Advanced Courses in Mathematics - CRM Barcelona

Managing Editor

David Romero i Sànchez, Centre de Rercerca Matemàtica, Barcelona, Spain

Since 1995 the Centre de Recerca Matemàtica (CRM) has organised a number of Advanced Courses at the post-doctoral or advanced graduate level on forefront research topics in Barcelona. The books in this series contain revised and expanded versions of the material presented by the authors in their lectures.

Mario Garcia-Fernandez • David Iglesias Ponte •
Eva Miranda • Cédric Oms • Roberto Rubio
Editors

Advances in Poisson Geometry

Editors
Mario Garcia-Fernandez
Universidad Autónoma de Madrid and ICMAT
Madrid, Spain

David Iglesias Ponte
Universidad de La Laguna
La Laguna, Spain

Eva Miranda
Universitat Politècnica de Catalunya
Barcelona, Spain

Cédric Oms
BCAM Bilbao Center of Applied Mathematics
Bilbao, Spain

Roberto Rubio
Universitat Autònoma de Barcelona
Barcelona, Spain

ISSN 2297-0304 ISSN 2297-0312 (electronic)
Advanced Courses in Mathematics - CRM Barcelona
ISBN 978-3-031-86388-2 ISBN 978-3-031-86657-9 (eBook)
https://doi.org/10.1007/978-3-031-86657-9

Mathematics Subject Classification: 53Dxx, 53D20, 53D17, 53C12, 53D50

© The Editor(s) (if applicable) and The Author(s), under exclusive license to Springer Nature Switzerland AG 2025

This work is subject to copyright. All rights are solely and exclusively licensed by the Publisher, whether the whole or part of the material is concerned, specifically the rights of translation, reprinting, reuse of illustrations, recitation, broadcasting, reproduction on microfilms or in any other physical way, and transmission or information storage and retrieval, electronic adaptation, computer software, or by similar or dissimilar methodology now known or hereafter developed.
The use of general descriptive names, registered names, trademarks, service marks, etc. in this publication does not imply, even in the absence of a specific statement, that such names are exempt from the relevant protective laws and regulations and therefore free for general use.
The publisher, the authors and the editors are safe to assume that the advice and information in this book are believed to be true and accurate at the date of publication. Neither the publisher nor the authors or the editors give a warranty, expressed or implied, with respect to the material contained herein or for any errors or omissions that may have been made. The publisher remains neutral with regard to jurisdictional claims in published maps and institutional affiliations.

This book is published under the imprint Birkhäuser, www.birkhauser-science.com by the registered company Springer Nature Switzerland AG
The registered company address is: Gewerbestrasse 11, 6330 Cham, Switzerland

If disposing of this product, please recycle the paper.

Preface

The study of Poisson geometry can be traced back to the pioneering works of Siméon Denis Poisson, Carl Gustav Jacob Jacobi, and Sophus Lie in the nineteenth century. However, it was not until the 1970s and 1980s, with the groundbreaking contributions of André Lichnerowicz, that Poisson manifolds gained significant attention. Since then, Poisson geometry has evolved into an independent and a rapidly growing field, significantly shaped by the contributions of Alan Weinstein.

Poisson geometry is now a vibrant and fast-developing area, enriched by its connections to differential geometry, Lie theory, quantization, noncommutative geometry, representation theory, quantum groups, geometric mechanics, and integrable systems. The influence of Poisson geometry extends beyond pure mathematics, finding direct applications in the theory of geometric phases, string theory, nonlinear control theory, nonholonomic mechanics, robotics (particularly in locomotion generation), and even planetary mission design.

Poisson manifolds possess a rich and versatile structure, offering multiple perspectives for exploration. On the one hand, they lead to, generally singular, symplectic foliations, positioning Poisson geometry as a natural generalization of symplectic geometry. On the other hand, Poisson manifolds can be approached through the lens of Lie groupoids and algebroids. Another important link between symplectic and Poisson geometry is reduction theory, where often the quotient of a symplectic manifold by a group action results in a Poisson space.

This volume is dedicated to the Summer School on Poisson Geometry, hosted at the Centre de Recerca Matemàtica (CRM) in Barcelona in July 2022. It formed part of the 12th edition of the biennial Poisson activities series, alongside the internationally renowned Poisson conference. The Summer School on Poisson Geometry aims to introduce participants to Poisson geometry and related fields while also presenting recent developments and open problems in the field. The school targeted advanced undergrad students, PhD candidates, and young researchers. This volume contains an enhanced compilation of notes from three courses presented at the Summer School on Poisson Geometry 2022:

- Henrique Bursztyn: *A Brief Introduction to Poisson Geometry*
- Ana Bălibanu: *Wonderful Varieties with a View Towards Poisson Geometry*

- Camille Laurent-Gengoux, Ruben Louis, and Leonid Ryvkin: *An Invitation to Singular Foliations*

This volume aims to introduce mathematicians to the fundamental concepts of Poisson manifolds while also providing an in-depth understanding of active, cutting-edge research in the field. We hope that readers, both newcomers and experienced researchers, will enjoy exploring the many facets of Poisson geometry through the lectures contained in this volume.

Madrid, Spain	Mario Garcia-Fernandez
La Laguna, Spain	David Iglesias Ponte
Barcelona, Spain	Eva Miranda
Bilbao, Spain	Cédric Oms
Barcelona, Spain	Roberto Rubio
October 3, 2024	

Contents

Part I   Introduction to Poisson Geometry

A Brief Introduction to Poisson Geometry 3
Henrique Bursztyn

Part II   Wonderful Varieties

Wonderful Varieties with a View Towards Poisson Geometry 85
Ana Bălibanu

Part III   An Invitation to Singular Foliations

What Is a Singular Foliation? 123
Camille Laurent-Gengoux, Ruben Louis, and Leonid Ryvkin

Canonical Geometric and Algebraic Structures Hidden Behind a Singular Foliation 263
Camille Laurent-Gengoux, Ruben Louis, and Leonid Ryvkin

State of the Art and Open Questions 391
Camille Laurent-Gengoux, Ruben Louis, and Leonid Ryvkin

Part I
Introduction to Poisson Geometry

A Brief Introduction to Poisson Geometry

Henrique Bursztyn

1 Introduction

These notes arose from a minicourse given at Centre de Reserca Matemàtica, Barcelona, in July 2022, as part of the *2022 Advanced Poisson School*. Versions of this minicourse were subsequently delivered at IMPA in 2023 and 2024, leading to a slight expansion of the original notes.

Keeping the original spirit of the minicourse, these notes focus on basic aspects of Poisson geometry. After a short introductory section below recalling the historical origins of Poisson structures, the notes are divided into four main parts: Sect. 2 covers foundations, where we define the central objects of interest in Poisson geometry; Sect. 3 presents some of the most relevant classes of examples of Poisson structures; Sect. 4 concerns basic structural results and classical problems, mostly going back to Weinstein's seminal paper [86], but with an effort to highlight refreshed perspectives and recent developments; Sect. 5 briefly discusses additional topics, such as symplectic groupoids and Dirac structures, that gained interest since [86] and led to noteworthy advances in the field.

Very little is actually proven in these notes; instead, many results and examples are presented through 132 exercises that permeate the main text with the purpose of guiding the reader toward the desired conclusions.

Despite the intention to offer a panoramic view of the subject, the selection of material in these notes has an inherent bias, and several important topics and applications are omitted. Various textbooks on the subject, such as [16, 31, 37, 60, 70], provide additional

---

H. Bursztyn (✉)
Instituto de Matemática Pura e Aplicada, Rio de Janeiro, Brazil
e-mail: henrique@impa.br

material not covered here, as well as further details on topics that we only briefly mention or sketch. In particular, these notes only treat Poisson geometry of C^∞-manifolds, despite the importance of Poisson structures in the holomorphic and algebraic contexts (see e.g. [77] and references therein).

Although not strictly necessary, readers would benefit from familiarity with basic concepts in symplectic geometry.

"The" Poisson Bracket Poisson geometry originated in the work of S. D. Poisson [76], who introduced the so-called *Poisson bracket* as a tool for the geometric formulation of classical mechanics in the nineteenth century; see [59] for more on the history. In its simplest form, the description of a classical mechanical system consists of a *phase space*

$$\mathbb{R}^{2n} = \{(q_i, p_i),\ i = 1, \ldots, n\},$$

with coordinates q_i representing positions and p_i representing momenta, along with a distinguished function $H \in C^\infty(\mathbb{R}^{2n})$, called the *Hamiltonian*, that gives rise to a *Hamiltonian vector field*:

$$X_H := \begin{pmatrix} 0 & \mathrm{Id} \\ -\mathrm{Id} & 0 \end{pmatrix} \begin{pmatrix} \frac{\partial H}{\partial q_i} \\ \frac{\partial H}{\partial p_i} \end{pmatrix} = \sum_i \frac{\partial H}{\partial p_i} \frac{\partial}{\partial q_i} - \frac{\partial H}{\partial q_i} \frac{\partial}{\partial p_i}.$$

The time evolution of the system is given by integral curves of X_H, i.e., curves $\sigma(t) = (q_i(t), p_i(t))$ in phase space satisfying $\dot\sigma(t) = X_H(\sigma(t))$ or, equivalently,

$$\dot q_i = \frac{\partial H}{\partial p_i},\quad \dot p_i = -\frac{\partial H}{\partial q_i},$$

which are *Hamilton's equations*.

In this context, Poisson introduced a skew-symmetric, $\mathbb{R}$-bilinear operation

$$\{\cdot, \cdot\} : C^\infty(\mathbb{R}^{2n}) \times C^\infty(\mathbb{R}^{2n}) \to C^\infty(\mathbb{R}^{2n}),$$

known as the *Poisson bracket*, given by

$$\{f, g\} = \sum_i \frac{\partial f}{\partial p_i}\frac{\partial g}{\partial q_i} - \frac{\partial f}{\partial q_i}\frac{\partial g}{\partial p_i}. \tag{1.1}$$

The Poisson bracket has a dynamical meaning as the rate of change of a given function with respect to the Hamiltonian vector field of another function:

$$\{H, f\} = \mathcal{L}_{X_H} f.$$

A Brief Introduction to Poisson Geometry

It is immediate that any function is preserved by its own Hamiltonian flow,

$$\mathcal{L}_{X_f} f = \{f, f\} = 0. \qquad (1.2)$$

In order to gain insight on (and some times explicitly solve) Hamilton's equations, there was a special interest around Poisson's time in finding *conserved quantities* (also called *first integrals*) of the system, i.e., functions f such that $\{H, f\} = \mathcal{L}_{X_H} f = 0$. (The link between the existence of conserved quantities and finding solutions to Hamilton's equation has led to the important notion of *complete integrability*, see e.g. [6, Chp. 10]). Poisson used his brackets to verify the following fact:

$$\mathcal{L}_{X_H} f = 0, \quad \mathcal{L}_{X_H} g = 0 \implies \mathcal{L}_{X_H} \{f, g\} = 0, \qquad (1.3)$$

i.e., the Poisson bracket can be used to produce new conserved quantities from old ones. This result was clarified by C. G. Jacobi in 1842, who observed that Poisson's theorem was a consequence of a fundamental identity satisfied by the Poisson bracket, now known as the *Jacobi identity*:

$$\{h, \{f, g\}\} + \{g, \{h, f\}\} + \{f, \{g, h\}\} = 0, \qquad f, g, h \in C^\infty(\mathbb{R}^{2n}).$$

The notion of "Poisson structure" appeared later in S. Lie's treatise on transformation groups of 1880; there Lie recognized the importance of the Jacobi identity when setting the foundations of the theory of Lie algebras and Lie groups.

Despite their important role in physics (not only in classical mechanics but also in the passage to quantum mechanics [32]) Poisson brackets were not systematically studied in mathematics until the 1970s. Two fundamental papers marking the beginning of the "modern era" of Poisson geometry are due to Lichnerowicz [62] and Weinstein [86]. The general study of Poisson structures since then has been stimulated by such diverse areas as geometric mechanics and field theories, integrable systems (in both finite and infinite dimensions), representation theory and quantum groups, singularity theory, noncommutative geometry and deformation quantization, among others; more information on these topics and further references can be found in the textbooks [6, 16, 37, 43, 53, 55, 57, 70].

2 Foundations

2.1 Poisson Brackets

A *Poisson bracket* on a smooth manifold M is an $\mathbb{R}$-bilinear operation $\{\cdot, \cdot\} : C^\infty(M) \times C^\infty(M) \to C^\infty(M)$ satisfying

(1) $\{f, g\} = -\{g, f\}$,

(2) $\{f, \{g, h\}\} + \{h, \{f, g\}\} + \{g, \{h, f\}\} = 0,$ (Jacobi identity)
(3) $\{f, gh\} = \{f, g\}h + \{f, h\}g,$ (Leibniz rule)

for $f, g, h \in C^\infty(M)$. The pair $(M, \{\cdot, \cdot\})$ is called a *Poisson manifold*.

Conditions (1) and (2) say that $(C^\infty(M), \{\cdot, \cdot\})$ is a Lie algebra, and the third condition is a compatibility of the Lie bracket with the commutative product on $C^\infty(M)$. One can use the same axioms to define a Poisson bracket $\{\cdot, \cdot\}$ on any commutative algebra $\mathcal{A}$, and the pair $(\mathcal{A}, \{\cdot, \cdot\})$ is called a *Poisson algebra*. If only conditions (1) and (3) are satisfied, one refers to an *almost Poisson bracket*.

A *Poisson map* from $(M_1, \{\cdot, \cdot\}_1)$ to $(M_2, \{\cdot, \cdot\}_2)$ is a smooth map $\varphi : M_1 \to M_2$ such that $\varphi^* : C^\infty(M_2) \to C^\infty(M_1)$ preserves brackets, i.e.,

$$\{f, g\}_2 \circ \varphi = \{f \circ \varphi, g \circ \varphi\}_1, \qquad \forall f, g \in C^\infty(M_2).$$

Poisson diffeomorphisms provide a natural notion of equivalence for Poisson manifolds, though there are other weaker notions of equivalence of interest (see Sect. 5.1).

For an almost Poisson bracket, the Leibniz rule implies that any $f \in C^\infty(M)$ determines a unique vector field X_f such that $\mathcal{L}_{X_f} g = \{f, g\}$, called the *Hamiltonian vector field* of f. Since

$$\{f, g\} = dg(X_f) = -df(X_g),$$

it is clear that, for any open subset $U \subseteq M$, $f|_U = 0$ implies that $\{f, g\}|_U = 0$ for all $g \in C^\infty(M)$; hence almost Poisson brackets are local, in the sense that they can be restricted to open subsets.

By the skew symmetry of the bracket, $\mathcal{L}_{X_f} f = \{f, f\} = 0$. In case of a Poisson bracket, the Jacobi identity ensures that (1.3) holds.

We say that $f, g \in C^\infty(M)$ are in *involution* if $\{f, g\} = 0$. A function f is in involution with any other function g if and only if $X_f = 0$, in which case f is called a *Casimir*.

We list a few initial examples.

Example 2.1 Any manifold carries the trivial Poisson bracket $\{\cdot, \cdot\} \equiv 0$. A Poisson map from a Poisson manifold $(M, \{\cdot, \cdot\})$ to $(\mathbb{R}^k, 0)$ is same as a collection of functions $f_1, \ldots, f_k \in C^\infty(M)$ in pairwise involution, $\{f_i, f_j\} = 0$. ◇

Example 2.2 A direct calculation shows that the original Poisson bracket (1.1) is a Poisson bracket on $M = \mathbb{R}^{2n}$ (see Example 2.4 below to avoid any calculation).

More generally, let (M, ω) be a symplectic manifold, i.e., $\omega \in \Omega^2(M)$ is closed and nondegenerate (the map $\omega^\flat : TM \to T^*M$, $X \mapsto i_X\omega$, is an isomorphism). Any $f \in C^\infty(M)$ defines a Hamiltonian vector field X_f by the condition $i_{X_f}\omega = df$, and

$$\{f, g\} := \omega(X_g, X_f) = dg(X_f) = \mathcal{L}_{X_f} g \tag{2.1}$$

defines a Poisson bracket on M (see Sect. 2.4). This can be verified by noticing that "the" Poisson bracket (1.1) on $\mathbb{R}^{2n}$ corresponds to the symplectic form $\omega = \sum_i dq_i \wedge dp_i$, so by Darboux's theorem for symplectic forms, any bracket as in (2.1) is locally equivalent to Poisson's original bracket. ◇

Exercise 2.1 Consider Poisson manifolds $(M_1, \{\cdot, \cdot\}_1)$ and $(M_2, \{\cdot, \cdot\}_2)$, and let $M = M_1 \times M_2$. Show that the formula

$$\{f, g\}(x_1, x_2) = \{f(\cdot, x_2), g(\cdot, x_2)\}_1(x_1) + \{f(x_1, \cdot), g(x_1, \cdot)\}_2(x_2) \tag{2.2}$$

defines a Poisson structure on M, in such a way that the projections $p_i : M \to M_i$ are Poisson maps and $\{p_1^* C^\infty(M_1), p_2^* C^\infty(M_2)\} = 0$.

For a Poisson manifold M, note that the diagonal map $M \to M \times M$ is not Poisson (unless the Poisson bracket on M is trivial); considering the category of Poisson manifolds and Poisson maps, conclude that the product of Poisson manifolds is not categorical[1] (cf. Exercise 2.3 (b)).

Example 2.3 Let S be a manifold with a smooth family of symplectic structures $\omega_t \in \Omega^2(S)$, $t \in \mathbb{R}$. Denote by $\{\cdot, \cdot\}_t$ the Poisson bracket defined by ω_t. Then there is a Poisson bracket on $M = S \times \mathbb{R}$ given by

$$\{f, g\}(x, t) = \{f(\cdot, t), g(\cdot, t)\}_t(x).$$

(More generally, in this example $\mathbb{R}$ could be replaced by any other manifold, and $\{\cdot, \cdot\}_t$ could be any smooth family of Poisson brackets, not necessarily symplectic.) ◇

2.2 Tensorial Viewpoint: Bivector Fields

Given a manifold M, we will use the notation $\mathfrak{X}^k(M) := \Gamma(\wedge^k TM)$. In particular, the space of vector fields on M will be denoted by $\mathfrak{X}^1(M)$.

For an almost Poisson bracket $\{\cdot, \cdot\}$ on M, since

$$\{f, g\} = dg(X_f) = -df(X_g),$$

---

[1] A product in a category satisfies the following universal property: given a product of objects $M_1 \times M_2$ with projections $p_i : M_1 \times M_2 \to M_i$, $i = 1, 2$, then for any object N and morphisms $\psi_i : N \to M_i$, $i = 1, 2$, there us a unique morphism $\psi : N \to M_1 \times M_2$ such that $p_i \circ \psi = \psi_i$, $i = 1, 2$.

we see that the bracket of two functions only depends on their differentials. It follows that there is a unique bivector field $\pi \in \mathcal{X}^2(M)$ such that

$$\{f, g\} = \pi(df, dg). \tag{2.3}$$

By means of the previous formula one obtains a bijective correspondence between almost Poisson brackets on $C^\infty(M)$ and bivector fields π on M. We will refer to a bivector field whose corresponding bracket satisfies the Jacobi identity as a *Poisson bivector field*. One can therefore think of a *Poisson structure* on M as either a Poisson bracket $\{\cdot, \cdot\}$ on $C^\infty(M)$, or as a Poisson bivector field $\pi \in \mathcal{X}^2(M)$.

In local coordinates $(x_1, \ldots, x_n)$ on M, a bivector field π is written as

$$\pi = \frac{1}{2} \sum_{i,j} \pi_{ij}(x) \frac{\partial}{\partial x_i} \wedge \frac{\partial}{\partial x_j} = \sum_{i<j} \pi_{ij}(x) \frac{\partial}{\partial x_i} \wedge \frac{\partial}{\partial x_j},$$

and the associated almost Poisson bracket takes the form

$$\{f, g\}(x) = \sum_{i,j} \pi_{ij}(x) \frac{\partial f}{\partial x_i} \frac{\partial g}{\partial x_j}, \qquad \pi_{ij} = \{x_i, x_j\}.$$

A bivector field π on M can be equivalently described by a vector-bundle map $\pi^\sharp : T^*M \to TM$ (over the identity map on M) satisfying $(\pi^\sharp)^* = -\pi^\sharp$ via

$$\pi(\alpha, \beta) = \beta(\pi^\sharp(\alpha)).$$

Note that, for $f \in C^\infty(M)$,

$$\pi^\sharp(df) = X_f.$$

Exercise 2.2 Let (M_1, π_1) and (M_2, π_2) be Poisson manifolds. Show that the following are equivalent.

(a) $\varphi : M_1 \to M_2$ is a Poisson map;
(b) $X_{\varphi^* f}$ is φ-related to X_f for all $f \in C^\infty(M_2)$;
(c) $\pi_2^\sharp|_{\varphi(x)} = d_x\varphi \circ \pi_1^\sharp|_x \circ (d_x\varphi)^*$ for all $x \in M_1$.

Exercise 2.3

(a) For Poisson manifolds (M_1, π_1) and (M_2, π_2), show that the Poisson bracket on $M_1 \times M_2$ in Exercise 2.1 corresponds to the natural direct product of bivector fields $\pi_1 \times \pi_2 \in \mathcal{X}^2(M_1 \times M_2)$.

(b) Suppose that $\varphi_1 : M \to M_1$ and $\varphi_2 : M \to M_2$ are Poisson maps. Show that $M \overset{(\varphi_1,\varphi_2)}{\longrightarrow} M_1 \times M_2$ is a Poisson map if and only if $\{\varphi_1^* C^\infty(M_1), \varphi_2^* C^\infty(M_2)\} = 0$.

Given a bivector field π on M, the failure of the Jacobi identity of the corresponding bracket $\{\cdot, \cdot\}$ is measured by the *Jacobiator*, i.e., the trilinear operation $\operatorname{Jac}_\pi : C^\infty(M) \times C^\infty(M) \times C^\infty(M) \to C^\infty(M)$,

$$\operatorname{Jac}_\pi(f, g, h) = \{f, \{g, h\}\} + \{h, \{f, g\}\} + \{g, \{h, f\}\}.$$

We will simply write Jac whenever π is clear from the context.

Exercise 2.4 Verify that

$$\operatorname{Jac}(f, g, h) = \mathcal{L}_{[X_f, X_g]} h - \mathcal{L}_{X_{\{f,g\}}} h = (\mathcal{L}_{X_f} \pi)(dg, dh).$$

It follows from the previous exercise that, for a Poisson structure, the map $C^\infty(M) \to \mathfrak{X}^1(M)$, $f \mapsto X_f$ is a Lie algebra homomorphism,

$$[X_f, X_g] = X_{\{f,g\}}, \qquad \forall\, f, g \in C^\infty(M), \tag{2.4}$$

as well as the fact that Poisson bivector fields are preserved by Hamiltonian flows. Another direct consequence of Exercise 2.4 is that the Jacobiator only depends on the differentials of functions, so there exists a unique trivector field $\Upsilon_\pi \in \mathfrak{X}^3(M) = \Gamma(\wedge^3 TM)$ such that, for all $f, g, h \in C^\infty(M)$,

$$\operatorname{Jac}(f, g, h) = \Upsilon_\pi(df, dg, dh). \tag{2.5}$$

Hence the vanishing of Jac can be verified by checking the vanishing of Υ_π on local frames of T^*M. In particular, in a coordinate chart $(x_1, \ldots, x_n)$, it suffices to check that $\operatorname{Jac}(x_i, x_j, x_k) = 0$, $\forall\, i, j, k$.

Example 2.4 On any open subset of $\mathbb{R}^m$, a constant bivector field is Poisson. In coordinates $(x_1, \ldots, x_m)$, the fact that $\pi_{ij} = \{x_i, x_j\}$ is constant implies that $\{x_i, \{x_j, x_k\}\} = \mathcal{L}_{X_{x_i}} \{x_j, x_k\} = 0$, so $\operatorname{Jac}(x_i, x_j, x_k) = 0$ for all i, j, k. This is the case for Poisson's original bracket (1.1), which corresponds to the bivector field

$$\pi_{can} = \sum_i \frac{\partial}{\partial p_i} \wedge \frac{\partial}{\partial q_i}$$

in $\mathbb{R}^{2n}$. ◇

Example 2.5 On a 2-dimensional manifold M any trivector field is trivial, so any bivector field is Poisson. E.g. any $\varphi \in C^\infty(\mathbb{R}^2)$ defines a Poisson structure $\pi = \varphi(x, y)\partial_x \wedge \partial_y$ on $\mathbb{R}^2$. ◇

Example 2.6 On $M = \mathbb{R}^3 = \{\xi = (x, y, z)\}$, we have the Poisson bracket

$$\{f, g\}(\xi) = \langle \xi, (\nabla f|_\xi \times \nabla g|_\xi)\rangle,$$

with corresponding bivector field $\pi = z\partial_x \wedge \partial_y + x\partial_y \wedge \partial_z + y\partial_z \wedge \partial_x$. In this case, integral curves of the Hamiltonian vector field of the function $H(x, y, z) = \frac{x^2}{2I_x} + \frac{y^2}{2I_y} + \frac{z^2}{2I_z}$ are the same as solutions of the Euler equations of a rigid body with moments of inertia I_x, I_y, I_z (see e.g. [70, § 15]). ◇

2.3 The Schouten Bracket

For a bivector field π on M, there is a more intrinsic way to express the trivector field Υ_π in terms of π through the so-called *Schouten bracket*. For each $k = 0, 1, \ldots$, consider the space $\mathfrak{X}^k(M) = \Gamma(\wedge^k TM)$ of k-vector fields on M, and the (graded commutative) algebra

$$\mathfrak{X}^\bullet(M) = \oplus_{k=0}^{\dim(M)} \mathfrak{X}^k(M)$$

of multivector fields (here $\mathfrak{X}^0(M) = C^\infty(M)$). The Schouten bracket extends the usual Lie bracket of vector fields as follows: it is the unique $\mathbb{R}$-bilinear bracket

$$[\cdot, \cdot] : \mathfrak{X}^k(M) \times \mathfrak{X}^l(M) \to \mathfrak{X}^{k+l-1}(M)$$

such that, for $X \in \mathfrak{X}^k(M), Y \in \mathfrak{X}^l(M)$,

- $[X, Y] = -(-1)^{(k-1)(l-1)}[Y, X]$,
- $[X, Y \wedge Z] = [X, Y] \wedge Z + (-1)^{(k-1)l} Y \wedge [X, Z]$,
- If Z is a vector field, then $[Z, \cdot] = \mathcal{L}_Z$.

It follows from these properties that $[\cdot, \cdot]$ satisfies a graded version of the Jacobi identity,

$$(-1)^{(k-1)(m-1)}[X, [Y, Z]] + (-1)^{(m-1)(l-1)}[Z, [X, Y]] + (-1)^{(l-1)(k-1)}[Y, [Z, X]] = 0,$$

for $X \in \mathfrak{X}^k(M), Y \in \mathfrak{X}^l(M), Z \in \mathfrak{X}^m(M)$, so the Schouten bracket makes $\mathfrak{X}(M)$ into a Gerstenhaber algebra (i.e., $[\cdot, \cdot]$ is a graded Poisson bracket of degree -1).

Exercise 2.5 Given $\pi \in \mathfrak{X}^2(M)$, check that

(a) $[\pi, f] = -X_f$, for $f \in C^\infty(M)$;
(b) $[\pi, [\pi, \cdot]] = \frac{1}{2}[[\pi, \pi], \cdot]$.

Remark 2.7 In local coordinates $(x_1, \ldots, x_n)$ and using the notation $\xi_i = \frac{\partial}{\partial x_i}$, one can write $X \in \mathfrak{X}^k(M)$ and $Y \in \mathfrak{X}^l(M)$ as

$$X = \sum_{i_1 < \ldots < i_k} a_{i_1 \cdots i_k}(x) \xi_{i_1} \ldots \xi_{i_k}, \quad Y = \sum_{i_1 < \ldots < i_l} b_{i_1 \cdots i_l}(x) \xi_{i_1} \ldots \xi_{i_l},$$

so we can regard them as "functions" of x_i and ξ_i. With this notation, the Schouten bracket of X and Y has the following local expression:

$$[X, Y] = \sum_i \frac{\partial X}{\partial \xi_i} \frac{\partial Y}{\partial x_i} - (-1)^{(k-1)(l-1)} \frac{\partial X}{\partial x_i} \frac{\partial Y}{\partial \xi_i}.$$

By formally interpreting x_i, ξ_j as local coordinates, one can regard the Schouten bracket as an analogue of Poisson's original bracket on the "shifted" cotangent bundle $T^*[1]M$ (the notation "[1]" indicates that fiber coordinates are assigned degree 1), see e.g. [81, § 2]. ◇

One can verify (see e.g. [37]) that

$$\Upsilon_\pi = \frac{1}{2}[\pi, \pi].$$

So a Poisson bivector field is characterized by the condition $[\pi, \pi] = 0$.

Exercise 2.6 Consider vector fields X, Y on M and let $\pi = X \wedge Y \in \mathfrak{X}^2(M)$. Verify that π is Poisson if and only if X, Y and $[X, Y]$ are linearly dependent at all points.

2.4 The Characteristic Distribution

By a *distribution* on a manifold M we will mean a subset $D \subseteq TM$ such that $D_x := D \cap T_x M$ is a vector subspace of $T_x M$. A distribution D is called *smooth* if for any $x \in M$ and $v \in D_x$, there exists a vector field $X \in \mathfrak{X}^1(M)$ such that $X|_x = v$ and $X|_y \in D_y$ for all $y \in M$. The dimension of D_x is called the *rank* of D at x, and a distribution with constant rank is called *regular*. A distribution that is smooth and regular is the same as a vector subbundle of TM.

Exercise 2.7 Let $\omega \in \Omega^2(M)$ and $\omega^\flat : TM \to T^*M$, $\omega^\flat(X) = \omega(X, \cdot)$. Consider the distribution

$$K = \ker(\omega^\flat) = \{X \in TM \mid \omega(X, \cdot) = 0\} \subseteq TM.$$

Show that K is smooth if and only if it is locally regular (i.e., regular on connected components). (It might be helpful to notice that the rank of a smooth distribution is a lower semi-continuous function on M.)

For a bivector field $\pi \in \mathfrak{X}^2(M)$, the image of the map

$$\pi^\sharp : T^*M \to TM, \quad \alpha \mapsto \pi(\alpha, \cdot), \tag{2.6}$$

defines a distribution on M,

$$R := \pi^\sharp(T^*M) \subseteq TM,$$

called the *characteristic distribution* of π. Note that $R_x = \{X_f|_x, \ f \in C^\infty(M)\}$.

Exercise 2.8 For any bivector field π, show that R is smooth.

As the next exercise shows, each R_x naturally acquires the structure of a symplectic vector space, and the collection of these symplectic vector spaces completely codifies the bivector field π. (In case of Poisson structures, much more will be true: the characteristic distribution will give rise to "symplectic leaves", see Sect. 4.1.)

Exercise 2.9 Let V be a (real) vector space, and let $\pi \in \wedge^2 V$. Consider $\pi^\sharp : V^* \to V$ defined by $\beta(\pi^\sharp(\alpha)) = \pi(\alpha, \beta)$, and let $R = \pi^\sharp(V^*) \subseteq V$. Show that there is a unique nondegenerate skew-symmetric bilinear form Ω on R given by $\Omega(u, v) = \pi(\beta, \alpha)$, for $u = \pi^\sharp(\alpha)$ and $v = \pi^\sharp(\beta)$. Conversely, show that given a pair (R, Ω), where $R \subseteq V$ is a subspace and $\Omega \in \wedge^2 R^*$ is nondegenerate, there is a unique $\pi \in \wedge^2 V$ such that $R = \pi^\sharp(V^*)$ and Ω is defined as before.

We define the *rank* of π at $x \in M$ as the rank of R at x, and we say that π is *regular* if so is R. We say that a bivector field π is *nondegenerate* if $R = TM$, or, equivalently, if $\pi^\sharp : T^*M \to TM$ is an isomorphism.

Any nondegenerate 2-form $\omega \in \Omega^2(M)$ defines a nondegenerate bivector field π by $\pi^\sharp = (\omega^\flat)^{-1}$, or

$$\pi(\beta, \alpha) = \omega((\omega^\flat)^{-1}(\alpha), (\omega^\flat)^{-1}(\beta)),$$

and this establishes a bijective correspondence

$$\text{nondegenerate 2-forms} \rightleftharpoons \text{nondegenerate bivector fields.}$$

Exercise 2.10 Suppose that π is a nondegenerate bivector field with corresponding 2-form ω, so that, for $f \in C^\infty(M)$, $X_f = \pi^\sharp(df)$ if and only if $i_{X_f}\omega = df$. For $f, g, h \in C^\infty(M)$, compare $d\omega(X_f, X_g, X_h)$ and $\text{Jac}(f, g, h)$, and verify that π is Poisson if and only if ω is symplectic.

It follows from the previous exercise that symplectic structures are equivalent to nondegenerate Poisson structures:

$$\text{Symplectic structures} \rightleftharpoons \text{nondegenerate Poisson structures.}$$

In spite of this equivalence, maps that preserve Poisson brackets are not the same as maps that preserve symplectic forms, as explained in the next exercise.

Exercise 2.11 Consider symplectic manifolds (M_i, ω_i), with corresponding Poisson brackets $\{\cdot, \cdot\}_i$, $i = 1, 2$, and let $\varphi : M_1 \to M_2$ be a smooth map.

(a) Show that if φ preserves symplectic forms (i.e., $\varphi^*\omega_2 = \omega_1$), then it must be an immersion, and that if φ is a Poisson map, then it must be a submersion.
(b) Prove that, if φ is a (local) diffeomorphism, then it is a Poisson map if and only if it preserves symplectic forms.
(c) Find examples of M_1, M_2 and $\varphi : M_1 \to M_2$ such that (1) φ is a Poisson map but does not preserve symplectic forms; (2) φ preserves symplectic forms but is not a Poisson map.

2.5 Poisson, Coisotropic and Cosymplectic Submanifolds

We briefly discuss some important types of submanifolds in Poisson geometry. By a submanifold of a manifold M we mean a manifold N together with an injective immersion $\iota : N \hookrightarrow M$; we will often omit ι and identify N with a subset of M.

Let ζ be a k-vector field on M. We say that ζ is *tangent* to a submanifold $N \hookrightarrow M$ if $\zeta_x \in \wedge^k T_x N \subseteq \wedge^k T_x M$ for all $x \in N$. In this case N inherits a k-vector field ζ_N by restriction of ζ.

Exercise 2.12 Check that a k-vector field ζ is tangent to $N \hookrightarrow M$ if and only if, for any open subset $U \subseteq N$, $(i_{df}\zeta)|_U = 0$ for all $f \in C^\infty(M)$ such that $f|_U = 0$.

Exercise 2.13

(a) Show that a bivector field π on M is tangent to a submanifold N if and only if $R|_N = \pi^\sharp(T^*M|_N) \subseteq TN$ if and only if $\pi^\sharp(\mathrm{Ann}(TN)) = 0$. In this case, check that the characteristic distribution of the induced bivector field π_N on N is $R|_N$.

(b) Consider the Jacobiator 3-vector field Υ_π, see (2.5). Show that if π is tangent to N, then, for any open subset $U \subseteq N$, $\mathrm{Jac}_\pi(f, g, h)|_U = \mathrm{Jac}_{\pi_N}(f|_U, g|_U, h|_U)$ for all $f, g, h \in C^\infty(M)$, and hence Υ_π is also tangent to N (see Exercise 2.12).

A submanifold N of a Poisson manifold M is a *Poisson submanifold* if it is equipped with a Poisson structure for which the inclusion map $N \hookrightarrow M$ is a Poisson map. (It is clear that there is at most one Poisson structure on a submanifold making it into a Poisson submanifold.)

Exercise 2.14 Let N be a submanifold of a Poisson manifold (M, π). Show that N is a Poisson submanifold if and only if π is tangent to N, if and only if every Hamiltonian vector field on M is tangent to N.

Exercise 2.15 Let N be a submanifold and $I_N = \{f \in C^\infty(M) \mid f|_N = 0\} \subseteq C^\infty(M)$ its vanishing ideal. Check that if N is Poisson then I_N is a Lie-ideal, i.e., $\{C^\infty(M), I_N\} \subseteq I_N$, and that the converse holds if N is embedded.

Exercise 2.16 Let (M, π) be a Poisson manifold and $\Psi = (\psi_1, \ldots, \psi_k) : M \to \mathbb{R}^k$ a map such that $r \in \mathbb{R}^k$ is a regular value, and consider the submanifold $N = \Psi^{-1}(r)$. Show that N is Poisson if and only if $X_{\psi_i}|_N = 0$ for all $i = 1, \ldots, k$. (In particular, level sets of Casimirs are Poisson submanifolds.)

Exercise 2.17 Consider $\mathbb{R}^3$ with the Poisson structure $\pi = z\partial_x \wedge \partial_y + x\partial_y \wedge \partial_z + y\partial_z \wedge \partial_x$. Show that the unit sphere $\mathbb{S}^2 \subseteq \mathbb{R}^3$ is a Poisson submanifold, and that the Poisson structure it inherits corresponds to the symplectic structure given by the negative of its area form.

Exercise 2.18 On $\mathbb{R}^{n+1} = \{(x, y, z) \mid x, y \in \mathbb{R}, z = (z_1, \ldots, z_{n-1}) \in \mathbb{R}^{n-1}\}$, consider the bivector field π defined by the bracket relations

$$\{x, y\} = |z|^2, \quad \{x, z_i\} = -yz_i, \quad \{y, z_i\} = xz_i, \quad \{z_i, z_j\} = 0,$$

where $|z|^2 = z_1^2 + \ldots + z_{n-1}^2$. Show that π is a Poisson structure on $\mathbb{R}^{n+1}$ and the unit sphere $\mathbb{S}^n \subseteq \mathbb{R}^{n+1}$ is a Poisson submanifold.

The Poisson submanifolds of a symplectic manifold are just its open subsets. We will now see ways to extend the notions of coisotropic and symplectic submanifolds to the

Poisson setting. (In physics, these are known as *first-class* and *second-class* constraints, respectively.)

Given a submanifold N of a Poisson manifold (M, π), we define

$$TN^\pi = \pi^\sharp(\mathrm{Ann}(TN)), \tag{2.7}$$

which is the Poisson analogue of symplectic orthogonals[2] in symplectic geometry. With this notation, Poisson submanifolds are defined by the condition $TN^\pi = 0$ (Exercise 2.13).

Exercise 2.19 Check that $TN^\pi = \mathrm{Ann}((\pi^\sharp)^{-1}(TN))$. Use this fact to show that, if $f, g \in C^\infty(M)$ are such that $df|_N, dg|_N \in \Gamma(\mathrm{Ann}(TN^\pi))$, then $(d\{f, g\})|_N \in \Gamma(\mathrm{Ann}(TN^\pi))$.

Exercise 2.20

(a) Show that the following holds pointwise: $TN^\pi = (TN \cap R)^\Omega \subseteq R$ (the symplectic orthogonal of $TN \cap R$ in R), where we regard the characteristic distribution $R = \pi^\sharp(T^*M)$ as a family of symplectic vector spaces (see Exercise 2.9). In particular, $(TN^\pi)^\pi = TN \cap R$.
(b) Check that $\mathrm{rank}(TN^\pi) \leq \dim(M) - \dim(N)$, and hence $\mathrm{rank}(TN) + \mathrm{rank}(TN^\pi) \leq \dim(M)$.

A submanifold N is called *coisotropic* if $TN^\pi \subseteq TN$. An immediate consequence of the previous exercise is that N is a coisotropic submanifold if and only if, at each point, $TN \cap R$ is a coisotropic subspace of R.

Exercise 2.21 Following the notation of Exercise 2.15, show that if N is coisotropic then $\{I_N, I_N\} \subseteq I_N$, and that the converse holds as long as N is embedded.

Exercise 2.22 Consider the submanifold N of M as in Exercise 2.16. Show that N is coisotropic if and only if $\{\psi_i, \psi_j\}|_N = 0$ for all $i, j = 1, \ldots, k$.

The following three exercises illustrate the role of coisotropic manifolds in Poisson geometry, see [88].

---

[2] If W is a subspace of a symplectic vector space (V, Ω), its symplectic orthogonal is $W^\Omega := \{v \in V \mid \Omega(v, w) = 0 \,\forall w \in W\} = (\Omega^\flat)^{-1}(\mathrm{Ann}(W))$. Recall that a subspace $W \subseteq V$ is *coisotropic* if $W^\Omega \subseteq W$ and *symplectic* if $W \cap W^\Omega = \{0\}$ (equivalently, if $V = W \oplus W^\Omega$).

Exercise 2.23 Show that a map $M_1 \to M_2$ between Poisson manifolds is a Poisson map if and only if its graph is a coisotropic submanifold of $M_1 \times \overline{M_2}$, where $\overline{M_2}$ has minus the Poisson structure of M_2.

Exercise 2.24 Consider a Poisson map $\varphi : M_1 \to M_2$ and a coisotropic submanifold $N \hookrightarrow M_2$ such that $\varphi^{-1}(N)$ is a submanifold of M_1 with $T(\varphi^{-1}(N)) = (d\varphi)^{-1}(TN)$ (this is the case e.g. when φ is transverse to N). Show that $\varphi^{-1}(N)$ is coisotropic.

Exercise 2.25 Let $\varphi : M \to B$ be a surjective submersion, and consider the submanifold $M \times_B M = \{(x, y) \in M \times M \mid \varphi(x) = \varphi(y)\} \subseteq M \times M$.

Suppose that M is a Poisson manifold. Show that B carries a Poisson structure for which φ is a Poisson map if and only if $M \times_B M$ is a coisotropic submanifold of $M \times \overline{M}$. (Exercises 2.21 and 2.24 can be helpful.)

Symplectic submanifolds can be generalized to Poisson geometry in different ways; we will discuss one possibility now (see Sect. 5.4 for a more thorough discussion).

Given a Poisson manifold (M, π), a submanifold N is called *cosymplectic* if

$$TN \oplus TN^\pi = TM|_N. \tag{2.8}$$

By Exercise 2.20 (b), this last condition is equivalent to $TN + TN^\pi = TM|_N$.

Exercise 2.26 Let N be a submanifold of (M, π), and consider the (fiberwise) skew-symmetric bilinear form on $\mathrm{Ann}(TN)$ defined by the restriction of π. Show that N is cosymplectic if and only if this bilinear form is nondegenerate on each fiber.

It follows from the previous exercise that $\mathrm{Ann}(TN)$ is a symplectic vector bundle, and this motivates the name "cosymplectic" for this class of submanifolds;[3] cosymplectic submanifolds are alternatively called *Poisson transversals* [40] (see Example 5.10 (b) for further comments on terminology).

Exercise 2.27 Let N be a submanifold of (M, π) and $R = \pi^\sharp(T^*M)$ (see Exercise 2.20 (a)). Show that

(a) pointwise, $TN \cap TN^\pi = \{0\}$ if and only if $TN \cap R \subseteq R$ is a symplectic subspace;

---

[3] The term "cosymplectic" in this context is not to be confused with the concept of "cosymplectic structure" due to Libermann, defined on an odd-dimensional manifold M^{2n+1} by a closed 1-form η and a closed 2-form ω such that $\eta \wedge \omega^n$ is a volume form; see [31, Prop. 4.19] for a description of how such structures are actually related to Poisson structures.

(b) a submanifold N is cosymplectic if and only if $TN \cap R \subseteq R$ is a symplectic subspace pointwise and N is transverse to R, i.e., $TN + R|_N = TM|_N$.

Exercise 2.28 Let N be a submanifold of (M, π). Show that if the cosymplectic condition (2.8) holds at a given point in N, then it holds locally (in N) around this point. Use this fact to show that, if $x \in N$ is such that $T_x N \oplus R_x = T_x M$, then there is a neighborhood of x in N that is a cosymplectic submanifold.

A cosymplectic submanifold N is not a Poisson (or, more generally, coisotropic) submanifold unless $TN = TM|_N$, i.e., N is an open subset of M. Nevertheless, a key property of cosymplectic submanifolds is that they always inherit a natural Poisson structure (a further discussion of how this Poisson structure relates to the ambient Poisson structure on M is presented in Sect. 5.4).

If N is a cosymplectic submanifold of (M, π), we can write

$$T^* M|_N = T^* N \oplus (TN^\pi)^*,$$

using the identifications

$$T^* N \xrightarrow{\sim} \mathrm{Ann}(TN^\pi) \subseteq T^* M|_N, \quad (TN^\pi)^* \xrightarrow{\sim} \mathrm{Ann}(TN) \subseteq T^* M|_N,$$

given by the dual maps to the projections $TM|_N \twoheadrightarrow TN$ and $TM|_N \twoheadrightarrow TN^\pi$. Since $\pi^\sharp(\mathrm{Ann}(TN)) = TN^\pi$ and $\pi^\sharp(\mathrm{Ann}(TN^\pi)) = (TN^\pi)^\pi = TN \cap R \subseteq TN$ (Exercise 2.20 (a)), we see that $\pi^\sharp|_N : T^* M|_N \to TM|_N$ decomposes as maps

$$T^* N \to TN, \quad \text{and} \quad (TN^\pi)^* \to TN^\pi. \tag{2.9}$$

The first map defines a bivector field π_N on N, while the second map defines an element $\chi \in \Gamma(\wedge^2 TN^\pi)$, so that

$$\pi|_N = \pi_N + \chi.$$

Note that χ is nondegenerate, cf. Exercise 2.26.

Exercise 2.29 Following Exercise 2.27, check that the characteristic distribution of π_N is $TN \cap R$, with pointwise symplectic form given by pullback of the one on R.

The next exercise indicates how to verify that the bivector field π_N is indeed a Poisson structure on N. We assume that N is embedded for simplicity (otherwise one can work locally on N).

Exercise 2.30

(a) Show that the bracket $\{\cdot,\cdot\}_N$ on $C^\infty(N)$ corresponding to π_N can be calculated as follows:

$$\{f,g\}_N = \{\widehat{f},\widehat{g}\}|_N,$$

where $\widehat{f},\widehat{g} \in C^\infty(M)$ are extensions of $f,g \in C^\infty(N)$ satisfying $d\widehat{f}|_{TN^\pi} = d\widehat{g}|_{TN^\pi} = 0$. (Note that such extensions always exist.)

(b) Use (a) and Exercise 2.19 to show that

$$\text{Jac}_{\pi_N}(f,g,h) = \text{Jac}_\pi(\widehat{f},\widehat{g},\widehat{h})|_N, \qquad f,g,h \in C^\infty(N).$$

Therefore $\{\cdot,\cdot\}_N$ satisfies the Jacobi identity, so π_N is a Poisson structure on N.

The following exercise gives a general local formula for the bracket $\{\cdot,\cdot\}_N$ in terms of $\{\cdot,\cdot\}$, known as the *Dirac bracket*.

Exercise 2.31 (Dirac Bracket) Consider a Poisson manifold (M,π), a map $\Psi = (\psi_1,\ldots,\psi_k) : M \to \mathbb{R}^k$ such that $r \in \mathbb{R}^k$ is a regular value, and the submanifold $N = \Psi^{-1}(r)$. Let (c^{ij}) be the matrix with entries $c^{ij} = \{\psi_i,\psi_j\}|_N$. (a) Show that N is cosymplectic if and only if the matrix (c^{ij}) is invertible. Let (c_{ij}) be the inverse matrix. (b) Prove that the Poisson bracket on N is given by

$$\{f,g\}_N = (\{\widetilde{f},\widetilde{g}\} - \sum_{i,j}\{\widetilde{f},\psi_i\}c_{ij}\{\psi_j,\widetilde{g}\})|_N,$$

where $\widetilde{f}$ and $\widetilde{g}$ are *arbitrary* extensions of f and g to M (cf. Exercise 2.30 (a)).

Exercise 2.32 Let $M = S \times N$ be the product of two Poisson manifolds S and N, with S symplectic.

(a) Show that, for all $x \in S$, $\{x\} \times N$ is a cosymplectic submanifold of M and the Poisson structure it inherits as such agrees with the original Poisson structure on N.
(b) Check that, for $y \in N$, $S \times \{y\}$ is a Poisson submanifold of M if and only if $\pi_N|_y = 0$.

2.6 Some Poisson Invariants

Let (M,π) be a Poisson manifold. We can use the Schouten bracket $[\cdot,\cdot]$ on multivector fields to define an operator

$$d_\pi := [\pi,\cdot] : \mathfrak{X}^\bullet(M) \to \mathfrak{X}^{\bullet+1}(M).$$

By the graded Jacobi identity for the Schouten bracket, we have that $d_\pi^2 = \frac{1}{2}[[\pi, \pi], \cdot] = 0$ (see Exercise 2.5). The cohomology of the cochain complex $(\mathfrak{X}^\bullet(M), d_\pi)$, denoted by $H_\pi^\bullet(M)$, is called the *Poisson cohomology* of M.

Recall the bundle map $\pi^\sharp : T^*M \to TM$, $\pi^\sharp(\alpha) = \pi(\alpha, \cdot)$, and consider the induced $C^\infty(M)$-linear homomorphism $(\pi^\sharp)^* : \Omega^\bullet(M) \to \mathfrak{X}^\bullet(M)$ given by

$$(\pi^\sharp)^* \eta(\alpha_1, \ldots, \alpha_k) = \eta(\pi^\sharp(\alpha_1), \ldots, \pi^\sharp(\alpha_k)),$$

for $\eta \in \Omega^k(M)$ and $\alpha_i \in \Omega^1(M)$, $i = 1, \ldots, k$.

Exercise 2.33 Show that, for any $\eta \in \Omega^k(M)$,

$$(\pi^\sharp)^* d\eta = d_\pi (\pi^\sharp)^* \eta,$$

i.e., $(\pi^\sharp)^*$ is a morphism from the de Rham complex to the Poisson complex. (Exercise 2.5 checks this equality when η has degree 0; the general result can be reduced to this case and the case where η is an exact 1-form.)

It follows that $(\pi^\sharp)^* : \Omega^\bullet(M) \to \mathfrak{X}^\bullet(M)$ induces a homomorphism in cohomology,

$$H_{dR}^\bullet(M) \to H_\pi^\bullet(M).$$

When π is nondegenerate (i.e., symplectic), this map is an isomorphism. In general, this map is neither surjective nor injective, and Poisson cohomology tends to be very different from de Rham cohomology. For example, when $\pi = 0$, we have $H_\pi^k(M) = \mathfrak{X}^k(M)$, so Poisson cohomology groups can be infinite-dimensional even when de Rham cohomology groups are trivial.

As with other types of cohomologies, Poisson cohomology groups have meaningful interpretations in low degrees.

- The zeroth Poisson cohomology group consists of functions $f \in C^\infty(M)$ such that (see Exercise 2.5)

$$d_\pi f = -X_f = 0,$$

  i.e., $H_\pi^0(M)$ is the space of Casimir functions.
- A *Poisson vector field* on (M, π) is a vector field X such that

$$\mathcal{L}_X \pi = -d_\pi X = 0,$$

i.e., X is an infinitesimal automorphism of the Poisson structure (its flow is by Poisson automorphisms). Equivalently, a Poisson vector field X is a derivation of the Poisson bracket,

$$\mathcal{L}_X\{f, g\} = \{\mathcal{L}_X f, g\} + \{f, \mathcal{L}_X g\},$$

for all $f, g \in C^\infty(M)$. The space of Poisson vector fields, denoted by $\mathfrak{X}_\pi(M)$, contains the space $\mathfrak{X}_{\text{Ham}}(M)$ of Hamiltonian vector fields (cf. Exercise 2.4), which are the inner derivations of the bracket.

Exercise 2.34 Note that $\mathfrak{X}_\pi(M)$ is closed under the Lie bracket of vector fields. Check that, if $X \in \mathfrak{X}_\pi(M)$ and $X_f \in \mathfrak{X}_{\text{Ham}}(M)$, then $[X, X_f] = X_{\mathcal{L}_X f} \in \mathfrak{X}_{\text{Ham}}(M)$, and hence $\mathfrak{X}_{\text{Ham}}(M) \subseteq \mathfrak{X}_\pi(M)$ is a Lie ideal.

The first Poisson cohomology group is given by

$$H^1_\pi(M) = \mathfrak{X}_\pi(M)/\mathfrak{X}_{\text{Ham}}(M),$$

i.e., it is the Lie algebra of outer derivations of the Poisson bracket.

- The second Poisson cohomology group on (M, π) has an interpretation in terms of deformations of the Poisson structure π. For $\pi_1 \in \mathfrak{X}^2(M)$ and denoting by ε a formal parameter, the condition

$$d_\pi \pi_1 = [\pi, \pi_1] = 0$$

is equivalent to the property that $\pi + \varepsilon \pi_1$ satisfies the Jacobi identity up to order ϵ^2,

$$[\pi + \varepsilon \pi_1, \pi + \varepsilon \pi_1] = 0 \mod \varepsilon^2.$$

If this holds we view π_1 as defining an infinitesimal deformation of π. This infinitesimal deformation is called *trivial* if $\pi_1 = d_\pi X = -\mathcal{L}_X \pi$ for a vector field X, since in this case π and $\pi + \varepsilon \pi_1$ are isomorphic up to order ε^2 via the time-ε flow of X:

$$(\varphi_X^\varepsilon)_* \pi = \pi + \varepsilon \pi_1 \mod \varepsilon^2.$$

Therefore $H^2_\pi(M)$ is the space of infinitesimal deformations of π modulo trivial deformations, so it could be intuitively regarded as the tangent space at π to the moduli space of Poisson structures on M up to Poisson automorphisms. This indicates the importance of degree 2 Poisson cohomology in the study of normal forms of Poisson structures. Note that π itself defines a class in $H^2_\pi(M)$; when this class vanishes, the Poisson structure is called *exact* (see Exercise 3.15 for an example).

Explicit computations of Poisson cohomology groups are notoriously difficult; see [37, Ch. 2] for many of the known results and calculation methods. See also Sects. 3.2, 3.4, and 3.5 for further references in specific examples.

Exercise 2.35 Consider $M = \mathbb{R}^2$ with Poisson structure $\pi = x\partial_x \wedge \partial_y$.

(a) Verify that the restriction of a Poisson vector field to the y-axis is a constant multiple of $\frac{\partial}{\partial y}$, and that a Poisson vector field is Hamiltonian if and only if it vanishes on the y-axis.
(b) Show that $H_\pi^0(M) = \mathbb{R}$, $H_\pi^1(M) = \mathbb{R}$ (with generator given by the class of $\frac{\partial}{\partial y}$), and $H_\pi^2(M) = 0$.

A Poisson manifold (M, π) has a canonical class in $H_\pi^1(M)$, known as the *modular class*. Let us assume for simplicity that M is orientable. Given a volume form $\eta \in \Omega^{top}(M)$, we recall that the *divergence* of a vector field Y on M with respect to η is the function $\mathrm{div}_\eta(Y) \in C^\infty(M)$ defined by

$$\mathcal{L}_Y \eta = \mathrm{div}_\eta(Y)\eta.$$

Exercise 2.36 Check that the map $C^\infty(M) \to C^\infty(M)$, $f \mapsto \mathrm{div}_\eta(X_f)$, is a derivation of the pointwise product of functions.

Therefore each volume form η defines a vector field X_η on (M, π) by the condition

$$\mathcal{L}_{X_\eta} f = \mathrm{div}_\eta(X_f), \quad \forall f \in C^\infty(M).$$

We call X_η the *modular vector field* with respect to η.

Exercise 2.37

(a) Show that X_η is a Poisson vector field.
(b) Let $\eta' = f\eta$ be another volume form, for $f \in C^\infty(M)$ nowhere vanishing. Show that
$$X_\eta - X_{\eta'} = X_{\log|f|}.$$

By the previous exercise, X_η defines a 1-cocycle in the Poisson complex of (M, π), and its cohomology class

$$[X_\eta] \in H_\pi^1(M)$$

is independent of the choice of volume form. We refer to $[X_\eta]$ as the *modular class* of (M, π). A Poisson manifold is called *unimodular* when its modular class is trivial.

Exercise 2.38 Show that a Poisson manifold is unimodular if and only if there exists a volume form that is invariant by all Hamiltonian vector fields. Conclude that a symplectic manifold is unimodular.

Exercise 2.39 Show that the modular class of the Poisson manifold in Exercise 2.35 is nontrivial, given by $[-\frac{\partial}{\partial y}]$.

See also Exercise 3.7.

Remark 2.8 We may drop the orientability requirement on M by replacing volume forms by smooth densities in the definition of modular vector fields. ◇

The modular class of a Poisson manifold goes back to the work of Koszul, see [58] for original references; it was rediscovered in [90], where the terminology was introduced motivated by analogies with the modular theory of von Neumann algebras.

3 Examples

We have seen some initial examples of Poisson manifolds in Sects. 2.1 and 2.2. We now discuss some broader classes of examples.

3.1 Quotients by Symmetries

Let (M, π) be a Poisson manifold carrying an action of a Lie group G by Poisson diffeomorphisms. Then the space of invariant functions on M,

$$C^\infty(M)^G := \{f \in C^\infty(M) \,|\, \sigma^* f = f \;\forall \sigma \in G\} \subseteq C^\infty(M),$$

is a Poisson subalgebra, i.e.,

$$\{C^\infty(M)^G, C^\infty(M)^G\} \subseteq C^\infty(M)^G,$$

as shown by the following (more general) exercise.

Exercise 3.1 Let K and M be Poisson manifolds. An immediate consequence of (2.2) is that a map $\psi : K \times M \to M$ is Poisson if and only if

$$\{f, g\}_M(\psi(k, x)) = \{f(\psi(k, \cdot)), g(\psi(k, \cdot))\}_M(x) + \{f(\psi(\cdot, x)), g(\psi(\cdot, x))\}_K(k).$$

(When $\{\cdot,\cdot\}_K \equiv 0$, it follows that ψ is Poisson if and only if $\psi(k,\cdot) : M \to M$ is a Poisson map for each $k \in K$.) Show that if $f \circ \psi(k,\cdot) = f$ and $g \circ \psi(k,\cdot) = g$ for all $k \in K$, then $\{f,g\}_M \circ \psi(k,\cdot) = \{f,g\}_M$ for all $k \in K$.

Suppose that the G-action on M is free and proper, so that $B = M/G$ is a smooth manifold and the quotient map $p : M \to B$ is a surjective submersion. Then $C^\infty(B)$ inherits a Poisson bracket from the identification $p^* : C^\infty(B) \xrightarrow{\sim} C^\infty(M)^G$, and in this way B acquires a Poisson structure with the property that $p : M \to B$ is a Poisson map.

A special case of this construction is when (M,ω) is symplectic and G acts on M by symplectomorphisms. For instance, whenever a Lie group G acts on a manifold Q, its cotangent lift[4] induces a G-action on $M = T^*Q$ preserving the canonical symplectic form (cf. Exercise 3.19 (a)). If the G-action on Q is free and proper, so is the lifted action on T^*Q, and hence T^*Q/G inherits a Poisson structure. As an example, consider the action of a Lie group G on itself by right multiplication, $\sigma \mapsto (r_{\sigma^{-1}} : G \to G)$, and the lift of this action to T^*G, $\sigma \mapsto ((dr_\sigma)^* : T^*G \to T^*G)$. In this case the map $T^*G \to \mathfrak{g}^*$ defined by right translations gives rise to an identification of T^*G/G with $\mathfrak{g}^*$.

Exercise 3.2 Check that the induced Poisson structure on $\mathfrak{g}^* = T^*G/G$ is given by

$$\{f,g\}(\xi) := \xi([df|_\xi, dg|_\xi]),$$

for $f,g \in C^\infty(\mathfrak{g}^*)$ and $\xi \in \mathfrak{g}^*$ (you may get a sign depending on your conventions). (Here $df|_\xi$ and $dg|_\xi$ are viewed as elements in $\mathfrak{g}$ through the identification $T_\xi^*\mathfrak{g}^* = (\mathfrak{g}^*)^* = \mathfrak{g}$.)

3.2 Linear Poisson Structures I: Lie Algebras

Let V be a (real, finite-dimensional) vector space. Consider $V^* = C^\infty_{lin}(V) \subseteq C^\infty(V)$, the subspace of linear functions on V. A Poisson structure on V is called *linear* if

$$\{C^\infty_{lin}(V), C^\infty_{lin}(V)\} \subseteq C^\infty_{lin}(V).$$

In this case the restriction of $\{\cdot,\cdot\}$ makes $V^* = C^\infty_{lin}(V)$ into a Lie algebra.

Conversely, let $(\mathfrak{g}, [\cdot,\cdot])$ be a Lie algebra. As a vector space, we can identify it with the space of linear functions on $\mathfrak{g}^*$,

$$\mathfrak{g} = C^\infty_{lin}(\mathfrak{g}^*) \subseteq C^\infty(\mathfrak{g}^*).$$

---

[4] The cotangent lift of a diffeomorphism $\psi : Q \to Q$ is $(d\psi^{-1})^* : T^*Q \to T^*Q$.

Exercise 3.3 Check that (a) there exists at most one almost Poisson bracket on $C^\infty(\mathfrak{g}^*)$ such that $\{\cdot,\cdot\}|_\mathfrak{g} = [\cdot,\cdot]$, and (b) an almost Poisson bracket on $C^\infty(\mathfrak{g}^*)$ with this property is Poisson.

By the previous exercise, the bracket on $C^\infty(\mathfrak{g}^*)$ given in Exercise 3.2,

$$\{f,g\}(\xi) := \xi([df|_\xi, dg|_\xi]),$$

is the unique Poisson structure on $\mathfrak{g}^*$ satisfying $\{\cdot,\cdot\}|_\mathfrak{g} = [\cdot,\cdot]$. This bracket is characterized by the fact that, for $u \in \mathfrak{g}$ (viewed as a linear function on $\mathfrak{g}^*$), the corresponding Hamiltonian vector field on $\mathfrak{g}^*$ is

$$X_u|_\xi = \xi(\mathrm{ad}_u(\cdot)) = \mathrm{ad}_u^*(\xi) \in \mathfrak{g}^* = T_\xi \mathfrak{g}^*. \tag{3.1}$$

In conclusion, we have a natural bijection between linear Poisson structures on a vector space $V = \mathfrak{g}^*$ and Lie algebra structures on its dual $V^* = \mathfrak{g}$.

With respect to linear coordinates $(\xi_1, \ldots, \xi_n)$ on V (corresponding to a basis on V^*), a linear Poisson structure is determined by the linear functions $\{\xi_i, \xi_j\} = \sum_k c_{ijk} \xi_k$, so it has the form

$$\pi = \sum_{i<j} (\sum_k c_{ijk} \xi_k) \frac{\partial}{\partial \xi_i} \wedge \frac{\partial}{\partial \xi_j},$$

and c_{ijk} are the structure constants of the corresponding Lie bracket on V^*.

Note that Example 2.6 is the linear Poisson structure on the dual of the Lie algebra $\mathfrak{so}(3)$.

Exercise 3.4 Let π be a Poisson structure on a vector space V. Show that π is linear if and only if the addition map $+ : V \times V \to V$ is Poisson.

Exercise 3.5 Let $\mathfrak{g}$ and $\mathfrak{h}$ be Lie algebras. Show that a linear map $\mathfrak{g} \to \mathfrak{h}$ is a Lie algebra homomorphism if and only if the dual map $\mathfrak{h}^* \to \mathfrak{g}^*$ is Poisson.

Exercise 3.6 Let $\mathfrak{g}$ be a Lie algebra and $\mathfrak{h} \subseteq \mathfrak{g}$ be a subspace. Show that $\mathfrak{h}$ is an ideal (resp. Lie subalgebra) if and only if $\mathrm{Ann}(\mathfrak{h}) \subseteq \mathfrak{g}^*$ is a Poisson (resp. coisotropic) submanifold. Moreover, when $\mathfrak{h}$ is an ideal, check that the Poisson structure on $\mathrm{Ann}(\mathfrak{h})$ coincides with the one on the dual of the Lie algebra $\mathfrak{g}/\mathfrak{h}$ under the natural identification $\mathrm{Ann}(\mathfrak{h}) = (\mathfrak{g}/\mathfrak{h})^*$.

Exercise 3.7 Let $\mathfrak{g}$ be a Lie algebra. Show that the modular vector field on $\mathfrak{g}^*$ with respect to any constant volume form is the constant vector field

$$\mathfrak{g}^* \ni \xi \mapsto \chi \in T_\xi \mathfrak{g}^* = \mathfrak{g}^*,$$

where χ is the *modular character* of $\mathfrak{g}$, given by $\chi(u) = \mathrm{Tr}(\mathrm{ad}_u)$.

The Poisson cohomology of linear Poisson structures associated to compact Lie algebras is described in [47]. For a description of the Poisson cohomology of all three-dimensional linear Poisson structures, see [52].

The next two exercises concern a generalization of linear Poisson structures.

Exercise 3.8

(a) Let $\mathfrak{g}$ be a Lie algebra with linear Poisson structure $\pi_{\mathfrak{g}^*}$ on $\mathfrak{g}^*$. Let λ be a constant Poisson structure on $\mathfrak{g}^*$ (viewed as an element in $\wedge^2 \mathfrak{g}^*$). Show that $\pi_{\mathfrak{g}^*} + \lambda$ is a Poisson structure on $\mathfrak{g}^*$ if and only if

$$\lambda(u_1, [u_2, u_3]) + \lambda(u_3, [u_1, u_2]) + \lambda(u_2, [u_3, u_1]) = 0, \quad \forall u_1, u_2, u_3 \in \mathfrak{g},$$

i.e., λ is a *2-cocycle* on $\mathfrak{g}$.

(b) A 2-cocycle λ defines a Lie algebra structure on $\mathfrak{g} \oplus \mathbb{R}$ via $[(u,t),(v,s)] = ([u,v], \lambda(u,v))$ (a central extension of $\mathfrak{g}$ by $\mathbb{R}$). Check that $(\mathfrak{g}^*, \pi_{\mathfrak{g}^*} + \lambda)$ sits in $(\mathfrak{g} \oplus \mathbb{R})^* = \mathfrak{g}^* \oplus \mathbb{R}$ as a Poisson submanifold via $\xi \mapsto (\xi, 1)$.

Exercise 3.9 Let $C^\infty(V)_{aff} \subseteq C^\infty(V)$ denote the subspace of affine functions on a vector space V. Show that a Poisson structure on V is *affine*, in the sense that $\{C^\infty_{aff}(V), C^\infty_{aff}(V)\} \subseteq C^\infty_{aff}(V)$, if and only if it is of the form $\pi + \lambda$, where π is a linear Poisson structure on V and $\lambda \in \wedge^2 V$ is a 2-cocycle on the Lie algebra V^*.

Many important examples of linear and affine Poisson structures on duals of Lie algebras occur in infinite dimensions, providing the Hamiltonian description of various PDEs related e.g. to fluid dynamics, see e.g. [55].

Linear (and affine) Poisson structures play a central role in the theory of Hamiltonian actions. Let M be a Poisson manifold and $\mathfrak{g}$ a Lie algebra. A $\mathfrak{g}$-action (i.e. Lie-algebra homomorphism) $\psi : \mathfrak{g} \to \mathfrak{X}^1(M)$ is called *weakly Hamiltonian* if there exists a linear map $\mu^* : \mathfrak{g} \to C^\infty(M)$ such that

$$\psi(u) = X_{\mu^* u}, \text{ for all } u \in \mathfrak{g}. \tag{3.2}$$

The $\mathfrak{g}$-action is said to be *Hamiltonian* when there is such a μ^* that is, in addition, a Lie algebra homomorphism.

Any linear map $\mu^* : \mathfrak{g} \to C^\infty(M)$ is equivalent to a smooth map $\mu : M \to \mathfrak{g}^*$ via

$$\langle \mu(x), u \rangle = \mu^*(u)(x),$$

for $x \in M$ and $u \in \mathfrak{g}$. For a weakly Hamiltonian $\mathfrak{g}$-action $\psi : \mathfrak{g} \to \mathfrak{X}^1(M)$, a map $\mu : M \to \mathfrak{g}^*$ for which (3.2) holds is called a *moment map*.

Exercise 3.10 Check that a linear map $\mu^* : \mathfrak{g} \to C^\infty(M)$ is a Lie algebra homomorphism if and only if $\mu : M \to \mathfrak{g}^*$ is a Poisson map.

An example of a Hamiltonian action is the coadjoint $\mathfrak{g}$-action on $\mathfrak{g}^*$; in this case the identity map $\mathfrak{g}^* \to \mathfrak{g}^*$ is a moment map (the corresponding Lie-algebra homomorphism $\mathfrak{g} \to C^\infty(\mathfrak{g}^*)$ is the natural inclusion of $\mathfrak{g}$ as linear functions on $\mathfrak{g}^*$), and generating vector fields are given in (3.1).

Exercise 3.11 If $\mu : M \to \mathfrak{g}^*$ is a moment map for a weakly Hamiltonian $\mathfrak{g}$-action, check that μ is a Poisson map if and only if it is $\mathfrak{g}$-equivariant with respect to the coadjoint action (cf. Exercise 2.2 (b)).

For Hamiltonian actions, moment maps will always be assumed to be equivariant (or equivalently, by the previous exercise, to be Poisson maps). Given a Hamiltonian $\mathfrak{g}$-action ψ on (M, π) with moment map $\mu : M \to \mathfrak{g}^*$, the data (M, π, ψ, μ) is called a *Hamiltonian $\mathfrak{g}$-space*. Note that the action ψ is determined by μ via (3.2); on the other hand, any Poisson map $\mu : M \to \mathfrak{g}^*$ defines a Hamiltonian action on M by the same formula (3.2) (see (2.4) and Exercise 3.10). It follows that Hamiltonian $\mathfrak{g}$-spaces are the same as Poisson maps into $\mathfrak{g}^*$,

$$\text{Hamiltonian } \mathfrak{g}\text{-spaces} \rightleftharpoons \text{Poisson maps into } \mathfrak{g}^*.$$

More generally, given a Poisson manifold M, a Lie algebra $\mathfrak{g}$, and a smooth map $\mu : M \to \mathfrak{g}^*$, define the skew-symmetric bilinear map $\lambda : \mathfrak{g} \times \mathfrak{g} \to C^\infty(M)$,

$$\lambda(u, v) = \{\mu^* u, \mu^* v\} - \mu^*[u, v].$$

Exercise 3.12 Show that μ is the moment map for a weakly Hamiltonian $\mathfrak{g}$-action on M if and only if λ takes values in Casimirs of M. If M is symplectic and connected, conclude that λ defines an element in $\wedge^2 \mathfrak{g}^*$ and verify that it is a 2-cocycle on $\mathfrak{g}$ (cf. Exercise 3.8).

Suppose that (M, ω) is a connected symplectic manifold carrying a weakly Hamiltonian $\mathfrak{g}$-action with moment map $\mu : M \to \mathfrak{g}^*$ and associated 2-cocycle $\lambda \in \wedge^2 \mathfrak{g}^*$, as in the previous exercise. Then $\mathfrak{g}^*$ carries an affine Poisson structure $\pi_{\mathfrak{g}^*} + \lambda$, cf. Exercises 3.8

and 3.9, and an affine $\mathfrak{g}$-action $\psi_\lambda : \mathfrak{g} \to \mathfrak{X}^1(\mathfrak{g}^*)$, where

$$\psi_\lambda(u)|_\xi = \mathrm{ad}_u^*(\xi) + i_u \lambda \in \mathfrak{g}^* = T_\xi \mathfrak{g}^*.$$

(The affine action ψ_λ is simply the Hamiltonian action on $(\mathfrak{g}^*, \pi_{\mathfrak{g}^*} + \lambda)$ with moment map given by the identity map $\mathfrak{g}^* \to \mathfrak{g}^*$, cf. (3.1).)

Exercise 3.13 Verify that the moment map μ is a Poisson map $(M, \omega) \to (\mathfrak{g}^*, \pi_{\mathfrak{g}^*} + \lambda)$, noticing that μ is Poisson if and only if it is $\mathfrak{g}$-equivariant (with respect to the affine action on $\mathfrak{g}^*$).

3.3 Linear Poisson Structures II: Lie Algebroids

The notion of linear Poisson structure makes sense, more generally, on vector bundles.

For a vector bundle $E \xrightarrow{q} M$, denote by $C_{bas}^\infty(E) = q^* C^\infty(M) \subseteq C^\infty(E)$ the subalgebra of basic functions, and by $C_{lin}^\infty(E) \subseteq C^\infty(E)$ the subspace of fiberwise linear functions. We use the fact that $C_{bas}^\infty(E) \cdot C_{lin}^\infty(E) \subseteq C_{lin}^\infty(E)$ to regard $C_{lin}^\infty(E)$ as a $C^\infty(M)$-module, so that we have a natural identification $\Gamma(E^*) = C_{lin}^\infty(E)$ (as $C^\infty(M)$-modules).

A Poisson structure on the total space E is called *linear* if

$$\{C_{lin}^\infty(E), C_{lin}^\infty(E)\} \subseteq C_{lin}^\infty(E).$$

We denote by

$$[\cdot, \cdot] : \Gamma(E^*) \times \Gamma(E^*) \to \Gamma(E^*) \qquad (3.3)$$

the Lie bracket on $\Gamma(E^*) = C_{lin}^\infty(E)$ obtained by restriction of $\{\cdot, \cdot\}$.

Exercise 3.14 Check that (by the Leibniz rule) a linear Poisson structure on $E \to M$ satisfies

$$\{C_{lin}^\infty(E), C_{bas}^\infty(M)\} \subseteq C_{bas}^\infty(E) \qquad \{C_{bas}^\infty(E), C_{bas}^\infty(E)\} = 0.$$

Show also that there is a bundle map $\rho : E^* \to TM$ such that

$$\{\xi, q^* f\} = q^* \mathcal{L}_{\rho(\xi)} f, \qquad (3.4)$$

for $\xi \in \Gamma(E^*)$ and $f \in C^\infty(M)$, and that ρ and the Lie bracket $[\cdot, \cdot]$ on $\Gamma(E^*)$ satisfy $[\xi, f\eta] = (\mathcal{L}_{\rho(\xi)} f)\eta + f[\xi, \eta]$, for all $\xi, \eta \in \Gamma(E^*)$ and $f \in C^\infty(M)$.

From the previous exercise, in local bundle coordinates $(x_1, \ldots, x_n, \xi_1, \ldots, \xi_r)$ on E (defined by a local frame on E^*), a linear Poisson structure satisfies

$$\{x_i, x_j\} = 0, \quad \{x_i, \xi_j\} = -\mathcal{L}_{\rho(\xi_j)} x_i = -\rho_{ij}(x), \quad \{\xi_i, \xi_j\} = \sum_k c_{ijk}(x) \xi_k,$$

so the corresponding bivector field $\pi \in \mathcal{X}^2(E)$ locally has the form

$$\pi = \sum_{i<j} (\sum_k c_{ijk}(x) \xi_k) \frac{\partial}{\partial \xi_i} \wedge \frac{\partial}{\partial \xi_j} - \sum_{i,j} \rho_{ij}(x) \frac{\partial}{\partial x_i} \wedge \frac{\partial}{\partial \xi_j}.$$

Exercise 3.15 Let $\kappa_t : E \to E$ denote the fiberwise scalar multiplication by $t \in \mathbb{R}$, and denote by $\mathcal{E} \in \mathcal{X}^1(E)$ the Euler vector field (in local bundle coordinates (x_i, ξ_j), $\mathcal{E} = \sum_j \xi_j \frac{\partial}{\partial \xi_j}$). Show that a Poisson structure π on E is linear if and only if $(\kappa_t)_* \pi = t\pi$ for all $t > 0$, if and only if $\mathcal{L}_\mathcal{E} \pi = -\pi$.

A *Lie algebroid* is a vector bundle $A \to M$ equipped with a Lie bracket $[\cdot, \cdot]$ on $\Gamma(A)$ and a vector bundle morphism $\rho : A \to TM$ (over the identity map in M), called the *anchor*, such that

$$[u, fv] = (\mathcal{L}_{\rho(u)} f) v + f[u, v]$$

for all $u, v \in \Gamma(A)$ and $f \in C^\infty(M)$. One refers to this last condition as the *Leibniz identity*.

Exercise 3.16 Show that, on a Lie algebroid, the anchor map $\rho : \Gamma(A) \to \mathcal{X}^1(M)$ preserves Lie brackets.

Examples of Lie algebroids include Lie algebras (when M is a point) and tangent bundles $A = TM$ (with Lie bracket given by the usual commutator bracket of vector fields and ρ the identity map) or, more generally, involutive subbundles $F \subseteq TM$. We will see other examples later.

For now, the point to be made is that, given a linear Poisson structure on $E \to M$, the Lie bracket (3.3) on $\Gamma(E^*)$ and the map ρ defined in (3.4) make (by Exercise 3.14) E^* into a Lie algebroid. Conversely, a Lie algebroid structure on $A \to M$ gives rise to a linear Poisson structure on A^*, uniquely determined by the conditions

$$\{u, v\} = [u, v], \qquad \{u, q^* f\} = q^* (\mathcal{L}_{\rho(u)} f) \tag{3.5}$$

for $u, v \in \Gamma(A) = C^\infty_{lin}(A^*)$, $f \in C^\infty(M)$, and $q : A^* \to M$ the bundle projection. These constructions are mutually inverse and establish a bijection between linear Poisson structures on vector bundles $E = A^*$ and Lie algebroid structures on $E^* = A$.

Exercise 3.17 Check that conditions (3.5) indeed define a unique linear Poisson bracket on A^*.

Exercise 3.18

(a) Let $A \to M$ and $B \to M$ be Lie algebroids. Show that a vector bundle map $\varphi : A \to B$ (over the identity map) is a Lie algebroid morphism (i.e., it intertwines anchor maps and the induced map $\Gamma(A) \to \Gamma(B)$ preserves brackets) if and only if $\varphi^* : B^* \to A^*$ is a Poisson map.
(b) Let $B \to M$ be a vector subbundle of a Lie algebroid $A \to M$. Show that B is a Lie subalgebroid (i.e., $\Gamma(B) \subseteq \Gamma(A)$ is a Lie subalgebra) if and only if $\text{Ann}(B) \subseteq A^*$ is a coisotropic submanifold.

Cf. Exercises 3.5 and 3.6.

Exercise 3.19

(a) Check that the Poisson bracket defined by the canonical symplectic form on T^*M is linear, and that the dual Lie algebroid structure on TM is the canonical one, i.e., with $\rho = \text{Id}$. (It follows that $\mathfrak{X}^1(M)$ sits in $C^\infty(T^*M)$ as a Lie subalgebra, and hence any action $\mathfrak{g} \to \mathfrak{X}^1(M) \subseteq C^\infty(T^*M)$ automatically defines a Hamiltonian $\mathfrak{g}$-action on T^*M, known as its *cotangent lift*.)
(b) Let π be a linear Poisson structure on $E \to M$. Show that π is nondegenerate if and only if the anchor $E^* \to TM$ is an isomorphism, if and only if E is isomorphic to T^*M equipped with its canonical Poisson structure.

Remark 3.1 For any Lie algebroid $A \to M$ there is a Schouten bracket on $\Gamma(\wedge^\bullet A)$, defined analogously to the one on Sect. 2.3 when $A = TM$, extending the Lie bracket on $\Gamma(A)$. When $A = \mathfrak{g}$ is a Lie algebra, the bracket on $\wedge^\bullet \mathfrak{g}$ agrees with the one induced from the Schouten bracket of left-invariant multivector fields on a Lie group with Lie algebra $\mathfrak{g}$.

◇

3.4 Poisson-Lie Groups and Their Homogeneous Spaces

Let G be a Lie group. A bivector field π on G is called *multiplicative* if

$$\pi_{\sigma_1 \sigma_2} = (l_{\sigma_1})_* \pi_{\sigma_2} + (r_{\sigma_2})_* \pi_{\sigma_1}, \quad \forall \sigma_1, \sigma_2 \in G,$$

where $l_\sigma, r_\sigma : G \to G$ denote left and right multiplication by $\sigma \in G$. It follows that $\pi_e = 0$, so any nontrivial example of multiplicative bivector field has varying rank.

Exercise 3.20 Check that a Poisson structure π on G is multiplicative if and only if the multiplication map $m : G \times G \to G$ is a Poisson map and that, in this case, the inversion map is anti-Poisson (i.e., it is a Poisson map from (G, π) to $(G, -\pi)$)).

A Lie group equipped with a multiplicative Poisson structure is called a *Poisson Lie group*.

Exercise 3.21 A bivector field π on a Lie group G is called *affine* if

$$\pi_{\sigma_1 \sigma_2} = (l_{\sigma_1})_* \pi_{\sigma_2} + (r_{\sigma_2})_* \pi_{\sigma_1} - (r_{\sigma_2})_* (l_{\sigma_1})_* \pi_e, \quad \forall \sigma_1, \sigma_2 \in G,$$

so that π is multiplicative if and only if it is affine and $\pi_e = 0$. Prove that if π is affine, then $\mathcal{L}_{u^r} \pi$ is right invariant for all $u \in \mathfrak{g}$ (equivalently, $\mathcal{L}_{u^l} \pi$ is left invariant for all $u \in \mathfrak{g}$), and the converse holds if G is connected.

Any Lie group is a Poisson Lie group with the trivial Poisson structure. Multiplicative Poisson structures on vector spaces, viewed as abelian groups, are the same as linear Poisson structures (see Exercise 3.4).

Exercise 3.22 Let $G = \mathbb{R} \ltimes \mathbb{R}$ be the semi-direct product Lie group with respect to the action of $\mathbb{R}$ on itself via $y \mapsto e^x y$; i.e., G is $\mathbb{R}^2 = \{(x, y)\}$ as a manifold with group structure $(x_1, y_1) \cdot (x_2, y_2) = (x_1 + x_2, y_1 + e^{x_1} y_2)$. For $\varphi \in C^\infty(\mathbb{R}^2)$, check that the Poisson structure $\varphi(x, y)\partial_x \wedge \partial_y$ (see Exercise 2.5) is multiplicative if and only if $\varphi(x_1 + x_2, y_1 + e^{x_1} y_2) = \varphi(x_1, y_1) + e^{x_1} \varphi(x_2, y_2)$. In particular, each of the Poisson structures $(e^x - 1)\partial_x \wedge \partial_y$ and $y \partial_x \wedge \partial_y$ makes $\mathbb{R} \ltimes \mathbb{R}$ into a Poisson Lie group.

An important class of examples of Poisson Lie groups arises as follows. Let G be a Lie group with Lie algebra $\mathfrak{g}$. For $\mathfrak{r} \in \wedge^2 \mathfrak{g}$, consider the bivector field on π on G given by

$$\pi_\sigma = (r_\sigma)_* \mathfrak{r} - (l_\sigma)_* \mathfrak{r} = (\mathfrak{r}^r - \mathfrak{r}^l)|_\sigma. \tag{3.6}$$

The next exercise explains when such π makes G into a Poisson Lie group. We will make use of the Schouten bracket $[\cdot, \cdot]$ on $\wedge^\bullet \mathfrak{g}$, see Remark 3.1

Exercise 3.23

(a) Check that any bivector field π of the form (3.6) is multiplicative.
(b) Verify that $[\pi, \pi] = 0$ if and only if $[\mathfrak{r}, \mathfrak{r}]^r = [\mathfrak{r}, \mathfrak{r}]^l$, i.e., $[\mathfrak{r}, \mathfrak{r}]$ is Ad-invariant.

When $\pi = \mathfrak{r}^r - \mathfrak{r}^l$ is a Poisson structure, i.e., $[\mathfrak{r}, \mathfrak{r}]$ is Ad-invariant, $\mathfrak{r}$ is called an *r-matrix* and the corresponding Poisson-Lie group is called *exact* or *coboundary*; when $\mathfrak{r}$ satisfies $[\mathfrak{r}, \mathfrak{r}] = 0$ (known as the classical Yang-Baxter equation), $\mathfrak{r}$ is called a *triangular r-matrix*.

If G is connected and semisimple, or if G is compact, any multiplicative Poisson structure is defined by an r-matrix (see [67, Thm. 1.11]).

Example 3.2 Let G be the Lie group

$$SU(2) = \left\{ \begin{pmatrix} a & b \\ -\bar{b} & \bar{a} \end{pmatrix}, \ a = x+iy, \ b = z+iw \in \mathbb{C}, \ |a|^2 + |b|^2 = 1 \right\}.$$

Its Lie algebra $\mathfrak{g} = \mathfrak{su}(2)$ has a basis $\{e_1, e_2, e_3\}$, where

$$e_1 = \frac{1}{2}\begin{pmatrix} i & 0 \\ 0 & -i \end{pmatrix}, \quad e_2 = \frac{1}{2}\begin{pmatrix} 0 & 1 \\ -1 & 0 \end{pmatrix}, \quad e_3 = \frac{1}{2}\begin{pmatrix} 0 & i \\ i & 0 \end{pmatrix}, \tag{3.7}$$

satisfying $[e_1, e_2] = e_3$, $[e_3, e_1] = e_2$, $[e_2, e_3] = e_1$. Notice that $e_1 \wedge e_2 \wedge e_3$ is Ad-invariant; since it generates the 1-dimensional space $\wedge^3 \mathfrak{g}$, any $\mathfrak{r} \in \mathfrak{g} \wedge \mathfrak{g}$ is such that $[\mathfrak{r}, \mathfrak{r}]$ is a multiple of $e_1 \wedge e_2 \wedge e_3$, and hence Ad-invariant as well. The choice $\mathfrak{r} = 2e_2 \wedge e_3$ defines the so-called "standard" Poisson structure on $SU(2)$.

◇

The next exercise relates the standard Poisson structure on $SU(2)$ with the Poisson structure on spheres given in Exercise 2.18.

Exercise 3.24 Consider $\mathbb{S}^3 = \{(x, y, z, w) \in \mathbb{R}^4 \mid x^2 + y^2 + z^2 + w^2 = 1\}$ with the Poisson structure π defined by the bracket relations

$$\{x, y\} = z^2 + w^2, \ \{x, z\} = -yz, \ \{x, w\} = -yw, \ \{y, z\} = xz, \ \{y, w\} = wx, \ \{z, w\} = 0,$$

as in Exercise 2.18, and Lie group structure coming from the quaternionic multiplication in $\mathbb{R}^4 = \{x + y\mathbf{i} + z\mathbf{j} + w\mathbf{k}\}$. With respect to the usual identification of Lie groups $SU(2) = \mathbb{S}^3$,

$$\begin{pmatrix} x+iy & z+iw \\ -z+iw & x-iy \end{pmatrix} \mapsto x + y\mathbf{i} + z\mathbf{j} + w\mathbf{k},$$

we have that $e_1 = \mathbf{i}/2$, $e_2 = \mathbf{j}/2$ and $e_3 = \mathbf{k}/2$ (see (3.7)). Verify that

$$2(e_2^r \wedge e_3^r - e_2^l \wedge e_3^l) = \pi,$$

i.e., π agrees with the standard Poisson structure.

A multiplicative Poisson structure on a Lie group G admits an infinitesimal description in terms of the Lie algebra $\mathfrak{g}$. Any bivector field π on G such that $\pi_e = 0$ gives rise to a map

$$\delta : \mathfrak{g} \to \wedge^2 \mathfrak{g}, \quad u \mapsto (\mathcal{L}_X \pi)_e, \tag{3.8}$$

where X is any vector field on G satisfying $X|_e = u$ (check that this is well defined). Let $[\cdot, \cdot]_* : \wedge^2 \mathfrak{g}^* \to \mathfrak{g}^*$ be the dual map. One can prove that (see [67, Thm. 1.6])

- if π is multiplicative, then

$$\delta([u, v]) = [\delta u, v] + [u, \delta v], \quad \forall u, v \in \mathfrak{g}; \tag{3.9}$$

- if π is Poisson, then $[\cdot, \cdot]_*$ defines a Lie bracket on $\mathfrak{g}^*$.

Exercise 3.25 Let V be a vector space equipped with a linear Poisson structure π, regarded as an abelian Poisson Lie group. Check that the map $\wedge^2 V^* \to V^*$ dual to (3.8) coincides with the Lie bracket on V^* corresponding to π (in the sense of Sect. 3.2).

A *Lie bialgebra* is a pair of Lie algebras $(\mathfrak{g}, [\cdot, \cdot])$ and $(\mathfrak{g}^*, [\cdot, \cdot]_*)$ such that the map δ dual to $[\cdot, \cdot]_*$, referred to as the *cobracket*, satisfies the compatibility condition (3.9). We will use the notation $(\mathfrak{g}, \mathfrak{g}^*)$ or $(\mathfrak{g}, \delta)$ for a Lie bialgebra. A *Lie-bialgebra morphism* from $(\mathfrak{g}_1, \mathfrak{g}_1^*)$ to $(\mathfrak{g}_2, \mathfrak{g}_2^*)$ is a Lie algebra morphism $T : \mathfrak{g}_1 \to \mathfrak{g}_2$ satisfying one (and hence all) of the following equivalent conditions: T intertwines δ_1 and δ_2; $T^* : \mathfrak{g}_2^* \to \mathfrak{g}_1^*$ is a Lie algebra morphism; T is a Poisson map (with respect to the Poisson structures on $\mathfrak{g}_1$ and $\mathfrak{g}_2$ dual to the Lie brackets on $\mathfrak{g}_1^*$ and $\mathfrak{g}_2^*$, respectively).

As explained above, any Poisson Lie group G gives rise to a Lie bialgebra via (3.8). Conversely, any Lie bialgebra $(\mathfrak{g}, \delta)$ arises in this way from a unique multiplicative Poisson structure on a connected, simply-connected Lie group G integrating $\mathfrak{g}$ [67, Thm. 1.8].

A key fact about Lie bialgebras is that if $(\mathfrak{g}, \delta)$ is a Lie bialgebra, then so is $(\mathfrak{g}^*, \delta_*)$ (where δ_* is the cobracket corresponding to the Lie bracket on $\mathfrak{g}$), called the Lie bialgebra *dual* to $(\mathfrak{g}, \delta)$. For a Poisson Lie group G with Lie bialgebra $(\mathfrak{g}, \delta)$, the connected, simply-connected Lie group integrating the dual Lie bialgebra $(\mathfrak{g}^*, \delta_*)$ is a Poisson Lie group G^* called the *dual* of G. (More generally, we refer to any Poisson Lie group integrating the dual Lie bialgebra $(\mathfrak{g}^*, \delta_*)$ as a *dual* to G.)

As an example, for a Lie group G, regarded as a Poisson Lie group with the trivial Poisson structure, the dual Poisson Lie group is $\mathfrak{g}^*$ (see Exercise 3.25).

Exercise 3.26 Consider the Lie group $G = \mathbb{R} \ltimes \mathbb{R}$ of Exercise 3.22. Check that the Poisson Lie groups $(G, y\partial_x \wedge \partial_y)$ and $(G, (e^x - 1)\partial_x \wedge \partial_y)$ are both self-dual, i.e., in each case there is an isomorphism $(\mathfrak{g}, \mathfrak{g}^*) \cong (\mathfrak{g}^*, \mathfrak{g})$.

Exercise 3.27 Consider $SU(2) = \mathbb{S}^3$ with the standard Poisson structure (see Exercise 3.24). Show that the dual Lie algebra $\mathfrak{su}(2)^*$ is isomorphic to $\mathbb{R}^3$ with Lie bracket defined by

$$[e_1, e_2] = e_2, \quad [e_1, e_3] = e_3, \quad [e_2, e_3] = 0,$$

called the "book" Lie algebra $\mathfrak{sb}(2, \mathbb{C})$ (since its regular coadjoint orbits look like the pages of an open book, see Exercise 4.10 (c)). In particular, as a Lie group, the dual of $SU(2)$ can be identified with $SB(2, \mathbb{C}) = \left\{ \begin{pmatrix} a & b+ic \\ 0 & a^{-1} \end{pmatrix}, a > 0, b, c \in \mathbb{R} \right\}$.

If (G, π_G) is a Poisson Lie group and (M, π) is a Poisson manifold, we say that an action of G on M is *Poisson* if the action map $\Psi : G \times M \to M$, $(\sigma, x) \mapsto \sigma.x$, is a Poisson map; equivalently,

$$\pi_{\sigma.x} = (\Psi_\sigma)_* \pi_x + (\Psi_x)_* (\pi_G)_\sigma, \tag{3.10}$$

where $\Psi_\sigma = \Psi(\sigma, \cdot) : M \to M$ and $\Psi_x = \Psi(\cdot, x) : G \to M$, for all $\sigma \in G$, $x \in M$. Note that π is generally not G-invariant (cf. Exercise 3.28 below), but it directly follows from (3.10) that if a G-action on M is Poisson for two Poisson structures π_1 and π_2, then $\pi_1 - \pi_2$ is a G-invariant bivector field.

Even though a Poisson action is not an action by Poisson diffeomorphisms, it still follows from Exercise 3.1 that the space of G-invariant functions $C^\infty(M)^G \subseteq C^\infty(M)$ is a Poisson subalgebra. Hence if the action is free and proper, the orbit space M/G acquires a Poisson structure for which the quotient map $M \to M/G$ is Poisson (generalizing the discussion in Sect. 3.1). The theory of Hamiltonian actions and moment maps extends to Poisson actions of Poisson Lie groups—but in this more general setting moment maps are Poisson maps into dual Poisson Lie groups [65].

Exercise 3.28 Let G be a Poisson Lie group with Lie bialgebra $(\mathfrak{g}, \delta)$. Suppose that G acts on a Poisson manifold (M, π_M) with infinitesimal action denoted by $\psi : \mathfrak{g} \to \mathfrak{X}(M)$. Verify that if the G-action on M is Poisson, then $\mathcal{L}_{\psi(u)} \pi_M = \psi(\delta(u))$ for all $u \in \mathfrak{g}$, and that the converse holds if G is connected.

A class of examples of Poisson manifolds closely related to Poisson Lie groups is given by their Poisson homogeneous spaces [34]: For a Poisson Lie group G, a *Poisson homogeneous G-space* is a Poisson manifold equipped with a transitive Poisson G-action.

Any manifold M carrying a transitive action of a Lie group G is of the form G/H for a closed subgroup H (we consider H acting on G by right multiplication). So a Poisson homogeneous G-space is equivalently described by a closed subgroup $H \subseteq G$ and a Poisson structure on G/H for which the map

$$G \times G/H \to G/H, \qquad (\sigma_1, \sigma_2 H) \mapsto \sigma_1 \sigma_2 H$$

is a Poisson map. A special situation is when H is a Poisson subgroup: in this case the H-action on G is Poisson and the Poisson structure on G descends to a homogeneous Poisson structure on G/H. In general, if a Poisson structure π makes G/H into a Poisson homogeneous space, it is *not* true that the quotient map $G \to G/H$ is a Poisson map. The next exercise explains when this happens.

Exercise 3.29 Let H be a closed subgroup of a Poisson Lie group G.

(a) Suppose that H has the property that the Poisson bracket of any two H-invariant functions on G vanishes on H. Use the multiplicativity of the Poisson structure on G to show that if f and g are H-invariant, then so is $\{f, g\}$.
(b) Show that G/H carries a Poisson structure for which $G \to G/H$ is a Poisson map if and only if H is coisotropic. Moreover, in this case G/H is a Poisson homogeneous space.

Exercise 3.30 Let G be a Poisson Lie group with Lie bialgebra $(\mathfrak{g}, \mathfrak{g}^*)$, let $H \subseteq G$ be a Lie subgroup. Show that if H is coisotropic, then $\mathrm{Ann}(\mathfrak{h}) \subseteq \mathfrak{g}^*$ is a Lie subalgebra; similarly, if H is Poisson then $\mathrm{Ann}(\mathfrak{h})$ is a Lie ideal of $\mathfrak{g}^*$ (cf. Exercise 3.6). Moreover, the converses of both statements hold when H is connected.

Exercise 3.31 For a Poisson Lie group (G, π_G), let π be a Poisson homogeneous structure on G itself, with respect to the action by left multiplication ($H = \{e\}$). Show that $\pi - \pi_G = (\pi_e)^l$, and that π is an affine Poisson structure on G (see Exercise 3.9).

(More is true: given any affine Poisson structure π on G, $\pi - (\pi_e)^l$ is a multiplicative Poisson structure on G with respect to which (G, π) is a Poisson homogeneous space.)

Hence affine Poisson structures and quotients of Poisson Lie groups by closed coisotropic subgroups provide many examples of Poisson homogeneous spaces.

Remark 3.3 (Drinfeld's Classification) Given a Lie bialgebra $(\mathfrak{g}, \mathfrak{g}^*)$, there is a natural Lie (bi)algebra structure on $\mathfrak{d} = \mathfrak{g} \oplus \mathfrak{g}^*$, known as the *Drinfeld double*; the Lie bracket on $\mathfrak{d}$ is characterized by the properties that its natural symmetric pairing is ad-invariant and $\mathfrak{g}$ and $\mathfrak{g}^*$ are Lie subalgebras. For a closed Lie subgroup H of a Poisson Lie group G, Poisson homogeneous structures on G/H are classified by suitable lagrangian subalgebras of $\mathfrak{d}$ [35], see also [71,72]; for coisotropic subgroups, the corresponding lagrangian subalgebras are of the form $\mathfrak{h} \oplus \mathrm{Ann}(\mathfrak{h})$, see Exercise 3.30. ◇

A Brief Introduction to Poisson Geometry

As an explicit example of Poisson homogeneous space, consider the standard Poisson structure π on $SU(2) = \mathbb{S}^3$ described in Exercise 3.24. The subgroup $H = \{(x, y, 0, 0) \mid x^2 + y^2 = 1\} \subseteq \mathbb{S}^3$ is such that $\pi|_H = 0$, so H is a Poisson subgroup. The quotient space of $\mathbb{S}^3$ by the action of $H = \mathbb{S}^1$ by right multiplication is $\mathbb{C}P^1 = \mathbb{S}^2 \subseteq \mathbb{R}^3 = \{(x_1, x_2, x_3)\}$, with quotient map $\mathbb{S}^3 \to \mathbb{S}^2$ given by the Hopf map

$$x_1 = 2(xz - yw), \quad x_2 = 2(yz + xw), \quad x_3 = x^2 + y^2 - z^2 - w^2.$$

As explained above, $\mathbb{S}^2$ inherits a quotient Poisson structure that makes it into a Poisson homogeneous space. We describe it in the next exercise.

Exercise 3.32 Show that

$$\{x_1, x_2\} = 2(x_3 - 1)x_3, \quad \{x_2, x_3\} = 2(x_3 - 1)x_1, \quad \{x_1, x_3\} = -2(x_3 - 1)x_2.$$

Following Exercise 2.17, we see that this Poisson structure on $\mathbb{S}^2$ agrees with

$$2(1 - x_3)\pi_{area}, \tag{3.11}$$

where π_{area} is the Poisson structure corresponding to the area form. So it is nondegenerate at all points except for $(0, 0, 1)$, where it vanishes. The next exercise provides suitable coordinates around this singular point.

Exercise 3.33

(a) Consider the stereographic projection (from the south pole $\mathbf{s} = (0, 0, -1)$),

$$\mathbb{S}^2 \setminus \mathbf{s} \to \mathbb{R}^2, \quad (x_1, x_2, x_3) \mapsto \left(y_1 = \frac{x_1}{1 + x_3}, y_2 = \frac{x_2}{1 + x_3} \right),$$

with inverse

$$x_1 = \frac{2y_1}{(1 + y_1^2 + y_2^2)}, \quad x_2 = \frac{2y_2}{(1 + y_1^2 + y_2^2)}, \quad x_3 = \frac{(1 - (y_1^2 + y_2^2))}{(1 + y_1^2 + y_2^2))}.$$

Show that $\{y_2, y_1\} = (y_1^2 + y_2^2)(1 + y_1^2 + y_2^2)$.

(b) With the additional change of coordinates given by $z_1 = y_1/\sqrt{1 + y_1^2 + y_2^2}$ and $z_2 = y_2/\sqrt{1 + y_1^2 + y_2^2}$, show that the Poisson structure becomes

$$(z_1^2 + z_2^2)\partial_{z_2} \wedge \partial_{z_1}.$$

Exercise 3.34 Show that any Poisson structure on $\mathbb{S}^2 = \{(x_1, x_2, x_3) \mid x_1^2 + x_2^2 + x_3^2 = 1\}$ that makes it into a Poisson homogeneous $SU(2)$-space is of the form $2(\lambda - x_3)\pi_{area}$, for $\lambda \in \mathbb{R}$.

The Poisson cohomology of the family of Poisson structures in the previous exercise is computed in [80, Thm. 3.2].

The Poisson structure (3.11) on $\mathbb{S}^2$, obtained as a quotient of the standard Poisson structure on $SU(2)$, is an example of a *Bruhat* Poisson structure. Much more generally [67, Thm. 4.7], any connected, compact, semi-simple Lie group carries a nontrivial "standard" Poisson structure making it into a Poisson Lie group, and each of its coadjoint orbits carries a "Bruhat" Poisson structure making it into a Poisson homogeneous space.

Further details on this subsection can be found e.g. in [57, Chp. 1].

3.5 Log-Symplectic Manifolds and Symplectic Lie Algebroids

Log-symplectic structures are special types of Poisson structures that are very close to being symplectic, in the sense that they are generically nondegenerate. Their systematic study began in [50, 78].

Let M be a $2n$-dimensional manifold. Any bivector field π defines a section $\pi^n := \wedge^n \pi$ of the line bundle $\wedge^{2n} TM$; note that π^n is nonzero at points where π is nondegenerate, and it vanishes at points where the rank of π is less than $2n$.

A Poisson structure π on M is called *log-symplectic* (or *b-symplectic*) if π^n vanishes transversally, i.e., π^n is transverse to the zero section of $\wedge^{2n} TM$. It follows that the zero set $Z := (\pi^n)^{-1}(0)$ is a hypersurface (i.e., a codimension 1 submanifold, possibly disconnected) of M, called the *singular* or *degeneracy locus* of π.

In local coordinates $(x_1, \ldots, x_{2n})$ on $U \subseteq M$, we have $\pi^n = f \partial_{x_1} \wedge \ldots \wedge \partial_{x_{2n}}$ for a smooth function f on U, and $f^{-1}(0) = Z \cap U$. In these coordinates the transversality condition is equivalent to 0 being a regular value of f.

The next exercise shows how 2-dimensional log-symplectic structures look like around points in the degeneracy locus.

Exercise 3.35

(a) Consider a nondegenerate Poisson structure on $\mathbb{R}^2$. Let $f \in C^\infty(\mathbb{R}^2)$ be such that $f(m) = 0$ and $df|_m \neq 0$. Prove that, in a neighborhood of m, one can find local coordinates (y_1, y_2), such that $y_1 = f$ and $\{y_1, y_2\} = 1$. (Use the straightening theorem for X_f.)

(b) Let π be a log-symplectic structure on $\mathbb{R}^2$ with degeneracy locus Z. Show that any $m \in Z$ admits a neighborhood with coordinates (y_1, y_2) such that $\pi = y_1 \partial_{y_1} \wedge \partial_{y_2}$. (Write $\pi = f\bar{\pi}$, where $\bar{\pi}$ is nondegenerate, and use (a).)

More generally, as we will see (Exercise 4.19), given a log-symplectic structure π on a $2n$-dimensional manifold M with degeneracy locus $Z \subseteq M$, any point in Z admits a neighborhood U with coordinates $(y_1, y_2, q_1, \ldots, q_{n-1}, p_1, \ldots, p_{n-1})$ such that

$$\pi|_U = y_1 \partial_{y_1} \wedge \partial_{y_2} + \sum_{i=1}^{n-1} \partial_{p_i} \wedge \partial_{q_i}. \tag{3.12}$$

Since $U \cap Z = \{(0, y_2, q_1, \ldots, q_{n-1}, p_1, \ldots, p_{n-1})\}$, it is clear that π is tangent to Z, i.e., Z is a Poisson submanifold, and the rank of the restricted Poisson structure on Z is $2n - 2$. Away from the degeneracy locus, π is symplectic; in the coordinates above, the corresponding symplectic form is

$$\frac{1}{y_1} dy_2 \wedge dy_1 + \sum_{i=1}^{n-1} dq_i \wedge dp_i = dy_2 \wedge d(\log |y_1|) + \sum_{i=1}^{n-1} dq_i \wedge dp_i,$$

which is a 2-form with logarithmic singularities along Z.

Log-symplectic manifolds have been the object of much study is recent years. Existence results, constructions and obstructions for log-symplectic structures are discussed e.g. in [18, 42]. A full set of invariants classifying them in dimension 2 was described in [78] (partially generalized in [50] to higher dimensions). Their Poisson cohomology is described in [78] and [50, 69].

A fruitful approach to Poisson structures that are "generically symplectic" relies on viewing them as symplectic structures on suitable Lie algebroids [56, 75]. We will briefly recall this viewpoint to log-symplectic structures, following [50].

As seen in Sect. 3.3, tangent bundles of manifolds are special cases of Lie algebroids. Let us consider a Lie algebroid $A \to M$, with anchor $\rho : A \to TM$ and Lie bracket $[\cdot, \cdot]$ on $\Gamma(A)$, and think of it as a replacement for TM. Then sections of $\wedge^k A$ are regarded as analogues of k-vector fields. The space of A-multivector fields $\Gamma(\wedge^\bullet A)$ inherits a Schouten bracket extending the Lie bracket on $\Gamma(A)$ (see Remark 3.1) and the action of A-vector fields $u \in \Gamma(A)$ on functions $f \in C^\infty(M)$ via $\mathcal{L}_{\rho(u)} f$. Similarly, the space of A-forms $\Gamma(\wedge^\bullet A^*)$ carries a differential

$$d_A : \Gamma(\wedge^\bullet A^*) \to \Gamma(\wedge^{\bullet+1} A^*) \tag{3.13}$$

satisfying

$$d_A f(u) = \mathcal{L}_{\rho(u)} f, \qquad d_A \xi(u, v) = \mathcal{L}_{\rho(u)} \xi(v) - \mathcal{L}_{\rho(v)} \xi(u) - \xi([u, v]),$$

for $f \in C^\infty(M)$, $u, v \in \Gamma(A)$, and $\xi \in \Gamma(A^*)$. The corresponding cohomology is called the *Lie algebroid cohomology* of A.

An *A-Poisson structure* is an A-bivector field $\pi_A \in \Gamma(\wedge^2 A)$ such that $[\pi_A, \pi_A] = 0$. Since the anchor map preserves Lie brackets, it extends to a bracket-preserving map $\rho : \Gamma(\wedge^\bullet A) \to \mathcal{X}^\bullet(M)$; as a consequence,

$$\pi := \rho(\pi_A) \in \mathcal{X}^2(M) \qquad (3.14)$$

is an ordinary Poisson structure on M. A situation of interest is when π_A is nondegenerate, in the sense that it induces an isomorphism $A^* \to A$, or, equivalently, π_A^n never vanishes. In analogy with the discussion in Sect. 2.4, π_A is nondegenerate if and only if it is the inverse of an *A-symplectic structure*, i.e., a d_A-closed 2-form $\omega_A \in \Gamma(\wedge^2 A^*)$ that is nondegenerate, in the sense that $\omega_A^\flat : A \to A^*$ is an isomorphism. Log-symplectic structures are examples of Poisson structures defined from A-symplectic structures via (3.14), for an appropriate choice of Lie algebroid A.

Given a manifold M and a hypersurface $Z \hookrightarrow M$, denote by $\mathcal{X}_Z^1(M)$ the space of vector fields on M that are tangent to Z. In a neighborhood U of a point in Z with adapted coordinates $(z, x_2, \ldots, x_l)$ (so that $U \cap Z$ is defined by $z = 0$), a vector field tangent to Z has the form $X = a \partial_z + \sum_i b_i \partial_{x_i}$, for $a, b_i \in C^\infty(U)$ and $a(0, x_1, \ldots, x_l) = 0$. Then $a = zb$, for $b \in C^\infty(U)$, and $X = bz\partial_z + \sum_i b_i \partial_{x_i}$, showing that the vector fields

$$e_1 = z\partial_z, \ e_2 = \partial_{x_2}, \ \ldots, \ e_l = \partial_{x_l}$$

form a basis for the $C^\infty(U)$-module of vector fields on U that are restrictions of vector fields in $\mathcal{X}_Z^1(M)$. In other words, as a sheaf of $C^\infty(M)$-modules, $\mathcal{X}_Z^1(M)$ is locally freely generated, and therefore there is a vector bundle $T_Z M$ over M such that $\mathcal{X}_Z^1(M) = \Gamma(T_Z M)$ [73].

Exercise 3.36 Take $M = \mathbb{S}^1$ and Z to be a point. Check that $T_Z M$ is a nontrivial line bundle (hence isomorphic to the Möbius line bundle over $\mathbb{S}^1$), as opposed to TM, which is trivial. What if Z is the union of 2 points?

The Lie bracket of vector fields defines a Lie bracket on $\mathcal{X}_Z^1(M)$, and the inclusion $\mathcal{X}_Z^1(M) \to \mathcal{X}^1(M)$ induces an anchor map $\rho : T_Z M \to TM$ making $T_Z M$ into a Lie algebroid, called the *log-tangent bundle* (or *b-tangent bundle*) associated with $Z \subseteq M$, see [16, §17.4]. Note that the anchor map is an isomorphism away from Z, but its kernel has rank 1 (locally spanned by e_1) over Z.

Exercise 3.37

(a) Let $A = T_Z M$ be the log-tangent bundle associated with a hypersurface $Z \subseteq M$, and let π_A be an A-Poisson structure. Let $\pi = \rho(\pi_A)$ be the induced Poisson structure on M. Show that π_A is nondegenerate if and only if π is log-symplectic.

(b) Let π be a log-symplectic structure with degenerate locus $Z \subseteq M$, and let $A = T_Z M$. Show that there is a (unique) A-Poisson structure π_A such that $\pi = \rho(\pi_A)$ (it may help to use that Z is a Poisson submanifold, see (3.12)).

In conclusion, log-symplectic structures on a manifold M with degeneracy locus Z are equivalent to symplectic structures on the Lie algebroid $T_Z M$ (see [50, Prop. 20]).

For more general discussions on how to view other types of generically symplectic Poisson structures by means of symplectic structures on appropriate Lie algebroids, see e.g. [56, 75] and references therein.

4 Local Structure and Symplectic Foliation

Let M be a manifold and $D \subseteq TM$ a smooth distribution (see Sect. 2.4). An *integral submanifold* of D is a connected submanifold $S \hookrightarrow M$ such that $D|_S = TS$. We say that D is *integrable* if any point in M is contained in an integral submanifold. An integrable distribution D gives rise to a partition of M into maximal integral submanifolds (which are immersed, not necessarily embedded), called *leaves*; the collection of leaves will be referred to as a *foliation*, see e.g. [37, Sec. 1.5]. When D has constant rank, the classical Frobenius theorem asserts that D is integrable if and only if it is involutive; the corresponding foliation in this case is called *regular*.

4.1 Integrability of the Characteristic Distribution and Symplectic Leaves

Let π be a bivector field on M, and let $R = \pi^\sharp(T^*M) \subseteq TM$ be its characteristic distribution. Suppose that $S \subseteq M$ is an integral submanifold of R,

$$R|_S = TS.$$

Then

- π is tangent to S, and the induced bivector field π_S on S is nondegenerate (Exercise 2.13 (a));
- π_S is equivalent to a nondegenerate 2-form $\omega_S \in \Omega^2(S)$, see Sect. 2.4.

It follows that a bivector field π whose characteristic distribution R is integrable gives rise to a foliation in which each leaf $\iota : \mathcal{O} \hookrightarrow M$ carries a nondegenerate 2-form $\omega_\mathcal{O}$ such that

$$\{f, g\}(x) = \omega_\mathcal{O}(X_{\iota^* g}, X_{\iota^* f})(x), \qquad \forall\, x \in \mathcal{O}. \tag{4.1}$$

Note that π is uniquely determined by the foliation and leafwise 2-forms.

Exercise 4.1 Let π be a bivector field on M with integrable characteristic distribution. Check that π is Poisson if and only if each leafwise 2-form $\omega_\mathcal{O}$ is closed, hence symplectic (see Exercises 2.10 and 2.13 (b)). In particular, π is Poisson if leaves have dimension at most 2.

As we will see in Sect. 4.2 below, when π is Poisson the integrability of R always holds, as a consequence of Weinstein's splitting theorem (see also Remark 5.6). In the regular case, this can be checked more directly using Frobenius theorem.

Exercise 4.2 Let (M, π) be a regular Poisson manifold. Use (2.4) to show that the characteristic distribution $R \subseteq TM$ is involutive, and hence integrable.

In summary, when π is a Poisson structure, its characteristic distribution is integrable (see Sect. 4.2) and each leaf $\mathcal{O} \hookrightarrow M$ is equipped with a symplectic form $\omega_\mathcal{O}$ uniquely determined by the property that the inclusion map is Poisson; the collection of all such symplectic leaves is called the *symplectic foliation* of π, and π is uniquely characterized by its symplectic foliation.

Exercise 4.3 Consider $M = \mathbb{T}^3 = \mathbb{S}^1 \times \mathbb{S}^1 \times \mathbb{S}^1 = \{(\theta_1, \theta_2, \theta_3)\}$, and for each $\lambda \in \mathbb{R}$, let $\pi_\lambda = \frac{\partial}{\partial \theta_1} \wedge (\frac{\partial}{\partial \theta_2} + \lambda \frac{\partial}{\partial \theta_3})$. Verify that its characteristic distribution has constant rank equal to 2 and is involutive, so π_λ is Poisson (by Exercises 4.1 and 4.2). Check that all its symplectic leaves are embedded (and isomorphic to $\mathbb{T}^2$) when $\lambda \in \mathbb{Q}$; if λ is irrational, leaves are isomorphic to cylinders and are not embedded.

Exercise 4.4 Use Exercises 2.27 and 2.29 to check that a submanifold N of a Poisson manifold M is cosymplectic if and only if it intersects each symplectic leaf $\mathcal{O} \hookrightarrow M$ transversally in a symplectic submanifold of $\mathcal{O}$, and, in this case, the symplectic foliation of the induced Poisson structure on N is given by the connected components of the intersections of N with the symplectic leaves of M.

Exercise 4.5 Let M_1 and M_2 be Poisson manifolds. Verify that a diffeomorphism $\varphi : M_1 \to M_2$ is Poisson if and only if, for every $x \in M_1$, φ restricts to a symplectomorphism from the symplectic leaf through x to the symplectic leaf containing $\varphi(x) \in M_2$.

As indicated by the next exercise, Casimir functions give useful information about symplectic leaves.

Exercise 4.6 Let (M, π) be a regular m-dimensional Poisson manifold with $\text{rank}(\pi) = k$. Suppose that $f_1, \ldots, f_{m-k}$ are Casimir functions such that $df_1, \ldots, df_{m-k}$ are linearly independent at all points. Show that the symplectic leaves are given by connected components of the level sets of $(f_1, \ldots, f_{m-k}) : M \to \mathbb{R}^{m-k}$.

The next exercise illustrates a way to obtain bivector fields with integrable characteristic distributions that are not Poisson.

Exercise 4.7 Let (M, π) be a Poisson manifold and $B \in \Omega^2(M)$. Let $B^\flat : TM \to T^*M$, $B^\flat(X) = i_X B$, and consider the map $\mathrm{Id} + B^\flat \circ \pi^\sharp : T^*M \to T^*M$.

(a) Check that $\mathrm{Id} + B^\flat \circ \pi^\sharp$ is an isomorphism if and only if, on each leaf $\mathcal{O}$ of (M, π), the 2-form $\omega_\mathcal{O} + B_\mathcal{O}$ is nondegenerate, where $B_\mathcal{O}$ denotes the pullback of B to $\mathcal{O}$.
(b) If $\mathrm{Id} + B^\flat \circ \pi^\sharp$ is an isomorphism, note that there is a bivector field π_B on M such that $\pi_B^\sharp = \pi^\sharp \circ (\mathrm{Id} + B^\flat \circ \pi^\sharp)^{-1}$, and whose characteristic distribution coincides with that of π, so they have the same leaves. Verify that, for each leaf $\mathcal{O}$, the 2-form on $\mathcal{O}$ induced by π_B is given by $\omega_\mathcal{O} + B_\mathcal{O}$. Conclude (see Exercise 4.1) that π_B is Poisson if and only if the pullback of dB to each leaf vanishes.

For a closed 2-form B, we call the Poisson structure π_B of the previous exercise the *gauge transformation* of π by B, and say that the two Poisson structures are *gauge equivalent*.

As an application, we can use gauge transformations and the decomposition of a Poisson manifold into symplectic leaves to extend the well-known Moser's trick from symplectic geometry to the Poisson setting, see e.g. [2]. We briefly recall the setup in symplectic geometry. For a smooth 1-parameter family of symplectic forms $\omega_t \in \Omega^2(M)$, let $B_t = \omega_t - \omega_0$, and suppose that there exists a 1-parameter family of 1-forms $\alpha_t \in \Omega^1(M)$ such that $\frac{dB_t}{dt} = -d\alpha_t$. Let X_t be the time dependent vector field defined by $i_{X_t}\omega_t = \alpha_t$, and let ϕ_t be its flow. Then

$$\phi_t^* \omega_t = \omega_0,$$

which follows from the equality

$$\frac{d}{dt}\phi_t^* \omega_t = \phi_t^* \left(\mathcal{L}_{X_t}\omega_t + \frac{d}{dt}\omega_t \right) = \phi_t^* d \left(i_{X_t}\omega_t - \alpha_t \right).$$

Exercise 4.8 Let π_t be a 1-parameter family of Poisson structures on M for which there exists a family of closed 2-forms $B_t \in \Omega^2(M)$ with $B_0 = 0$ such that $\pi_t = (\pi_0)_{B_t}$ (see Exercise 4.7, part (b), for the notation). Suppose that there exists a 1-parameter family of 1-forms $\alpha_t \in \Omega^2(M)$ such that $\frac{dB_t}{dt} = -d\alpha_t$. Consider the time dependent vector field $X_t = \pi_t^\sharp(\alpha_t)$, and let ϕ_t be its flow. Then

$$(\phi_t)_* \pi_0 = \pi_t.$$

4.2 The Splitting Theorem

The central result describing the local structure of a Poisson manifold is *Weinstein's splitting theorem* [86]: *for a Poisson manifold* (M, π), *any point m admits a neighborhood with coordinates* $(q_1, \ldots, q_k, p_1, \ldots, p_k, y_1, \ldots, y_l)$ *(centered at m) where* π *takes the form*

$$\pi = \sum_{i=1}^{k} \frac{\partial}{\partial p_i} \wedge \frac{\partial}{\partial q_i} + \frac{1}{2} \sum_{i,j=1}^{l} \varphi_{ij}(y) \frac{\partial}{\partial y_i} \wedge \frac{\partial}{\partial y_j}, \quad \text{and} \quad \varphi_{ij}(0) = 0. \tag{4.2}$$

Since φ_{ij} vanishes at m, the rank of π at m is $2k$.

The following are consequences of this result.

- Since the functions φ_{ij} only depend on the coordinates $y_1, \ldots, y_l$, the theorem gives a local decomposition of M as the direct product of two Poisson manifolds,

$$(S, \pi_S) \times (N, \pi_N),$$

where S, defined by the coordinates (q_i, p_i), has a nondegenerate Poisson structure $\pi_S = \sum_i \frac{\partial}{\partial p_i} \wedge \frac{\partial}{\partial q_i}$, and N, defined by the coordinates $(y_1, \ldots, y_l)$, has Poisson structure $\pi_N = \sum_{i<j} \varphi_{ij}(y) \frac{\partial}{\partial y_i} \wedge \frac{\partial}{\partial y_j}$ whose rank vanishes at the point $y = 0$.

- When π has constant rank $2k$ around m, it locally looks like Poisson's original bracket (1.1),

$$\pi = \sum_{i=1}^{k} \frac{\partial}{\partial p_i} \wedge \frac{\partial}{\partial q_i}.$$

When π is nondegenerate, this local normal form recovers the Darboux theorem for symplectic structures.

- The submanifold $S \hookrightarrow M$, $(q_i, p_i) \mapsto ((q_i, p_i), 0)$ satisfies $TS = R|_S$, i.e., it is an integral submanifold of the characteristic distribution R containing $m = (0, 0)$. It follows that R is integrable, and its leaves define the symplectic foliation of π, as explained in Sect. 4.1. In Weinstein's splitting coordinates, the leafwise symplectic forms have the canonical expression $\sum_i dq_i \wedge dp_i$.

- Leaves of (M, π) can be characterized by the following equivalence relation: $x, x' \in M$ belong to the same leaf if and only if there are functions $f_1, \ldots, f_r$ on M such that

$$x' = \phi_{X_{f_1}}^{t_1} \circ \ldots \circ \phi_{X_{f_r}}^{t_r}(x), \tag{4.3}$$

where ϕ_X^t denotes the flow of a vector field X at time t. (This can be directly verified if x, x' belong to a neighborhood where the splitting theorem holds, and in general by covering any path from x to x' with finitely many of those.)

Exercise 4.9 Let $\varphi : M_1 \to M_2$ be a Poisson map that is *complete*, in the sense that whenever a Hamiltonian vector field X_f on M_2 is complete, X_{φ^*f} is also complete. Show that if $y = \varphi(x)$, then the entire symplectic leaf through y is contained in the image of φ. Hence the image of a complete Poisson map is a union of symplectic leaves.

The following steps outline the proof of the splitting theorem.

(1) Assuming that the rank of π at m is not zero, let p_1 be a function such that $X_{p_1}|_m \neq 0$. By the straightening theorem one can find a function q_1 around m such that $\{p_1, q_1\} = 1$.
(2) The vector fields X_{p_1} and X_{q_1} are linearly independent around m and commute: $[X_{p_1}, X_{q_1}] = X_{\{p_1,q_1\}} = X_1 = 0$. It follows that there are coordinates $(y_1, \ldots, y_n)$ around m satisfying $X_{q_1} = \partial_{y_{n-1}}$, $X_{p_1} = \partial_{y_n}$, in such a way that $dy_j(X_{q_1}) = 0$ and $dy_j(X_{p_1}) = 0$ for $j = 1, \ldots, n-2$.
(3) The differentials $dq_1, dp_1, dy_1, \ldots dy_{n-2}$ are linearly independent, so the functions $q_1, p_1, y_1, \ldots y_{n-2}$ form a local coordinate system around m, satisfying

$$\{p_1, q_1\} = 1, \quad \{p_1, y_j\} = \{q_1, y_j\} = 0.$$

Hence $X_{p_1} = \partial_{q_1}$ and $X_{q_1} = -\partial_{p_1}$.
(4) By the Jacobi identity,

$$\partial_{p_1}\{y_i, y_j\} = -\{q_1, \{y_i, y_j\}\} = 0, \quad \partial_{q_1}\{y_i, y_j\} = \{p_1, \{y_i, y_j\}\} = 0,$$

so the functions $\varphi_{ij} = \{y_i, y_j\}$ only depend on $y_1, \ldots, y_{n-2}$. Hence

$$\pi = \frac{\partial}{\partial p_1} \wedge \frac{\partial}{\partial q_1} + \frac{1}{2}\sum_{i,j} \varphi_{ij}(y) \frac{\partial}{\partial y_i} \wedge \frac{\partial}{\partial y_j}.$$

If $\varphi_{ij}(0) = 0$, the theorem is proven. If not, one repeats the argument for the Poisson structure $\frac{1}{2}\sum_{i,j} \varphi_{ij}(y) \frac{\partial}{\partial y_i} \wedge \frac{\partial}{\partial y_j}$.

Different proofs of the splitting theorem, leading to more general results, can be found in [13, 40].

4.3 Symplectic Leaves in Examples

4.3.1 Poisson Surfaces

For a Poisson structure π on a 2-dimensional manifold M (Example 2.5), zero-dimensional leaves are given by points where the rank of π is zero, and 2-dimensional leaves are given by the connected components of the open subset where π is nondegenerate

4.3.2 Duals of Lie Algebras

For a Lie algebra $\mathfrak{g}$, the infinitesimal coadjoint action of $\mathfrak{g}$ on $\mathfrak{g}^*$ sends $u \in \mathfrak{g}$ to the vector field ad^*_u on $\mathfrak{g}^*$ given by

$$\mathfrak{g}^* \ni \xi \mapsto \mathrm{ad}^*_u(\xi) = \xi([u, \cdot]) \in \mathfrak{g}^* = T_\xi \mathfrak{g}^*.$$

It follows from (3.1) that the characteristic distribution of the linear Poisson structure on $\mathfrak{g}^*$ is tangent to the orbits of the coadjoint G-action, where G is any connected Lie group with Lie algebra $\mathfrak{g}$. Following (4.1), the symplectic form on a coadjoint orbit $\mathcal{O}$ is given, at $\xi \in \mathcal{O}$, by

$$\omega_\mathcal{O}(\mathrm{ad}^*_u(\xi), \mathrm{ad}^*_v(\xi)) = \xi([v, u]).$$

In the next exercise, we will consider $\mathbb{R}^3 = \{(x, y, z)\}$ with linear Poisson structures corresponding to each of the 3-dimensional Lie algebras $\mathfrak{su}(2)$, $\mathfrak{sl}(2, \mathbb{R})$ and $\mathfrak{sb}(2, \mathbb{C})$ (see Exercise 3.27), respectively.

Exercise 4.10 Verify the following (Exercise 4.6 should be useful).

(a) The only zero-dimensional leaf of $\pi = z\partial_x \wedge \partial_y + x\partial_y \wedge \partial_z + y\partial_z \wedge \partial_x$ is the origin, and its 2-dimensional leaves are given by the fibers of the map $f : \mathbb{R}^3 \setminus \{(0, 0, 0)\} \to \mathbb{R}$, $f(x, y, z) = x^2 + y^2 + z^2$.
(b) The only zero-dimensional leaf of $\pi = -z\partial_x \wedge \partial_y + x\partial_y \wedge \partial_z + y\partial_z \wedge \partial_x$ is the origin, and its 2-dimensional leaves are the (connected components of the) fibers of the map $f : \mathbb{R}^3 \setminus \{(0, 0, 0)\} \to \mathbb{R}$, $f(x, y, z) = x^2 + y^2 - z^2$.
(c) The zero-dimensional leaves of $\pi = x\partial_x \wedge \partial_z + y\partial_y \wedge \partial_z$ are points along the z-axis, and its 2-dimensional leaves are the fibers of the map

$$f : \mathbb{R}^3 \setminus \{(0, 0, z), z \in \mathbb{R}\} \to \mathbb{S}^1, \quad f(x, y, z) = \left(\frac{x}{\sqrt{x^2 + y^2}}, \frac{y}{\sqrt{x^2 + y^2}} \right).$$

In the more general case of Poisson-Lie groups, symplectic leaves are given by orbits of the so-called *dressing action*, see e.g. [57, § 3.4] and [67, Sec. 2], that generalize the coadjoint action on the dual of a Lie algebra.

The next exercise describes some leaves of duals of Lie algebroids.

Exercise 4.11 Let $A \to M$ be a Lie algebroid with surjective anchor map $\rho : A \twoheadrightarrow TM$. Show that, if M is connected, then ρ^* identifies (T^*M, ω_{can}) with a symplectic leaf of A^* (cf. Exercises 3.18 and 4.9). More generally, a Lie algebroid A defines a foliation on M whose tangent distribution is $\rho(A) \subseteq TM$ (see Remark 5.6); show that, for each leaf $\mathcal{O} \hookrightarrow M$, the cotangent bundle $(T^*\mathcal{O}, \omega_{can})$ is identified with a symplectic leaf of A^* via $\rho^*|_{\mathcal{O}} : T^*\mathcal{O} \to A^*|_{\mathcal{O}} \subseteq A^*$.

4.3.3 Quotients of Symplectic Manifolds

We will now discuss the symplectic foliation of Poisson manifolds obtained as quotients of symplectic manifolds. We start with some general considerations.

Exercise 4.12 Let (S, ω) be a symplectic manifold and $\varphi : S \to M$ a surjective submersion. Consider the "vertical" bundle $\mathcal{V} = \ker(d\varphi) \subseteq TS$, and let $\mathcal{V}^\omega \subseteq TS$ be the symplectic orthogonal subbundle. Let $C^\infty_\mathcal{V}(S) = \{g \in C^\infty(S) \,|\, dg \in \Gamma(\mathrm{Ann}(\mathcal{V}))\}$, i.e., the space of functions that are locally constant on φ-fibers. Notice that

- $g \in C^\infty_\mathcal{V}(S)$ if and only if $X_g \in \Gamma(\mathcal{V}^\omega)$,
- $(d_x\varphi)^* : T^*_{\varphi(x)}M \xrightarrow{\sim} \mathrm{Ann}(\mathcal{V}_x)$, and

$$\mathcal{V}^\omega_x = \{X_{\varphi^*f}|_x \,,\, f \in C^\infty(M)\}, \quad \forall x \in S. \tag{4.4}$$

Verify that

(a) If $\varphi^*C^\infty(M) \subseteq C^\infty(S)$ is a Poisson subalgebra (equivalently, there is a Poisson structure on M for which φ is a Poisson map), then $\mathcal{V}^\omega$ is involutive.
(b) If $\mathcal{V}^\omega$ is involutive, then $C^\infty_\mathcal{V}(S) \subseteq C^\infty(S)$ is a Poisson subalgebra.

Let (M, π) be a Poisson manifold, and suppose that there is a symplectic manifold (S, ω) and a Poisson surjective submersion $\varphi : S \to M$. We will use φ to give a description of the symplectic leaves of M.

Consider the vector subbundle $\mathcal{V}^\omega \subseteq TS$, the symplectic orthogonal of the distribution $\mathcal{V}$ tangent to φ-fibers. By the previous exercise (part (a)), $\mathcal{V}^\omega$ is involutive, so it is the tangent distribution to a regular foliation on S. Since the Hamiltonian vector fields X_{φ^*f} and X_f are φ-related for all $f \in C^\infty(M)$ (see Exercise 2.2), (4.4) implies that

$$d\varphi(\mathcal{V}^\omega_x) = R_{\varphi(x)}, \quad \forall x \in S,$$

where R is the characteristic distribution of π. As a consequence, denoting by $\mathcal{L} \stackrel{\iota_\mathcal{L}}{\hookrightarrow} S$ the leaf through a given point $x \in S$ and $\mathcal{O}$ the symplectic leaf of π containing $\varphi(x)$, then $\varphi(\mathcal{L}) \subseteq \mathcal{O}$; moreover, the map $\varphi_\mathcal{L} := \varphi|_\mathcal{L} : \mathcal{L} \to \mathcal{O}$ is a submersion.

By Exercise 4.9, if we additionally assume the Poisson map φ to be *complete*, then

$$\varphi(\mathcal{L}) = \mathcal{O},$$

showing that the symplectic leaves in M are the images of the leaves of $\mathcal{V}^\omega \subseteq TS$ under φ. The discussion is summarized by the following commutative diagram of horizontal inclusions and vertical surjective submersions,

$$\begin{array}{ccc} \mathcal{L} & \stackrel{\iota_\mathcal{L}}{\longrightarrow} & (S, \omega) \\ \varphi_\mathcal{L} \downarrow & & \downarrow \varphi \\ (\mathcal{O}, \omega_\mathcal{O}) & \longrightarrow & (M, \pi). \end{array} \qquad (4.5)$$

The next exercise shows that the symplectic forms ω and $\omega_\mathcal{O}$ are related by

$$\varphi_\mathcal{L}^* \omega_\mathcal{O} = \iota_\mathcal{L}^* \omega. \qquad (4.6)$$

Exercise 4.13 For $f, g \in C^\infty(M)$, check that

$$\varphi_\mathcal{L}^* \omega_\mathcal{O}(X_{\varphi^* g}(x), X_{\varphi^* f}(x)) = \{f, g\}(\varphi(x)), \qquad \iota_\mathcal{L}^* \omega(X_{\varphi^* g}(x), X_{\varphi^* f}(x)) = \{\varphi^* f, \varphi^* g\}(x).$$

A special case of the previous setup was considered in Sect. 3.1: (S, ω) is a symplectic manifold carrying an action of a Lie group G by symplectomorphisms that is free and proper, and $M = S/G$ is equipped with the Poisson structure such that the quotient map $\varphi : S \to S/G$ is Poisson. We keep the notation $\mathcal{V} \subseteq TS$ for the distribution tangent to the G-orbits.

Exercise 4.14 Show that the Poisson map $\varphi : S \to M = S/G$ is complete.

Suppose that the G-action on S is Hamiltonian (cf. Sect. 3.2) with moment map $\mu : S \to \mathfrak{g}^*$; recall that this means that μ is G-equivariant (with respect to the coadjoint action on $\mathfrak{g}^*$) and satisfies

$$i_{\psi(u)} \omega = d(\mu^* u), \qquad \forall u \in \mathfrak{g},$$

where $\psi : \mathfrak{g} \to \mathfrak{X}^1(M)$ is the infinitesimal action and $\mu^* u(x) = \langle \mu(x), u \rangle$.

For $\xi \in \mathfrak{g}^*$, denote by $G_\xi \subseteq G$ the stabilizer group of ξ.

Exercise 4.15 Verify that

(a) $\mu : S \to \mathfrak{g}^*$ is a submersion,
(b) $\ker(d\mu) = \mathcal{V}^\omega$,
(c) for $\xi \in \mathfrak{g}^*$ and $x \in \mu^{-1}(\xi)$, $\sigma x \in \mu^{-1}(\xi)$ if and only if $\sigma \in G_\xi$.

It follows from (a) and (b) of the previous exercise that, when the G-action on S is Hamiltonian, the leaves of $\mathcal{V}^\omega$ are the connected components of the level sets of μ. From part (c) of the exercise, along with (4.5) and (4.6), it follows that the symplectic leaves of S/G are identified with the connected components of the symplectic reduced spaces $\mu^{-1}(\xi)/G_\xi$,

$$\begin{array}{ccccc} \mathcal{L} & \longrightarrow & \mu^{-1}(\xi) & \longrightarrow & (S, \omega) \\ \downarrow & & \downarrow & & \downarrow \varphi \\ \mathcal{O} & \longrightarrow & \mu^{-1}(\xi)/G_\xi & \longrightarrow & (S/G, \pi). \end{array}$$

4.3.4 Standard Poisson Structure on $\mathbb{S}^3$

Let us consider the standard Poisson structure on $\mathbb{S}^3 = \{(x, y, z, w) \in \mathbb{R}^4 \mid x^2 + y^2 + z^2 + w^2 = 1\}$, given by

$$\begin{aligned} \pi &= (z^2 + w^2)\partial_x \wedge \partial_y + \left(-y\partial_x + x\partial_y\right) \wedge (z\partial_z + w\partial_w) \\ &= (-z\partial_x + x\partial_z) \wedge (-z\partial_y + y\partial_z) + (-w\partial_x + x\partial_w) \wedge (-w\partial_y + y\partial_w), \end{aligned}$$

see Exercise 3.24 (note that the second expression for π makes apparent the fact that it is tangent to $\mathbb{S}^3$). The zero-dimensional leaves of $(\mathbb{S}^3, \pi)$ are the points in the unit circle $C = \{(x, y, 0, 0) \mid x^2 + y^2 = 1\}$. Hence the restriction of π to the open subset $\mathbb{S}^3 \setminus C$ is a regular Poisson structure of rank 2.

Consider $\mathbb{S}^1 \subseteq \mathbb{R}^2$ and the map

$$F : \mathbb{S}^3 \setminus C \to \mathbb{S}^1, \qquad F(x, y, z, w) = \left(\frac{z}{\sqrt{z^2 + w^2}}, \frac{w}{\sqrt{z^2 + w^2}}\right).$$

Denoting by $\mathbb{D}^2 \subseteq \mathbb{R}^2$ the open unit disk, we have a diffeomorphism

$$\mathbb{S}^3 \setminus C \xrightarrow{\sim} \mathbb{D}^2 \times \mathbb{S}^1, \quad (x, y, z, w) \mapsto ((x, y), F(x, y, z, w)), \tag{4.7}$$

with inverse map given by $((x, y), c) \mapsto (x, y, \sqrt{1 - x^2 - y^2}\, c)$.

Exercise 4.16

(a) Check that the functions $F_1(x, y, z, w) = \frac{z}{\sqrt{z^2+w^2}}$ and $F_2(x, y, z, w) = \frac{w}{\sqrt{z^2+w^2}}$ are Casimirs for π on $\mathbb{S}^3 \setminus C$.

(b) Show that the diffeomorphism (4.7) identifies $(\mathbb{S}^3 \setminus C, \pi)$ with the direct product of Poisson manifolds $(\mathbb{D}^2, \pi_{\mathbb{D}^2}) \times (\mathbb{S}^1, 0)$, where $\pi_{\mathbb{D}^2} = (1 - x^2 - y^2)\partial_x \wedge \partial_y$.

It follows from the previous exercise that the 2-dimensional leaves of π agree with the fibers of the map F. In more detail, the set of 2-dimensional leaves of $(\mathbb{S}^3, \pi)$ is in bijection with $\mathbb{S}^1$: for each $c \in \mathbb{S}^1$ the corresponding leaf is the image of

$$\mathbb{D}^2 \to \mathbb{S}^3, \quad (x, y) \mapsto (x, y, \sqrt{1 - x^2 - y^2}\, c),$$

with symplectic form in these coordinates given by $\frac{1}{x^2+y^2-1} dx \wedge dy$.

Exercise 4.17 Give an analogous description of the symplectic leaves of the higher-dimensional Poisson spheres of Exercise 2.18.

Recall from Sect. 3.4 that $\mathbb{S}^2$ has a natural Poisson homogeneous structure such that the Hopf map $\mathbb{S}^3 \to \mathbb{S}^2$ is a Poisson map (see (3.11)). This Poisson structure has two symplectic leaves: one singular point (the image of $C \subseteq \mathbb{S}^3$), and an open symplectic leaf given by the complement of the singular point. The Hopf map takes each 2-dimensional symplectic leaf in $\mathbb{S}^3 \setminus C$ symplectomorphically onto the open symplectic leaf of $\mathbb{S}^2$.

Exercise 4.18 Describe the symplectic leaves of the Poisson structures on $\mathbb{S}^2$ of Exercise 3.34.

4.3.5 Log-Symplectic Structures

The following exercise describes the symplectic leaves of log-symplectic manifolds (Sect. 3.5).

Exercise 4.19 Let (M^{2n}, π) be a log-symplectic manifold with degeneracy locus Z, and $m \in Z$. Check that, in Weinstein's splitting coordinates $(q_1, \ldots, q_k, p_1, \ldots, p_k, y_1, \ldots, y_l)$ around m, one must have $l = 2$. Use Exercise 3.35 to conclude that there are local coordinates around m such that

$$\pi = y_1 \frac{\partial}{\partial y_1} \wedge \frac{\partial}{\partial y_2} + \sum_{i=1}^{n-1} \frac{\partial}{\partial p_i} \wedge \frac{\partial}{\partial q_i}.$$

As a Poisson submanifold, Z inherits a regular Poisson structure of corank one. So the symplectic leaves of (M, π) are either open subsets given by connected components of $M \setminus Z$, or are contained in Z and have dimension $2n - 2$.

4.4 Transverse Structure and Linearization

Given a Poisson manifold (M, π), recall that Weinstein's local splitting theorem (Sect. 4.2) asserts that any point in M admits a neighborhood that can be written as a direct product of Poisson manifolds,

$$(S, \pi_S) \times (N, \pi_N),$$

where π_S is nondegenerate and π_N vanishes at the given point. The symplectic factor S is an open subset of a symplectic leaf $\mathcal{O} \subseteq M$. The degenerate factor N sits in M as a cosymplectic submanifold (see Exercise 2.32) and π_N is called the *transverse Poisson structure* to the symplectic leaf $\mathcal{O}$; although (N, π_N) is not intrinsically defined, the terminology is justified by the following uniqueness result [86]: *let $m_0, m_1 \in \mathcal{O}$ and, for each $i = 0, 1$, let N_i be a cosymplectic submanifold containing m_i such that*

$$T_{m_i}\mathcal{O} \oplus T_{m_i} N_i = T_{m_i} M; \tag{4.8}$$

then there are open neighborhoods U_0 of m_0 in N_0 and U_1 of m_1 in N_1 with a Poisson diffeomorphism $(U_0, \pi_{N_0}) \cong (U_1, \pi_{N_1})$ taking m_0 to m_1.

In the previous result, one can think of the transverse cosymplectic submanifolds as obtained from local Weinstein splittings around points in the symplectic leaf. More generally, any submanifold N containing $m \in \mathcal{O}$ and satisfying the transversality condition (4.8) will be cosymplectic in a neighborhood of m, see Exercise 2.28.

We outline the proof of the uniqueness of the transverse structure.

- Since m_0 and m_1 belong to the same symplectic leaf, they are connected by a composition of Hamiltonian flows as in (4.3); this composition defines a Poisson automorphism of M taking m_0 to m_1, and as a consequence we can reduce the problem to the case where $m_0 = m_1$. So we henceforth assume that we have a Weinstein splitting chart $S \times N$ around a point $m = (0, 0)$ and $N' \subseteq S \times N$ is a cosymplectic submanifold, with induced Poisson structure denoted by $\pi_{N'}$, containing m and satisfying

$$T_m S \oplus T_m N' = T_m M.$$

- By the previous transversality condition at m, the map $N' \to N$ given by restriction of the natural projection $S \times N \to N$ is a local diffeomorphism around m. By shrinking N, we can assume that N' projects diffeomorphically onto N, and hence there is a smooth map $\phi : N \to S$ such that

$$N' = \{(\phi(y), y) \mid y \in N\}.$$

Let π' be the Poisson structure on N corresponding to $\pi_{N'}$ via the projection $N' \xrightarrow{\sim} N$, or, equivalently, such that the inverse diffeomorphism

$$\iota_\phi : N \xrightarrow{\sim} N', \qquad \iota_\phi(y) = (\phi(y), y),$$

is a Poisson map. The remainder of the proof consists in showing that π_N and π' are locally isomorphic around m. We start by checking that they are gauge equivalent (see Exercise 4.7).

- The symplectic leaves of $(N', \pi_{N'})$ are obtained by intersecting N' with the leaves of $(S, \pi_S) \times (N, \pi_N)$ (Exercise 4.4). So each leaf is given by

$$(S \times \mathcal{O}) \cap N' = \{(\phi(y), y) \mid y \in \mathcal{O}\},$$

with symplectic form given by the pullback of $\omega_S \times \omega_\mathcal{O}$, where $(\mathcal{O}, \omega_\mathcal{O})$ is a symplectic leaf of (N, π_N) and $\omega_S = \pi_S^{-1}$. We know that the leaves of π' correspond to the leaves of $\pi_{N'}$ under the Poisson diffeomorphism $\iota_\phi : (N, \pi') \to (N', \pi_{N'})$, so they are given by

$$\iota_\phi^{-1}((S \times \mathcal{O}) \cap N') = \mathcal{O},$$

with symplectic form given by the pullback of $\omega_S \times \omega_\mathcal{O}$ by $\iota_\phi : \mathcal{O} \to S \times \mathcal{O}$,

$$\iota^*\phi^*\omega_S + \omega_\mathcal{O},$$

where $\iota : \mathcal{O} \hookrightarrow N$ is the inclusion. Following Exercise 4.7, we conclude that π' is the gauge transformation of π_N by the closed 2-form $B = \phi^*\omega_S$,

$$(\pi')^\sharp = \pi_N^\sharp \circ (\mathrm{Id} + B^\flat \circ \pi_N^\sharp)^{-1}.$$

- In a neighborhood admitting Weinstein's splitting coordinates, the symplectic form ω_S is exact, and the local isomorphism between π_N and π' around m can be obtained by Moser's trick (Exercise 4.8), as explained in the next exercise.

Exercise 4.20 Let π be a Poisson structure on N vanishing at $y_0 \in N$. Let π' be a Poisson structure gauge equivalent to π via an exact 2-form $B = -d\alpha$. The following items show that there are open neighborhoods U and U' of y_0 and a Poisson diffeomorphism $(U, \pi) \xrightarrow{\sim} (U', \pi')$.

(a) Let $B_t = tB$. Show that there is an open neighborhood of y_0 where $(\mathrm{Id} + B_t^\flat \circ \pi^\sharp)$ is invertible for all $t \in [0, 1]$, so there we can consider the family of Poisson structures $\pi_t := \pi_{B_t}$ (note that they all vanish at y_0).

(b) Let $X_t = \pi_t^\sharp(\alpha)$. Show that there is a neighborhood of y_0 where the flow ϕ_t of X_t is defined for all $t \in [0, 1]$. By Exercise 4.8,

$$(\phi_t)_* \pi_t = \pi_0 = \pi,$$

so ϕ_1 gives the desired Poisson diffeomorphism.

At any point m of a Poisson manifold (M, π) there is a natural Lie algebra structure on the conormal space to the symplectic leaf $\mathcal{O}$ through m. Denoting by $\nu_m = T_m M / T_m \mathcal{O}$ the normal space to $\mathcal{O}$ at m, the Lie bracket on the conormal space

$$\nu_m^* = (T_m M / T_m \mathcal{O})^* = \mathrm{Ann}(T_m \mathcal{O}) = \ker(\pi_m^\sharp)$$

is defined as follows: for $\alpha, \beta \in \mathrm{Ann}(T_m \mathcal{O})$, take functions $f, g \in C^\infty(M)$ such that $\alpha = df|_m$ and $\beta = dg|_m$ and set

$$[\alpha, \beta] := d(\{f, g\})|_m.$$

Note that $d(\{f, g\})|_m \in \mathrm{Ann}(T_m \mathcal{O})$ as a consequence of the Jacobi identity:

$$d(\{f, g\})|_m(X_h) = dg|_m(X_{\{h, f\}}) - df|_m(X_{\{h, g\}}) = 0$$

for any $h \in C^\infty(M)$.

Exercise 4.21 Check that, by the Leibniz identity, the value of $[\alpha, \beta]$ is well defined, independent of the choices of f and g.

The Lie algebra $(\nu_m^*, [\cdot, \cdot])$ is called the *transverse Lie algebra*, or the *isotropy Lie algebra* of π at m. The corresponding linear Poisson structure on the normal space ν_m is closely related to the transverse Poisson structure. To see that, we consider Weinstein's splitting coordinates centered at m, with respect to which the transverse Poisson structure on $N = \{(y_j)\}$ is written as

$$\pi_N = \sum_{i<j} \varphi_{ij}(y) \frac{\partial}{\partial y_i} \wedge \frac{\partial}{\partial y_j},$$

with $\varphi_{ij}(0) = 0$. Then the Lie bracket on $\nu_m^* = T_0^* N$ satisfies

$$[dy_i, dy_j] = d(\{y_i, y_j\})|_0 = \sum_k \frac{\partial \varphi_{ij}}{\partial y_k}(0) dy_k,$$

i.e., the structural constants in this basis are $c_{ijk} = \frac{\partial \varphi_{ij}}{\partial y_k}(0)$. The corresponding linear Poisson structure on $\nu_m = T_0 N$ is given by

$$\sum_{i<j} \left(\sum_k \frac{\partial \varphi_{ij}}{\partial y_k}(0) y_k \right) \frac{\partial}{\partial y_i} \wedge \frac{\partial}{\partial y_j},$$

which is the first-order term of the Taylor expansion of π_N around m. So the linear Poisson structure dual to the transverse Lie algebra at a given point is the *linear approximation* to the transverse Poisson structure at that point.

A consequence of Weinstein's splitting theorem is that, in the local study of Poisson manifolds, it suffices to consider Poisson structures in the neighborhoods of points where they vanish. Given a Poisson manifold (M, π) and $m \in M$ such that $\pi_m = 0$, a natural question is whether π is locally isomorphic around m to its linear approximation on $\nu_m = T_m M$. In the affirmative case, π is said to be *linearizable* at m. A simple example of a Poisson structure that is not linearizable is $(x^2 + y^2)\partial_x \wedge \partial_y$ at the origin: its linear approximation is the zero Poisson structure, while π is nondegenerate away from the origin.

Exercise 4.22 Show that, in dimension 2, any Poisson structure whose isotropy Lie algebra at a vanishing point is nonabelian is linearizable at that point (use Exercise 3.35).

The systematic study of linearization of Poisson structures began with Weinstein in [86] and led to an active line of research in Poisson geometry; see [37, Ch. 4] for many of the important results and references. A guiding question has been that of characterizing Lie algebras $\mathfrak{g}$ with the property that any Poisson structure having $\mathfrak{g}$ as its isotropy Lie algebra at a vanishing point is linearizable at that point. Lie algebras with this property are called *nondegenerate*, and *degenerate* otherwise. One of the main theorems in this direction, obtained by Conn in [19], is that any semisimple Lie algebra of compact type[5] is nondegenerate (as conjectured in [86, § 6]). Conn's proof of this linearization result was analytic and relied on Nash-Moser type techniques; a soft geometric proof was later obtained in [27]. There is also a "semi-local" version of the linearization problem in the neighborhood of a general symplectic leaf (rather than just vanishing points of the Poisson

---

[5] I.e., a Lie algebra with negative definite Killing form, or equivalently such that all of its integrating Lie groups are compact.

structure) as well as an extension of Conn's theorem to this setting [29] (see [38] for further generalizations and references).

The next exercise shows that $\mathfrak{sl}(2,\mathbb{R})$ is degenerate by giving an explicit example of a nonlinearizable Poisson structure with $\mathfrak{sl}(2,\mathbb{R})$ as its isotropy Lie algebra [86, Prop. 6.3].

Exercise 4.23 Consider the linear Poisson structure on $\mathbb{R}^3$ dual to $\mathfrak{sl}(2,\mathbb{R})$,

$$\pi_1 = -z\partial_x \wedge \partial_y + x\partial_y \wedge \partial_z + y\partial_z \wedge \partial_x = X \wedge Y,$$

where $X = x\partial_y - y\partial_x$ and $Y = \partial_z + \frac{z}{x^2+y^2}(x\partial_x + y\partial_y)$. Let $f \in C^\infty(\mathbb{R})$ be such that $f(t) = 0$ for $t \leq 0$, and $f(t) > 0$ for $t > 0$, consider the vector field

$$Z = \frac{f(x^2 + y^2 - z^2)}{x^2 + y^2}(x\partial_x + y\partial_y),$$

and let $\pi = (X - Z) \wedge Y$.

(a) Show that $[X, Y] = 0$ and $[Z, Y] = 0$, and conclude that $[\pi, \pi] = 0$ (see Exercise 2.6).
(b) Verify that the linear approximation of π at the origin is π_1.
(c) Notice that $X - Z$ is the Hamiltonian vector field of $h(x, y, z) = -z$ with respect to π. Check that its integral curves in the region $\{x^2 + y^2 - z^2 > 0\}$ spiral in towards the cone $\{x^2 + y^2 - z^2 = 0\}$. By comparison with the symplectic leaves of π_1 (see Exercise 4.10, part (b)), conclude that π is not linearizable at the origin.

The linearization problem has been extensively studied in the realm of Poisson Lie groups. Most compact Lie groups with the standard Poisson structure turn out to be nonlinearizable at the group unit; in the case of a compact simple Lie group, the only exception is $SU(2)$ [15]. By contrast, the dual Poisson Lie group K^* to any compact, semisimple Lie group K is globally linearizable, i.e., it is globally Poisson diffeomorphic to its linear approximation on $T_eK^* = \mathfrak{k}^*$ [47]. See [2] for further results and references.

5 Further Topics

5.1 Symplectic Realizations and Dual Pairs

As previously seen, many interesting examples of Poisson manifolds arise as quotients of symplectic manifolds. Given the degenerate nature of Poisson structures and their singular symplectic foliations, it is often convenient to study Poisson manifolds by looking at symplectic manifolds that realize them as quotients.

Given a symplectic manifold (S, ω) and a surjective submersion $\varphi : S \to M$, a natural question is whether the symplectic structure on S descends to a Poisson structure on M, or in other words if one can endow M with a Poisson structure for which φ is a Poisson map. The conditions for this to happen are described in Exercise 4.12, from where the so-called *Libermann's lemma* directly follows: *If M carries a Poisson structure such that φ is a Poisson map then the symplectic orthogonal distribution to the subbundle $\mathcal{V} = \ker(d\varphi) \subseteq TM$ is integrable, and the converse holds provided the fibers of φ are connected.*

The necessity of the connectedness assumption on φ-fibers is illustrated by the next exercise.

Exercise 5.1 Consider the 2-sphere $\mathbb{S}^2$ equipped with a symplectic structure and the natural projection $\varphi : \mathbb{S}^2 \to \mathbb{R}P^2$. Verify that φ cannot be made into a Poisson map.

A Poisson map $S \to M$ from a symplectic manifold S to a Poisson manifold M is called a *symplectic realization* of M. We call a symplectic realization *full* when it is a surjective submersion. We list below some examples that we have already encountered.

Example 5.1

(a) Inclusions of symplectic leaves into Poisson manifolds are symplectic realizations, generally not full.
(b) The cotangent bundle projection $T^*M \to M$ is a full symplectic realization of the zero Poisson structure on M.
(c) As explained in Sect. 3.2, symplectic realizations $\mu : S \to \mathfrak{g}^*$ of the dual of a Lie algebra $\mathfrak{g}$ are the same as moment maps of Hamiltonian $\mathfrak{g}$-actions on symplectic manifolds. In case of a free $\mathfrak{g}$-action, μ is a submersion, and hence a full symplectic realization of the open subset $\mu(S) \subseteq \mathfrak{g}^*$.
(d) When a Lie group G acts freely and properly on a symplectic manifold (S, ω) by symplectomorphisms, the quotient map $S \to S/G$ is a full symplectic realization of the induced Poisson structure on S/G (see Sect. 3.1). In particular, if $\mathfrak{g}$ is the Lie algebra of G, then T^*G is a full symplectic realization of $\mathfrak{g}^* = T^*G/G$ (see Exercise 3.2).

$\diamond$

Exercise 5.2 Let $\varphi : S \to M$ be a full symplectic realization with connected fibers. Let $\mathcal{V} = \ker(d\varphi)$, and consider the (smooth) distribution $\mathcal{V} + \mathcal{V}^\omega \subseteq TS$. Show that if $\mathcal{O}$ is a leaf of the symplectic foliation of π, then $\varphi^{-1}(\mathcal{O})$ is a leaf of $\mathcal{V} + \mathcal{V}^\omega$ (in particular, $\mathcal{V} + \mathcal{V}^\omega$ is integrable). Moreover, this map induces a homeomorphism between the leaf spaces of the corresponding foliations on M and S. (Cf. Sect. 4.3.3.)

Let $\varphi : S \to M$ be a full symplectic realization, and suppose that the foliation tangent to $\mathcal{V}^\omega$ is *simple*, in the sense that its leaves are given by fibers of a surjective submersion

$\varphi' : S \to M'$. By Libermann's lemma, M' carries a Poisson structure such that φ' is also a full symplectic realization.

Given a symplectic manifold S and Poisson manifolds M_1 and M_2, a diagram

 (5.1)

of full symplectic realizations satisfying $(\ker(d\varphi_1))^\omega = \ker(d\varphi_2)$ is called a *dual pair*.

For a key example, consider a free and proper Hamiltonian G-action on a symplectic manifold S, with moment map $\mu : S \to \mathfrak{g}^*$. Then (see Exercise 4.15) $S/G \leftarrow S \xrightarrow{\mu} \mu(\mathfrak{g}^*)$ is a dual pair.

Poisson manifolds M_1 and M_2 fitting into a dual pair (5.1) turn out to share many features.

- If φ_1- and φ_2-fibers are connected, then there is an induced bijection between symplectic leaves of M_1 and M_2 (in fact, a homeomorphism of leaf spaces) with respect to which $\mathcal{O}_1 \hookrightarrow M_1$ corresponds to $\mathcal{O}_2 \hookrightarrow M_2$ if and only if

$$\varphi_1^{-1}(\mathcal{O}_1) = \varphi_2^{-1}(\mathcal{O}_2).$$

This follows from Exercise 5.2. In particular, there is an identification of Casimir functions on M_1 and M_2 (i.e., $H^0_\pi(M_1) \cong H^0_\pi(M_2)$).
- For any $x \in S$, the transverse Poisson structures at $\varphi_1(x)$ and $\varphi_2(x)$ are anti-isomorphic (i.e., isomorphic up to a sign); see Exercise 5.32 for a proof.
- If φ_1- and φ_2-fibers are 1-connected (i.e., connected and simply-connected), then there is an induced isomorphism $H^1_\pi(M_1) \cong H^1_\pi(M_2)$ preserving modular classes [24,45,46].

A dual pair $M_1 \xleftarrow{\varphi_1} S \xrightarrow{\varphi_2} M_2$ with the additional requirements that φ_1 and φ_2 have 1-connected fibers and are complete Poisson maps (see Exercise 4.9) defines a *Morita equivalence* between M_1 and $\overline{M_2}$ [93]. The notion of Morita equivalence of Poisson manifolds bears close resemblance with the original concept of Morita equivalence in algebra, see e.g. [11]; following this analogy, the set of self-Morita equivalences of a Poisson manifold is called its *Picard group* [9,10]. Various examples of Morita equivalent Poisson manifolds and Picard groups are described in [9,11]; see [10,79,85] for the study of Morita equivalence and Picard groups of log-symplectic manifolds.

The fundamental question as to whether any Poisson manifold admits a full symplectic realization was answered in the affirmative by Weinstein [21, 86] and Karasev [54]. The original proof consisted in gluing local symplectic realizations using a uniqueness property. More direct, global arguments to prove this result were given in [28] (see also [41]).

One can take as the starting point to construct a full symplectic realization $S \to M$ of an arbitrary Poisson manifold M the fact that the zero Poisson structure on M has a canonical symplectic realization given by the cotangent bundle projection $q : (T^*M, \omega_{can}) \to M$. For a given Poisson structure π on M, the intuitive idea is to obtain a symplectic realization by "deforming" ω_{can} using the flow of a suitably chosen vector field Y on T^*M. This vector field is defined with the aid of the horizontal lift hor $: q^*TM = T^*M \times_M TM \to T(T^*M)$ of a linear connection on TM by

$$Y|_\xi = \text{hor}(\xi, \pi^\sharp(\xi)), \qquad \xi \in T^*M.$$

(The vector field Y is more conceptually understood as a *Poisson spray* [28, § 1], i.e., the Poisson-geometric analogue of a classical geodesic spray, see [31, Chp. 11].)

Denoting the local flow of Y by ϕ_t, the fact that Y vanishes on the zero section of T^*M ensures that ϕ_t is defined for $t \in [0, 1]$ in a sufficiently small neighborhood of $M \subseteq T^*M$. The main result [28] is that *there is a neighborhood $U \subseteq T^*M$ of the zero section where the 2-form*

$$\omega = \int_0^1 (\phi_t)^* \omega_{can} dt \tag{5.2}$$

is symplectic and the restriction of the projection $q|_U : (U, \omega) \to (M, \pi)$ is a (full) symplectic realization.

One can refine the problem of existence of full symplectic realizations of Poisson manifolds by the additional requirement that the symplectic realization be *complete*. For instance, the realizations of Example 5.1 (d) are all complete (Exercise 4.14). As it turns out [26], it is *not* the case that any Poisson manifold admits a complete full symplectic realization.

To hint at what is involved in this problem, consider the case of the dual of a Lie algebra. As mentioned in Example 5.1 (d), any Lie group G integrating a Lie algebra $\mathfrak{g}$ gives rise to a natural complete full symplectic realization of $M = \mathfrak{g}^*$ by T^*G. More generally, the existence of a complete full symplectic realization of a Poisson manifold is equivalent to the existence of a certain global object "integrating" the Poisson manifold (discussed in Sect. 5.2 below); the issue is that, as opposed to Lie algebras, not every Poisson manifold is "integrable" in this sense.

The fact that complete full symplectic realizations do not always exist implies that the notion of Morita equivalence mentioned above is not defined for all Poisson manifolds; on the other hand, among "integrable" Poisson manifolds, Morita equivalence is indeed an equivalence relation [93] (see also [11]).

5.2 Symplectic Groupoids

The full symplectic realization of a Poisson manifold (M, π) defined by the symplectic form (5.2) on a neighborhood $U \subseteq T^*M$ of the zero section via the cotangent bundle projection $\mathsf{t} := q|_U : U \to M$ comes naturally equipped with more structure:

- the map $\mathsf{s} := q \circ \phi_1 : U \to M$ is anti-Poisson, and $M \xleftarrow{\mathsf{t}} U \xrightarrow{\mathsf{s}} \overline{M}$ is a dual pair;
- the zero-section $M \hookrightarrow U$ is a lagrangian submanifold.

There is, in addition, a group-like multiplication on U, and understanding how these structures fit together leads to the notion of (local) *symplectic groupoid* [21, Thm. 1.4].

Recall that a *Lie groupoid* consists of manifolds $\mathcal{G}$ and M equipped with structure maps given by surjective submersions $\mathsf{t}, \mathsf{s} : \mathcal{G} \to M$ (called *source and target maps*), a smooth partial *multiplication* map

$$m : \mathcal{G}^{(2)} \to \mathcal{G}, \qquad (g, h) \mapsto gh,$$

defined on the submanifold $\mathcal{G}^{(2)} = \{(g, h) \mid \mathsf{s}(g) = \mathsf{t}(h)\} \subseteq \mathcal{G} \times \mathcal{G}$ of composable pairs, a diffeomorphism $i : \mathcal{G} \to \mathcal{G}, g \mapsto g^{-1}$, called *inversion* map, and an embedding $\varepsilon : M \to \mathcal{G}, x \mapsto 1_x$, called *unit* map, satisfying

(i) $\mathsf{s}(gh) = \mathsf{s}(h), \mathsf{t}(gh) = \mathsf{t}(g)$,
(ii) $(gh)k = g(hk)$,
(iii) $\mathsf{s}(1_x) = \mathsf{t}(1_x) = x$, and $g 1_{\mathsf{s}(g)} = 1_{\mathsf{t}(g)} g = g$,
(iv) $\mathsf{s}(g^{-1}) = \mathsf{t}(g), \mathsf{t}(g^{-1}) = \mathsf{s}(g)$ and $g^{-1} g = 1_{\mathsf{s}(g)}, gg^{-1} = 1_{\mathsf{t}(g)}$.

We usually identify M with its image in $\mathcal{G}$ under ε. It is convenient to think of an element $g \in \mathcal{G}$ as an "arrow" $\mathsf{t}(g) \xleftarrow{g} \mathsf{s}(g)$ from its source to its target, so that e.g. the composition law (i) for the multiplication looks like

$$m(z \xleftarrow{g} y, y \xleftarrow{h} x) = z \xleftarrow{gh} x.$$

We use the notation $\mathcal{G} \rightrightarrows M$ for a Lie groupoid. Morphisms of Lie groupoids are smooth maps preserving all groupoid structural maps.

A Lie group G is the same as a Lie groupoid for which M is a point, $G \rightrightarrows \{*\}$. Any manifold M can be viewed as a Lie groupoid $M \rightrightarrows M$ with source and target maps being the identity. We will give more examples below.

A Lie groupoid $\mathcal{G} \rightrightarrows M$ defines an equivalence relation on M where $x \sim y$ if there exists an arrow $y \xleftarrow{g} x$. The equivalence class of $x \in M$,

$$\mathcal{O}_x = \{y \in M \mid \exists\, y \xleftarrow{g} x\} = \mathsf{t}(\mathsf{s}^{-1}(x)) \subseteq M,$$

is called the *orbit* of x, and it is a (immersed) submanifold of M. To each $x \in M$ there also corresponds a Lie group

$$\mathcal{G}_x = \{g \in \mathcal{G} \mid x \xleftarrow{g} x\} = \mathsf{t}^{-1}(x) \cap \mathsf{s}^{-1}(x) \subseteq \mathcal{G},$$

called the *isotropy group* at x. The decomposition of M into orbits and the collection of isotropy groups are essential ingredients of a Lie groupoid.

Exercise 5.3 On a Lie groupoid $\mathcal{G} \rightrightarrows M$, consider the graph of the multiplication map,

$$\Gamma_m := \{(g, h, gh),\ (g, h) \in \mathcal{G}^{(2)}\}.$$

Check that $\dim(\Gamma_m) = \dim(\mathcal{G}^{(2)}) = 2\dim(\mathcal{G}) - \dim(M)$.

Example 5.2

(a) Given a manifold M, the *pair groupoid* $M \times M \rightrightarrows M$ has arrows $y \xleftarrow{(y,x)} x$ and multiplication given by $(z, y)(y, x) = (z, x)$. This Lie groupoid has M itself as the only orbit and trivial isotropy groups. A closely related Lie groupoid is the *fundamental groupoid* of M, $\Pi(M) \rightrightarrows M$, whose arrows are homotopy classes of paths on M,

$$\gamma(1) \xleftarrow{[\gamma]} \gamma(0),$$

with multiplication given by concatenation. The orbits of $\Pi(M)$ are the connected components of M, and its isotropy group at x is the fundamental group of the connected component containing x. The map $\Pi(M) \to M \times M$, $[\gamma] \mapsto (\gamma(1), \gamma(0))$ is a groupoid morphism, which is an isomorphism when M is 1-connected (i.e., connected with trivial fundamental group).

(b) A Lie groupoid for which source and target maps coincide is a *bundle of Lie groups*. Any vector bundle is a particular example of a Lie groupoid of this kind, with multiplication given by fiberwise addition.

(c) An action of a Lie group G on a manifold M gives rise to an *action groupoid* $G \times M \rightrightarrows M$, with arrows

$$\sigma x \xleftarrow{(\sigma,x)} x$$

and multiplication $(\sigma', y)(\sigma, x) = (\sigma'\sigma, x)$. The orbits and isotropy groups of the action groupoid coincide with those of the action.

On a Lie groupoid $\mathcal{G} \rightrightarrows M$, right and left multiplication by an arrow $y \stackrel{g}{\leftarrow} x$ define diffeomorphisms

$$r_g : \mathsf{s}^{-1}(y) \to \mathsf{s}^{-1}(x), \qquad l_g : \mathsf{t}^{-1}(x) \to \mathsf{t}^{-1}(y).$$

Exercise 5.4 Use properties (i), (iii) and (iv) of Lie groupoids to verify the following identities.

(a) For $V \in T_g \mathcal{G}$,

$$V = (dm)_{(\mathsf{t}(g),g)}((d\mathsf{t})_g(V), V) = (dm)_{(g,\mathsf{s}(g))}(V, (d\mathsf{s})_g(V)).$$

In particular, $X = dm_{(x,x)}(X, X)$ for $X \in T_x M$.

(b) For $u \in \ker(d\mathsf{s})_y$,

$$(dr_g)_y(u) = dm_{(y,g)}(u, 0), \quad \text{and} \quad (d\mathsf{t})_g((dr_g)_y(u)) = (d\mathsf{t})_y(u).$$

(c) For $V \in T_g \mathcal{G}$, $(d\mathsf{t})_g(V) = dm_{(g,g^{-1})}(V, (di)_g(V))$.

To define symplectic groupoids, we need to consider differential forms on Lie groupoids $\mathcal{G} \rightrightarrows M$ suitably compatible with the groupoid structure. A 2-form $\omega \in \Omega^2(\mathcal{G})$ is called *multiplicative* if

$$m^*\omega = \text{pr}_1^*\omega + \text{pr}_2^*\omega, \tag{5.3}$$

where $\text{pr}_1, \text{pr}_2 : \mathcal{G}^{(2)} \to \mathcal{G}$ are the natural projections. (The same definition applies to differential forms of any degree.) A *symplectic groupoid* is a Lie groupoid equipped with a multiplicative symplectic form.

Exercise 5.5 Let ω be a multiplicative 2-form on $\mathcal{G} \rightrightarrows M$.

(a) Use Exercise 5.4, (a) and (b), to show that, for $y \stackrel{g}{\leftarrow} x$, $u \in (\ker(d\mathsf{s}))_y$, and $V \in T_g\mathcal{G}$,

$$\omega((dr_g)_y(u), V) = \omega(u, (d\mathsf{t})_g(V)).$$

Using the isomorphism $(dr_g)_y : \ker(d\mathsf{s})_y \stackrel{\sim}{\to} \ker(d\mathsf{s})_g$, conclude that

$$\ker(d\mathsf{s}) \subseteq \ker(d\mathsf{t})^\omega.$$

(b) For $X, Y \in T_x M$, check that $\omega(X, Y) = 0$ (see Exercise 5.4 (a)).

(c) Use the previous item and Exercise 5.4 (c) to show that, for $V, W \in T_g \mathcal{G}$,

$$\omega(V, W) + \omega((di)_g(V), (di)_g(W)) = 0.$$

On a symplectic groupoid $(\mathcal{G}, \omega)$, the compatibility condition (5.3) is equivalent to requiring that the graph of the multiplication map, $\Gamma_m := \{(g, h, gh), (g, h) \in \mathcal{G}^{(2)}\}$, is an isotropic submanifold of $(\mathcal{G}, \omega) \times (\mathcal{G}, \omega) \times (\mathcal{G}, -\omega)$. On the other hand, by Exercise 5.5 (b), the image of $M \overset{\varepsilon}{\hookrightarrow} \mathcal{G}$ is isotropic. Therefore (see Exercise 5.3)

$$\dim(\Gamma_m) = 2\dim(\mathcal{G}) - \dim(M) \leq \frac{3}{2}\dim(\mathcal{G}), \qquad \dim(M) \leq \frac{1}{2}\dim(\mathcal{G}).$$

Combining these two inequalities, we have that

$$\dim(M) = \frac{1}{2}\dim(\mathcal{G}), \quad \text{and} \quad \dim(\Gamma_m) = \frac{3}{2}\dim(\mathcal{G}).$$

As a consequence, we see that *a symplectic form ω on a Lie groupoid $\mathcal{G}$ is multiplicative if and only if Γ_m is a lagrangian submanifold of $(\mathcal{G}, \omega) \times (\mathcal{G}, \omega) \times (\mathcal{G}, -\omega)$*; this last condition is the original definition of symplectic groupoid in [87].

The following are some important properties of a symplectic groupoid $(\mathcal{G}, \omega)$.

(SG1) *The submanifold of units $M \hookrightarrow \mathcal{G}$ is lagrangian*, as shown above.
(SG2) *The inversion map satisfies $i^*\omega = -\omega$*, by Exercise 5.5 (c).
(SG3) $\ker(ds) = \ker(dt)^\omega$ (by Exercise 5.5 (a), since the subbundles $\ker(ds)$ and $\ker(dt)^\omega$ have the same rank).

Another key property, linking symplectic groupoids to Poisson structures, will follow from the next exercise (see (SG4) below).

Exercise 5.6 Let $(\mathcal{G}, \omega)$ be a symplectic groupoid, and consider the vector bundle map

$$\mu_\omega : \ker(ds)|_M \to T^*M, \qquad \mu_\omega(u) = \omega(u, \cdot)|_{TM}.$$

(a) Check that μ_ω is an isomorphism.
(b) Let $y \overset{g}{\leftarrow} x$, $u \in \ker(ds)|_y$ and $f \in C^\infty(M)$ such that $df|_y = \mu_\omega(u)$. Conclude from Exercise 5.5 (a) that

$$X_{t^*f}|_g = (dr_g)_y(u).$$

(c) For $f_1, f_2 \in C^\infty(M)$, show that

$$\{\mathsf{t}^* f_1, \mathsf{t}^* f_2\}(g) = \omega(u_2, (d\mathsf{t})_y(u_1)),$$

where $y = \mathsf{t}(g)$, $u_i \in (\ker(d\mathsf{s}))_y$, and $\mu_\omega(u_i) = df_i|_y$, $i = 1, 2$. Note in particular that the function $\{\mathsf{t}^* f_1, \mathsf{t}^* f_2\} \in C^\infty(\mathcal{G})$ is constant along t-fibers.

A consequence of the previous exercise (part (c)) is that, on a symplectic groupoid,

$$\{\mathsf{t}^* C^\infty(M), \mathsf{t}^* C^\infty(M)\} \subseteq \mathsf{t}^* C^\infty(M),$$

and therefore M inherits a Poisson structure such that t is a Poisson map. (In the special case of symplectic groupoids with t-connected fibers, this property can also be deduced from (SG3) and Libermann's lemma, see Sect. 5.1).

Exercise 5.7 Let $(\mathcal{G}, \omega)$ be a symplectic groupoid and $f \in C^\infty(M)$. It follows from Exercise 5.6 (b) that the Hamiltonian vector field $X_{\mathsf{t}^* f}$ is tangent to the s-fibers, and, for $y \overset{g}{\leftarrow} x$,

$$(r_g)_*(X_{\mathsf{t}^* f}|_{\mathsf{s}^{-1}(y)}) = X_{\mathsf{t}^* f}|_{\mathsf{s}^{-1}(x)}.$$

Use this last property to show that the Poisson map $\mathsf{t} : \mathcal{G} \to M$ is complete.

In conclusion, for a symplectic groupoid $(\mathcal{G}, \omega)$,

(SG4) *there exists a (unique) Poisson structure on M making $\mathsf{t} : \mathcal{G} \to M$ into a complete, full symplectic realization.*

Note that, by the discussion in Sect. 4.3.3 (see (4.5)), the symplectic leaves of the Poisson structure on M are the connected components of the orbits of $\mathcal{G} \rightrightarrows M$.

Example 5.3

(a) Given a symplectic manifold (S, ω), the pair groupoid $S \times S \rightrightarrows S$ is a symplectic groupoid with respect to $\mathrm{pr}_1^* \omega - \mathrm{pr}_2^* \omega$. The induced Poisson structure on S is the one defined by ω. Since the morphism $\Pi(S) \to S \times S$ is a local diffeomorphism, the pullback of $\mathrm{pr}_1^* \omega - \mathrm{pr}_2^* \omega$ makes $\Pi(S)$ into a symplectic groupoid over S.
(b) Let $E \to M$ be a vector bundle, viewed as a Lie groupoid with multiplication given by fiberwise addition (see Example 5.2 (b)). A symplectic form $\omega \in \Omega^2(E)$ is multiplicative if and only if it defines a linear Poisson structure, in the sense of Sect. 3.3. It follows from Exercise 3.19 that (E, ω) is isomorphic to T^*M with its canonical symplectic form ω_{can}. The induced Poisson structure on M is trivial.

(c) For a Lie group G, the action groupoid $(G \times \mathfrak{g}^*) \rightrightarrows \mathfrak{g}^*$ (with respect to the coadjoint action) becomes a symplectic groupoid when equipped with the symplectic form obtained from the identification $T^*G \simeq G \times \mathfrak{g}^*$ via left (or right) translations. The induced Poisson structure on $\mathfrak{g}^*$ is the linear Poisson structure dual to the Lie algebra $\mathfrak{g}$ (up to a sign, depending on the chosen identification).

$\diamond$

A Poisson manifold is called *integrable* if it is isomorphic to the manifold of units of a symplectic groupoid; in such a case one refers to the symplectic groupoid as an *integration* of the Poisson manifold. The examples above show that Poisson structures that are nondegenerate (symplectic), trivial or linear are integrable, and also that integrations are not unique. On the other hand, as first observed by Weinstein [87], there are Poisson manifolds that fail to be integrable; a concrete example is the Poisson structure on $M = \mathbb{S}^2 \times \mathbb{R}$ defined by the family of symplectic structures $\omega_t = (1 + t^2)\omega_{area}$ on $\mathbb{S}^2$, as in Example 2.3 (see [17, 26, 87]). This phenomenon can be understood in light of the broader integrability problem for Lie algebroids [25], briefly recalled in Sect. 5.3 below.

We mention other important classes of integrable Poisson manifolds.

- If M is an integrable Poisson manifold endowed with a free and proper action of a Lie group G by Poisson automorphisms, then the Poisson manifold M/G (see Sect. 3.1) is also integrable. As shown in [39], M admits an integration by a symplectic groupoid $\mathcal{G}$ carrying a Hamiltonian G-action lifting the G-action on M, and its symplectic reduction at zero is a symplectic groupoid over M/G. The simplest instance of this construction is when M has the trivial Poisson structure and $\mathcal{G} = T^*M$; in this case, it is well known that the canonical lift of the G-action on M to T^*M is Hamiltonian (see Exercise 3.19) and its reduction at zero is $T^*(M/G)$.
- Any Poisson Lie group (G, π) is integrable [66]. The symplectic groupoids integrating G described in [66] have the property of carrying a second groupoid structure making them into a symplectic groupoid over the dual Poisson Lie group G^*,

$$\begin{array}{ccc} (\mathcal{G}, \omega) & \rightrightarrows & G^* \\ \Downarrow & & \Downarrow \\ G & \rightrightarrows & *. \end{array}$$

These are examples of *symplectic double groupoids*. The simplest case is given by (T^*G, ω_{can}), which can be regarded as a symplectic groupoid over G (with the trivial Poisson structure) and over $\mathfrak{g}^*$, see Example 5.3, (b) and (c).

More generally, any Poisson homogeneous space is integrable, see [14].
- Log-symplectic Poisson manifolds are integrable. Their symplectic groupoids were constructed and classified in [49] by means of blow-up and gluing operations.

For an integrable Poisson manifold (M, π), there is a canonical symplectic groupoid $\Pi(M, \pi) \rightrightarrows M$ that generalizes the fundamental groupoid of a symplectic manifold (Example 5.3 (a)), obtained as a quotient of "cotangent paths" on M by "cotangent homotopies" [17, 26]. We refer to $\Pi(M, \pi)$ as the *Poisson fundamental groupoid* of (M, π). For a general Poisson manifold (M, π), the construction of $\Pi(M, \pi)$ produces a *topological* groupoid which is always smooth on an open neighborhood of the units $M \subseteq \Pi(M, \pi)$, yielding a "local" symplectic groupoid (its global smoothness occurs if and only if M is integrable [26]).

We outline the key points relating symplectic groupoids to the study of symplectic realizations in Sect. 5.1. The general symplectic realization of a Poisson manifold (M, π) defined by the 2-form (5.2) on a neighborhood of the zero section of T^*M is isomorphic to the symplectic realization given by the target map on a smooth neighborhood of the units in $\Pi(M, \pi)$. On the other hand, by (SG4) any integrable Poisson manifold has a complete, full symplectic realization given by the target map of a symplectic groupoid. More generally, as proven in [26], *the existence of a complete, full symplectic realization of a Poisson manifold is equivalent to the existence of a symplectic groupoid that integrates it* (see also [5]).

Remark 5.4 (More on Symplectic Groupoids)

⋄ Symplectic groupoids were originally introduced as part of a quantization scheme for Poisson manifolds [54, 89, 91] (see [7, 51] for more recent developments) based on the fact that groupoids are rich sources of noncommutative algebras via their convolution algebras.
⋄ The Poisson fundamental groupoid of a Poisson manifold (M, π) can be regarded as an infinite-dimensional symplectic quotient, and it has an interpretation in terms of (classical) topological field theory as the phase space of the so-called *Poisson sigma model* with target M [17].
⋄ Symplectic groupoids and their actions lead to a broad extension of the theory of Hamiltonian actions on symplectic manifolds, which allows symplectic realizations to be regarded as "moment maps" [74] (see also [31, § 14.4]). Symplectic groupoid actions are also key elements in the theory of Morita equivalence of Poisson manifolds [92, 93] (see also [11]).
⋄ A symplectic groupoid gives rise to a dual pair

(5.4)

see properties (SG2), (SG3), and (SG4). Moreover, if $\mathcal{G}$ has 1-connected s-fibers, this dual pair defines a self Morita equivalence of the Poisson manifold M, which is the unit element of its Picard group.
⋄ Symplectic groupoids serve as a tool e.g. for the calculation of Poisson cohomology [94] and linearization problems [27]. Compactness properties of symplectic groupoids have many relevant effects on their underlying Poisson manifolds; their systematic study has recently begun in [30].

⋄

Remark 5.5 *Poisson groupoids* [88] provide a common framework for symplectic groupoids and Poisson Lie groups (Sect. 3.4). The extension of Lie bialgebras and their Drinfeld doubles (see Remark 3.3) to this context leads to the important notions of *Lie bialgebroids* and *Courant algebroids* [63]. ⋄

5.3 The Lie Algebroid of a Poisson Manifold

Lie algebroids (see Sect. 3.3) are infinitesimal versions of Lie groupoids, extending the way Lie algebras arise from Lie groups. A key fact explaining the link between Poisson manifolds and symplectic groupoids is that any Poisson manifold has an associated Lie algebroid. From this perspective, the correspondence between Poisson manifolds and symplectic groupoids is analogous to that of Lie algebras and Lie groups.

The Lie algebroid of a Lie groupoid $\mathcal{G} \rightrightarrows M$ is defined by as follows:[6]

- As a vector bundle, $A = \ker(ds)|_M \to M$;
- The anchor map is the restriction of $dt : T\mathcal{G} \to TM$ to $A \subseteq T\mathcal{G}|_M$,

$$\rho = dt|_A : A \to TM.$$

- The Lie bracket on the space of sections $\Gamma(A)$ arises from its identification with vector fields X on $\mathcal{G}$ that are tangent to s-fibers and right-invariant, i.e., $dr_g(X_h) = X_{hg}$.

A Lie algebroid $A \to M$ is *integrable* if it is isomorphic to the Lie algebroid of a Lie groupoid $\mathcal{G} \rightrightarrows M$, as above. In this case, we will say that $\mathcal{G}$ *integrates* A.

Unlike Lie algebras, which can be always integrated to Lie groups, Lie algebroids are not necessarily integrable (see [25] for a description of obstructions to integrability and examples). Nonetheless, other fundamental results in classical Lie theory are still

---

[6] An alternative convention reverses the roles of source and target maps, and uses left instead of right-invariant vector fields.

valid in this broader context; e.g., the construction of the Lie algebroid of a Lie groupoid gives rise to a *Lie functor* from Lie groupoids to Lie algebroids that defines an *equivalence of categories between source-simply connected Lie groupoids[7] and integrable Lie algebroids*. (In particular, an integrable Lie algebroid has a unique source-simply connected integration, up to isomorphism.)

Given a Poisson manifold (M, π), its cotangent bundle $A = T^*M$ carries a Lie algebroid structure with anchor map $\pi^\sharp : T^*M \to TM$ and Lie bracket on $\Gamma(A) = \Omega^1(M)$ given by

$$[\alpha, \beta]_\pi := \mathcal{L}_{\pi^\sharp(\alpha)}\beta - \mathcal{L}_{\pi^\sharp(\beta)}\alpha - d(\pi(\alpha, \beta)),$$

so that $[df, dg]_\pi = d\{f, g\}$.

Exercise 5.8 Show that a Lie algebroid structure on T^*M, with anchor $\rho : T^*M \to TM$ and bracket $[\cdot, \cdot]$ on $\Omega^1(M)$, is induced by a Poisson structure on M as above if and only if $\rho^* = -\rho$ and $[\cdot, \cdot]$ preserves the subspace $\Omega^1_{cl}(M)$ of closed 1-forms. (As a key step, verify that if $\Lambda : \Omega^1(M) \times \Omega^1(M) \to \Omega^1(M)$ is $C^\infty(M)$-bilinear, skew-symmetric, and $\Lambda(\Omega^1_{cl}(M), \Omega^1_{cl}(M)) \subseteq \Omega^1_{cl}(M)$, then $\Lambda = 0$.)

Exercise 5.9 Given a Lie algebroid $A \to M$, a subbundle $B \to N$ is a *Lie subalgebroid* if $\rho(B) \subseteq TN$ and $[u, v]|_N \in \Gamma(B)$ whenever $u|_N, v|_N \in \Gamma(B)$.

Let (M, π) be a Poisson manifold, and consider the corresponding Lie algebroid T^*M. Show that a submanifold $N \hookrightarrow M$ is coisotropic if and only if $\mathrm{Ann}(TN) \subseteq T^*M$ is a Lie subalgebroid. Hence there is a natural bijective correspondence between coisotropic submanifolds of M and lagrangian Lie subalgebroids of T^*M (i.e., Lie subalgebroids that are lagrangian submanifolds with respect to the canonical symplectic form).

Remark 5.6 Many properties of Poisson manifolds can be understood in terms of general features of Lie algebroids. For a Lie algebroid $A \to M$,

- the distribution $\rho(A) \subseteq TM$ given by the image of the anchor map is integrable; the leaves of the corresponding (singular) foliation on M are called *orbits* of A;
- for each $m \in M$, $\ker(\rho|_m) \subseteq A|_m$ acquires a Lie algebra structure by restriction of the Lie bracket on $\Gamma(A)$, called the *isotropy Lie algebra* at m.

When $A = T^*M$ is the cotangent Lie algebroid of a Poisson structure π on M, orbits of A are the leaves of the symplectic foliation (see Sect. 4.1), and the isotropy Lie algebra of A at m is the transverse Lie algebra of π at m (see Sect. 4.4). The Lie algebroid cohomology of A (see Sect. 3.5) agrees with the Poisson cohomology of (M, π). ◇

---

[7] We say that a Lie groupoid is source-simply connected if its s-fibers are 1-connected.

Remark 5.7 For a Poisson manifold (M, π), the Lie algebroid structure on T^*M gives rise to a linear Poisson structure π_{TM} on TM (see Sect. 3.3) uniquely characterized by the property that

$$\{df, dg\}_{TM} = d\{f, g\},$$

for all $f, g \in C^\infty(M)$, where here we view df, dg as (linear) functions on TM. The Poisson structure π_{TM} on TM is called the *tangent lift* of π.

Exercise 5.10 Check that if (M, π) is symplectic, then so is its tangent lift π_{TM} on TM. How is this symplectic structure on TM related to the canonical symplectic structure on T^*M?

◇

Symplectic groupoids are closely related to integrations of cotangent Lie algebroids of Poisson manifolds.

First, suppose that (M, π) is the Poisson manifold of units of a symplectic groupoid $(\mathcal{G}, \omega)$, and let A be the Lie algebroid of $\mathcal{G}$. The next exercise shows that the Lie algebroid T^*M is naturally identified with A.

Exercise 5.11 Following Exercise 5.6, consider the vector bundle isomorphism

$$\mu_\omega : A \to T^*M, \qquad \mu_\omega(u) = \omega(u, \cdot)|_{TM},$$

and note that the Poisson structure π on M satisfies $\pi^\sharp = \rho \circ \mu_\omega^{-1} : T^*M \to TM$. Verify that μ_ω is an isomorphism of Lie algebroids, i.e., it intertwines anchor maps and preserves brackets.

Therefore any symplectic groupoid integrates the cotangent Lie algebroid of the Poisson structure on its units. In the converse direction, if M is a Poisson manifold with integrable cotangent Lie algebroid, then any source-simply-connected Lie groupoid integrating T^*M is naturally a symplectic groupoid [68, Thm. 5.2] (but this may not be the case for other integrations, see Remark 5.8 below); moreover, the Poisson fundamental groupoid $\Pi(M, \pi)$ gives a concrete model for source-simply-connected integrations of T^*M [26].

We conclude that *a Poisson manifold is integrable (in the sense of Sect. 5.2) if and only if its cotangent Lie algebroid is integrable.*

Remark 5.8 Consider the 3-sphere $\mathbb{S}^3$ equipped with the zero Poisson structure. As a Poisson manifold, it is integrated by $T^*\mathbb{S}^3 = \mathbb{S}^3 \times \mathbb{R}^3$, viewed as a symplectic groupoid with respect to fiberwise addition and canonical symplectic structure (see Example 5.3

(b)). In this case the cotangent Lie algebroid of $\mathbb{S}^3$ has trivial anchor and bracket. The Lie groupoid $\mathbb{S}^3 \times \mathbb{T}^3$ (viewed as a bundle of abelian Lie groups over $\mathbb{S}^3$) is another integration of this Lie algebroid, but it cannot be a symplectic groupoid (since the manifold $\mathbb{S}^3 \times \mathbb{T}^3$ carries no symplectic form). ◇

5.4 Dirac Structures

On a symplectic manifold M, any submanifold $N \hookrightarrow M$ inherits a closed 2-form (usually not symplectic) via pullback of the ambient symplectic form. Dirac structures are geometric objects that give a similar description of the intrinsic geometry inherited by submanifolds of Poisson manifolds. The slogan one should have in mind is that *Dirac structures are to Poisson structures what closed 2-forms are to symplectic structures*. In what follows, we will also use the term "presymplectic" to refer to closed 2-forms.

Dirac structures have several applications to Poisson geometry and related areas, some of which are mentioned at the end of this subsection. Here we will mostly focus on their role in the study of submanifolds of Poisson manifolds. Besides Courant's original paper [22], general references include [8] and [31, Chp. 7].

A peculiarity of Dirac structures is that, in contrast with Poisson and symplectic structures, their general definition is not in terms of tensors on a manifold M, but rather as special types of vector subbundles of the direct sum

$$\mathbb{T}M := TM \oplus T^*M.$$

The vector bundle $\mathbb{T}M$ carries a natural nondegenerate, symmetric, fiberwise bilinear pairing $\langle \cdot, \cdot \rangle$ given by

$$\langle (X, \alpha), (Y, \beta) \rangle = \beta(X) + \alpha(Y).$$

With respect to this pairing, a subbundle $L \subseteq \mathbb{T}M$ is *lagrangian* (i.e., $L = L^\perp$) if and only if $\langle \cdot, \cdot \rangle$ vanishes on L and $\mathrm{rank}(L) = \dim(M)$. As shown in the next exercise, 2-forms and bivector fields on M can be characterized by appropriate lagrangian subbundles of $\mathbb{T}M$.

Exercise 5.12 Verify the following.

(a) For a 2-form $\omega \in \Omega^2(M)$, the graph of $\omega^\flat : TM \to T^*M$ is a lagrangian subbundle of $\mathbb{T}M$, and this establishes a bijection between 2-forms on M and lagrangian subbundles of $\mathbb{T}M$ that intersect $0 \oplus T^*M$ trivially.
(b) The map taking a bivector field $\pi \in \mathcal{X}^2(M)$ to the graph of $\pi^\sharp : T^*M \to TM$ is a bijection between bivector fields on M and lagrangian subbundles of $\mathbb{T}M$ intersecting $TM \oplus 0$ trivially.

To identify 2-forms that are closed and bivector fields that are Poisson, one considers the *Courant-Dorfman bracket* $[\![\cdot,\cdot]\!]$ on $\Gamma(\mathbb{T}M) = \mathfrak{X}^1(M) \oplus \Omega^1(M)$ [22, 33],

$$[\![(X,\alpha),(Y,\beta)]\!] = ([X,Y], \mathcal{L}_X\beta - i_Y d\alpha). \tag{5.5}$$

A *Dirac structure* on M [22, 23] is a lagrangian subbundle $L \subset \mathbb{T}M$ satisfying

$$[\![\Gamma(L), \Gamma(L)]\!] \subset \Gamma(L),$$

referred to as the *integrability condition*. The pair (M, L) is a *Dirac manifold*.

Exercise 5.13 Given a lagrangian subbundle $L \subseteq \mathbb{T}M$, verify that the operation

$$(e_1, e_2, e_3) \mapsto \langle [\![e_1, e_2]\!], e_3 \rangle, \quad \text{for } e_1, e_2, e_3 \in \Gamma(L),$$

is skew-symmetric and $C^\infty(M)$-trilinear, so it defines an element $\Upsilon_L \in \Gamma(\wedge^3 L^*)$, called the *Courant tensor* of L. Moreover, the integrability of L is equivalent to the vanishing condition $\Upsilon_L = 0$.

Exercise 5.14

(a) Let $L \subseteq \mathbb{T}M$ be given by the graph of $\omega^\flat$, for a 2-form $\omega \in \Omega^2(M)$. Given $e_i = (X_i, \omega^\flat(X_i)) \in \Gamma(L)$, $i = 1, 2, 3$, check that $\Upsilon_L(e_1, e_2, e_3) = d\omega(X_1, X_2, X_3)$. Therefore L is integrable (i.e., it is a Dirac structure) if and only if $d\omega = 0$.

(b) Let $L \subseteq \mathbb{T}M$ be the graph of $\pi^\sharp$, for a bivector field $\pi \in \mathfrak{X}^2(M)$. For $e_i = (X_{f_i}, df_i) \in \Gamma(L)$, $i = 1, 2, 3$, check that $\Upsilon_L(e_1, e_2, e_3) = \text{Jac}(f_1, f_2, f_3)$. Hence L is integrable if and only if π is Poisson.

(c) Any subbundle $D \subseteq TM$ gives rise to a lagrangian subbundle $L = D \oplus \text{Ann}(D) \subseteq \mathbb{T}M$. Show that L is a Dirac structure if and only if D is involutive (and hence the tangent distribution to a regular foliation on M).

We define the *kernel* distribution of a Dirac structure L on M by

$$\ker(L) = L \cap (TM \oplus 0) \subseteq TM,$$

so that Poisson structures on M are equivalent to Dirac structures with trivial kernel, see Exercise 5.12 (b). (For Dirac structures defined by closed 2-forms, those with trivial kernel correspond to symplectic forms.) As a consequence, a general Dirac structure L defines a Poisson structure on the (possibly empty) open subset of M where its kernel is trivial.

The next exercise illustrates how "singular" Poisson structure may be encoded by (smooth!) Dirac structures.

Exercise 5.15 On $M = \mathbb{R}^3 = \{(x, y, z)\}$, check that the subbundle of $\mathbb{T}M$ spanned by the sections $(\frac{\partial}{\partial y}, zdx)$, $(\frac{\partial}{\partial x}, -zdy)$, $(0, dz)$ is a Dirac structure. Verify that this Dirac structure has trivial kernel on the open subset $\{(x, y, z) \,|\, z \neq 0\}$, where it corresponds to the Poisson structure $\pi = \frac{1}{z} \frac{\partial}{\partial x} \wedge \frac{\partial}{\partial y}$.

The exercise below shows how the abelian group of closed 2-forms on a manifold acts on the set of Dirac structures by means of an operation called *gauge transformation* [84].

Exercise 5.16 For a 2-form $B \in \Omega^2(M)$, consider the operation
$$\tau_B : \mathbb{T}M \to \mathbb{T}M, \quad (X, \alpha) \mapsto (X, i_X B + \alpha).$$

(a) Check that τ_B preserves the natural symmetric pairing on $\mathbb{T}M$ and that it preserves the Courant-Dorfman bracket if and only if $dB = 0$. In this case, if L is a Dirac structure, so is $\tau_B(L)$.

(b) Suppose that π is a Poisson structure on M and $L_\pi = \text{graph}(\pi^\sharp)$ is the corresponding Dirac structure. For a closed 2-form B, show that the Dirac structure $\tau_B(L_\pi)$ is again given by a Poisson structure if and only if $(\text{Id} + B^\flat \circ \pi^\sharp) : T^*M \to T^*M$ is an isomorphism, in which case $\tau_B(L_\pi) = L_{\pi_B}$ (see Exercise 4.7).

Remark 5.9 (More on the Courant-Dorfman Bracket) The Courant-Dorfman bracket (5.5) is an interesting object in its own right. The following are its main properties:

(i) $[\![e_1, e_2]\!] + [\![e_2, e_1]\!] = d\langle e_1, e_2 \rangle$,
(ii) $[\![e_1, [\![e_2, e_3]\!]]\!] = [\![[\![e_1, e_2]\!], e_3]\!] + [\![e_2, [\![e_1, e_3]\!]]\!]$,
(iii) $[\![e_1, fe_2]\!] = (\mathcal{L}_{X_1} f)e_2 + f[\![e_1, e_2]\!]$,
(iv) $\mathcal{L}_{X_1}\langle e_2, e_3 \rangle = \langle [\![e_1, e_2]\!], e_3 \rangle + \langle e_2, [\![e_1, e_3]\!] \rangle$,

for $e_1, e_2, e_3 \in \Gamma(\mathbb{T}M)$, and $e_1 = (X_1, \alpha_1)$. The Courant-Dorfman bracket is not skew-symmetric, hence not a Lie bracket, but we see that it satisfies versions of the Jacobi and Leibniz identities (as well as a natural compatibility with the pairing). This is an example of a *Courant algebroid*, called the *standard Courant algebroid*. One can alternatively consider the skew-symmetrization of this bracket (which is Courant's original bracket in [22]), but this has the effect of introducing "anomalies" in properties (ii), (iii) and (iv) (that can be suitably understood in terms of "Lie 2-algebras" [82, 83]). By (i) the Courant-Dorfman bracket coincides with its skew symmetrization on sections of lagrangian subbundles $L \subseteq \mathbb{T}M$, so the integrability condition of Dirac structures can be phrased using either one of the brackets. ◇

We mention some of the key features of Dirac manifolds extending properties of Poisson manifolds.

- *(Lie algebroid)* For a Dirac structure L on M, the vector bundle $L \to M$ acquires a Lie-algebroid structure with anchor given by the natural projection $L \to TM$ and bracket on $\Gamma(L)$ given by the restriction of the Courant-Dorfman bracket. When L corresponds to a Poisson structure, the projection $\mathbb{T}M \to T^*M$ induces a Lie algebroid isomorphism $L \xrightarrow{\sim} T^*M$.
- *(Characteristic distribution and presymplectic foliation)* The image of a lagrangian subbundle $L \subseteq \mathbb{T}M$ under the projection $\mathrm{pr}_{TM} : \mathbb{T}M \to TM$ defines a distribution

$$R := \mathrm{pr}_{TM}(L) \subseteq TM$$

that generalizes the characteristic distribution of a bivector field, see Sect. 2.4; additionally, over each point $x \in M$, L defines a skew-symmetric bilinear form Ω_x on $R|_x$ by

$$\Omega_x(X, Y) = \alpha(Y),$$

for $X, Y \in R|_x$ and α any element in T_x^*M such that $(X, \alpha) \in L|_x$.

Exercise 5.17 Verify that $\mathrm{Ann}(R) = L \cap (0 \oplus T^*M)$, and check that Ω_x is well defined.

When L is a Dirac structure, the distribution R is integrable and each leaf $\mathcal{O} \hookrightarrow M$,

$$T\mathcal{O} = R|_\mathcal{O},$$

carries a closed 2-form $\omega_\mathcal{O}$ defined pointwise by the 2-forms Ω_x on $R|_x$, $x \in \mathcal{O}$. The collection of all such presymplectic leaves defines the *presymplectic foliation* of L, which uniquely characterizes L. Note that $\omega_\mathcal{O}$ is nondegenerate at $x \in M$ if and only if $\ker(L)|_x = \{0\}$, so the presymplectic foliation of L has symplectic leaves if and only if L has trivial kernel at all points, which happens if and only if L is defined by a Poisson structure.

- *(Integration)* The "differentiation/integration" type of correspondence relating Poisson manifolds and symplectic groupoids also extends to Dirac manifolds [12]. The global objects arising in this more general setting are known as *presymplectic groupoids* (or *quasi-symplectic groupoids* [95]) and consist of Lie groupoids $\mathcal{G} \rightrightarrows M$ carrying a multiplicative, closed 2-form ω satisfying a weaker version of nondegeneracy.

Exercise 5.18 Verify that two Dirac structures on a manifold M related by a gauge transformation τ_B (as in Exercise 5.16) have the same leaves, and the corresponding closed 2-forms on a given leaf $\mathcal{O} \hookrightarrow M$ differ by the pullback of B to $\mathcal{O}$ (cf. Exercise 4.7).

The next exercise shows that any Dirac manifold gives rise to a Poisson algebra; building on this fact, the subsequent exercise indicates that a Dirac structure on a manifold M can be also thought of as a Poisson structure on the leaf space of a foliation on M.

Exercise 5.19 For a Dirac manifold (M, L), we say that a vector field X is a *Hamiltonian vector field* for a function $f \in C^\infty(M)$ if $(X, df) \in \Gamma(L)$; functions that admit a Hamiltonian vector field are called *admissible* (note that it may happen that a function admits many or no Hamiltonian vector fields). Show that the set of admissible functions, denoted by $C^\infty_{adm}(M)$, is a Poisson algebra with bracket given by

$$\{f, g\} = dg(X),$$

where X is any Hamiltonian vector field of f.

Exercise 5.20 Let (M, L) be a Dirac manifold, and let $\varphi : M \to B$ be a surjective submersion with connected fibers and such that $\ker(d\varphi) = \ker(L)$. Show that $\varphi^* : C^\infty(B) \to C^\infty(M)$ is an isomorphism onto $C^\infty_{adm}(M)$, and therefore (by Exercise 5.19) B acquires a Poisson structure.

Regarding "morphisms" of Dirac manifolds, there are two natural ways in which Dirac manifolds can be related by a map, one extending the notion of Poisson map, and the other generalizing the pullback operation on 2-forms.

For Dirac manifolds (M_1, L_1) and (M_2, L_2), we say that a map $\varphi : M_1 \to M_2$ is a *forward Dirac map* (or a *Dirac map*, for simplicity) if

$$L_2|_{\varphi(x)} = \{(d_x\varphi(X), \beta) \mid (X, (d_x\varphi)^*\beta) \in L_1|_x\} \quad \forall x \in M_1.$$

We also write $L_2 = \varphi_! L_1$ to denote that L_1 and L_2 are related as above.

Exercise 5.21 In the context of Exercise 5.20, check that the Poisson structure on B, denoted by π, satisfies $\varphi_! L = L_\pi$.

Exercise 5.22

(a) For Dirac structures L_1 and L_2 defined by Poisson structures, check that $\varphi : M_1 \to M_2$ is a Dirac map if and only if it is a Poisson map.
(b) For Dirac structures L_1 and L_2 defined by closed 2-forms ω_1 and ω_2, show that $\varphi : M_1 \to M_2$ is a Dirac map if and only if φ is a submersion and $\varphi^*\omega_2 = \omega_1$.

As for pullbacks, consider a Dirac manifold (M, L) and a smooth map $\varphi : N \to M$. The *backward image* of L under φ is defined by

$$\varphi^! L := \{(X, (d\varphi)^*\beta) \mid (d\varphi(X), \beta) \in L\} \subset TN \oplus T^*N.$$

Over each point $y \in N$, the backward image of L is a lagrangian subspace of $\mathbb{T}N|_y$, but it may happen that this family of lagrangian subspaces does not fit into a smooth vector subbundle of $\mathbb{T}N$ (see Exercise 5.26 below); but if that is the case, then $\varphi^!L$ indeed defines a Dirac structure on N [8, Prop. 1.10], that we call the *pullback* of L.

We collect some properties of pullbacks of Dirac structures and Dirac maps in the following three exercises.

Exercise 5.23 Check that the characteristic distribution of the pullback $\varphi^!L$ is $(d\varphi)^{-1}(R)$, where R is the characteristic distribution of L; moreover, the pointwise 2-forms on $(d\varphi)^{-1}(R)$ are given by the pullbacks of those in R.

Exercise 5.24 Let $\varphi : N \to M$ be a smooth map.

(a) Check that $\varphi^!TM = TN$, and $\varphi^!T^*M = \ker(d\varphi) \oplus \mathrm{Ann}(\ker(d\varphi))$.
(b) Let L be a Dirac structure on M, and let $\omega \in \Omega^2(M)$ be a closed 2-form. Check that

$$\varphi^!(\tau_\omega(L)) = \tau_{\varphi^*\omega}(\varphi^!L).$$

(In particular, $\varphi^!L_\omega = \varphi^!(\tau_\omega(TM)) = \tau_{\varphi^*\omega}(TN) = L_{\varphi^*\omega}$, where L_ω is the Dirac structure defined by ω.)

Exercise 5.25 Consider Dirac manifolds (M_1, L_1) and (M_2, L_2), and a smooth map $\varphi : M_1 \to M_2$.

(a) Suppose that $\varphi_! L_1 = L_2$. Check that $\varphi^!L_2 = L_1$ if and only if $\ker(d\varphi) \subseteq \ker(L_1)$.
(b) Suppose that $\varphi^!L_2 = L_1$. Check that $\varphi_!L_1 = L_2$ if and only if $R_2|_{\varphi(x)} \subseteq d\varphi(T_xM_1)$ for all $x \in M_1$ (here R_2 is the projection of L_2 on TM).
(c) If φ is a diffeomorphism, check that $\varphi_!L_1 = L_2$ if and only if $L_1 = \varphi^!L_2$.

The next exercise provides a simple example where the backward image of a Dirac structure is not smooth.

Exercise 5.26 Let L be the Dirac structure in $\mathbb{R}^2$ defined by the Poisson structure $x\partial_x \wedge \partial_y$. Show that the backward image of L under the inclusion $\varphi : \mathbb{R} \to \mathbb{R}^2$, $x \mapsto (x, 0)$, is given by $T_x\mathbb{R} \oplus 0 \subseteq \mathbb{T}_x\mathbb{R}$ for $x \neq 0$, and for $x = 0$ it is $0 \oplus T_0^*\mathbb{R}$.

A sufficient (but not necessary) condition to ensure the smoothness of the family $\varphi^! L$ is that [8, Prop. 1.10]

- the subspaces $\ker((d_x\varphi)^*) \cap L|_{\varphi(x)} \subseteq T^*M|_{\varphi(x)}$, for $x \in N$, have the same dimension, or equivalently (by considering their annihilators) that the subspaces $\mathrm{Im}(d_x\varphi) + R|_{\varphi(x)} \subseteq TM|_{\varphi(x)}$ have the same dimension.

We will refer to this property as the *co-regularity condition*, following the terminology in [44]. In particular, Dirac structures can be always pulled back by submersions (or, more generally, by maps that are transverse to their characteristic distributions).

Whenever the co-regularity condition holds, *the presymplectic leaves of the Dirac structure $\varphi^! L$ coincide with the connected components of $\varphi^{-1}(\mathcal{O})$, equipped with the 2-form $\varphi^*\omega_{\mathcal{O}}$, for each presymplectic leaf $(\mathcal{O}, \omega_{\mathcal{O}})$ of L* [31, Thm. 7.33]. (In spite of what Exercise 5.23 may suggest, this description of presymplectic leaves is not valid in general, without additional "cleanness" conditions, see [20] for examples.)

Exercise 5.27

(a) Let $\varphi : M \to B$ be a surjective submersion with connected fibers. Verify that the assignment $\pi \mapsto \varphi^! L_\pi$ establishes a bijection between Poisson structures on B and Dirac structures L on M satisfying $\ker(d\varphi) = \ker(L)$ (cf. Exercises 5.20 and 5.21).
(b) Suppose that M carries a G-action that is free and proper, and $\varphi : M \to B = M/G$ is the quotient map. Show that $\pi \mapsto \varphi^! L_\pi$ yields a one-to-one correspondence between Poisson structures on B and G-invariant Dirac structures L on M for which $\ker(L)$ coincides with the distribution tangent to G-orbits (here G-invariance means that $\sigma^! L = L$, or equivalently $\sigma_! L = L$, for all $\sigma \in G$).

The next exercise uses pullbacks and gauge transformations to give a Dirac-geometric characterization of dual pairs [41].

Exercise 5.28 Consider a symplectic manifold (S, ω), Poisson manifolds (M_1, π_1) and (M_2, π_2), and let $\varphi_i : S \to M_i$, $i = 1, 2$, be surjective submersions. Show that $M_1 \xleftarrow{\varphi_1} S \xrightarrow{\varphi_2} M_2$ is a dual pair (see (5.1)) if and only if $\dim(S) = \dim(M_1) + \dim(M_2)$ and

$$\varphi_1^!(L_{\pi_1}) = \tau_\omega(\varphi_2^! L_{\pi_2}).$$

We now suppose that $N \xhookrightarrow{\iota} M$ is a submanifold. We know from the discussion so far that, for a Dirac structure L on M, the backward image $\iota^! L$ defines a Dirac structure on N whenever it is smooth. The co-regularity condition assuring the smoothness of $\iota^! L$ in this case is that $\mathrm{Ann}(TN) \cap L|_N$ has constant rank [22, § 3.1], or equivalently that

$$TN + R|_N \subseteq TM|_N \text{ has constant rank.}$$

We refer to a submanifold satisfying these conditions as *co-regular*. If $TN + R|_N = TM|_N$, we call the submanifold N a *transversal*.

Exercise 5.29 Check that the kernel of the pullback Dirac structure $\iota^! L$ on N is the distribution given by the projection on TN of $(TN \oplus \mathrm{Ann}(TN)) \cap L$.

Exercise 5.30 Let (M, L) be a Dirac manifold, and suppose that $\iota : N \hookrightarrow M$ is a submanifold containing $x \in M$ and such that $T_x M = R|_x \oplus T_x N = T_x \mathcal{O} \oplus T_x N$, where $\mathcal{O}$ is the presymplectic leaf through x. Show that there exists a neighborhood of x in N where $\iota^! L$ is given by a Poisson structure.

In the special case where M is a Poisson manifold, the Poisson structure of the previous exercise is just the transverse Poisson structure of Sect. 4.4, as given in Exercise 2.28 (compare Exercises 2.29 and 5.23). A uniqueness theorem for these more general "transverse Poisson structures" is proven in [36, Thm. 4.5], see also [13, § 5.6].

The next exercise shows that transverse Poisson structures are preserved under pullbacks by submersions. This fact will be used in Exercise 5.32 to prove the invariance of transverse Poisson structures in dual pairs stated in Sect. 5.1.

Exercise 5.31 Let (M, L) be a Dirac manifold, $\varphi : S \to M$ a submersion, and consider the pullback Dirac structure $\varphi^! L$ on S. Pick $m \in S$, let $\mathcal{O}$ be the presymplectic leaf through $x = \varphi(m)$ and $\mathcal{O}' := \varphi^{-1}(\mathcal{O})$. Let $N' \hookrightarrow S$ be a submanifold containing m such that $T_m S = T_m(\mathcal{O}') \oplus T_m N'$. Check that, by shrinking N' and using Exercise 5.30, one can assume that

- N' inherits a Poisson structure by the pullback of $\varphi^! L$;
- φ restricts to an embedding $N' \to M$, with image denoted by $N \subseteq M$;
- N inherits a Poisson structure by the pullback of L.

Verify that φ restricts to a Poisson isomorphism $N' \to N$.

Exercise 5.32 Let $(M_1, \pi_1) \xleftarrow{\varphi_1} (S, \omega) \xrightarrow{\varphi_2} (M_2, -\pi_2)$ be a dual pair (see (5.1)). For any $m \in S$, show that the transverse Poisson structures at $\varphi_1(m)$ and $\varphi_2(m)$ are isomorphic. The following are the key steps:

- Use the Darboux theorem to find a submanifold $N' \hookrightarrow S$ such that $T_m S = T_m N' \oplus R'|_m$, where $R' = \ker(d\varphi_1) + \ker(d\varphi_2)$, that is isotropic.
- By Exercises 5.24 (b) and 5.28, $\varphi_1^! L_{\pi_1}$ and $\varphi_2^! L_{\pi_2}$ have the same pullback on N'.
- By shrinking N', this pullback is a Poisson structure that is isomorphic to the transverse Poisson structures at $\varphi_1(m)$ and $\varphi_2(m)$, by Exercise 5.31.

Let us now specialize to the case where (M, π) is a Poisson manifold and $\iota : N \hookrightarrow M$ is a submanifold. As we have seen, modulo smoothness issues (cf. Exercise 5.26), N generally carries a Dirac structure $\iota^! L_\pi$ via pullback of the ambient Poisson structure.

Recalling that $TN^\pi = \pi^\sharp(\text{Ann}(TN))$, note that

- $\ker(\iota^! L_\pi) = TN \cap TN^\pi$ (see Exercise 5.29);
- N is co-regular if and only if $\text{Ann}(TN) \cap \ker(\pi^\sharp)$ has constant rank, which is equivalent to the condition that

$$TN^\pi \text{ has constant rank,}$$

since $\text{Ann}(TN) \cap \ker(\pi^\sharp) \hookrightarrow \text{Ann}(TN) \twoheadrightarrow TN^\pi$ is exact.

A submanifold $\iota : N \hookrightarrow M$ for which the backward image $\iota^! L_\pi$ is smooth (hence Dirac) and has trivial kernel ($TN \cap TN^\pi = 0$) is called a *Poisson-Dirac submanifold* of M. Equivalently, a Poisson-Dirac submanifold N is characterized by the fact that it carries a Poisson structure π_N uniquely determined by the condition

$$L_{\pi_N} = \iota^! L_\pi.$$

Example 5.10

(a) A submanifold $N \xhookrightarrow{\iota} M$ is Poisson if and only if $TN^\pi = 0$ (see Exercise 2.13 (a)), so Poisson submanifolds are co-regular Poisson-Dirac submanifolds. Recall that, being a Poisson submanifold, N carries a Poisson structure π_N for which the inclusion is a Poisson map, which means that $L_\pi = \iota_! L_{\pi_N}$ (Exercise 5.22 (a)). But then π_N satisfies $L_{\pi_N} = \iota^! L_\pi$ by Exercise 5.25 (a).

(b) Since cosymplectic submanifolds satisfy $TM|_N = TN \oplus TN^\pi$, they are co-regular Poisson-Dirac submanifolds, and the Poisson structure π_N that N inherits as a cosymplectic submanifold indeed satisfies $L_{\pi_N} = \iota^! L_\pi$ (compare its characterization in Exercise 2.29 with Exercise 5.23). Note that, by Exercise 2.27, cosymplectic submanifolds are the same as Poisson-Dirac submanifolds that are also transversals.[8]

(c) Poisson-Dirac submanifolds of a symplectic manifold are the same as symplectic submanifolds (see Exercise 5.24). Note that in a symplectic manifold any submanifold is co-regular.

◇

---

[8] This fact suggests "Poisson-Dirac transversals", rather than "Poisson transversals", as a more accurate alternative terminology for "cosymplectic submanifolds".

It is shown in [44] that any co-regular Poisson-Dirac submanifold of M is a Poisson submanifold in a cosymplectic submanifold of M (see also [31, Thm. 8.44]).

On a co-regular Poisson-Dirac submanifold $N \hookrightarrow M$, symplectic leaves are given by the connected components of the intersections of N with the symplectic leaves in M, equipped with the pullback of the symplectic forms. The description of the induced Poisson bracket on N in the next exercise is entirely analogous to the one for cosymplectic submanifolds in Exercise 2.30.

Exercise 5.33 For a co-regular Poisson-Dirac submanifold $N \hookrightarrow M$ (embedded, for simplicity), verify that the induced Poisson bracket $\{\cdot,\cdot\}_N$ on N is given by

$$\{f,g\}_N = \{\widehat{f},\widehat{g}\}|_N,$$

where $\widehat{f}, \widehat{g} \in C^\infty(M)$ are extensions of $f, g \in C^\infty(N)$ satisfying $d\widehat{f}|_{TN^\pi} = d\widehat{g}|_{TN^\pi} = 0$.

For a detailed account of Poisson-Dirac submanifolds with different types of regularity conditions, see [20]. For a broader discussion including other types of submanifolds, see e.g. [31, Chp. 8] and [96], as well as [44].

Besides offering an effective framework for the study of submanifolds of Poisson manifolds, Dirac structures are essential in several other aspects of Poisson geometry, especially due to their flexible pullbacks and enlarged group of symmetries (see Exercise 5.16). They can be used to prove, clarify, and often extend, several fundamental results in Poisson geometry, see [72]. Those include the classification of Poisson homogeneous spaces [63, 71] (see Remark 3.3), the integration of Poisson homogeneous spaces by symplectic groupoids [14], the construction of symplectic realizations [41], linearization of Poisson-Lie groups [2], and normal forms of Poisson structures around transversals (extending Weinstein's splitting theorem) [13, 40].

We have only discussed Dirac structures in the "standard" Courant algebroid $\mathbb{T}M$. Many situations of interest involve other types of Courant algebroids, such as "exact" ones, where the Courant-Dorfman bracket (5.5) is twisted by a closed 3-form [84]. In this extended setting Dirac structures and Dirac maps provide a geometric framework for quasi-Hamiltonian geometry [3] (see [4, 12]) and become natural ingredients in the construction of Poisson structures on moduli spaces of flat bundles over surfaces and their variants [61]. Dirac structures are also studied in complexified Courant algebroids, especially in connection with generalized complex structures [48], see [1].

Acknowledgments I thank David Iglesias Ponte, Eva Miranda, Cedric Oms, and Roberto Rubio for their encouragement in having these notes written up. I am particularly indebted to David and Roberto for their feedback on preliminary versions of the manuscript. While preparing these notes, I benefited from discussions with many people, including Daniel Alvarez, Alejandro Cabrera, Pedro H. Carvalho, Matías del Hoyo, and Hudson Lima. I am grateful to all the students from the courses on which these notes are based; I am especially thankful to Andrés Rodriguez and Tianhao Ye for carefully reading the manuscript and providing corrections and suggestions. *I am delighted to dedicate these notes to A. Weinstein, whose scientific leadership and fundamental contributions have shaped Poisson geometry as we know it.*

References

1. D. Aguero, R. Rubio, Complex Dirac structures: invariants and local structure. Comm. Math. Phys. **396**, 623–646 (2022)
2. A. Alekseev, E. Meinrenken, Linearization of Poisson Lie group structures. J. Symplectic Geom. **14**, 227–267 (2016)
3. A. Alekseev, A. Malkin, E. Meinrenken, Lie group valued moment maps. J. Differential Geom. **48**, 445–495 (1998)
4. A. Alekseev, H. Bursztyn, E. Meinrenken, Pure spinors on Lie groups. Astérisque **327**, 131–199 (2009)
5. D. Álvarez, Complete Lie algebroid actions and the integrability of Lie algebroids. Proc. Amer. Math. Soc. **149**, 4923–4930 (2021)
6. V.I. Arnold, *Mathematical Methods of Classical Mechanics*. Grad. Texts in Math., vol. 60 (Springer, New York, 1995), xvi+516 pp.
7. F. Bonechi, N. Ciccoli, J. Qiu, M. Tarlini, Quantization of Poisson manifolds from the integrability of the modular function. Comm. Math. Phys. **331**, 851–885 (2014)
8. H. Bursztyn, A brief introduction to Dirac manifolds, in *Geometric and Topological Methods for Quantum Field Theory* (pp. 4–38) (Cambridge University Press, Cambridge, 2013)
9. H. Bursztyn, R. Fernandes, Picard groups of Poisson manifolds. J. Differential Geom. **109**, 1–38 (2018)
10. H. Bursztyn, A. Weinstein, Picard groups in Poisson geometry. Mosc. Math. J. **4**, 39–66, 310 (2004)
11. H. Bursztyn, A. Weinstein, Poisson geometry and Morita equivalence, in *Poisson Geometry, Deformation Quantisation and Group Representations*. London Math. Soc. Lecture Note Ser., vol. 323 (Cambridge University Press, Cambridge, 2005), pp. 1–78
12. H. Bursztyn, M. Crainic, A. Weinstein, C. Zhu, Integration of twisted Dirac brackets. Duke Math. J. **123**, 549–607 (2004)
13. H. Bursztyn, H. Lima, E. Meinrenken, Splitting theorems for Poisson and related structures. J. Reine Angew. Math. **754**, 281–312 (2019)
14. H. Bursztyn, D. Iglesias-Ponte, J.-H. Lu, Dirac geometry and integration of Poisson homogeneous spaces. J. Differential Geom. **126**, 939–1000 (2024)
15. M. Cahen, S. Gutt, J. Rawnsley, Nonlinearizability of the Iwasawa Poisson Lie structure. Lett. Math. Phys. **24**, 79–83 (1992)
16. A. Cannas da Silva, A. Weinstein, *Geometric Models for Noncommutative Algebras*. Berkeley Mathematics Lecture Notes, vol. 10 (American Mathematical Society/Berkeley Center for Pure and Applied Mathematics, Providence, RI/Berkeley, CA, 1999)

17. A. Cattaneo, G. Felder, Poisson sigma models and symplectic groupoids, in *Quantization of Singular Symplectic Quotients*. Progr. Math., vol. 198 (Birkhäuser Verlag, Basel, 2001), pp. 61–93
18. G. Cavalcanti, Examples and counter-examples of log-symplectic manifolds. J. Topol. **10**, 1–21 (2017)
19. J. Conn, Normal forms for smooth Poisson structures. Ann. Math. **121**, 565–593 (1985)
20. L. Cordeiro Brambila, P. Frejlich, D. Martínez Torres, Coregular submanifolds and Poisson submersions. Rev. Mat. Iberoam. **40**, 1419–1468 (2024)
21. A. Coste, P. Dazord, A. Weinstein, *Groupoïdes Symplectiques*. Publ. Dép. Math. Nouvelle Sér. A **2**(i–ii), 1–62 (1987), 87-2, Univ. Claude-Bernard, Lyon
22. T. Courant, Dirac manifolds. Trans. Amer. Math. Soc. **319**, 631–661 (1990)
23. T. Courant, A. Weinstein, *Beyond Poisson Structures*. Séminaire sudrhodanien de géométrie VIII. Travaux en Cours, vol. 27 (Hermann, Paris, 1988), pp. 39–49
24. M. Crainic, Differentiable and algebroid cohomology, van Est isomorphisms, and characteristic classes. Comment. Math. Helv. **78**, 681–721 (2003)
25. M. Crainic, R. Fernandes, Integrability of Lie brackets. Ann. Math. **157**, 575–620 (2003)
26. M. Crainic, R. Fernandes, Integrability of Poisson brackets. J. Differential Geom. **66**, 71–137 (2004)
27. M. Crainic, R.L. Fernandes, A geometric approach to Conn's linearization theorem. Ann. Math. **173**, 1121–1139 (2011)
28. M. Crainic, I. Marcut, On the existence of symplectic realizations. J. Symplectic Geom. **9**, 435–444 (2011)
29. M. Crainic, I. Marcut, A normal form theorem around symplectic leaves. J. Differential Geom. **92**, 417–461 (2012)
30. M. Crainic, R. Fernandes, D. Martínez Torres, Poisson manifolds of compact types (PMCT 1). J. Reine Angew. Math. **756**, 101–149 (2019)
31. M. Crainic, R. Fernandes, I. Marcut, *Lectures on Poisson Geometry*. Grad. Stud. Math., vol. 217 (American Mathematical Society, Providence, RI, 2021), xix+479 pp.
32. P. Dirac, *Lectures on Quantum Mechanics*, Belfer Graduate School of Science (Yeshiva University, New York, 1964)
33. I. Dorfman, Dirac structures of integrable evolution equations. Phys. Lett. A **125**, 240–246 (1987)
34. V.G. Drinfel'd, Hamiltonian structures on Lie groups, Lie bialgebras and the geometric meaning of classical Yang-Baxter equations. Dokl. Akad. Nauk SSSR **268**, 285–287 (1983)
35. V.G. Drinfel'd, On Poisson homogeneous spaces of Poisson-Lie groups. Theoret. Math. Phys. **95**, 524–525 (1993)
36. J.-P. Dufour, A. Wade, On the local structure of Dirac manifolds. Compos. Math. **144**, 774–786 (2008)
37. J.-P. Dufour, N.-T. Zung, *Poisson Structures and Their Normal Forms*, Progress in. Mathematics, vol. 242 (Birkhauser Boston, 2005)
38. R. Fernandes, I. Mărcut, Poisson geometry around Poisson submanifolds. J. Eur. Math. Soc. (2024), published online first
39. R. Fernandes, J.-P. Ortega, T. Ratiu, The momentum map in Poisson geometry. Amer. J. Math. **131**, 1261–1310 (2009)
40. P. Frejlich, I. Marcut, The normal form theorem around Poisson transversals. Pacific J. Math. **287**, 371–391 (2017)
41. P. Frejlich, I. Marcut, On dual pairs in Dirac geometry. Math. Z. **289**, 171–200 (2018)
42. P. Frejlich, D. Martinez Torres, E. Miranda, A note on the symplectic topology of b-manifolds. J. Symplectic Geom. **15**, 719–739 (2017)

43. M. Gekhtman, M. Shapiro, A. Vainshtein, *Cluster Algebras and Poisson Geometry*. Math. Surveys Monogr., vol. 167 (American Mathematical Society, Providence, RI, 2010), xvi+246 pp.
44. S. Geudens, The Poisson saturation of coregular submanifolds. Int. Math. Res. Not. IMRN **2023**(11), 9667–9710 (2023)
45. V.L. Ginzburg, Grothendieck groups of Poisson vector bundles. J. Symplectic Geom. **1**, 121–169 (2001)
46. V.L. Ginzburg, J.-H. Lu, Poisson cohomology of Morita-equivalent Poisson manifolds. Int. Math. Res. Not. **1992**(10), 199–205 (1992)
47. V.L. Ginzburg, A. Weinstein, Lie-Poisson structure on some Poisson Lie groups. J. Amer. Math. Soc. **5**, 445–453 (1992)
48. M. Gualtieri, Generalized complex geometry. Ann. Math. (2) **174**, 75–123 (2011)
49. M. Gualtieri, S. Li, Symplectic groupoids of log symplectic manifolds. Int. Math. Res. Not. IMRN **2014**(11), 3022–3074 (2014)
50. V. Guillemin, E. Miranda, A.R. Pires, Symplectic and Poisson geometry on b-manifolds. Adv. Math. **264**, 864–896 (2014)
51. E. Hawkins, A groupoid approach to quantization. J. Symplectic Geom. **6**, 61–125 (2008)
52. Hoekstra, D., Zeiser, F., Poisson cohomology of 3D Lie algebras. J. Geom. Phys. **191**, Paper No. 104862, 38 pp. (2023)
53. V. Kac, Introduction to vertex algebras, Poisson vertex algebras, and integrable Hamiltonian PDE, in *Perspectives in Lie Theory*. Springer INdAM Ser., vol. 19 (Springer, Cham, 2017), pp. 3–72
54. M.V. Karasev, Analogues of objects of the theory of Lie groups for nonlinear Poisson brackets. Izv. Akad. Nauk SSSR Ser. Mat. **50**(3), 508–538, 638 (1986)
55. B. Khesin, R. Wendt, The geometry of infinite-dimensional groups. Ergeb. Math. Grenzgeb. (3), 51 [Results in Mathematics and Related Areas. 3rd Series. A Series of Modern Surveys in Mathematics] Springer-Verlag, Berlin (2009), xii+304 pp.
56. Klaasse, R., Poisson structures of divisor-type. arXiv:1811.04226 (2018)
57. L.I. Korogodski, Y.S. Soibelman, *Algebras of Functions on Quantum Groups. Part I.*, Math. Surveys Monogr., vol. 56 (American Mathematical Society, Providence, RI, 1998), x+150 pp.
58. Y. Kosmann-Schwarzbach, Poisson manifolds, Lie algebroids, modular classes: a survey. SIGMA Symmetry Integrability Geom. Methods Appl. **4**, Paper 005, 30 pp. (2008)
59. Y. Kosmann-Schwarzbach, La géométrie de Poisson, création du XXe siécle, in *Siméon-Denis Poisson*. Hist. Math. Sci. Phys. (Éditions de l'École Polytechnique, Palaiseau, 2013), pp. 129–172
60. C. Laurent-Gengoux, A. Pichereau, P. Vanhaecke, *Poisson Structures, Grundlehren der Mathematischen Wissenschaften [Fundamental Principles of Mathematical Sciences]*, vol. 347 (Springer, Heidelberg, 2013)
61. D. Li-Bland, P. Ševera, Symplectic and Poisson geometry of the moduli spaces of flat connections over quilted surfaces, in *Mathematical Aspects of Quantum Field Theories*. Math. Phys. Stud. (Springer, Cham, 2015), pp. 343–411
62. A. Lichnerowicz, Les variétés de Poisson et leurs algèbres de Lie associées. J. Differential Geometry **12**, 253–300 (1977)
63. Z.-J. Liu, A. Weinstein, P. Xu, Manin triples for Lie bialgebroids. J. Differential Geom. **45**, 547–574 (1997)
64. Z.-J. Liu, A. Weinstein, P. Xu, Dirac structures and Poisson homogeneous spaces. Comm. Math. Phys. **192**, 121–144 (1998)

65. J.-H. Lu, Momentum mappings and reduction of Poisson actions, in *Symplectic Geometry, Groupoids, and Integrable Systems* (Berkeley, CA, 1989). Math. Sci. Res. Inst. Publ., vol. 20 (Springer, New York, 1991), pp. 209–226
66. J.-H. Lu, A. Weinstein, Groupoïdes symplectiques doubles des groupes de Lie-Poisson. C. R. Acad. Sci. Paris Sér. I Math. **309**, 951–954 (1989)
67. J.-H. Lu, A. Weinstein, Poisson Lie groups, dressing transformations, and Bruhat decompositions. J. Differential Geom. **31**, 501–526 (1990)
68. K.C.H. Mackenzie, P. Xu, Integration of Lie bialgebroids. Topology **39**, 445–467 (2000)
69. I. Marcut, B. Osorno Torres, Deformations of log-symplectic structures. J. Lond. Math. Soc. (2) **90**, 197–212 (2014)
70. J. Marsden, T. Ratiu, *Introduction to Mechanics and Symmetry*, Text in Applied Mathematics, vol. 17 (Springer, 1994)
71. E. Meinrenken, Dirac actions and Lu's Lie algebroid. Transform. Groups **22**, 1081–1124 (2017)
72. E. Meinrenken, Poisson geometry from a Dirac perspective. Lett. Math. Phys. **108**, 447–498 (2018)
73. R. Melrose, *The Atiyah-Patodi-Singer Index Theorem*. Res. Notes Math., vol. 4 (A K Peters, Ltd., Wellesley, MA, 1993), xiv+377 pp.
74. K. Mikami, A. Weinstein, Moments and reduction for symplectic groupoid actions. Publ. RIMS Kyoto Univ. **24**, 121–140 (1988)
75. E. Miranda, G. Scott, The geometry of E-manifolds. Rev. Mat. Iberoam. **37**, 1207–1224 (2021)
76. S.-D. Poisson, Sur la variation des constantes arbitraires dans les questions de mecanique. J. Ecole Polytech. **8**, 266–344 (1809)
77. B. Pym, Constructions and classifications of projective Poisson varieties. Lett. Math. Phys. **108**, 573–632 (2018)
78. O. Radko, A classification of topologically stable Poisson structures on a compact oriented surface. J. Symplectic Geom. **1**, 523–542 (2002)
79. O. Radko, D. Shlyakhtenko, Picard groups of topologically stable Poisson structures. Pacific J. Math. **224**, 151–183 (2006)
80. D. Roytenberg, Poisson cohomology of SU(2)-covariant "necklace" Poisson structures on S^2. J. Nonlinear Math. Phys. **9**, 347–356 (2002)
81. D. Roytenberg, On the structure of graded symplectic supermanifolds and Courant algebroids, in *Quantization, Poisson Brackets and Beyond* (Manchester, 2001). Contemp. Math., vol. 315 (American Mathematical Society, Providence, RI, 2002), pp. 169–185
82. D. Roytenberg, On weak Lie 2-algebras, in *XXVI Workshop on Geometrical Methods in Physics*. AIP Conf. Proc., vol. 956 (American Institute of Physics, Melville, NY, 2007), pp. 180–198
83. D. Roytenberg, A. Weinstein, Courant algebroids and strongly homotopy Lie algebras. Lett. Math. Phys. **46**, 81–93 (1998)
84. P. Ševera, A. Weinstein, Poisson geometry with a 3-form background. Prog. Theoret. Phys. Suppl. **144**, 145–154 (2001)
85. J. Villatoro, Picard groups of b-symplectic manifolds. J. Symplectic Geom. **19**, 723–775 (2021)
86. A. Weinstein, The local structure of Poisson manifolds. J. Differential Geom. **18**, 523–557 (1983)
87. A. Weinstein, Symplectic groupoids and Poisson manifolds. Bull. Amer. Math. Soc. (N.S.) **16**, 101–104 (1987)
88. A. Weinstein, Coisotropic calculus and Poisson groupoids. J. Math. Soc. Japan **40**, 705–727 (1988)
89. A. Weinstein, Noncommutative geometry and geometric quantization, in *Symplectic Geometry and Mathematical Physics* (Aix-en-Provence, 1990). Progr. Math., vol. 99 (Birkhäuser Boston, Boston, MA, 1991), pp. 446–461

90. A. Weinstein, The modular automorphism group of a Poisson manifold. J. Geom. Phys. **23**, 379–394 (1997)
91. A. Weinstein, P. Xu, Extensions of symplectic groupoids and quantization. J. Reine Angew. Math. **417**, 159–189 (1991)
92. P. Xu, Morita equivalent symplectic groupoids, in *Symplectic Geometry, Groupoids, and Integrable Systems* (Berkeley, CA, 1989). Math. Sci. Res. Inst. Publ., vol. 20 (Springer, New York, 1991), pp. 291–311
93. P. Xu, Morita equivalence of Poisson manifolds. Comm. Math. Phys. **142**, 493–509 (1991)
94. P. Xu, Poisson cohomology of regular Poisson manifolds. Ann. Inst. Fourier (Grenoble) **42**, 967–988 (1992)
95. P. Xu, Momentum maps and Morita equivalence. J. Differential Geom. **67**, 289–333 (2004)
96. M. Zambon, Submanifolds in Poisson geometry: a survey, in *Complex and Differential Geometry*. Springer Proc. Math., vol. 8 (Springer, Heidelberg, 2011), pp. 403–420

Part II
Wonderful Varieties

Wonderful Varieties with a View Towards Poisson Geometry

Ana Bălibanu

A homogeneous space is an algebraic variety $\mathring{X}$ on which an algebraic group G acts transitively. To study $\mathring{X}$, it is often useful to study its equivariant compactifications—that is, those compactifications in which the action of G extends to the boundary. These model the equivariant behavior of $\mathring{X}$ "at infinity," where many interesting features of its geometry become apparent. For several large classes of homogeneous spaces, such compactifications can be described entirely by discrete combinatorial data. The most famous example of such a classification is the theory of toric varieties, in which equivariant compactifications of a torus are indexed by fans or polytopes.

The goal of these notes is to give an introduction to an important class of equivariant compactifications of homogeneous spaces known as wonderful varieties. We begin with an overview that places these objects into the broader setting of spherical varieties, and explains some of the motivation behind the development of this theory. Then we construct the wonderful compactification of a semisimple algebraic group, and we describe some features of its remarkable structure. In the final three sections, we connect these constructions to varieties of Lagrangian subalgebras and to Poisson geometry.

A Note on Background A main goal of the minicourse on which these notes are based was to make this classical topic accessible to an audience without a broad knowledge of algebraic geometry. These notes assume only some basic notions of algebraic groups, Lie algebras, group actions, and algebraic varieties, and provide a long list of references for detailed introductions of these subjects. We also give concrete examples and intuitive

---

A. Bălibanu (✉)
Department of Mathematics, Harvard University, Cambridge, MA, USA
e-mail: ana@math.harvard.edu

© The Author(s), under exclusive license to Springer Nature Switzerland AG 2025
M. Garcia-Fernandez et al. (eds.), *Advances in Poisson Geometry*,
Advanced Courses in Mathematics - CRM Barcelona,
https://doi.org/10.1007/978-3-031-86657-9_2

heuristics to illustrate the geometry of the constructions that follow. In particular, readers who are not very familiar with algebraic varieties will lose nothing by thinking of algebraic groups as complex Lie groups, and of smooth algebraic varieties as holomorphic manifolds.

1   Introduction

Let H be a connected algebraic group over the complex numbers. A H-*homogeneous space* is an algebraic variety $\mathring{X}$ equipped with a transitive action of H. By fixing a point $x \in \mathring{X}$ we obtain an identification

$$\mathring{X} \cong H/K,$$

where $K = \operatorname{Stab}_H(x)$ is the stabilizer of x in H.

Definition 1.1 An *equivariant embedding* of the homogeneous space $\mathring{X}$ is a normal H-variety X together with an H-equivariant embedding

$$\mathring{X} \hookrightarrow X$$

whose image is open and dense. An *equivariant compactification* of $\mathring{X}$ is an equivariant embedding whose target is a projective variety.

For details on the study of equivariant embeddings, we refer the reader to the extensive introduction [27] by Timashev.

Remark While equivariant embeddings may fail to be smooth, in this exposition we require them to be *normal*. This means that the local ring at every point is an integrally closed domain. One consequence of normality is that algebraic local functions on the smooth locus X^{sm} of X extend to all of X. In particular, as in Hartog's theorem, this implies that X has no singularities in codimension one.

Given a homogeneous space, one can ask for a classification of all its equivariant embeddings. The most well-known answer to this type of question is the theory of *toric varieties*, which are equivariant embeddings of the complex torus

$$H = (\mathbb{C}^*)^n.$$

These are parametrized by discrete combinatorial objects known as *fans*—collections of convex rational polyhedral cones in $\mathbb{R}^n$ closed under taking intersections and faces. All the geometric features of toric varieties, like their orbit structure, their cohomology, the morphisms between them, etc., are indexed by combinatorial data encoded in their fans. For more background on toric varieties, we refer to the classical book [12] by Fulton.

When H is a more general algebraic group, the collection of all equivariant embeddings of a given homogeneous space is often intractably large. For example, even the equivariant compactifications of the apparently simple additive group $\mathbb{C}^n$ are not "discrete"—instead, they are parametrized by moduli spaces [13]. However, a good theory of equivariant embeddings does exist for certain relatively large classes of homogeneous spaces.

The first such class that we will be interested in consists of symmetric spaces. To define them, we need the notion of an *involution* of H, which is a nontrivial algebraic group automorphism $\sigma : H \longrightarrow H$ that satisfies $\sigma \circ \sigma = \mathrm{Id}_H$.

Definition 1.2 A homogeneous space H/K is a *symmetric space* if K is an open algebraic subgroup of the fixed-point subgroup

$$H^\sigma := \{h \in H \mid \sigma(h) = h\}$$

of an involution σ. Since algebraic subgroups are always closed, this condition is equivalent to requiring that

$$(H^\sigma)^\circ \subset K \subset H^\sigma.$$

Remark If H was a compact real Lie group, Definition 1.2 would coincide with the usual "Riemannian" definition of a symmetric space, as a Riemannian manifold which has an inversion symmetry centered at every point.

Example 1.3 Suppose that $H = G \times G$ is a direct product of two copies of a group G, and consider the involution

$$\sigma : G \times G \longrightarrow G \times G$$

$$(g, h) \longmapsto (h, g).$$

Its locus of fixed points is

$$H^\sigma = G_\Delta := \{(g, g) \in G \times G \mid g \in G\},$$

which is the stabilizer of the identity element under the transitive action of $G \times G$ on G given by

$$(g, h) \cdot a = gah^{-1}.$$

Therefore we get an identification

$$G \cong (G \times G)/G_\Delta$$

which shows that the group G itself is a symmetric $G \times G$-space.

Exercise 1 Suppose that $U \subset H$ is a connected normal unipotent subgroup of H.

(a) Let σ be an involution of H. Show that $U \subset (H^\sigma)^\circ$.
(b) Conclude that, if $\mathring{X}$ is a symmetric space for the action of H, then $\mathring{X}$ is also a symmetric space for the action of a quotient of H which has no nontrivial connected normal unipotent subgroups.

Groups that have no nontrivial connected normal unipotent subgroups are called *reductive* groups. One can show that a group H is reductive if and only if all its finite-dimensional representations are completely reducible. Complex tori, the general linear group GL_n, and the complex algebraic groups SL_n, SO_n, and Sp_{2n} are all examples of reductive groups, and we refer to [14] for more background on their structure theory.

A special class of reductive groups consists of the *semisimple* groups, which are groups with no nontrivial connected normal solvable subgroups. Equivalently, a reductive group is semisimple if and only if the identity component of its center is trivial.

Exercise 2 In this exercise we'll see some simple examples.

(a) As a sanity check, give two examples of reductive groups that are not semisimple, two examples of semisimple groups with nontrivial center, and two examples of semisimple groups with trivial center.
(b) Note that the additive group $\mathbb{C}^n$ is not reductive (because it is unipotent.) To see that its representations are generally not semisimple, find a representation of $\mathbb{C}^n$ that is not completely reducible.

Exercise 1 shows that any symmetric space $\mathring{X}$ is homogeneous for the action of a reductive group H. In the early 1980s, de Concini and Procesi [4] observed that, if H has trivial center, then $\mathring{X}$ admits a distinguished equivariant compactification with remarkable properties.

Definition 1.4 A *wonderful compactification* of $\mathring{X}$ is an equivariant embedding X such that

(1) X is smooth and projective,
(2) the boundary divisor

$$D := X \setminus \mathring{X}$$
$$= D_1 \cup \ldots \cup D_l$$

has simple normal crossings and nonempty partial intersections, and
(3) two points $a, b \in X$ are in the same H-orbit if and only if they are contained in exactly the same irreducible components of D—that is,

$$\{j \mid a \in D_j\} = \{j \mid b \in D_j\}.$$

Theorem 1.5 ([4, Theorem 3.1]) *Suppose that the center of H is trivial and that $\sigma : H \longrightarrow H$ is an involution. Then the symmetric space*

$$H/H^\sigma$$

has a wonderful compactification.

Exercise 3 Suppose that X is a wonderful variety.

(a) Show that, for any proper subset $J \subset \{1, \ldots, l\}$, the partial intersection

$$\bigcap_{j \notin J} D_j$$

is the closure of a single H-orbit on D.
(b) Conclude that X contains a unique closed H-orbit of minimal dimension, given by

$$\mathcal{O}_\emptyset = D_1 \cap \ldots \cap D_l.$$

Not all homogeneous spaces have compactifications that satisfy the wonderful properties of Definition 1.4. For example, using some more advanced facts in algebraic geometry, the next exercise shows that there is no such thing as a "wonderful toric variety."

Exercise 4 Suppose that X is a toric variety for the action of a complex torus T which has a unique closed orbit. Show that X is an affine variety. (Hint: Use the fact, due to Sumihiro, that any normal T-variety is covered by T-invariant affine open subsets [26, Corollary 2].)

It is therefore natural to ask which homogeneous spaces have wonderful compactifications. The answer to this question has its roots in a wide-ranging research program of Luna and Vust that began in the 1980s, which gave a systematic classification of a class of equivariant embeddings known as spherical varieties. To formulate their definition, we will assume from now on that H is a reductive group. A maximal solvable subgroup of H is called a *Borel subgroup*, and all such subgroups are conjugate.

Example 1.6 Suppose that $H = GL_n$. Then any Borel subgroup is conjugate to the subgroup

$$B = \left\{ \begin{pmatrix} * & \cdots & * \\ & \ddots & \vdots \\ & & * \end{pmatrix} \right\}$$

of upper-triangular matrices.

Definition 1.7 The homogeneous space $\mathring{X}$ is *spherical* if a Borel subgroup of H acts on $\mathring{X}$ with an open dense orbit. Equivariant embeddings of such spaces are called *spherical varieties*.

Example 1.8

(a) Since any torus T is a solvable reductive group, all toric varieties are spherical.
(b) Let $H = G \times G$ for some reductive group G, and let $B \subset G$ be a Borel subgroup. Fix a maximal torus T contained in B, let

$$W = N_G(T)/T$$

be the associated Weyl group, and for each element $w \in W$ fix a coset representative $\dot{w}$ in $N_G(T)$. Then the Bruhat decomposition tells us that G can be written as a disjoint union

$$G = \bigsqcup_{w \in W} B\dot{w}B$$

of finitely many $B \times B$-orbits, indexed by elements of W and called *Bruhat cells*. In particular, when $w_0 \in W$ is the longest element of the Weyl group, the Bruhat cell

$$B\dot{w}_0 B$$

is open and dense in G. Since this is a single $B \times B$-orbit, it follows that G itself is a spherical $G \times G$-space.

(c) Let G be a reductive group and let B be a Borel subgroup. Then the projective homogeneous space

$$G/B$$

is called the *flag variety* of G. By the Bruhat decomposition of part (b), it is spherical relative to the left action of G.

Vust showed in [28, Theorem 1] that all symmetric spaces are spherical. (The converse is not true—for example, the flag variety of Example 1.8(c) is not a symmetric space.) Luna and Vust [22] then developed a classification of equivariant embeddings of spherical spaces, using combinatorial objects known as *colored fans*.

There are two distinguished classes of equivariant compactifications that appear in this classification. The first is the class of *toroidal compactifications*, which behave locally like toric varieties. Such compactifications correspond to colorless fans, and their boundaries are "large," in the sense that the boundary divisor has simple normal crossings. Every equivariant compactification of $\mathring{X}$ is dominated by a toroidal one through a sequence blow-ups.

The second class is the class of *simple compactifications*, which are those with a unique closed orbit. As we have seen in Exercise 4, not all homogeneous spaces admit simple compactifications. In fact, $\mathring{X}$ has a simple compactification if and only if the stabilizer K of a point of $\mathring{X}$ has finite index in its normalizer [2, Corollary 5.3].

If $\mathring{X}$ admits simple compactifications, there is a unique simple compactification X which is also toroidal. It dominates all other simple compactifications and is dominated by all toroidal ones. In other words, it has the universal property that, for every toroidal compactification X' and every simple compactification X'' of $\mathring{X}$, there exist proper G-equivariant morphisms

$$X' \longrightarrow X \longrightarrow X''$$

which restrict to the identity along the open dense orbit $\mathring{X}$.

If the distinguished compactification X is smooth, then it is a wonderful variety in the sense of Definition 1.4. Smoothness is guaranteed, for example, if the stabilizer K is self-normalizing [17, Corollary 7.2]. This occurs in particular when K is the fixed-point subgroup of an involution, in which case $\mathring{X}$ is a symmetric space and the compactification X is the wonderful compactification constructed by de Concini and Procesi in Theorem 1.5.

2 The Wonderful Compactification of a Semisimple Algebraic Group

Let G be a reductive group whose center is trivial—we will often consider the example of the projectivized general linear group PGL_n. In this section we describe one way to construct the wonderful compactification of G, viewed as a symmetric space for the action of the reductive group $G \times G$. This construction is an adaptation of the original one given by de Concini and Procesi [4], and is explained in detail in the now-classical survey [8] by Evens and Jones.

Fix a Borel subgroup B of G containing a maximal torus T. In the simply-connected cover $\widetilde{G}$ of G, we write $\widetilde{T}$ and $\widetilde{B}$ for the preimages of T and B. The irreducible representations of $\widetilde{G}$ can then be indexed by weights of the torus $\widetilde{T}$ which are dominant with respect to the Borel $\widetilde{B}$, and a representation is called *regular* if its dominant weight does not lie on any walls of the corresponding Weyl chamber.

Exercise 5 Let λ be a dominant weight of $\widetilde{T}$ and let V be the corresponding irreducible representation. If $v_\lambda \in V$ is a highest weight vector, show that the following are equivalent:

(a) λ is regular.
(b) The stabilizer of the subspace $\mathbb{C}v_\lambda$ in G is B.
(c) The stabilizer of λ in W is trivial.

If V is a regular irreducible representation of $\widetilde{G}$, we can construct the following diagram:

$$\begin{array}{ccc} \widetilde{G} & \xrightarrow{\rho} & \operatorname{End} V \setminus \{0\} \\ \downarrow & & \downarrow \\ G & \dashrightarrow{\psi} & \mathbb{P}(\operatorname{End} V). \end{array}$$

The top arrow is the representation map, the left vertical arrow is the quotient by the center of $\widetilde{G}$, and the right vertical arrow is the quotient by scalars.

Exercise 6 Use Exercise 5 to show that the representation map ρ descends to a well-defined morphism

$$\psi : G \hookrightarrow \mathbb{P}(\operatorname{End} V),$$

and that this morphism is an embedding.

Definition 2.1 The *wonderful compactification* of G is the closure

$$\overline{G} := \overline{\psi(G)}$$

of the image of the embedding ψ.

The compactification $\overline{G}$ is a smooth projective variety which is independent of the choice of regular representation V. It is also stable under the $G \times G$-action on $\mathbb{P}(\text{End } V)$ given by pre- and post-composition of projectivized endomorphisms, which extends the action of G on itself by left- and right-multiplication—therefore, it is an equivariant compactification in the sense of Definition 1.1.

The boundary divisor

$$D := \overline{G} \backslash G$$
$$= D_1 \cup \ldots \cup D_l$$

is a union of smooth irreducible components whose number is equal to the rank l of G. They have simple normal crossings and nonempty partial intersections—in other words, $\overline{G}$ is a wonderful compactification of G in the sense of Definition 1.4.

The $G \times G$-orbits on D are indexed by proper subsets of the simple roots $\{1, \ldots, l\}$. More precisely, to each proper subset $J \subset \{1, \ldots, l\}$ we can associate the data of a parabolic subgroup P_J, generated by the positive Borel B and the negative simple roots

$$\{\alpha_j \mid j \in J\}.$$

It has an opposite parabolic P_J^-, and we write $U_J^\pm$ for their unipotent radicals and $L_J := P_J \cap P_J^-$ for their common Levi component. The Levi decomposition then gives

$$P_J^\pm = U_J^\pm L_J.$$

Example 2.2 Let $G = PGL_4$ be the projectivized general linear group of rank 3, and let $J = \{1, 3\}$. The corresponding parabolic data is

$$P_J = \left\{ \begin{pmatrix} * & * & * & * \\ * & * & * & * \\ 0 & 0 & * & * \\ 0 & 0 & * & * \end{pmatrix} \right\} \qquad P_J^- = \left\{ \begin{pmatrix} * & * & 0 & 0 \\ * & * & 0 & 0 \\ * & * & * & * \\ * & * & * & * \end{pmatrix} \right\}$$

$$U_J = \left\{ \begin{pmatrix} 0 & 0 & * & * \\ 0 & 0 & * & * \\ 0 & 0 & 0 & 0 \\ 0 & 0 & 0 & 0 \end{pmatrix} \right\} \quad U_J^- = \left\{ \begin{pmatrix} 0 & 0 & 0 & 0 \\ 0 & 0 & 0 & 0 \\ * & * & 0 & 0 \\ * & * & 0 & 0 \end{pmatrix} \right\}$$

$$L_J = \left\{ \begin{pmatrix} * & * & 0 & 0 \\ * & * & 0 & 0 \\ 0 & 0 & * & * \\ 0 & 0 & * & * \end{pmatrix} \right\}.$$

The index set J corresponds to a $G \times G$-orbit $\mathcal{O}_J$ whose closure is the partial intersection

$$\overline{\mathcal{O}_J} = \bigcap_{j \notin J} D_j$$

of divisor components not indexed by J. This orbit contains a distinguished basepoint z_J whose stabilizer is

$$\text{Stab}_{G \times G}(z_J) = \left\{ (ul_1, vl_2) \mid u \in U_J, v \in U_J^-, l_1 l_2^{-1} \in Z_{L_J} \right\}. \tag{2.3}$$

Exercise 7 Show that the orbit $\mathcal{O}_J$ is a fibration

$$\begin{array}{c} G_J \hookrightarrow \mathcal{O}_J \\ \downarrow \\ G/P_J \times G/P_J^- \end{array} \tag{2.4}$$

over the product of partial flag varieties $G/P_J \times G/P_J^-$, with fiber is isomorphic to the quotient group $G_J := L_J / Z_{L_J}$.

The group G_J is itself a reductive group with trivial center, and when we take the closure of the orbit $\mathcal{O}_J$ we obtain a fibration

$$\begin{array}{c} \overline{G_J} \hookrightarrow \overline{\mathcal{O}_J} \\ \downarrow \\ G/P_J \times G/P_J^-, \end{array} \tag{2.5}$$

whose fiber is the wonderful compactification of G_J. In particular, when $J = \emptyset$ is the empty set, we obtain the unique closed $G \times G$-orbit

$$\mathcal{O}_\emptyset = D_1 \cap \ldots \cap D_l.$$

In this case $P_\emptyset = B$ is a Borel subgroup, $P_\emptyset^- = B^-$ is the opposite Borel, and $L_\emptyset = T$ is the maximal torus of G, which is abelian. Therefore the subgroup $G_\emptyset$ is trivial, and diagrams (2.4) and (2.5) imply that this closed orbit

$$\mathcal{O}_\emptyset = \overline{\mathcal{O}_\emptyset} \cong G/B \times G/B^-$$

is isomorphic to the product of two copies of the full flag variety of G.

Example 2.6 Let $G = PGL_2$. Then $\tilde{G} = SL_2$, and the weights of the maximal torus of are given by the integers. The irreducible representations are indexed by the nonnegative integers, and all nonzero weights are regular. We can therefore choose $V = \mathbb{C}^2$ to be the standard representation. In this case

$$\psi : PGL_2 \hookrightarrow \mathbb{P}(M_{2\times 2})$$

is the standard embedding with image

$$\psi(PGL_2) = \left\{ \begin{bmatrix} a & b \\ c & d \end{bmatrix} \in \mathbb{P}(M_{2\times 2}) \mid ad - bc \neq 0 \right\}.$$

The wonderful compactification of PGL_2 is its closure

$$X = \mathbb{P}(M_{2\times 2}) \cong \mathbb{P}^3.$$

The boundary of this compactification is a smooth irreducible divisor with a transitive action of $PGL_2 \times PGL_2$. It is given cut out by the zero locus

$$D = \left\{ \begin{bmatrix} a & b \\ c & d \end{bmatrix} \in \mathbb{P}(M_{2\times 2}) \mid ad - bc = 0 \right\}$$

of the determinant polynomial, and consists exactly of the projectivized 2-by-2 matrices of rank 1. Exercise 8 shows that D is isomorphic to a product of two copies of $\mathbb{P}^1$, which is the flag variety of the group PGL_2.

Exercise 8 Construct an isomorphism

$$\mathbb{P}^1 \times \mathbb{P}^1 \xrightarrow{\sim} \left\{ \begin{bmatrix} a & b \\ c & d \end{bmatrix} \in \mathbb{P}(M_{2\times 2}) \mid ad - bc = 0 \right\}.$$

(You answer is probably an example of the *Segre embedding*.)

Example 2.6 does not generalize to larger ranks—when $n \geq 3$, the standard representation of SL_n is not regular, and the wonderful compactification of PGL_n is not simply the projective space $\mathbb{P}^{n^2-1}$. While this projective space is still a simple equivariant compactification of PGL_n, its boundary divisor is the vanishing locus

$$\{A \in \mathbb{P}(M_{n \times n}) \mid \det(A) = 0\}$$

of the determinant polynomial, which is irreducible and has singularities along the subvarieties of $\mathbb{P}(M_{n \times n})$ where the rank of the matrix A drops.

Exercise 9 Check by hand that the compactification $\mathbb{P}(M_{3 \times 3})$ of PGL_3 is not a wonderful variety.

3 Logarithmic Geometry of $\overline{G}$

Let X be a smooth algebraic variety containing a simple normal crossings divisor D, and write $\mathring{X}$ for the complement of D in X. The *logarithmic tangent sheaf* or *log-tangent sheaf* of X with respect to D is the subsheaf

$$\mathcal{T}_{X,D} \subset \mathcal{T}_X$$

consisting of vector fields on X that are tangent to D. In other words, the sections of $\mathcal{T}_{X,D}$ are those derivations which preserve the ideal sheaf of local functions that vanish along D.

Since D has simple normal crossings, the sheaf $\mathcal{T}_{X,D}$ is locally free—if $x_1, \ldots, x_n$ are local coordinates on X in which the divisor D is given by

$$D = \{(x_1, \ldots, x_n) \mid x_1 \ldots x_k = 0\},$$

the sheaf $\mathcal{T}_{X,D}$ is locally generated by the vector fields

$$x_1 \partial_1, \ldots, x_k \partial_k, \partial_{k+1}, \ldots, \partial_n.$$

This implies that $\mathcal{T}_{X,D}$ is the sheaf of sections of a vector bundle $T_{X,D}$, called the *log-tangent bundle* of the pair (X, D). Since the sheaves $\mathcal{T}_{X,D}$ and $\mathcal{T}_X$ have the same sections over the open dense locus $\mathring{X}$, the restriction of the bundle $T_{X,D}$ to $\mathring{X}$ is precisely the usual tangent bundle $T_{\mathring{X}}$.

Exercise 10 The inclusion $\mathcal{T}_{X,D} \hookrightarrow \mathcal{T}_X$ induces a morphism of vector bundles

$$T_{X,D} \longrightarrow T_X. \tag{3.1}$$

Is this morphism injective? Surjective? Describe its kernel and its image.

The dual vector bundle of $T_{X,D}$ is the *log-cotangent bundle* $T^*_{X,D}$, whose sections are meromorphic forms with logarithmic poles along the divisor D. In the same coordinates, it is locally generated by the logarithmic 1-forms

$$\frac{dx_1}{x_1}, \ldots, \frac{dx_k}{x_k}, dx_{k+1}, \ldots, dx_n.$$

In particular, its restriction to $\mathring{X}$ is the cotangent bundle $T^*_{\mathring{X}}$.

The usual differential gives the sheaves

$$\wedge^\bullet T^*_{X,D}$$

the structure of a complex, and a global section $\omega \in \Gamma(\wedge^2 T^*_{X,D})$ is called *log-symplectic* if it is closed and logarithmically nondegenerate—that is, if

- $d\omega = 0$, and
- the induced map $\omega^\flat : T_{X,D} \longrightarrow T^*_{X,D}$ is an isomorphism.

In the next exercise, we will show that any log-cotangent bundle has a canonical log-symplectic structure.

Exercise 11 Let M be a smooth algebraic variety with a simple normal crossings divisor Z, and write $\mathring{M}$ for the complement of Z in M. Let $\pi : T^*_{M,Z} \longrightarrow M$ be the bundle map, and write $D := \pi^{-1}(Z)$ for the preimage of the boundary divisor.

(a) Check that D is a simple normal crossings divisor in $T^*_{M,Z}$.
(b) Show that the canonical symplectic structure on $T^*_{\mathring{M}}$ extends to a log-symplectic structure on the log-cotangent bundle $T^*_{M,Z}$. (Hint: Try to write a "logarithmic" version of the usual construction of the canonical symplectic structure on a cotangent bundle, by finding a logarithmic Liouville 1-form.)

Once again let H be a connected complex algebraic group that acts on X, and write $\mathfrak{h}$ for its Lie algebra. The infinitesimal action of $\mathfrak{h}$ gives a morphism of vector bundles

$$\rho : X \times \mathfrak{h} \longrightarrow T_X \qquad (3.2)$$

from the trivial $\mathfrak{h}$-bundle on X to the tangent bundle T_X, which associates to each Lie algebra element ξ the action vector field ξ_X. Writing $\mathcal{O}_X$ for the structure sheaf of X, on the level of sheaves the map ρ corresponds to a morphism

$$\mathcal{O}_X \otimes \mathfrak{h} \longrightarrow T_X$$

that takes values in the tangent sheaf $\mathcal{T}_X$.

Exercise 12 Show that X is homogeneous for the action of H if and only if the bundle map (3.2) is surjective.

If the simple normal crossings divisor D is stable under the action of H, the infinitesimal action map (3.2) factors through the morphism (3.1) via a logarithmic action map

$$\rho_D : X \times \mathfrak{h} \longrightarrow T_{X,D}.$$

In other words, there is a commutative diagram

$$\begin{array}{ccc} X \times \mathfrak{h} & \xrightarrow{\rho_D} & T_{X,D} \\ & \searrow{\rho} & \downarrow \\ & & T_X. \end{array}$$

Exercise 13 Returning to the notation of Exercise 11, suppose that H acts on M and preserves Z. Then the induced action on the cotangent bundle $T^*_{\overset{\circ}{M}}$ is Hamiltonian, and for any covector α on $\overset{\circ}{M}$ the moment map

$$\mu : T^*_{\overset{\circ}{M}} \longrightarrow \mathfrak{h}^*$$

is given by

$$\mu(\alpha)(\xi) = \alpha(\xi_M) \quad \text{for all } \xi \in \mathfrak{h}.$$

Show that this moment map extends to a moment map

$$\bar{\mu} : T^*_{M,Z} \longrightarrow \mathfrak{h}^*$$

with respect to the canonical log-symplectic structure on $T^*_{M,Z}$.

Definition 3.3 The pair (X, D) is called *log-homogeneous* if the morphism ρ_D is surjective.

In this case, by Exercise 12, the open dense locus $\overset{\circ}{X}$ is homogeneous for the action of H, and there is a short exact sequence of vector bundles

$$0 \longrightarrow \ker \rho_D \longrightarrow X \times \mathfrak{h} \xrightarrow{\rho_D} T_{X,D} \longrightarrow 0.$$

Log-homogeneous varieties were introduced and studied by Brion [1], who gave the following characterization of the kernel of the logarithmic action map ρ_D.

Proposition 3.4 ([1, Proposition 2.1.2]) *Suppose that (X, D) is log-homogeneous for the action of H, and fix a point $x \in X$ with H-orbit $\mathcal{O} = H \cdot x$. The kernel of $\rho_{D,x}$ is the Lie algebra of the kernel of the action of $\mathrm{Stab}_H(x)$ on the normal space*

$$N_{\mathcal{O},x} = T_{X,x}/T_{\mathcal{O},x}.$$

All wonderful varieties are log-homogeneous [23, Proposition 4.2]. In particular, in the special case when $H = G \times G$ and $X = \overline{G}$, the logarithmic action map gives a short exact sequence of vector bundles

$$0 \longrightarrow \ker \rho_D \longrightarrow \overline{G} \times \mathfrak{g} \times \mathfrak{g} \xrightarrow{\rho_D} T_{\overline{G},D} \longrightarrow 0.$$

Exercise 14

(a) Let $e \in G$ be the identity element. Show that the kernel of ρ_D at e is the diagonal subalgebra

$$\mathfrak{g}_\Delta = \{(x, x) \in \mathfrak{g} \times \mathfrak{g} \mid x \in \mathfrak{g}\}.$$

(b) Let $a \in G$ be an arbitrary group element. Find the kernel of ρ_D at a.
(c) Let z_J be the distinguished basepoint of the $G \times G$-orbit $\mathcal{O}_J$, with stabilizer given by (2.3). Show that the kernel of ρ_D above z_J is the subalgebra

$$\mathfrak{p}_J \times_{\mathfrak{l}_J} \mathfrak{p}_J^-,$$

where $\mathfrak{p}_J^\pm$ denotes the Lie algebra of the parabolic subgroup P_J and $\mathfrak{l}_J$ is the Lie algebra of the common Levi L_J.

Now we consider the Killing form

$$\kappa : \mathfrak{g} \times \mathfrak{g} \longrightarrow \mathbb{C}$$
$$(x, y) \longmapsto \mathrm{tr}(\mathrm{ad}_x \circ \mathrm{ad}_y),$$

and we use it to define a symmetric, nondegenerate, invariant bilinear form on $\mathfrak{g} \times \mathfrak{g}$ by

$$\langle (x_1, x_2), (y_1, y_2) \rangle = \kappa(x_1, y_1) - \kappa(x_2, y_2). \tag{3.5}$$

With respect to this form, the diagonal subalgebra $\mathfrak{g}_\Delta$ is Lagrangian. Since κ is G-invariant, this implies that the subalgebra

$$\ker \rho_{D,a} = (a, e) \cdot \mathfrak{g}_\Delta$$

of Exercise 14 is also Lagrangian for any $a \in G$. Moreover, since the condition of being Lagrangian is closed, this means that the bundle

$$\ker \rho_D \subset \overline{G} \times \mathfrak{g} \times \mathfrak{g}$$

is a Lagrangian subbundle.

We therefore obtain a natural isomorphism

$$\ker \rho_D \cong \left((\overline{G} \times \mathfrak{g} \times \mathfrak{g})/\ker \rho_D\right)^* \cong T^*_{\overline{G},D}.$$

In other words, the bundle $\ker \rho_D$ is isomorphic to the log-cotangent bundle of the wonderful compactification $\overline{G}$. Our short exact sequence becomes

$$0 \longrightarrow T^*_{\overline{G},D} \longrightarrow \overline{G} \times \mathfrak{g} \times \mathfrak{g} \longrightarrow T_{\overline{G},D} \longrightarrow 0, \qquad (3.6)$$

and this realizes the log-cotangent bundle as a bundle of Lie algebras over $\overline{G}$. Note that this vector bundle is not locally trivial as a bundle of Lie algebras—the fiber above the identity element is isomorphic to the semisimple Lie algebra $\mathfrak{g}_\Delta$, while the fibers over the $G \times G$-orbit $\mathcal{O}_J$ are not semisimple, as the next exercise shows.

Exercise 15 Check that the Lie algebra $\mathfrak{p}_J \times_{\mathfrak{l}_J} \mathfrak{p}_J^-$ is not reductive. (Hint: Reductive Lie algebras contain no nontrivial nilpotent ideals.)

Restricting the sequence (3.6) to the open dense locus G of $\overline{G}$, we obtain the commutative diagram

$$\begin{array}{ccccccccc}
0 & \longrightarrow & T^*_{\overline{G},D} & \longleftrightarrow & \overline{G} \times \mathfrak{g} \times \mathfrak{g} & \longrightarrow & T_{\overline{G},D} & \longrightarrow & 0 \\
 & & \uparrow & & \uparrow & & & & \\
 & & T^*_G & \longleftrightarrow & G \times \mathfrak{g} \times \mathfrak{g}, & & & &
\end{array}$$

in which the square is Cartesian. If we identify the cotangent bundle T^*_G with $G \times \mathfrak{g}$ using left-trivialization and the isomorphism between $\mathfrak{g}$ and $\mathfrak{g}^*$ given by the Killing form, the bottom inclusion is given by the map

$$T^*_G \cong G \times \mathfrak{g} \hookrightarrow G \times \mathfrak{g} \times \mathfrak{g}$$

$$(a, x) \longmapsto (a, \operatorname{Ad}_a x, x).$$

The last two components of this morphism recover the moment map

$$\mu : T^*_G \cong G \times \mathfrak{g} \longrightarrow \mathfrak{g} \times \mathfrak{g}$$

$$(a, x) \longmapsto (\operatorname{Ad}_a x, x)$$

for the Hamiltonian $G \times G$-action on T^*G given by

$$(g, h) \cdot (a, x) = (gah^{-1}, \mathrm{Ad}_h x).$$

Projection onto the fibers of the trivial bundle $\overline{G} \times \mathfrak{g} \times \mathfrak{g}$ then extends this to the moment map

$$\overline{\mu} : T^*_{\overline{G}, D} \longrightarrow \mathfrak{g} \times \mathfrak{g}$$

relative to the canonical log-symplectic structure on $T^*_{\overline{G}, D}$ constructed in Exercise 11.

Finally, we note that the short exact sequence (3.6) induces a morphism

$$\overline{G} \longrightarrow \mathrm{Gr}(\dim \mathfrak{g}, \mathfrak{g} \oplus \mathfrak{g}) \qquad (3.7)$$

$$a \longmapsto [\ker \rho_{D,a}],$$

from the wonderful compactification $\overline{G}$ into the Grassmannian of half-dimensional subspaces of $\mathfrak{g} \oplus \mathfrak{g}$. This morphism is an embedding, called the *Demazure embedding* [18]. By what we saw above, its image lies in the closed subvariety

$$\mathcal{L}(\mathfrak{g} \oplus \mathfrak{g}) \subset \mathrm{Gr}(\dim \mathfrak{g}, \mathfrak{g} \oplus \mathfrak{g})$$

consisting of subalgebras of $\mathfrak{g} \oplus \mathfrak{g}$ which are Lagrangian with respect to the inner product (3.5). This subvariety has many remarkable properties, which we now study in a slightly more general context.

4 Lagrangian Subalgebras and Poisson-Homogeneous Spaces

Let $\mathfrak{d}$ be a real or complex Lie algebra of dimension $2n$ that carries a nondegenerate, symmetric, invariant bilinear form $\langle \cdot, \cdot \rangle$. If

$$\mathfrak{d} = \mathfrak{l}_1 + \mathfrak{l}_2$$

is a splitting of $\mathfrak{d}$ into Lagrangian subalgebras and we write $\mathfrak{u} = \mathfrak{l}_1$, there is a natural identification of $\mathfrak{u}^*$ with $\mathfrak{l}_2$. Then

$$(\mathfrak{d}, \mathfrak{u}, \mathfrak{u}^*)$$

is a Manin triple, and the Lie bracket on $\mathfrak{u}^*$ induces a Poisson bracket π_U on the simply-connected Lie group U integrating $\mathfrak{u}$. This bracket has the property that the group multiplication map is a Poisson map—in other words, (U, π_U) is a *Poisson–Lie group*.

Equivalently, the Poisson bivector π_U satisfies the identity

$$\pi_{U,gh} = L_g \pi_{U,h} + R_h \pi_{U,g} \quad \text{for all } g, h \in U,$$

where L_g and R_h are the differentials of left multiplication by g and right multiplication by h, respectively.

Example 4.1 Let $\mathfrak{d} = \mathfrak{g}$ be a semisimple complex Lie algebra of complex dimension n, viewed as a $2n$-dimensional Lie algebra over the real numbers. Consider the nondegenerate, symmetric, invariant bilinear form

$$\langle \cdot, \cdot \rangle = \text{Im}(\kappa)$$

given by the imaginary part of the Killing form. The Iwasawa decomposition

$$\mathfrak{g} = \mathfrak{k} + \mathfrak{a} + \mathfrak{n}$$

gives a Lagrangian splitting of $\mathfrak{g}$ into the Lagrangian subalgebras $\mathfrak{u} = \mathfrak{k}$ and $\mathfrak{u}^* = \mathfrak{a} + \mathfrak{n}$, and

$$(\mathfrak{g}, \mathfrak{k}, \mathfrak{a} + \mathfrak{n})$$

is a Manin triple. In this way, the simply-connected Lie group K integrating $\mathfrak{k}$ becomes a Poisson–Lie group.

Example 4.2 Let $\mathfrak{g}$ be any real or complex Lie algebra, and view $\mathfrak{g}^*$ as an abelian Lie algebra equipped with the trivial Lie bracket. The semidirect product $\mathfrak{d} = \mathfrak{g} \ltimes \mathfrak{g}^*$ with respect to the coadjoint action is the Lie algebra, and its bracket is given by

$$[(x, \alpha), (y, \beta)] = ([x, y], \text{ad}_x \beta - \text{ad}_y \alpha) \quad \text{for all } (x, \alpha), (y, \beta) \in \mathfrak{g} \ltimes \mathfrak{g}^*.$$

It carries a natural nondegenerate symmetric invariant inner product

$$\langle (x, \alpha), (y, \beta) \rangle = \alpha(y) + \beta(x) \quad \text{for all } (x, \alpha), (y, \beta) \in \mathfrak{g} \ltimes \mathfrak{g}^*,$$

with respect to which the subalgebras $\mathfrak{g} \oplus 0$ and $0 \oplus \mathfrak{g}^*$ are Lagrangian. If G is a Lie group integrating $\mathfrak{g}$, the double $\mathfrak{g} \ltimes \mathfrak{g}^*$ integrates to the group $G \ltimes \mathfrak{g}^*$ with multiplication given by

$$(g, \alpha) \cdot (h, \beta) = (gh, \text{Ad}_{h^{-1}} \alpha + \beta) \quad \text{for all } (g, \alpha), (h, \beta) \in G \ltimes \mathfrak{g}^*.$$

The Manin triple

$$(\mathfrak{g} \ltimes \mathfrak{g}^*, \mathfrak{g}, \mathfrak{g}^*)$$

induces a trivial Poisson structure on the group G, and the Kostant–Kirillov–Souriau Poisson structure on its Poisson–Lie dual group $\mathfrak{g}^*$.

Example 4.3 Let $\mathfrak{g}$ be a complex semisimple Lie algebra and fix a maximal torus $\mathfrak{t}$ and a Borel subalgebra $\mathfrak{b}$ that contains it. Writing $\mathfrak{b}^-$ for the opposite Borel we obtain a Lagrangian splitting

$$\mathfrak{g} \oplus \mathfrak{g} = \mathfrak{g}_\Delta + \mathfrak{b} \times_\mathfrak{t} \mathfrak{b}^- \tag{4.4}$$

with respect to the nondegenerate invariant bilinear form

$$\langle (x_1, x_2), (y_1, y_2) \rangle = \kappa(x_1, y_1) - \kappa(x_2, y_2)$$

defined in (3.5). This produces a Manin triple

$$(\mathfrak{g} \oplus \mathfrak{g}, \mathfrak{g}_\Delta, \mathfrak{b} \times_\mathfrak{t} \mathfrak{b}^-),$$

and gives a canonical Poisson–Lie group structure on the group G integrating $\mathfrak{g}$.

Poisson–Lie groups were introduced by Drinfeld [5] and by Semenov-Tian-Shansky [25]. For background, we refer to [3, Section 1] or to [20, Section 1]. Given such a group (U, π_U), it is natural to consider its *Poisson actions*—that is, actions of U on a Poisson manifold (M, π) that have the property that the action map

$$U \times M \longrightarrow M$$

is a Poisson map. Equivalently, this means that the Poisson bivector π satisfies the identity

$$\pi_{gm} = g_* \pi_m + m_* \pi_{U,g} \quad \text{for all } g \in U, m \in M,$$

where g_* and m_* are the differentials of the maps

$$g : M \longrightarrow M \qquad \text{and} \qquad m : U \longrightarrow M$$
$$m' \longmapsto gm' \qquad\qquad\qquad g' \longmapsto g'm$$

respectively.

Definition 4.5 A (U, π_U)-*homogeneous space* is a Poisson manifold (M, π) equipped with an action of U that is transitive and Poisson.

Let (M, π) be a (U, π_U)-homogeneous space, and for any point m of M write U_m for the stabilizer of m and $\mathfrak{u}_m$ for its Lie algebra. Then there is a natural identification

$$T_{M,m} \cong \mathfrak{u}/\mathfrak{u}_m,$$

and we can view the Poisson bivector as π_m as an element of $\wedge^2(\mathfrak{u}/\mathfrak{u}_m)$. The induced map

$$\pi_m^\# : (\mathfrak{u}/\mathfrak{u}_m)^* \longrightarrow (\mathfrak{u}/\mathfrak{u}_m),$$

lets us define the subspace

$$\mathfrak{l}_m := \left\{ (x, \xi) \in \mathfrak{u} + \mathfrak{u}^* \mid \xi_{|\mathfrak{u}_m} = 0 \text{ and } \pi_m^\#(\xi) = x + \mathfrak{u}_m \right\}$$

of the Lie algebra $\mathfrak{d}$. In other words, $\mathfrak{l}_m$ consists of pairs (x, ξ) with the property that ξ descends to an element of $(\mathfrak{u}/\mathfrak{u}_m)^*$ whose image under $\pi_m^\#$ is the coset of x.

Exercise 16 ([6, Theorem 1]) Show that, for any point $m \in M$, the subspace $\mathfrak{l}_m$ is a Lagrangian subalgebra of $\mathfrak{d}$.

In this way we have obtained a map

$$\Psi : M \longrightarrow \mathrm{Gr}(n, \mathfrak{d}) \tag{4.5}$$
$$m \longmapsto [\mathfrak{l}_m],$$

known as the *Drinfeld map* [6], from the Poisson-homogeneous space (M, π) to the Grassmannian of half-dimensional subspaces of the Lie algebra $\mathfrak{d}$. Exercise 16 implies that its image is contained in the *variety of Lagrangian subalgebras*

$$\mathcal{L}(\mathfrak{d}) := \left\{ [\mathfrak{l}] \in \mathrm{Gr}(n, \mathfrak{d}) \mid [\mathfrak{l}, \mathfrak{l}] \subset \mathfrak{l} \text{ and } \mathfrak{l}^\perp = \mathfrak{l} \right\}.$$

Since both of the defining conditions on the right are algebraic, $\mathcal{L}(\mathfrak{d})$ is a closed subvariety of the Grassmannian. Moreover, it is possibly singular, possibly disconnected, and stabilized by the adjoint action of any algebraic group D that integrates $\mathfrak{d}$.

Theorem 4.7 ([6, Theorem 1]) *Let (M, π) be a (U, π_U)-homogeneous space.*

(a) *The Drinfeld map Ψ is U-equivariant—in other words, for any $m \in M$ and $u \in U$,*

$$\mathfrak{l}_{um} = \mathrm{Ad}_u(\mathfrak{l}_m).$$

(b) *The assignment* $(M, \pi) \longmapsto U \cdot [\mathfrak{l}_m]$ *defines an injection*

$$\{(U, \pi_U)\text{-homogeneous spaces with connected stabilizers}\} \hookrightarrow \{U\text{-orbits on } \mathcal{L}(\mathfrak{d})\}.$$

Its image consists of those Lagrangian subalgebras $\mathfrak{l}$ of $\mathfrak{d}$ with the property that $\mathfrak{l} \cap \mathfrak{u}$ integrates to a closed subgroup of U.

The variety $\mathcal{L}(\mathfrak{d})$ has a natural Poisson structure which is compatible with the map (4.5), and which was first introduced and studied by Evens and Lu [10, 11]. Before defining it we remark that, since $\mathcal{L}(\mathfrak{d})$ is a possibly singular variety, by a Poisson structure we mean a Lie bracket on the structure sheaf of $\mathcal{L}(\mathfrak{d})$ which is a derivative in each component. On the smooth locus of $\mathcal{L}(\mathfrak{d})$, this definition agrees with the usual notion of a Poisson bivector field.

Identify

$$\mathfrak{d}^* \cong \mathfrak{u}^* + \mathfrak{u}$$

and define a classical r-matrix $R \in \wedge^2 \mathfrak{d}$ by

$$R((\xi_1, x_1), (\xi_2, x_2)) = \xi_2(x_1) + \xi_1(x_2) \quad \text{for all } (\xi_1, x_i) \in \mathfrak{d}^*.$$

If D is an algebraic group integrating $\mathfrak{d}$, the adjoint action of D on $\mathrm{Gr}(n, \mathfrak{d})$ induces an infinitesimal action map

$$\rho : \mathfrak{d} \longrightarrow T_{\mathrm{Gr}(n, \mathfrak{d})}.$$

Via this action, the r-matrix R becomes a bivector field on $\mathrm{Gr}(n, \mathfrak{d})$ given by

$$\Pi := \rho(R).$$

Exercise 17 Consider the identification

$$\#: \mathfrak{d}^* \longrightarrow \mathfrak{d}$$
$$(\xi, x) \longmapsto (x, \xi).$$

(a) Show that the Schouten bracket $[R, R] \in \wedge^3 \mathfrak{d}$ is given by

$$[R, R](d_1, d_2, d_3) = \langle \#d_1, [\#d_2, \#d_3] \rangle \quad \text{for all } d_i \in \mathfrak{d}^*.$$

(b) Suppose that $\mathfrak{l}$ is a Lagrangian subalgebra of $\mathfrak{d}$. Use part (a) to show that

$$[\Pi, \Pi]_{[\mathfrak{l}]} = 0.$$

Since $\mathcal{L}(\mathfrak{d})$ is stable under the action of D, the skew-symmetric bracket induced by Π on the structure sheaf of $\mathrm{Gr}(n, \mathfrak{d})$ descends to a bracket on $\mathcal{L}(\mathfrak{d})$. Exercise 17 implies that this bracket satisfies the Jacobi identity, and therefore $\mathcal{L}(\mathfrak{d})$ has the structure of a Poisson algebraic variety. The D-orbits of $\mathcal{L}(\mathfrak{d})$ are Poisson submanifolds with respect to this structure.

Remark Note that the Poisson structure Π depends on the fixed Lagrangian splitting. There are therefore many possible Poisson structures on $\mathcal{L}(\mathfrak{d})$, given by different choices of Lagrangian splitting, and it is an interesting problem to classify them. In the case when $\mathfrak{d} = \mathfrak{g} \oplus \mathfrak{g}$, such a classification is carried out by Lu and Yakimov in [21].

Theorem 4.8 ([10, Proposition 2.16 and Theorem 2.17])

(a) *The action of (U, π_U) on $(\mathcal{L}(\mathfrak{d}), \Pi)$ is a Poisson action.*
(b) *Every U-orbit on $\mathcal{L}(\mathfrak{d})$ is a Poisson submanifold of $\mathcal{L}(\mathfrak{d})$, and therefore a (U, π_U)-homogeneous space.*

Let $\mathfrak{l}$ be a Lagrangian subalgebra of $\mathfrak{d}$. According to Theorem 4.8, the orbit

$$M = U \cdot [\mathfrak{l}] \subset \mathrm{Gr}(n, \mathfrak{d})$$

is a (U, π_U)-homogeneous space. The image of its basepoint under the Drinfeld map (4.5) is the Lagrangian subalgebra

$$\Psi([\mathfrak{l}]) = [\mathfrak{u}_{[\mathfrak{l}]} + (\mathfrak{u} + \mathfrak{u}_{[\mathfrak{l}]}^\circ) \cap \mathfrak{l}]$$

[10, Theorem 2.20]. In particular, the Drinfeld map in this case is not necessarily the identity map. In fact, viewed as a function from $\mathcal{L}(\mathfrak{d})$ to itself, the Drinfeld map is generally not even continuous [10, Remark 2.21].

Exercise 18

(a) Show that $\Psi([\mathfrak{l}]) = [\mathfrak{l}]$ if and only if

$$\mathfrak{l} \cap \mathfrak{u} = \mathfrak{u}_{[\mathfrak{l}]}.$$

The Lagrangian subalgebras satisfying this property are called *model points* of $\mathcal{L}(\mathfrak{d})$.
(b) Check that the set of model points of $\mathcal{L}(\mathfrak{d})$ is stable under the action of U.

(c) Suppose that (M, π) is a (U, π_U)-homogeneous space and that, for some $m \in M$, $[\mathfrak{l}_m]$ is a model point of $\mathcal{L}(\mathfrak{d})$. Show that the Drinfeld map

$$\Psi : M \longrightarrow U \cdot [\mathfrak{l}_m]$$

is a local diffeomorphism.

Theorem 4.9 ([10, Theorem 2.22]) *For every (U, π_U)-homogeneous space (M, π), the Drinfeld map defined by (4.5) is a Poisson map.*

Putting all of this together, Theorem 4.9 shows that the Drinfeld map (4.5) is a Poisson map from any given (U, π_U)-homogeneous space to a U-orbit in $\mathcal{L}(\mathfrak{d})$, which is itself a (U, π_U)-homogeneous space. When this homogeneous space consists of model points of $\mathcal{L}(\mathfrak{d})$, the Drinfeld map is a local Poisson diffeomorphism—in other words, model point U-orbits on the variety of Lagrangian subalgebras can be viewed as "local models" of Poisson homogeneous spaces.

Exercise 19 Consider the Manin triple $(\mathfrak{g}, \mathfrak{k}, \mathfrak{a} + \mathfrak{n})$ of Example 4.1, and let $\mathfrak{l} = \mathfrak{a} + \mathfrak{n}$.

(a) Show that $\mathfrak{l}$ is not a model point of $\mathcal{L}(\mathfrak{g})$.
(b) Show that the image of $\mathfrak{l}$ under the Drinfeld map Ψ is a model point of $\mathcal{L}(\mathfrak{d})$.

5 Poisson Structures on Wonderful Spaces

The Poisson geometry of the variety of Lagrangian subalgebras $\mathcal{L}(\mathfrak{d})$ has many unexpected connections to the representation theory of $\mathfrak{d}$, and many interesting equivariant embeddings of homogeneous spaces appear as closures of D-orbits or U-orbits on $\mathcal{L}(\mathfrak{d})$. We conclude this survey with three examples, following once again the work of Evens and Lu [10, 11].

Example 5.1 Let $\mathfrak{g}$ be a semisimple complex Lie algebra with the Lagrangian splitting given by the Iwasawa decomposition

$$\mathfrak{g} = \mathfrak{k} + (\mathfrak{a} + \mathfrak{n})$$

as in Example 4.1. The variety of Lagrangian subalgebras $\mathcal{L}(\mathfrak{g})$ and its induced Poisson structure were studied by Evens and Lu in [10], using the parametrization of Lagrangian subalgebras of $\mathfrak{g}$ given by Karolinsky [15].

In this work the authors classify the irreducible components of $\mathcal{L}(\mathfrak{g})$ and show that each one is a fibration over a partial flag variety with fiber the real points of the wonderful compactification associated to a Levi subgroup of $\mathfrak{g}$. A large number of familiar G- and K-homogeneous spaces appear as Poisson submanifolds of $\mathcal{L}(\mathfrak{g})$. Among them are the flag varieties of G, equipped with the Bruhat–Poisson structure whose symplectic leaves are Bruhat cells [20].

Example 5.2 Let $\mathfrak{g}$ be a real or complex Lie algebra, and consider the Manin triple

$$(\mathfrak{g} \ltimes \mathfrak{g}^*, \mathfrak{g}, \mathfrak{g}^*)$$

of Example 4.2. The normalizer of the Lagrangian subalgebra $\mathfrak{g}^*$ in $G \ltimes \mathfrak{g}^*$ is the subgroup

$$\mathrm{Stab}_{G \ltimes \mathfrak{g}^*}[\mathfrak{g}^*] = G \ltimes \{0\},$$

and therefore the $G \ltimes \mathfrak{g}^*$-orbit

$$(G \ltimes \mathfrak{g}^*) \cdot [\mathfrak{g}^*] \cong (G \ltimes \mathfrak{g}^*)/(G \times \{0\}) \cong \mathfrak{g}^*$$

is a Poisson submanifold of $\mathcal{L}(\mathfrak{g} \ltimes \mathfrak{g}^*)$. The induced Poisson structure on $\mathfrak{g}^*$ is precisely the Kostant–Kirillov–Souriau Poisson structure.

The variety $\mathcal{L}(\mathfrak{g} \ltimes \mathfrak{g}^*)$ is a reducible algebraic variety whose irreducible components are often singular, and generally its geometry is very complicated. However, in the case when $\mathfrak{g}$ is a complex semisimple Lie algebra, its points are indexed by a parametrization due to Karolinsky and Stolin [16]. Recently, Evens and Li [9] used this parametrization to show that the closed $G \ltimes \mathfrak{g}^*$-orbits of $\mathcal{L}(\mathfrak{g} \ltimes \mathfrak{g}^*)$ are in bijection with the abelian ideals of a maximal Borel of $\mathfrak{g}$.

Example 5.3 Let $\mathfrak{g}$ be a semisimple complex Lie algebra, and consider the Manin triple

$$(\mathfrak{g} \oplus \mathfrak{g}, \mathfrak{g}_\Delta, \mathfrak{b} \times_\mathfrak{t} \mathfrak{b}^-)$$

of Example 4.3, which can be viewed as a complexification of Example 4.1. This triple induces the standard Poisson–Lie group structure on the adjoint group G, and by the discussion in the previous section it gives the variety $\mathcal{L}(\mathfrak{g} \oplus \mathfrak{g})$ a Poisson structure for which the orbits of $G \times G$, G_Δ, and $B \times_T B^-$ are Poisson submanifolds. The geometry of this Poisson structure was studied by Evens and Lu in [11], and we devote the rest of this section to some of their results.

In [11], the authors classify the $G \times G$-orbits and the irreducible and connected components of $\mathcal{L}(\mathfrak{g} \oplus \mathfrak{g})$ using generalized Belavin–Drinfeld triples, which were introduced by Schiffman [24] and which we now briefly recall.

Definition 5.4 A *generalized Belavin–Drinfeld triple* is a triple (I, J, η), where

- I and J are subsets of the set Δ of simple roots, and
- $\eta: I \longrightarrow J$ is a bijection that preserves the inner product $\langle \cdot, \cdot \rangle$.

The subsets I and J correspond to parabolic subgroups P_I and P_J as illustrated in Example 2.2. We denote by P_I^- and P_J^- their opposite parabolics, and by L_I and L_J their respective Levi components. The isometry η induces a group isomorphism between the adjoint forms L_I/Z_{L_I} and L_J/Z_{L_J}, which we denote by G_I and G_J respectively.

Theorem 5.5 ([11, Proposition 2.17 and Proposition 2.25]) *Let $\mathcal{O}$ be a $G \times G$-orbit in $\mathcal{L}(\mathfrak{d})$.*

(a) *There is a generalized Belavin–Drinfeld triple (I, J, η) such that $\mathcal{O}$ is a fibration*

$$\begin{array}{ccc} G_I & \hookrightarrow & \mathcal{O} \\ & & \downarrow \\ & & G/P_I \times G/P_J^- \end{array}$$

over the partial flag variety $G/P_I \times G/P_J^-$ with fiber isomorphic to the adjoint group G_I.

(b) *The closure $\overline{\mathcal{O}}$ of $\mathcal{O}$ is a fibration*

$$\begin{array}{ccc} \overline{G_I} & \hookrightarrow & \overline{\mathcal{O}} \\ & & \downarrow \\ & & G/P_I \times G/P_J^- \end{array}$$

with fiber isomorphic to the wonderful compactification $\overline{G_I}$ of G_I.

Theorem 5.5 implies that the closures of $G \times G$-orbits in $\mathcal{L}(\mathfrak{g} \oplus \mathfrak{g})$ are smooth. Moreover, Evens and Lu show that each $G \times G$-orbit is a union of finitely many $B \times B^-$-orbits. Since one of these $B \times B^-$-orbits must be open and dense, this implies that every $G \times G$-orbit in $\mathcal{L}(\mathfrak{g} \oplus \mathfrak{g})$ is a spherical $G \times G$-homogeneous space [11, Corollary 2.22].

Example 5.6 The normalizer of the Lagrangian subalgebra $\mathfrak{g}_\Delta$ is the diagonal subgroup

$$\mathrm{Stab}_{G \times G}[\mathfrak{g}_\Delta] = G_\Delta.$$

This construction therefore gives a Poisson structure on the group G itself, via the isomorphism

$$(G \times G) \cdot [\mathfrak{g}_\Delta] \cong (G \times G)/G_\Delta \cong G,$$

which is different from the Poisson–Lie group structure on G. By the above discussion, this Poisson structure extends to the closure of the $G \times G$-orbit of $[\mathfrak{g}_\Delta]$, which by Exercise 14 is isomorphic to the wonderful compactification $\overline{G}$ via the map (3.7). Using this observation, the characterization of $G \times G$-orbits on $\overline{G}$ given in (2.4) and (2.5) can also be deduced from the statement of Theorem 5.5.

Now let $\mathcal{O}_\Delta$ be a G_Δ-orbit, let $\mathcal{O}_{B \times B^-}$ be a $B \times B^-$-orbit, and suppose that the two have nonempty intersection. If $\mathcal{O}$ is the $G \times G$-orbit that contains them both, the Lagrangian splitting (4.4) implies that $\mathcal{O}_\Delta$ and $\mathcal{O}_{B \times B^-}$ intersect transversally in $\mathcal{O}$. Therefore

$$\mathcal{O}_\Delta \cap \mathcal{O}_{B \times B^-}$$

is a smooth submanifold of $\mathcal{L}(\mathfrak{g} \oplus \mathfrak{g})$ and, since it is an intersection of Poisson submanifolds, it is in fact a Poisson submanifold.

Theorem 5.7 ([11, Theorem 4.5 and Theorem 4.14]) *The intersection $\mathcal{O}_\Delta \cap \mathcal{O}_{B \times B^-}$ is a regular Poisson manifold—that is, the Poisson bivector has constant rank, so that the symplectic foliation is a regular foliation. The diagonal torus T_Δ acts transitively on the symplectic leaves of this intersection.*

Remark The intersections $\mathcal{O}_\Delta \cap \mathcal{O}_{B \times B^-}$ are called the *T-leaves* of the Poisson structure on $\mathcal{L}(\mathfrak{g} \oplus \mathfrak{g})$, because they are T-saturations of symplectic leaves. In the more general setting of complex tori acting on Poisson manifolds, such leaves have recently come to play an important role in cluster geometry [7, 19].

Example 5.8 In particular, when we view the group G as a Poisson submanifold of $\mathcal{L}(\mathfrak{g} \oplus \mathfrak{g})$, the T-leaves are exactly the intersections of the conjugacy classes with the Bruhat cells.

We conclude with an important class of G_Δ-orbits on $\mathcal{L}(\mathfrak{g} \oplus \mathfrak{g})$ which can be identified with the symmetric G-spaces defined in Definition 1.2.

Exercise 20 Let

$$\sigma : \mathfrak{g} \longrightarrow \mathfrak{g}$$

be an algebraic involution of the Lie algebra $\mathfrak{g}$, and consider its graph

$$\mathfrak{l}_\sigma := \{(x, \sigma(x)) \in \mathfrak{g} \oplus \mathfrak{g} \mid x \in \mathfrak{g}\}.$$

(a) Show that $\mathfrak{l}_\sigma$ is a Lagrangian subalgebra of $\mathfrak{g} \oplus \mathfrak{g}$.
(b) Show that the stabilizer $\mathrm{Stab}_{G_\Delta}(\mathfrak{l}_\sigma)$ is isomorphic to the fixed-point subgroup G^σ.

It follows from Exercise 20 that the G_Δ-orbit of the Lagrangian subalgebra $[\mathfrak{l}_\sigma]$ in $\mathcal{L}(\mathfrak{g} \oplus \mathfrak{g})$ is isomorphic to the symmetric space

$$G_\Delta \cdot [\mathfrak{l}_\sigma] \cong G/G^\sigma.$$

Its closure in the variety of Lagrangian subalgebras is precisely the wonderful compactification of G/G^σ given by Theorem 1.5 [11, Proposition 3.21]. In this way, all de Concini–Procesi wonderful compactifications sit inside $\mathcal{L}(\mathfrak{g} \oplus \mathfrak{g})$ as Poisson submanifolds, and their resulting Poisson structure is characterized by the broader results of Evens and Lu.

Acknowledgments I would like to thank the organizers and the scientific committee of Poisson 2022 for the invitation to give the minicourse on which these notes are based. I am also grateful to Yu Li for his help in running this minicourse and for many interesting discussions.

References

1. M. Brion, Log-homogeneous varieties, in *XVI Coloquio Latinoamericano de Algebra* (2005)
2. M. Brion, F. Pauer, Valuations des espaces homogènes sphériques. Comment. Math. Helvetici **62**, 265–285 (1987)
3. V. Chari, A. Pressley. *A Guide to Quantum Groups* (Cambridge University Press, 1995)
4. C. de Concini, C. Procesi, Complete symmetric varieties, in *In Invariant Theory* (Springer, Berlin Heidelberg, 1983)
5. V. Drinfeld, Hamiltonian structures on Lie groups, Lie bialgebras and the geometric meaning of the classical Yang-Baxter equation. Soviet Math. Dokl. **27**, 68–71 (1983)
6. V. Drinfeld, On Poisson-homogeneous spaces of Poisson-Lie groups. Theor. Math. Phys. **95**(2), 226–227 (1993)
7. B. Elek, J. Lu, Bott-Samelson varieties and Poisson Ore extensions. Int. Math. Res. Not. **2021**(14), 10745–10797 (2021)
8. S. Evens, B.F. Jones. On the wonderful compactification. arXiv:0801.0456 (2008)
9. S. Evens, Y. Li. Abelian ideals and the variety of Lagrangian subalgebras. arXiv:2010.04358 (2020)
10. S. Evens, J. Lu, On the variety of Lagrangian subalgebras I. Ann. Sci. Éc. Norm. Sup. **34**(5), 631–668 (2001)
11. S. Evens, J. Lu, On the variety of Lagrangian subalgebras II. Ann. Sci. Éc. Norm. Sup. **39**(2), 347–379 (2006)
12. W. Fulton, *Introduction to Toric Varieties* (Princeton University Press, 1993)
13. B. Hassett, Y. Tschinkel, Geometry of equivariant compactifications of G_a^n. Int. Math. Res. Not. **22**, 1211–1230 (1999)
14. J. Humphreys, *Linear Algebraic Groups* (Springer, 1975)
15. E. Karolinsky, The classification of Poisson-homogeneous spaces of compact Poisson-Lie groups. Math. Phys. Anal. Geom. **3**(3), 274–289 (1996)

16. E. Karolinsky, A. Stolin, Classical dynamical r-matrices, Poisson homogeneous spaces, and Lagrangian subalgebras. Lett. Math. Phys. **60**, 257–274 (2002)
17. F. Knop, Automorphisms, root systems, and compactifications of homogeneous varieties. J. Amer. Math. Soc. **9**(1), 153–174 (1996)
18. I. Losev, Demazure embeddings are smooth. Int. Math. Res. Not. **2009**(14), 2588–2596 (2009)
19. J. Lu, V. Mouquin, On the t-leaves of some Poisson structures related to products of flag varieties. Adv. Math. **306**, 1209–1261 (2017)
20. J. Lu, A. Weinstein, Poisson-Lie groups, dressing transformations, and Bruhat decompositions. J. Differential Geom. **31**(2), 501–526 (1990)
21. J. Lu, M. Yakimov, Group orbits and regular partitions of Poisson manifolds. Comm. Math. Phys. **283**, 729–748 (2008)
22. D. Luna, Th. Vust, Plongements d'espaces homogènes. Comment. Math. Helv. **58**, 186–245 (1983)
23. G. Pezzini, Lectures on wonderful varieties. Acta Math. Sin. **34**, 417–438 (2018)
24. O. Schiffman, On classification of dynamical r-matrices. Math. Res. Lett. **5**, 13–31 (1998)
25. M.A. Semenov-Tian-Shansky, Dressing transformations and Poisson-Lie group actions. Publ. RIMS **21**, 1237–1260 (1985)
26. H. Sumihiro, Equivariant completion. J. Math. Kyoto Univ. **14**(1), 1–28 (1974)
27. D. Timashev, *Homogeneous Spaces and Equivariant Embeddings* (Springer, 2011)
28. T. Vust, Opération de groupes réductifs dans un type de cônes presque homogènes. Bull. Soc. Math. France **102**, 317–334 (1974)

Part III

An Invitation to Singular Foliations

Introduction

Flying Under Radar: Singular Foliations

Singular foliations are so common in mathematics that they often go unnoticed.

Regular foliations have been long studied; The Frobenius theorem [1, 19] is taught quite early in the differential geometry curriculum. Holonomy (or "first return") is a very classical notion [26]. In contrast, singular foliations have never been studied with such an intensity. Still, there is a long story behind foliations that have leaves which are not all the same dimension:

1. As pointed by Sylvain Lavau [22], the 1960s saw an intense debate about finding a correct definition of a singular foliation. The discussion led to some major discoveries by H. Hermann [20], T. Nagano [29], P. Stefan [33, 34], H. Sussmann [35, 36].[1]
2. Then the subject seems to have been slightly forgotten, or at least put aside. There were, still, important contributions to the linearization problem by Dominique Cerveau [11] (where "singular foliations" appear under the name of "involutive distributions") in 1977 and Pierre Dazord [12] who defined a holonomy map for a singular leaf in 1984. There were other contributions coming from complex geometry, in particular—but not only—about codimension 1 or dimension 1 singular foliations (see the excellent review [7]) and the theory of residues (see e.g. Paul Baum and Raoul Bott's [8], Ali Sinan Sertöz's [32], André Belotto da Silva and Daniel Panazzolo's [10] or Tatsuo Suwa's [37]). Also, Poisson geometers knew that symplectic leaves of a Poisson manifold, or

---

[1] We refer to Sylvain Lavau's excellent article [22] for the historical aspects. Lavau [22] can also be read as an introduction to the subject.

Lie algebroid leaves, were a sort of "singular foliation" [23, 42], but, to our knowledge, rarely developed it as such.

3. Then, starting in the 2000s, a "singular foliation's revival" arose from non-commutative geometry, with pioneering and fundamental works by Iakovos Androulidakis, Claire Debord, Georges Skandalis, and Marco Zambon in particular. It is unfair to summarize their contributions in one sentence, but since we have to do so, let us claim that, from the geometric point of view, a crucial feat is the construction, by Androulidakis and Skandalis [2], of a holonomy groupoid of a singular foliation, that extends holonomy groupoids of regular foliations [26], and a smooth groupoid previously constructed by Claire Debord for projective singular foliations [13]. A theorem of crucial importance was also established by Claire Debord: although Androulidakis-Skandalis holonomy groupoid is not smooth, it is longitudinally smooth [14]. Then, Omar Mohsen [27] introduced a quotient of the holonomy groupoid, now called Mohsen's groupoid. As an application, the so-called Helffer-Nourrigat conjecture was recently solved by Androulidalis et al. [6].

This holonomy groupoid, or more precisely its natural C^*-algebra, is used by this school to define and study elliptic pseudodifferential operators [41], analytic indexes [15], or to investigate its Baum-Connes conjecture [4], Boutet de Monvel calculus [16]—in one word, to do analysis of singular foliations and their differential operators, to define symbols [28]. It is now used to solve classical conjectures about hypo-elliptic operators [6]. We have no expertise in non-commutative geometry and will not speak much about that subject, although it is certainly the most quickly progressing one among those using singular foliations.

Singular foliations, especially those of dimension or codimension 1 are also currently used in complex dynamics, see, e.g. [9, 10, 31], and in theoretical physics for perturbation theories of moduli spaces, see e.g., [18, 21].

The purpose of the present invitation is not to do analysis of singular foliations, although it is certainly the most active topic at the moment. In particular, we are not competent in non-commutative geometry, in index theorem, and pseudo-differential operators. We do not claim to be highly competent neither in holomorphic dynamical systems nor in theoretical physics. Our purpose is to simply to introduce the geometry of a singular foliation—maybe we should even say the differential topology of singular foliations, since we will not speak much about metrics. In our opinion, geometry is the easiest entrance to singular foliations. We are also non-competent to include an overview of derived foliations, recently developed by Tony Pantev, Bertrand Toën and Gabriele Vezzosi (see, e.g. [30, 38]): there is certainly a link to be explored with what we called the universal Lie ∞-algebroid of a singular foliation in Sect. 7, while [39] may have some relation with the construction of the holonomy groupoid.

Let us go back to the initial debate—in a very anachronistic manner: Should singular foliations be seen:

(0) as level sets (called "leaves") of (maybe non-independent) functions?
♣ as a partition of a manifold into submanifolds?
◇ as the data, at each point, of sub-spaces of the tangent space satisfying an involutivity condition?
♡ as a regular foliation defined on some open subset of the manifold?
♠ or as an involutive $\mathcal{C}^\infty(M)$-module of vector fields (morally thought to be tangent to the leaves)?

Definition (0) (i.e., "level set of non-independent functions") is opposite to what we intend to study here: leaves would not be manifolds, and even if we work within the context of algebraic geometry (so that these level sets would be affine varieties), there is still a problem: exceptional leaves would be of bigger dimensions than the "regular" ones. We do not claim that such a geometry is not interesting by itself, but this is clearly opposite to what we are looking for.[2] ♡ is used in holomorphic geometry, where a singular foliation on a complex manifold M may be defined as a holomorphic regular foliation on a codimension ≥ 2 analytic subset of M: this definition, however, is essentially equivalent to the holomorphic equivalent of ♣, ◇, ♠ which are essentially equivalent one to the other, when made precise in the right way.

Now, in the smooth case, the three remaining points of views (♣, ◇, ♠) have to be made more precise to yield a reasonable definition of a singular foliation. As we shall see in the first chapter, all of them allow counter-examples to properties that we wish to be true. This does not mean that they have to be rejected, but they have to be made precise.

We may dare to say that after that debate took place in the late 1960s, only two definitions survived to the twenty-first century:

(♠⋆) A singular foliation is a sub-sheaf of the sheaf of vector fields stable under Lie bracket, stable under multiplication by a smooth function, and locally finitely generated as a module over smooth functions.[3]
(♣⋆) A singular foliation is a partition of a manifold into submanifolds called leaves, such that through any vector tangent to a leaf there is at least a vector field tangent to all leaves [17, 36].

---

[2] Jokingly, we say that we intend to study lasagna dishes with a few isolated spaghetti, but we do not wish to study isolated lasagnas in a spaghetti dish.
[3] For those unfamiliar with or hostile to sheaves, this definition can be equivalently stated as: a locally finitely generated involutive sub-$\mathcal{C}^\infty(M)$-module of the module of compactly supported vector fields.

We will work with the first of these definitions, for the following reasons:

1. Definition (♠⋆) implies definition (♣⋆): Singular foliations in the sense of (⋆) do admit leaves which are honest submanifolds and partition the manifold[4] and the henceforth obtained partition satisfies (♣⋆),
2. the tangent spaces of these leaves form a (singular) involutive distribution,
3. it is—according to us—general enough to contain most interesting examples,
4. but it is restrictive enough to be able to prove strong results, while, for instance, singular foliations as in (♣⋆) may not admit an AS-holonomy groupoid (at least, not a longitudinally smooth one).
5. Last, (♠⋆) is used by a now well-established community of non-commutative geometers (Androulidakis, Debord, Mohsen, Skandalis, Yuncken, Skandalis, Zambon—to cite a few) and some theoretical physicists (e.g., Kotov, Strobl), while (⋆ ⋆) seems to be less commonly used nowadays, although it is not abandoned [17, 24, 25].

For all these reasons, we will present the theory of singular foliations using Definition (♠⋆). Although we had no time or space to present it, we claim that these notes would in fact not present a fundamentally different theory had we decided to use Definition (♣⋆).

Are Singular Foliations Worth Studying?

Is there a point in studying singular foliations? It will depend on where you come from and where you want to go.

First, whoever studies Poisson geometry will encounter a highly non-trivial singular foliation: the symplectic leaves of a Poisson structure. But we claim more: whoever understands classical Poisson geometry has understood objects which are more or less analogous to those used in the geometry of singular foliations. Half of the way is behind you.

Below, we listed the classical notions of Poisson geometry on the left, and their equivalent objects in the SF-theory on the left:[5] if you know what the left-hand column is about, understanding the right-hand column should not be overly difficult. Also, we wrote $>, =, <$ to tell which side is, in our subjective opinion, harder to understand.

---

[4] and this is the least we can require to dare calling an object "singular foliation": leaves have to make sense!

[5] We use the abbreviations SF = Singular Foliations and AS = Androulidakis-Skandalis.

Notion in Poisson geometry		The equivalent notion in Singular foliation theory
Poisson manifold (M, π)	=	Singular foliation $\mathcal{F}$ on M
Hamiltonian flows are Poisson diffeo.		Vector fields tangent to $\mathcal{F}$ are symmetries of $\mathcal{F}$
(This is almost trivial)	<<	(This is really hard, at least in the smooth case, Many existing proofs have gaps…)
Weinstein's splitting theorem	=	Singular Foliations' splitting theorem
Partition into symplectic leaves	>	Partition into leaves
Transverse Poisson structure (of a leaf)	=	Transverse singular foliation (of a leaf)
Poisson-Dirac reduction	>	Induced SF on a transverse submanifold
Lie algebroid structure on T^*M	>	(easy) almost Lie algebroid structures generating $\mathcal{F}$
	<<	or (harder) the universal Lie ∞-algebroid of $\mathcal{F}$
Isotropy Lie algebra $\ker \pi_m^\#$ at $m \in M$	=	(easy) isotropy Lie algebra of $\mathcal{F}$ at m
	<<	(harder) isotropy Lie ∞-algebra of $\mathcal{F}$ at m
Poisson cohomology	>	Longitudinal cohomology (easy)
	<<	Cohomology of the universal Lie ∞-algebroid (harder)
Symplectic realization	=	bisubmersions
Morita equivalences	=	Equivalences of bisubmersions
Symplectic Groupoid		AS holonomy groupoid
(This is often a smooth groupoid at worst a stacky groupoid [40])	<<<	(This is almost never a smooth groupoid, not even a stacky groupoid)

In particular, the AS holonomy groupoid is not like any Lie groupoid Poisson geometry has so far produced. Its non-smoothness is at the origin of the subtle analysis developed by non-commutative geometers. Although the AS holonomy groupoid is certainly the most studied aspect of singular foliation at the present time [2–5, 14], we will construct it in detail.

To Which Area of Mathematics Do Singular Foliations Belong to?

As we will see, singular foliations shall be defined as a sub-algebra $\mathcal{F}$ of vector fields, stable under the Lie bracket and under multiplication by a function, and the leaf through a point $m \in M$ shall be the set of points reachable from m following the flows of vector fields in $\mathcal{F}$. Those vector fields in $\mathcal{F}$ are, heuristically, vector fields "tangent to all leaves". But there is an additional assumption in the definition that more or less makes consensus nowadays: we should require $\mathcal{F}$ to be "locally finitely generated". Also, vector fields in

$\mathcal{F}$ are often supposed to be compactly supported (see Definition 2.1). Before dealing with those technical points, we have to address a more fundamental question: in which area of mathematics are we?

The present manuscript is mainly written having in mind the universe of smooth differential geometry. But singular foliations do make sense in real analytic differential geometry, in complex geometry, and in algebraic geometry as well. And we will try to deal with all three aspects altogether. For that purpose, we will use the language of sheaves.[6] More precisely:

1. In real differential geometry, sheaves can be ignored, and singular foliations on a manifold M will be defined as a locally finitely generated sub-$\mathcal{C}^\infty(M)$-module of compactly supported vector fields stable under Lie brackets.
2. In real analytic or holomorphic or algebraic settings, global objects may not exist, or it may be that there are too few of them. One has to work with the sheaf of vector fields, and it does not make sense to consider compactly supported vector fields anymore. Moreover, the "locally finitely generated" condition is equivalent, in this context, to "coherent sheaf".
3. In smooth, real analytic or complex settings, singular foliations induce a partition of M into leaves which are smooth, real analytic or complex submanifolds respectively. This is not true anymore in algebraic geometry: the "leaves" are not algebraic sub-varieties. This is highly related to the well-known fact that the flow of a polynomial vector field is a real-analytic or holomorphic map, but not a polynomial map in general.

Again, although we will deal with real analytic or holomorphic or algebraic settings, we will mostly take the smooth differential geometry point of view. Also, we will assume that the reader knows everything about differential geometry: classical or less-classical theorems about flows of vector fields will often be admitted, and only those specific to singular foliations shall be detailed.

Acknowledgments

The authors would like to thank the organizers of the Poisson 2022 summer school for offering them the opportunity to present an overview of singular foliations, that eventually led to the present manuscript. We extend special thanks to Eva Miranda, and to Cédric Oms for the opportunity to publish these notes in the annals of the summer school.

We acknowledge several discussions with several mathematicians, in particular Georges Skandalis, who suggested several improvements, Iakovos Androulidakis, Camilo Angulo, Claire Debord, Simon Raphael Fischer, Anastasios Fotiadis, Noriaki Ikeda, Jun Jiang,

---

[6] The reader interested only in the smooth case may perfectly ignore the word "sheaf" and replace it by the corresponding global objects. For technical reasons, it is then better to use "compactly supported" objects, or to just replace "the sheaf of" by "the locally defined".

Oleksii Kotov, Sylvain Lavau, Hsuan-Yi Liao, Omar Mohsen, Hadi Nahari, Vladimir Salnikov, Yunhe Sheng, Bernd Stratmann, Thomas Strobl, Robert Yuncken, Chenchang Zhu. Last, we would like to thank Cédric Rigaud, who wrote a master degree memoir that we turned into examples.

All three authors have discussed the present manuscript during their stay at IHP in the summer of 2023.

C.L.-G. would like to thank the Tsing Hua University 國立清華大學 and the NCTS 國家理論科學研究 for their hospitality.

R.L. acknowledges the full financial support provided by Jilin University for his joint postdoctoral position with Göttingen University.

L.R. acknowledges the financial support of the DFG through the grant Higher Lie Theory and the LabEx MiLyon.

Conventions

Throughout the text: manifolds shall be separated and second countable. Vector fields on a manifold M shall be denoted by $\mathfrak{X}(M)$, or simply $\mathfrak{X}$ when there is no ambiguity, but, depending on the context, the previous notation may stand for global vector fields on M or for the sheaf of vector fields. We will always spell out our convention in due place. Sections of a vector bundle E shall be denoted by $\Gamma(E)$, but, again, we will sometimes make no difference between the notation of the sheaf of local sections or global sections. Sections over an open subset $\mathcal{U} \subset M$ of a vector bundle $E \to M$ shall be denoted by $\Gamma_{\mathcal{U}}(E)$. This non-constant notation was chosen to avoid using too heavy symbols coming from sheaf theory all along the text, especially when sheaves are not needed.

We also invite the reader to check the conventions that we use to denote singular foliations, at the end of Sect. 2.2. When we define them, we need to distinguish two notions of singular foliations, respectively denoted by $\mathcal{F}_c$ and $\mathcal{F}_\bullet$. This equivalence allows us to simply denote a singular foliation by an $\mathcal{F}$, and the rest of the text will simply use that notation. Compactly supported vector fields or sections shall be denoted by $\mathfrak{X}_c(M)$ and $\Gamma_c(E)$, respectively. In order to deal with holomorphic, real analytic and smooth settings simultaneously, we will often use the symbol $\mathcal{O}$ for the relevant sheaf of functions. Also, for X a vector field on M or e a section of a vector bundle $E \to M$, we denote by $X_{|m}$ and $e_{|m}$ their values at a point $m \in M$.

Also, to avoid having to repeat "Let M be a manifold equipped with a singular foliation", we will often say "Let $(M, \mathcal{F})$" be a foliated manifold.

Restrictions to an open $\mathcal{U} \subset M$ or "any-mathematical-notion-N-that-restricts" will mostly be denoted by $\mathfrak{i}_{\mathcal{U}}^* N$.

References

1. R. Abraham, J.E. Marsden, Foundations of Mechanics. Benjamin/Cummings Publishing Co., Inc., Advanced Book Program, Reading (1978). Second edition, revised and enlarged, With the

assistance of Tudor Raţiu and Richard Cushman
2. I. Androulidakis, G. Skandalis, The holonomy groupoid of a singular foliation. J. Reine Angew. Math. **626**, 1–37 (2009)
3. I. Androulidakis, G. Skandalis, The analytic index of elliptic pseudodifferential operators on a singular foliation. J. K-Theory **8**(3), 363–385 (2011)
4. I. Androulidakis, G. Skandalis, A Baum-Connes conjecture for singular foliations. Ann. K-Theory **4**(4), 561–620 (2019)
5. I. Androulidakis, M. Zambon, Smoothness of holonomy covers for singular foliations and essential isotropy. Math. Z. **275**(3–4), 921–951 (2013)
6. I. Androulidakis, O. Mohsen, R. Yuncken, A pseudodifferential calculus for maximally hypoelliptic operators and the Helffer-Nourrigat conjecture, arXiv.2201.12060 (2022)
7. C. Araujo, J.P. Figueredo, Foliations on complex manifolds. Not. Amer. Math. Soc. **69**(7), 1137–1144 (2022)
8. P. Baum, R. Bott, Singularities of holomorphic foliations. J. Differ. Geom. **7**, 279–342 (1972)
9. A. Belotto, Global resolution of singularities subordinated to a 1-dimensional foliation. J. Algebra **447**, 397–423 (2016)
10. A. Belotto da Silva, D. Panazzolo, Generalized flow-box property for singular foliations. Rev. R. Acad. Cienc. Exactas Fís. Nat. Ser. A Mat. RACSAM **113**(4), 3949–3965 (2019)
11. D. Cerveau, Distributions involutives singulières. Ann. Inst. Fourier (Grenoble) **29**(3), xii, 261–294 (1979)
12. P. Dazord, Feuilletages à singularités. Indag. Math. **47**, 21–39 (1985)
13. C. Debord, Holonomy groupoids of singular foliations. J. Differential Geom. **58**(3), 467–500 (2001)
14. C. Debord, Longitudinal smoothness of the holonomy groupoid. C. R. Math. Acad. Sci. Paris **351**(15–16), 613–616 (2013)
15. C. Debord, G. Skandalis, Lie groupoids, pseudodifferential calculus, and index theory, in *Advances in Noncommutative Geometry—on the Occasion of Alain Connes' 70th Birthday* (Springer, Cham, 2019), pp. 245–289
16. C. Debord, G. Skandalis, Blow-up constructions for Lie groupoids and a Boutet de Monvel type calculus. Münster J. Math. **14**(1), 1–40 (2021)
17. L.D. Drager, J.M. Lee, E. Park, K. Richardson, Smooth distributions are finitely generated. Ann. Global Anal. Geom. **41**(3), 357–369 (2012)
18. S.-R. Fischer, M.J. Farahani, H. Kim, C. Saemann, Topological classification of symmetry breaking and vacuum degeneracy, arXiv:2406.17634 (2024)
19. G. Frobenius, Ueber das Pfaffsche Problem. J. Reine Angew. Math. **82**, 230–315 (1877)
20. R. Hermann, The differential geometry of foliations. II. J. Math. Mech. **11**, 303–315 (1962)
21. A. Kotov, T. Strobl, Lie algebroids, gauge theories, and compatible geometrical structures. Rev. Math. Phys. **31**(4), 31 (2019). Id/No 1950015
22. S. Lavau, A short guide through integration theorems of generalized distributions. Differ. Geom. Appl. **61**, 42–58 (2018)
23. A. Lichnerowicz, Les variétés de Poisson et leurs algèbres de Lie associées. J. Differ. Geom. **12**, 253–300 (1977)
24. D. Miyamoto, The basic de Rham complex of a singular foliation. Int. Math. Res. Not. **2023**(8), 6364–6401 (2023)
25. D. Miyamoto, Singular foliations through diffeology, in *Recent Advances in Diffeologies and Their Applications. AMS-EMS-SMF Special Session, Université de Grenoble-Alpes, Grenoble, France, July 18–20, 2022* (American Mathematical Society (AMS), Providence, 2024), pp. 139–160

26. I. Moerdijk, J. Mrčun, *Introduction to Foliations and Lie Groupoids*, volume 91 of Camb. Stud. Adv. Math. (Cambridge University Press, Cambridge, 2003)
27. O. Mohsen, Blow-up groupoid of singular foliations. arXiv:2105.05201 (2021)
28. B. Toën, G. Vezzosi, Foliations and stable maps, in *Simons Symposia, Proceedings of the IMSA conference on "Recent Developments in Hodge Theory"* (Springer, 2025), pp. 1–35, arXiv:2202.09174, to appear
29. T. Nagano, Linear differential systems with singularities and an application to transitive Lie algebras. J. Math. Soc. Japan **18**, 398–404 (1966)
30. T. Pantev, B. Toën, Moduli of flat connections on smooth varieties. Algebr. Geom. **9**, 266–310 (2022)
31. B. Scárdua, *Holomorphic Foliations with Singularities. Key Concepts and Modern Results*. Lat. Am. Math. Ser. (Springer; Instituto de Matemática y Ciencias Afines (IMCA), Cham, 2021)
32. S. Sertöz, Residues of singular holomorphic foliations. Compos. Math. **70**(3), 227–243 (1989)
33. P. Stefan, Accessibility and foliations with singularities. Bull. Am. Math. Soc. **80**, 1142–1145 (1974)
34. P. Stefan, Integrability of systems of vector fields. J. Lond. Math. Soc. II. Ser. **21**, 544–556 (1980)
35. H.J. Sussmann, Orbits of families of vector fields and integrability of distributions. Trans. Am. Math. Soc. **180**, 171–188 (1973)
36. H.J. Sussmann, Orbits of families of vector fields and integrability of systems with singularities. Bull. Am. Math. Soc. **79**, 197–199 (1973)
37. T. Suwa, Residues of complex analytic foliation singularities. J. Math. Soc. Japan **36**, 37–45 (1984)
38. B. Toën, G. Vezzosi, Foliations and stable maps (2022)
39. B. Toën, G. Vezzosi, Analytic and algebraic integrability of quasi-smooth derived foliations, arXiv:2305.08212 (2023)
40. H.-H. Tseng, C. Zhu, Integrating Lie algebroids via stacks. Compos. Math. **142**(1), 251–270 (2006)
41. E. van Erp, R. Yuncken, A groupoid approach to pseudodifferential calculi. J. Reine Angew. Math. **756**, 151–182 (2019)
42. A. Weinstein, The local structure of Poisson manifolds. J. Differ. Geom. **18**, 523–557 (1983)

What Is a Singular Foliation?

Camille Laurent-Gengoux, Ruben Louis, and Leonid Ryvkin

1 Naive and Less Naive Attempts of a Definition of a Singular Foliation

In order to understand the geometric ideas behind the consensus definition of a singular foliation, let us make a list of definitions that are natural, but either turned out to be dead ends, or did not yet prevail so far for some reason.

This section is widely inspired by Sylvain Lavau's [45], and by Iakovos Androulidakis and Marco Zambon's [6].

C. Laurent-Gengoux (✉)
Université de Lorraine, CNRS, IECL, Metz, France
e-mail: camille.laurent-gengoux@univ-lorraine.fr

R. Louis
Department of Mathematics, Jilin University, Changchun, Jilin, China

Institut für Mathematik, Georg-August-Universität Göttingen, Göttingen, Germany
e-mail: ruben.louis@mathematik.uni-goettingen.de

L. Ryvkin
Université Claude Bernard Lyon 1, Villeurbanne, France

University of Göttingen, Göttingen, Germany
e-mail: ryvkin@math.univ-lyon1.fr

1.1 Partitionifolds: Is a Singular Foliation Simply a Partition by Smooth Manifolds?

Most differential geometers are used to hear the word "foliation" as referring to what we will call here "regular foliation". Such a "regular" foliation partitions a manifold into submanifolds, all of the same dimension. As a consequence, the most natural idea that comes to mind when trying to make up a definition of a singular foliation is to try to define them as being a disjoint union of submanifolds called "leaves"—now of varying dimension. This perfectly makes sense, but let us give it another name.

Unless otherwise specified, the discussion of this section makes sense in smooth, real analytic, or complex geometry.

> **Definition 1.1 (A First Attempt to Define Singular Foliations:Partitionifolds)**
> Let M be a manifold. A *partitionifold*[a] of M is a partition of M into connected immersed submanifolds,[b] called *leaves*.
>
> ---
> [a] We suggest the word *partitionniété* in French.
> [b] From now on, "submanifold" means by default "immersed submanifold".

Let us introduce a convenient notation.

> **Notation 1.2 (To a Point, We Associate Its Leaf)** *A partitionifold on a manifold M shall be denoted as a map:*
>
> $$L_\bullet \colon M \to \{Submanifolds\ of\ M\}$$
> $$m \mapsto L_m$$
>
> *that maps a point $m \in M$ to the submanifold in the partition to which m belongs. Also, for all $m \in M$, L_m shall be called the leaf through m.*

Below are two examples of partitionifolds that we not wish to allow as being decent "singular foliations".

Example 1.3 We call *bioriented partition* the partitionifold on $\mathbb{R}^2$ given by the following partition

1. the leaf L given by the straight line $\{y = 0\}$, called *central leaf*,
2. the half lines H_y given by $\{(x, y) \mid x \in \mathbb{R}_-^*\}$ for all $y \neq 0$,
3. the half lines V_x^+ given for all $x \geq 0$ by $\{(x, y) \mid y \in \mathbb{R}_+^*\}$, and
4. the half lines V_x^- given for all $x \geq 0$ by $\{(x, y) \mid y \in \mathbb{R}_-^*\}$.

Example 1.4 *The magnetic partition* is an example of partitionifold that behaves badly, although it is "regular" in the sense that all its leaves have the same dimension. It is given as follows:

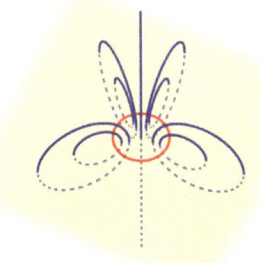

These can be seen as being the lines of a magnetic field generated by an electric current in the red circle, to which the red circle itself is added.

Remark 1.5 The reader used to regular foliations will notice that the Bioriented partition or the Magnetic partition are partitionifolds which are

1. similar to regular foliations in the sense that all leaves are submanifolds all of the same dimension 1,
2. but they are still not regular foliations in a neighborhood of the point $(0, 0)$, respectively the red circle, for the bioriented partition, respectively the magnetic partition.

Example 1.6 French speakers may also look at the Agrégation de Mathématiques of 1998, "Sujet de mathématiques générales": Its first part is dedicated to the construction of a partitionifold on $\mathbb{R}^3$ whose leaves are all circles of non-zero radius. It is of course not a regular foliation.

Example 1.7 *"Isolated lasagna in a dish of spaghettis"*. Consider the partitionifold on $M = \mathbb{R}^3$ with coordinates (x, y, z) whose leaves are defined to be:

1. The plane $z = 0$ (the "isolated lasagna" in red). This is the only leaf of dimension 2.

2. The straight lines parallel to the x-axis (the "spaghettis" in blue) and not contained in the plane $z = 0$.

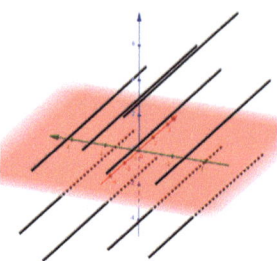

□

To any partitionifold, one can associate a very natural subspace of vector fields, namely those which are tangent to leaves.

> **Notation 1.8 (Vector Fields Tangent to Every Leaf)** Let $L_\bullet$ be a partitionifold on M. We denote by $\mathfrak{T}(L_\bullet) \subset \mathfrak{X}(M)$ the sub-sheaf[a] of vector fields tangent to all leaves, i.e., such that $X_{|\ell} \in T_\ell L_\ell$ for all ℓ in the open subset on which X is defined. We say that such a vector field is *tangent to the partitionifold* $L_\bullet$.
>
> ---
> [a]The reader unfamiliar or hostile to sheaves can define instead $\mathfrak{T}(L_\bullet)$ to be the sub-spaces of vector fields satisfying $X_{|\ell} \in T_\ell L_\ell$ for all $\ell \in M$. It is sometimes convenient to add the assumption "compactly supported" vector fields. Sheaves are only necessary while working within the framework of complex or real analytic geometry.

It is routine to check that $\mathfrak{T}(L_\bullet) \subset \mathfrak{X}(M)$ is a sub-module over the relevant algebra of functions. Also, it is a Lie subalgebra.

Consider a curve $\gamma : I \to M$ whose derivative is tangent to the leaf to which it belongs, i.e., such that for all $t \in I$:

$$\frac{d\gamma(t)}{dt} \in T_{\gamma(t)} L_{\gamma(t)}. \tag{1}$$

Is it true that this curve can not "jump from a leaf to another leaf"? The answer is no, and the next exercises give us counter examples.

Exercise 1.9 Consider the smooth partitionifold on $M = \mathbb{R}$ given by the three subsets

$$\mathbb{R}_-^*, \{0\}, \mathbb{R}_+^*.$$

Show that the curve $t \mapsto t^3$ satisfies (1) but is not contained in a single leaf.

Exercise 1.10 Consider the bioriented partition on $\mathbb{R}^2$ of Example 1.3. Show that the curve

$$t \mapsto \begin{cases} \left(e^{-1/t^2}, 0\right) & \text{if } t \leq 0 \\ \left(0, e^{-1/t^2}\right) & \text{if } t \geq 0 \end{cases}$$

satisfies (1) but is not contained in a single leaf.

The previous exercises show that curves may satisfy Eq. (1) and still jump from leaves to leaves.

But integral curves of vector fields in $\mathfrak{T}(L_\bullet)$ automatically satisfy Eq. (1), but can not jump from a leaf to a leaf, as we now see.

Proposition 1.11 (Not Jumping from Leaves to Leaves) *Let $L_\bullet$ be a partitionifold on M. An integral curve $\gamma(t)$ of a vector field $X \in \mathfrak{T}(L_\bullet) \subset \mathfrak{X}(M)$ tangent to the partitionifold is always contained in one fixed leaf.*

Proof The statement is not obvious. The difficulty is that a vector field X may be tangent to a submanifold $L \in M$, but its flow may not preserve it.[1] What is true, however, if that if a vector field X is tangent to a submanifold L, for any integral curve starting at $t = t_0$ from a point $\ell \in L$ is "locally in L", i.e., there is $\epsilon > 0$ such that $\gamma(t) \in L$ if $|t - t_0| < \epsilon$.

In the present situation, since we are given a vector field X is tangent to $L_\bullet$ at all points, this implies that an integral curve $t \mapsto \gamma(t)$ of X, defined on a connected open interval $I \subset \mathbb{R}$, "locally lies in the same leaf", i.e., for any $t_0 \in I$ there is $\epsilon > 0$ such that $L_{\gamma(t)} = L_{\gamma(t_0)}$ if $|t - t_0| < \epsilon$. Said otherwise, $\gamma^{-1}(L)$ is an open subset of I for any leaf L of $L_\bullet$. Since $L_\bullet$ form a partition of M, I is the disjoint union of the open sets $(\gamma^{-1}(L))_{L \in \mathcal{L}}$ with $\mathcal{L}$ the set of leaves of $L_\bullet$. Since I is connected, there exists a leaf L such that $\gamma^{-1}(L) = I$, i.e., the integral curve of $t \mapsto \gamma(t)$ must be in the same leaf on its full domain.

---

[1] For instance, consider the Euler vector field $X = \sum_{i=1}^n x_i \frac{\partial}{\partial x_i}$ on $M = \mathbb{R}^n$. The time-t flow is a homothety with multiplying factor e^t. In particular, no open ball centered at 0 is preserved under the flow, although they are submanifolds.

For any partitionifold $L_\bullet$ on M, and any open subset $\mathcal{U} \subset M$, a partitionifold on $\mathcal{U}$ is obtained by mapping $m \in \mathcal{U}$ to the connected component of m in $L_m \cap \mathcal{U}$. We denote by $i_\mathcal{U}^* L^\bullet$ this partitionifold and call it *restriction to $\mathcal{U}$* of $L_\bullet$.

Given partitionifolds $L_\bullet$ on M and $L'_\bullet$ on M', we call *isomorphism from $L_\bullet$ to $L'_\bullet$* a diffeomorphism $\phi \colon M \to M'$ such that $\phi(L_m) = \phi(L'_{\phi(m)})$ for all $m \in L$. When $M = M'$ and $L_\bullet = L'_\bullet$, we shall speak of a *symmetry of $L_\bullet$*.

Proposition 1.12 (Flows Are Symmetries) *Let M be a manifold equipped with a partitionifold $L_\bullet$. The flow at time t of a complete vector field $X \in \mathfrak{X}(L_\bullet)$ tangent to $L_\bullet$ is a symmetry of $L_\bullet$.*

More generally, for a maybe non-complete vector field $X \in \mathfrak{X}(L_\bullet)$ tangent to all leaves, its flow ϕ_t^X at time t, provided it is well-defined on some open subset $\mathcal{U} \subset L$, is an isomorphism from the restriction of $L_\bullet$ to $\mathcal{U}$ to the restriction of $L_\bullet$ to $\phi_t^X(\mathcal{U})$.

Proof The first part of Proposition 1.12 is a consequence of the second one. We therefore only prove the second part. Consider two points $m_0, m_1 \in \mathcal{U}$ that are in the same leaf of $i_\mathcal{U}^* L_\bullet$, and therefore in the same leaf L of $L_\bullet$. There is a smooth path $m \colon [0, 1] \to \mathcal{U}$ starting from m_0 and arriving at m_1 which is entirely contained in $L \cap \mathcal{U}$. Since integral curves can not jump from one leaf to another one by Proposition 1.11, for every $u \in [0, 1], s \in [0, t]$, the map $s \mapsto \phi_s^X(m(u))$ is valued in the leaf L. In particular, the curve

$$u \mapsto \phi_t^X(m(u))$$

is entirely contained in L. It is also contained in $\phi_t^X(\mathcal{U})$. Hence, $\phi_t^X(m_0)$ and $\phi_t^X(m_1)$ are in the same leaf of $i_{\phi_t^X(\mathcal{U})}^* L_\bullet$. This proves the claim.

Let $L_\bullet$ be a partitionifold on M. Consider $S \subset M$ a submanifold, we can associate to every $s \in S$ the connected component $(L_s \cap S)_0$ of s in the intersection $L_s \cap S$. The map

$$S \to \{\text{Connected subsets of } S\}$$
$$s \mapsto (L_s \cap S)_0$$

may not be a partitionifold: it is valued in connected subsets, but not in smooth manifolds. However, it is a classical result of differential geometry that if the intersection is clean, i.e., if for all $s \in S$:

$$T_s S + T_s L_s = T_s M, \tag{2}$$

then $L_s \cap S$ is a submanifold for every $s \in S$. It may not be connected, but the connected component $(L_s \cap S)_0$ of $s \in S$ in $L_s \cap S$ is now a non-empty connected submanifold of S. In particular,

$$S \to \{ \text{Connected submanifolds of } S\}$$
$$s \mapsto (L_s \cap S)_0 \tag{3}$$

is a partitionifold on S.

> **Notation 1.13 (How to Denote a Restriction?)** *Let M be a manifold equipped with a partitionifold $L_\bullet$. A manifold S satisfying Condition (2) shall be said to intersect $L_\bullet$ cleanly. We denote by $i_S^* L_\bullet$ the partitionifold on S given by Eq. (3), and call it the restriction of $L_\bullet$ to M*

Remark 1.14 Since open subsets of M intersect cleanly any partitionifold $L_\bullet$ on M, the terminology and notations of Notation 1.13 match the previous conventions.

Let us conclude this question.

> *Question 1.15 (Are Partitionifolds a Good Notion of Singular Foliations?)* No, it is not! It is too weak a notion to satisfy any significant theorem, beside the meager ones mentioned above.

1.2 Smooth Partitionifolds: Is a Singular Foliation a Smooth Partition by Submanifolds?

We now suggest a second notion, denoted as (∗∗) in the introduction, that we claim could be the definition of a singular foliation. In fact, it is used as such by several authors [20, 51, 67]. It is not the most popular definition, but it is a perfectly workable notion.

Of course, these authors call this notion "singular foliation", but for clarity, we prefer to give it another name. Unless otherwise specified, all results of this section are valid on smooth, real analytic and complex manifolds.

Definition 1.16 (A More Subtle Attempt: Smooth Partitionifolds) A partitionifold $L_\bullet$ is said to be *smooth* if for every $\ell \in M$ and every tangent vector $u \in T_\ell L_\ell$, there exists a vector field X through u, defined in a neighborhood $\mathcal{U}$ of m, which is tangent to all leaves.[a]

---

[a] I.e., $X|_m \in T_m L_m$ for all $m \in \mathcal{U}$.

Said differently, a partitionifold is smooth if and only if, for every $\ell \in M$, the evaluation map

$$\mathfrak{X}(L_\bullet) \to T_\ell L_\ell$$
$$X \mapsto X_{|\ell}$$

is a surjective linear map.

Remark 1.17 The word "smooth" in the expression "smooth partitionifold is confusing. We used it by analogy with smooth distributions [20, 64, 65]. But it is confusing because it is defined in the holomorphic or real analytic contexts as well. We kept the name, however, in order to be consistent with existing literature, but also because this confusion happens anyway quite often: a smooth affine variety is not a smooth variety. We hope that it will not cause further confusion.

Let us start by a few non-examples.

Exercise 1.18 Show that neither the "magnetic partition" (Example 1.4) nor the "isolated lasagna in a spaghetti dish" (Example 1.7) are smooth partitionifolds.

□

Exercise 1.19 *"Pinched curves"*. Consider the partitionifold on $\mathbb{R}^2$, with coordinates x, y whose leaves are the graph of the function $f_\lambda : x \mapsto \lambda \left(\text{th}(\sqrt[3]{x}) + 1 \right)$ with $\lambda \in \mathbb{R}$. For each value of λ, the graph of f_λ is a smooth[2] submanifold of dimension 1 in $\mathbb{R}^2$.

---

[2] Even if f_λ is not a smooth function at $x = 0$ for $\lambda \neq 0$, its graph is a smooth submanifold of dimension 1 in $\mathbb{R}^2$.

What Is a Singular Foliation?

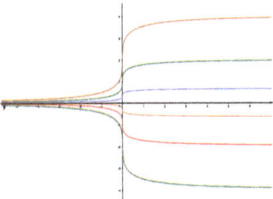

Show that this partitionifold of $\mathbb{R}^2$ is not smooth. *Hint:* consider a neighborhood of $(0, 0)$ and use the fact that the tangent space of $L_{(0,y)}$ is for all $y \neq 0$ a vertical straight line.

For a smooth partitionifold, the following proposition means that singular leaves (not defined yet!) have smaller dimensions than regular ones. This explains why "isolated lasagnas" of Example 1.7 are not smooth partitionifolds.

Proposition 1.20 *Let M be a manifold equipped with a smooth partitionifold $L_\bullet$, the function:*

$$M \to \mathbb{N}_0$$
$$m \mapsto \dim(L_m)$$

is lower semi-continuous.[3]

Proof Let us choose a point $m_0 \in M$, let r be the dimension of the leaf through L_{m_0}, and let $(e_1, \ldots, e_r)$ be a basis of $T_{m_0} L_{m_0}$. By assumption, there exist r vector fields $X_1, \ldots, X_r$ through $(e_1, \ldots, e_r)$, defined in a neighborhood $\mathcal{U}$ of m_0 and tangent to all leaves. They are therefore independent at each point of a sub-neighborhood $\mathcal{U}' \subset \mathcal{U}$, so that $\dim(L_m) \geq r$ for all $m \in \mathcal{U}'$.

Smooth partitionifolds behave much better than partitionifolds, as we will briefly show by giving several reasonable theorems that they satisfy. Their first nice property is that, for a smooth partitionifold $L_\bullet$, along a given leaf L, $L_\bullet$ "always looks locally the same". This idea that a "traveler going along a leaf will be bored" is an important one for singular foliations.

---

[3] I.e., for all $k \in \mathbb{N}_0$, $\{m \in M \mid \dim(L_m) \geq k\}$ is an open subset in M.

Theorem 1.21 (Along a Leaf, Landscape Is Always Identical) *Two points on the same leaf of a smooth partitionifold $L_\bullet$ have neighborhoods where the restrictions of $L_\bullet$ are isomorphic.*

We start with a lemma.

Lemma 1.22 *Given any two points ℓ_0, ℓ_1 on the same leaf L of a smooth partitionifold $L_\bullet$, there exists a finite number of vector fields $X_1, \ldots, X_n \in \mathfrak{T}(L_\bullet)$ (i.e., vector fields tangent to all leaves) such that if we apply successively the flows at time 1 of $X_1, \ldots, X_n$ to the point ℓ_0, we obtain the point ℓ_1.*

Proof Let us say that $x \in M$ and $y \in M$ are reachable one from the other when there exists vector fields as in the lemma. This relation clearly defines an equivalence relation, that we call the reachable relation. It follows from Proposition 1.11 that each equivalence class of the reachable relation is contained in a given leaf of $L_\bullet$. It suffices to check that equivalence classes are precisely the leaves of $L_\bullet$. Since each leaf is a connected set, it suffices to prove that the equivalence classes are open subsets of a leaf. Let $\ell \in L$ be a point. Let $X_1, \ldots, X_k$ be vector fields in $\mathcal{T}_{L_\bullet}$ whose values $e_1 := X_1|_\ell, \ldots, e_k := X_k|_\ell$ generate $T_\ell L$. Consider the map (Here $\mathbb{K}$ stands for the field of real of complex numbers, depending on the context.)

$$\mathbb{K}^k \to L$$
$$(\lambda_1, \ldots, \lambda_k) \mapsto \Phi_1^{\sum_{i=1}^k \lambda_i X_i}(\ell).$$

The differential of this map at the point $(0, \ldots, 0)$ is given for every $(\lambda_1, \ldots, \lambda_k) \in \mathbb{K}^k \simeq T_{(0,\ldots,0)}\mathbb{K}^k$ by

$$(\lambda_1, \ldots, \lambda_k) \mapsto \sum_{i=1}^k \lambda_i e_i$$

In particular, this differential is surjective. By the implicit function theorem, it means that the function is open in L, i.e., that every point in some neighborhood of ℓ is in the same equivalence class as ℓ is. This completes the proof.

Proof of Theorem 1.21 Lemma 1.22 implies that if two points ℓ_0, ℓ_1 are in the same leaves, then there are vector fields $X_1, \ldots, X_r$ tangent to all leaves whose flow at time 1 maps ℓ_0 to ℓ_1. Proposition 1.12 then implies that these flows are isomorphisms of partitionifolds. This completes the proof.

Let L be a leaf of a smooth partitionifold $L_\bullet$. A pointed submanifold (Σ, ℓ) that intersect L at ℓ (i.e., $\Sigma \subset M$ is a submanifold and $\ell \in \Sigma \cap L$) is said to be a $L_\bullet$-*cut* of L if

1. Σ is transverse to L at ℓ, i.e., $T_\ell \Sigma \oplus T_\ell L = T_\ell M$, and
2. Σ intersects $L_\bullet$-cleanly.

Lemma 1.23 *Any pointed submanifold (Σ, ℓ) transverse to L at ℓ admits a neighborhood Σ' of ℓ such that (Σ', ℓ) is a $L_\bullet$-cut of L.*

Proof Let $X^1, \ldots, X^k \in \mathfrak{T}(L_\bullet)$ be vector fields tangent to all leaves whose evaluations at ℓ form a basis of $T_\ell L$. There exists a neighborhood $\mathcal{V}$ of ℓ in Σ such that for all $\sigma \in \mathcal{V}$

$$T_\sigma \Sigma \oplus \langle X^1_{|\sigma}, \ldots, X^k_{|\sigma} \rangle = T_\sigma M.$$

This implies $T_\sigma \Sigma + T_\sigma L_\sigma = T_\sigma M$, which is precisely the definition of intersecting $L_\bullet$ cleanly.

For any $L_\bullet$-cut (Σ, ℓ) of a leaf L, $i_\Sigma^* L_\bullet$ is a smooth partitionifold, that we call a *transverse partitionifold* of the leaf L.

> **Corollary 1.24 (The Germ of a Slice Transverse to a Leaf)** *Let M be a manifold equipped with a smooth partitionifold $L_\bullet$. Any two $L_\bullet$-cuts of a given leaf L have neighborhoods on which their restrictions are isomorphic.*[a]
>
> ---
>
> [a]More precisely, for any two pointed submanifolds (Σ_1, ℓ_1) and (Σ_2, ℓ_2) transverse to the same leaf L, there exists neighborhoods $\mathcal{U}_1 \subset \Sigma_1, \mathcal{U}_2 \subset \Sigma_2$ of ℓ_1, ℓ_2 and an isomorphism
>
> $$i^*_{\Sigma_1 \cap \mathcal{U}_1} L_\bullet \xrightarrow{\sim} i^*_{\Sigma_2 \cap \mathcal{U}_2} L_\bullet.$$

Proof This comes from the fact that the vector fields $X_1, \ldots, X_d \in \mathfrak{T}(L_\bullet)$ in Lemma 1.22 can be chosen such that the composition of their flows (which are symmetries of $L_\bullet$) not only map ℓ_1 to ℓ_2 but also Σ_1 to Σ_2, at least in a neighborhood of ℓ_1. A similar argument is presented in the proof of Theorem 8.6: we refer the reader to that proof for more details.

Corollary 1.24 implies that it makes sense to speak of the transverse model of a leaf. To be more precise, it can be defined as follows. Let d be the dimension of the manifold and

r the dimension of the leaf L We call *representative of the transverse partitionifold of L* a pair $(\mathcal{U}, L^{\mathcal{U}})$ with $\mathcal{U}$ a neighborhood of 0 such that there exists a L-cut (Σ, ℓ) and an isomorphism isomorphic $(\Sigma, i_\Sigma^* L_\bullet) \simeq (\mathcal{U}, L^{\mathcal{U}})$ mapping ℓ to 0. More precisely, consider pairs $(\mathcal{U}, L^{\mathcal{U}})$ with $\mathcal{U}$ a neighborhood of 0 $\mathbb{K}^s$, and then identify two such pairs

$$(\mathcal{U}, L^{\mathcal{U}}) \sim (\mathcal{V}, L^{\mathcal{V}})$$

if they have neighborhoods $\mathcal{U}'$ and $\mathcal{V}'$ of 0 on which the restricted smooth partitionifolds are isomorphic, through an isomorphism that preserves 0. We call *germs of partitionifolds at 0 in dimension s* the equivalence classes of this equivalence relation. To any leaf L of a partitionifold, one associates a canonical germ of partitionifolds at 0 in dimension $d - k$ with $d = \dim(M)$ and $k = \dim(L)$: It is by definition the unique class which admits a representative isomorphic, as a smooth partitionifold, to the smooth partitionifold of one $L_\bullet$-cut of L. We will have very similar theorems for in Sect. 8.1 for the definition of singular foliation that we will choose.

A classical result of classical theorem in Control Theory called Nagano-Sussmann Theorem, which is a continuation of Rashevsky–Chow Theorem, see Chapter 5 in [2], justifies the introduction of smooth partitionifolds.

Let us describe it. Given a finite family of d vector fields $X_\bullet := X_1, \ldots, X_d$ on a manifold M, consider the equivalence relation on M generated by $m \sim m'$ if m and m' are on the same integral curve of $X_1, \ldots, X_d$. Two points in the same equivalence class of this relation are said to be *reachable one from the other*, and we call this class the *reachable equivalence relation*. Let us construct this class differently. Assume that the vector fields $X_1, \ldots, X_d$ are complete for the sake of simplicity. Consider the subgroup $\mathrm{Diff}_{X_\bullet}(M)$ of the group $\mathrm{Diff}(M)$ of diffeomorphisms of M generated by the flows at time $t \in \mathbb{R}$ of the vector fields $X_1, \ldots, X_d$. In equation, a diffeomorphism ϕ of M belongs to $\mathrm{Diff}_{X_\bullet}(M)$ if there exists an integer $n \geq 1$, a n-tuple $i_1 \ldots, i_n \in \{1, \ldots, d\}$ and a n-tuple $t_1 \ldots, t_n \in \{1, \ldots, d\}$ such that

$$\phi = \Phi_{X_{i_1}}^{i_1} \circ \ldots, \circ \Phi_{X_{i_n}}^{i_n}.$$

When the vector fields are not complete, $\mathrm{Diff}_{X_\bullet}(M)$ becomes what is called a pseudo-group, i.e., pairs made of an open subset of M and a diffeomorphism defined on that open subset. In any case, it makes sense to speak of the orbits of $\mathrm{Diff}_{X_\bullet}(M)$: two points $m, m' \in M$ belong to the same orbit of $(X_1, \ldots, X_d)$ if there exists $\phi \in \mathrm{Diff}_{X_\bullet}(M)$ such that $\phi(m) = m'$. The orbits are precisely the equivalence classes of the previous equivalence relation.

Theorem 1.25 (Nagano–Sussmann in Terms of Smooth Partitionifolds) *Let $X_\bullet = (X_1, \ldots, X_d)$ be a finite family of vector fields on a manifold M. The equivalence classes of the reachable equivalence relation above form a smooth partitionifold $L_\bullet$ of M.*

Nagano–Sussmann's theorem [55,66] says in fact something more. We say that a subspace $\mathcal{G} \subset \mathfrak{X}(M)$ of vector fields is *completely tangent to a smooth partitionifold $L_\bullet$* if for every $m \in m$, and every $u \in T_m L_m$ tangent to the leaf through L at m, there exists $X \in \mathcal{G}$ such that $X(m) = u$. Consider the vector fields obtained by considering:

1. the Lie subalgebra $\text{Lie}(X_\bullet)$ of $\mathfrak{X}(M)$ generated by the vector fields in $X_\bullet$, i.e., the space generated by all vector fields of the form

$$[\cdots [X_{i_1}, X_{i_2}] \ldots, X_{i_n}]$$

for all $n \geq 1, i_1, \ldots, i_n \in \{1, \ldots, d\}$
2. then consider all possible push-forwards through all the diffeomorphisms $\phi \in \text{Diff}_{X_\bullet}(M)$ of vector fields in $\text{Lie}(X_\bullet)$,
3. then all possible linear combinations, with coefficients in functions, of the previously defined vector fields.

In the real analytic or holomorphic setting, step 2 is not required. By proceeding as in steps 1,2,3 before, we obtain a sub-sheaf of $\mathfrak{X}(M)$ that we denote by $\text{Closure}(X_\bullet)$. It is a module over functions, and it is closed under Lie bracket. It is of course included into $\mathfrak{T}(L_\bullet)$.

Remark 1.26 In general, the inclusion $\text{Closure}(X_\bullet) \subset \mathfrak{T}(L_\bullet)$ is strict: take for instance on $M = \mathbb{R}$ the vector field $X_\bullet$ to be the family with one element, namely $X_1 = x^2 \frac{\partial}{\partial x}$. $\text{Closure}(X_\bullet)$ is the $\mathcal{C}^\infty(M)$-module generated by X_1 while $\mathfrak{T}(L_\bullet)$ is made of all vector fields on $\mathbb{R}$ vanishing at 0.

Nagano-Sussmann theorem is of course an extremely strong result, and would be an excellent reason to work with smooth partitionifolds as a definition of singular foliations. We will however not choose it for reasons that we discuss now.

Question 1.27 (Conclusion: Are Smooth Partitionifolds a Good Definition of Singular Foliations?) Our opinion is that it is fine as a definition. They are called "singular foliations" by several authors, even recently [51]. The Nagana-Sussmann Theorem above is an excellent reason to use them. But it not the definition which is most commonly used at the moment. We will see that they are not practical to deal with differential operators, which explains why non-commutative geometers do not use these. Most importantly, the theory one could develop out of smooth partitionifolds would not be so different from the one we will develop in Sect. 2: they are different, but parallel.

In addition, here is an oddity that we want our "singular foliations" to avoid, presented as an exercise. In Sect. 2, we will explain why having non-locally finitely generated modules is an issue.

Exercise 1.28 *"Vector fields tangent to the leaves are not finitely generated"*. On $M = \mathbb{R}$, consider the partitionifold whose 0-dimensional leaves are $\{1\}, \{\frac{1}{2}\}, \{\frac{1}{3}\}, \ldots, \{\frac{1}{n}\}, \ldots$ and $\{0\}$ and whose 1-dimensional leaves are the open intervals bounded by these points. Show that vector fields tangent to L are not a finitely generated module over $\mathcal{C}^\infty(M)$, and that there is no neighborhood $\mathcal{U}$ of 0 on which such vectors form a locally finitely generated $\mathcal{C}^\infty(\mathcal{U})$-module.

1.3 Is a Singular Foliation an Involutive Distribution?

Before presenting the definition of a singular foliation that we intend to use, let us introduce an alternative manner to define partitionifolds of non-constant dimensions. It is extremely classical in differential geometry [8, 53] that a regular foliation may be defined as being an integrable sub-vector bundle $D \subset TM$. It is therefore tempting to allow the fibers of the vector bundle D to be of non-constant dimension, as long as its sections are closed under the Lie bracket of vector fields:

Definition 1.29 (Integrable Singular Distributions) A singular distribution on a manifold M is a map $\mathfrak{D}$ associating to a point $m \in M$ a subspace $\mathfrak{D}_m \subset T_m M$. A singular distribution $\mathfrak{D}$ is said to be:

1. *involutive* when $[\Gamma(\mathfrak{D}), \Gamma(\mathfrak{D})] \subset \Gamma(\mathfrak{D})$, where $\Gamma(\mathfrak{D}) \subset \mathfrak{X}(M)$ is the $\mathcal{C}^\infty(M)$-module of vector fields X such that $X_m \in \mathfrak{D}_m$ for all $m \in M$.
2. *integrable* when there exists a partitionifold $L_\bullet$ such that for all $m \in M$, $T_m L_m = \mathfrak{D}_m$.

Exercise 1.30 Show that for any partitionifold $L_\bullet$ on M, the map

$$\mathfrak{D} : m \mapsto T_m L_m$$

is an involutive and integrable singular distribution.

Exercise 1.31 Let M be a manifold and m_0 a point. Show that the map

$$m \mapsto \begin{cases} T_{m_0} M & \text{if } m = m_0 \\ 0_{T_m M} & \text{if } m \neq m_0 \end{cases}$$

is an involutive but non-integrable singular distribution. (Here, 0_E stands for the zero element of a vector space E.)

For a given singular distribution, what would be the proper definition of a leaf? There is a natural manner to define leaves for a singular distribution $\mathfrak{D}$. Consider the equivalence relation on M generated by the relation $x_0 \sim x_1$ if there exists a path of class C^1 such that

$$x(0) = x_0, x(1) = x_1 \text{ and } \frac{d}{dt}x(t) \in \mathfrak{D}_{x(t)} \text{ and } \frac{d}{dt}x(t) \neq 0. \tag{4}$$

Equivalently, one could define an equivalence relation as follows: call integral submanifold of $\mathfrak{D}$ a submanifold Σ such that $T_\sigma \Sigma \subset \mathfrak{D}_\sigma$ for all $\sigma \in \Sigma$. We could then consider the equivalence relation generated by the relation $x_0 \sim x_1$ if there exists an integral submanifold containing both x_0 and x_1. The classes of this equivalence relation can not decently be called leaves, because they are not submanifolds, as seen in the following Exercise.

Exercise 1.32 Here is an example (the "trumpet foliation") of an involutive singular distribution for which one class of the equivalence definition (4) is not a manifold. Take $M = \mathbb{R}^2$ with coordinates (x, y). Let $k(x) = e^{-1/x}$ for $x > 0$ and $k(x) = 0$ for $x \leq 0$. Divide $\mathbb{R}^2$ in three zones:

$$North := \{y \geq k(x)\}, Middle := \{x > 0 \text{ and } -k(x) < y < k(x)\},$$
$$South := \{y \leq -k(x)\}$$

Define a singular distribution by:

$$\mathfrak{D}_m = \begin{cases} \langle (1, k'(x)) \rangle & \text{for } m \in North \\ T_m \mathbb{R}^2 & \text{for } m \in Middle \\ \langle (1, -k'(x)) \rangle & \text{for } m \in South \end{cases}$$

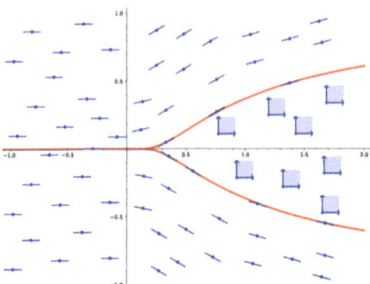

1. Show that $\mathfrak{D}$ is involutive
2. Show that $\mathfrak{D}$ is not integrable (*Hint* Show that the equivalence class of $(0, 0)$ is $\{y = 0\} \cup \overline{Middle}$, which is not a manifold).

It is clear that we have to avoid situations like the one in Exercise 1.31, as well as the tangent spaces of the partitionifolds of Examples 1.4 and 1.7 ("magnetic foliation" or "isolated lasagnas"). For that purpose, we will impose a second condition, similar to Definition 1.16 of smooth partitionifolds.

Definition 1.33 (Smooth Singular Distributions) A singular distribution $\mathfrak{D}$ is said to be smooth if for every point $m \in M$ and $u \in \mathfrak{D}_m$, there exists a vector field $X \in \Gamma(\mathfrak{D})$ through u.

Exercise 1.34 Let $L_\bullet$ be a smooth partitionifold of M. Consider the singular distribution $\mathfrak{D}_L : m \mapsto T_m L_m$.

1. Show that it is integrable and involutive,
2. and smooth
3. and that the flow of any section in $\Gamma(\mathfrak{D}_L)$ preserves $\mathfrak{D}_L$.

Exercise 1.35 For an involutive and integrable smooth distribution, show the classes of the equivalence relation (4) are precisely the leaves of $L_\bullet$.

The two exercises above seem to indicate that smooth involutive singular distributions are a "good" notion. There is however a type of example which is quite annoying:

Exercise 1.36 Here is a non-integrable distribution, the *"infinite comb"*, that will be a source of several counter-examples. Consider on $M = \mathbb{R}^2$ with variables (x, y) the singular distribution given by

$$\mathcal{D}_{(x,y)} = \begin{cases} \langle \frac{\partial}{\partial x} \rangle & \text{if } x \leq 0 \text{ i.e., "Dimension 1 in the black zone - and horizontal."} \\ \langle \frac{\partial}{\partial x}, \frac{\partial}{\partial y} \rangle & \text{if } x > 0 \text{ i.e., "Dimension 2 in the red zone."} \end{cases} \tag{5}$$

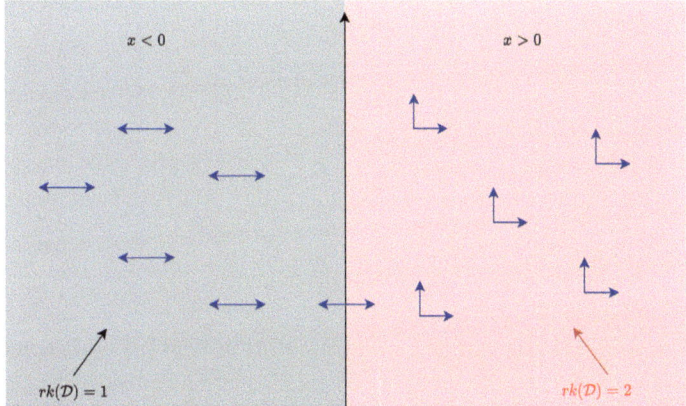

1. Show that the singular distribution $\mathcal{D}$ is smooth.
2. Show that the singular distribution $\mathcal{D}$ is involutive.
3. Show that any two points in $\mathbb{R}^2$ are in the same equivalence class of the relation (4).
4. Show that it is not integrable.

The last exercise shows that smooth involutive singular distributions may not be integrable. Another issue of the notion of smooth integrable distribution is that the flow of a complete vector field in $\Gamma(\mathcal{D})$ may not be a symmetry of $\mathcal{D}$. Again, the infinite comb is a counter-example:

Exercise 1.37 Show that the vector field $\frac{\partial}{\partial x}$ belongs to $\mathcal{D}$ but that its flow does not preserve $\mathcal{D}$.

In the context of smooth differential geometry, Stefan-Sussmann theorems are a way out of these counter-examples.

Theorem 1.38 (Stefan-Sussmann Theorem (1973) [20, 64–67]) Let $\mathfrak{D}$ be an involutive smooth distribution on a smooth manifold M. The following items are equivalent:

(i) $\mathfrak{D}$ is integrable.[a]
(ii) There exists a $C^\infty(M)$-module $\mathcal{F}$ of vector fields such that:
  (a) $\mathcal{F}$ generates $\mathfrak{D}$.[b]
  (b) The flow ϕ_t^X of any vector field $X \in \mathcal{F}$ preserves $\mathfrak{D}$.[c]

---

[a] I.e., there exists a smooth partitionifold $L_\bullet$ such that $T_m L_m = \mathfrak{D}_m$ at all points $m \in M$.
[b] I.e., for all $m \in M$ and $u \in \mathfrak{D}_m$, there exists $X \in \mathcal{F}$ with $X_{|m} = u$.
[c] I.e., $\phi_t^X(\mathfrak{D}_m) = \mathfrak{D}_{\phi_t^X(m)}$ for all $m \in M$, $X \in \mathcal{F}$, $t \in \mathbb{R}$ for which the flow is well-defined in a neighborhood of m.

In [20], it is proven that one can add a condition (c) in item *(ii)* of Theorem 1.38: one can also impose that $\mathcal{F}$ is locally finitely generated, and the equivalence of items *(i)* and *(ii)* still holds true. In general, however, we cannot assume $[\mathcal{F}, \mathcal{F}] \subset \mathcal{F}$, so $\mathcal{F}$ can not be assumed to be what we will call a "singular foliation" in the next section.

Let us conclude this section:

Question 1.39 (Are Involutive Smooth Singular Distributions a Good Notion of Singular Foliations?) No, they are not!

Foliations should have leaves, and there is an issue with the notion of leaves. See the discussion about infinite combs, which are not integrable.

However, with an additional condition on flows, Stefan-Sussmann Theorem 1.38 grants integrability. But this condition is hard to check in a concrete manner.

2 Singular Foliations Through Vector Fields: A Consensus Definition?

We now introduce the definition of singular foliations that we will use effectively. Essentially, the singular foliations will be for us a subspace $\mathcal{F}$ of the space of vector fields, a subspace supposed to "behave like" vector fields tangent to the leaves of a partition of M by sub-varieties. This is the view that seems to have prevailed for some time. It certainly predominates today if we count the number of publications. It is therefore tempting to say that there is a growing consensus around the definition that we present in the lines below. However, it is not universally used, see e.g. Miyamoto's recent Ph.D. [51,52]. Regardless, this vector field view is not *that* different from the "smooth partitionifold" view: both theories are in some way parallel. Although they are not so easily comparable, in the sense that there is no functorial relationship between them, and notably no inclusion, they are parallel in the sense that both theories have the same shape. There is a natural way to translate a theorem using our definition into a theorem about smooth partitionifolds, and the resulting statement is most likely true.

Note that if what we call a singular foliation is only a subspace of the space of vector fields satisfying several conditions, the leaves are not a priori given in the definition. The existence of leaves will now be a theorem, and in fact a very non-trivial theorem going back to Hermann's Theorem 7.8.

We will present the smooth setting apart from the real analytic and holomorphic ones, although there is indeed a manner to merge them all. We will also present a purely algebraic framework, which defines singular foliations on an affine variety, or even on an arbitrary commutative algebra.

2.1   Definition of a Singular Foliation: The Smooth Case

We now present a definition which has become the object of consensus in non-commutative geometry, see e.g. [3–5, 7, 17, 54] by Androulidakis, Debord, Mohsen, Skandalis, Yuncken, and Zambon - to quote a few. This branch of non-commutative geometry works essentially within the context of <u>smooth</u> differential geometry, so we start in this context.

Recall that the *support* of a smooth function or a smooth vector field on a manifold M is the closure of the subset of M of points where it is not equal to zero. We denote by $\mathcal{C}_c^\infty(M)$ and $\mathfrak{X}_c(M)$ the space of compactly supported smooth functions and vector fields, respectively. Recall that $\mathfrak{X}_c(M)$ is a module over the algebra $\mathcal{C}^\infty(M)$.

Definition 2.1 (Singular Foliations on Smooth Manifolds) A singular foliation on a smooth manifold M is a subspace $\mathcal{F} \subset \mathfrak{X}_c(M)$ which

(α) is involutive,
(β) is stable under multiplication by an element of $\mathcal{C}^\infty(M)$,
(γ) is locally finitely generated.

Remark 2.2 A subspace $\mathcal{F} \subset \mathfrak{X}_c(M)$ that satisfies only conditions (α) and (β) is a *Lie-Rinehart subalgebra*[4] of the Lie-Rinehart algebra of compactly supported vector fields. (Please ignore this remark is you never heard of Lie-Rinehart algebras).

2.1.1 Definition 2.1 Explained Item by Item

(α) "$\mathcal{F}$ is involutive" means that there is an inclusion

$$[\mathcal{F}, \mathcal{F}] \subset \mathcal{F}$$

where $[\cdot, \cdot]$ stands for the bracket of vector fields. In words, condition (α) means that $\mathcal{F}$ is a sub-Lie algebra of the Lie algebra $\mathfrak{X}_c(M)$ of compactly supported smooth vector fields on the manifold M.

(β) "$\mathcal{F}$ is stable under multiplication by an element of $\mathcal{C}^\infty(M)$" means that for all $F \in \mathcal{C}^\infty(M), X \in \mathcal{F}, FX \in \mathcal{F}$. In algebraic terminology, it means that $\mathcal{F}$ is a $\mathcal{C}^\infty(M)$-sub-module of the $\mathcal{C}^\infty(M)$-module $\mathfrak{X}_c(M)$ of compactly supported vector fields on M.

(γ) The meaning of "$\mathcal{F}$ is locally finitely generated" has to be made very precise. It means that for any point $m \in M$, there exists a finite family $X^1, \ldots, X^r \in \mathcal{F}$ and an open neighborhood $\mathcal{U}$ of x in M such that for every $X \in \mathcal{F}$, there exist $f_1, \ldots, f_r \in \mathcal{C}^\infty(M)$ satisfying for all $x \in \mathcal{U}$:

$$X_{|x} = \sum_{i=1}^{r} f_i(x) \, X^i_{|x}.$$

Exercise 2.3 Show that in Definition 2.1, $\mathcal{C}^\infty(M)$ could be replaced by $\mathcal{C}^\infty_c(M)$ without making any difference.

2.1.2 Definition 2.1 Justified Item by Item

Now that we have explained the meaning of the three items, let us justify these items. Recall that the general idea is that $\mathcal{F}$ shall behave like "vector fields tangent to the leaves".

To start with, why do we use the space $\mathfrak{X}_c(M)$ of compactly supported vector fields? We would first like to tell the reader that the reasons are above all technical, that there is nothing very deep in this choice. Let M be a non-compact manifold. The following spaces:

(a) compactly supported vector fields on M,
(b) all smooth vector fields on M (compactly supported or not),

---

[4] In fact, generic Lie-Rinehart subalgebras of vector fields behave very badly in the smooth case and there is not much to say about them. They are more interesting in the complex case. But we need the concept for pedagogical reasons, in particular in order to explain why we impose condition γ.

are *different* as modules over $C^\infty(M)$. But both of them should define the *same* singular foliation: we do not wish to distinguish them. They obviously have the same leaf: M itself (leaves have not been defined, but it is obvious in this case that M is the only leaf, whatever it means). If we did not impose "compactly supported", then we would have to say "and we will identify $\mathcal{F}$ with the subspace of compactly supported vector fields in $\mathcal{F}$". To avoid that two different subspaces of $\mathfrak{X}(M)$ define the same singular foliation, one directly requires that $\mathcal{F} \subset \mathfrak{X}_c(M)$.

One can get rid of all compactness assumption by using another possible definition that involves sheaves, as we will see in the next section.

1. Why do we assume (α), i.e., that $\mathcal{F}$ is involutive? If two vector fields X, Y are tangent to a submanifold L, so is its bracket. Since $\mathcal{F}$ must be thought of as being a replacement of vector fields tangent to the leaves, it makes sense to require $[\mathcal{F}, \mathcal{F}] \subset \mathcal{F}$.
2. Why do we assume (β), i.e., that $\mathcal{F}$ is a $C^\infty(M)$-module? Because, if a vector field X is tangent to all leaves, so is fX for all smooth functions $f \in C^\infty(M)$.
3. Why do we assume (γ), i.e., that $\mathcal{F}$ is "locally finitely generated"? The idea is to avoid weird counter-examples as the infinite comb (see Example 1.36). Imposing locally finitely generated guaranties that leaves will make sense. This is the topic of a subsequent section (Sect. 7).

Exercise 2.4 Show that the space of vector fields on $\mathbb{R}$ of the form $f(x)\frac{\partial}{\partial x}$ with f a function vanishing with all its derivatives at 0 satisfy items (α) and (β) by not item (γ). (*Hint:* Look at second item in Exercise 4.2).

Exercise 2.5 We call[5] *infinite comb*[6] the singular foliation on $\mathbb{R}^2$ made of all vector fields of the form

$$F(x, y)\frac{\partial}{\partial x} + G(x, y)\frac{\partial}{\partial y}$$

where $G(x, y)$ is a function with compact support in the half plane $\{(x, y) | x \geq 0\}$ and $F(x, y)$ is a function with compact support. Here (x, y) are the coordinates on $\mathbb{R}^2$. Show that these vector fields satisfy items (α) and (β) but not item (γ).

The next exercise is crucial, for quite a few singular foliations are defined as families $X_1, \ldots, X_r$ that satisfy one of the equivalent conditions listed there.

Exercise 2.6 Let M be a manifold, and let $X_1, \ldots, X_r \in \mathfrak{X}(M)$ be vector fields. Show that the following three items are equivalent:

---

[5] We owe this example to Thomas Strobl.
[6] This is if course related to Exercise 1.36, which describes the singular distribution obtained by taking the distribution associated to these vector fields.

(i) The $C_c^\infty(M)$-module[7] generated by $X_1, \ldots, X_r$ is a singular foliation,
(ii) There exist smooth functions $c_{i,j}^k \in C^\infty(M)$, with $i, j, k \in \{1, \ldots, r\}^3$, such that

$$[X_i, X_j] = \sum_{k=1}^n c_{ij}^k X_k \qquad (6)$$

for all $i, j \in 1, \ldots, r$.
(iii) There exists smooth functions $\tilde{c}_{i,j}^k \in C^\infty(M)$, with $i, j, k \in \{1, \ldots, r\}^3$ satisfying

$$\tilde{c}_{ij}^k = -\tilde{c}_{ji}^k \text{ and } [X_i, X_j] = \sum_{k=1}^n \tilde{c}_{ij}^k X_k \qquad (7)$$

for all possible indices.

Hint: Show that if a family c_{ij}^k of functions satisfies (6), then $\tilde{c}_{ij}^k := \frac{1}{2}\left(c_{ij}^k - c_{ji}^k\right)$ satisfies (7).

Definition 2.7 A singular foliation on M as in Exercise 2.6 is said to be *finitely generated*.

These singular foliations will be studied in detail in Sect. 2.5

2.1.3 Restriction to an Open Subset

Let $\mathcal{F}$ be a singular foliation on a manifold M, with $\mathcal{F}$ as in Definition 2.1. Let $\mathcal{U} \subset M$ be an open subset. Four natural spaces can be associated to it:

- ♣ The subspace of $\mathfrak{X}_c(\mathcal{U})$ of vector fields obtained by restriction to $\mathcal{U}$ of vector fields in $\mathcal{F}$ whose support lies within $\mathcal{U}$.
- ♢ The subspace of $\mathfrak{X}(\mathcal{U})$ of vector fields obtained by restriction to $\mathcal{U}$ of vector fields in $\mathcal{F}$.
- ♡ The $C^\infty(\mathcal{U})$-module generated by elements in ♢.
- ♠ The subspace of all vector fields $\mathfrak{X}(\mathcal{U})$ that coincide with a vector field in $\mathcal{F}$ in a neighborhood of every point $m \in \mathcal{U}$.

Remark 2.8 ♠ can alternatively be defined as the subspace of all vector fields $X \in \mathfrak{X}(\mathcal{U})$ such that for every compactly supported function f with support in $\mathcal{U}$, the vector field fX, extended by zero outside $\mathcal{U}$, in an element in $\mathcal{F}$.

Lemma 2.9 *There is a chain of inclusions:*

$$♣ \subset ♢ \subset ♡ \subset ♠.$$

---

[7] To be more explicit, this item means that the space of vector fields of the form $\sum_{i=1}^r f_i X_i$, with $f_1, \ldots, f_r$ compactly supported smooth functions on M, is a singular foliation.

Proof The only non-trivial inclusion is $\heartsuit \subset \spadesuit$. A vector field $Y \in \mathfrak{X}(\mathcal{U})$ lies in $\heartsuit$ if it reads

$$Y = \sum_{i=1}^{n} f_i X_i$$

with $X_i \in \mathcal{F}$ and $f_i \in C^\infty(\mathcal{U})$. Choose $m \in \mathcal{U}$. For each $i = 1, \ldots, n$, there exists a function $\tilde{f}_i \in C^\infty(M)$ that coincides with f_i on some neighborhood V_i of m in $\mathcal{U}$, but is now defined on the whole manifold M. The vector field $\sum_{i=1}^{n} \tilde{f}_i X_i$ coincides with Y on the neighborhood $\cap_{i=1}^{n} V_i$ of m. By construction, it belongs to $\mathcal{F}$. The vector field Y therefore belongs to $\spadesuit$.

Exercise 2.10 Show that for $M = \mathbb{R}$, $\mathcal{F} = \mathfrak{X}_c(M)$, and $\mathcal{U} = \mathbb{R}\setminus\{0\}$, each one of the inclusions in Lemma 2.9 is a strict inclusion.

Hint: Show that the support of every vector field in $\heartsuit$ is contained in $[-R, R]\setminus\{0\}$ for some $R > 0$.

Exercise 2.11 Show that $\clubsuit$ is a singular foliation on $\mathcal{U}$.

Exercise 2.12 Show that a $C^\infty(M)$-module $\mathcal{F} \subset \mathfrak{X}_c(M)$ stable under Lie bracket is a singular foliation (i.e., is finitely generated in the sense of Definition 2.1) if and only if every point admits a neighborhood in which $\clubsuit$ is a finitely generated singular foliation (see Definition 2.7).

Exercise 2.13 Let $\mathcal{F}$ be a globally finitely generated singular foliation with generators $X_1, \ldots, X_r$ as in Definition 2.7. Show that for every $\mathcal{U} \subset M$:

- $\clubsuit$ is the space of vector fields of the form $\sum_{i=1}^{r} f_i X_i$ with f_i smooth functions on $\mathcal{U}$ with compact support in $\mathcal{U}$.
- $\diamondsuit$ is the space of vector fields of the form $\sum_{i=1}^{r} f_i X_i$ with f_i smooth functions on $\mathcal{U}$ obtained by restriction of a compactly supported smooth function on M.
- $\heartsuit$ is the space of vector fields of the form $\sum_{i=1}^{r} f_i X_i$ with f_i smooth functions on $\mathcal{U}$ whose support is included into a compact subset of M.
- $\spadesuit$ is the space of vector fields of the form $\sum_{i=1}^{r} f_i X_i$ with f_i smooth functions on $\mathcal{U}$.

Warning: None of the four items are trivial, except the second one! For the last one, the proof goes as follows. Let X be in $\spadesuit$. For every $m \in \mathcal{U}$, there exists a r-tuple $f_1^r, \ldots, f_r^m$ in compactly supported functions on M such that X and $\sum_{j=1}^{r} f_j^r X_j$ coincide on a neighborhood $\mathcal{U}_m$ of m in $\mathcal{U}$. One extracts out of the open cover $(\mathcal{U}_m)_{m \in M}$ a sub-open-cover $(\mathcal{U}_{m_i})_{i \in I}$ such that a partition of unity $(\chi_i)_{i \in I}$ exists, and such that each $m \in \mathcal{U}$ has a neighborhood that belongs to finitely many of the open subsets $(\mathcal{U}_{m_i})_{i \in I}$. The vector fields

$\sum_{j=1}^{r}(\sum_{i\in I}\chi_i f^i_j)X_j$ and X coincide on the whole open subset $\mathcal{U}$. Since the function $\sum_{i\in I}\chi_i f^i_j$ is a smooth function on $\mathcal{U}$ for $j = 1,\ldots, r$, this completes the proof.

Exercise 2.14 Is it true that a vector field in ♠ which is compactly supported belongs to ♣?

It is therefore natural to ask which one of the four spaces ♣, ◊, ♡, ♠ is the "good" restriction of $\mathcal{F}$ to an open subspace $\mathcal{U}$. While the answer depends on the context, the one that we will need soon is ♠. Hence, we will change the notation:

Definition 2.15 Let $\mathcal{F}$ be a singular foliation on M. For every $\mathcal{U} \subset M$, we will denote by $\mathcal{F}(\mathcal{U})$ the subspace of vector fields on $\mathcal{U}$ that coincide with a vector field of $\mathcal{F}$ is a neighborhood of every point, i.e., the space denoted by ♠ in the above lines.

Exercise 2.16 Let $\mathcal{F}$ be a singular foliation on M, defined as in Definition 2.1. Show that by associating to an open subset $\mathcal{U} \subset M$ the space $\mathcal{F}(\mathcal{U})$, one defines a sheaf on M. Show that it is both a sheaf of Lie algebras, and a module over the sheaf of smooth functions on M.

We conclude this discussion by a technical but important lemma.

Lemma 2.17 *Let $\mathcal{F}$ be a singular foliation. For every point $m \in M$, there exist an open neighborhood $\mathcal{U}$ of m in M and $X_1, \ldots, X_r \in \mathcal{F}$ such that for any $\mathcal{V} \subset \mathcal{U}$, $\mathcal{F}(\mathcal{V})$ is generated over $\mathcal{C}^\infty(\mathcal{V})$ by the restriction to $\mathcal{V}$ of $X_1, \ldots, X_r$.*

Proof The proof is similar to the proof of item ♠ in Exercise 2.13. By Definition 2.1, there exists a neighborhood $\mathcal{U}$ and $X_1, \ldots, X_r \in \mathcal{F}$ such that for every $X \in \mathcal{F}$, there exists smooth functions $f_1, \ldots, f_r \in \mathcal{C}^\infty(M)$ satisfying that the vector fields X and $\sum_{j=1}^{r} f_j X_j$ coincide on $\mathcal{U}$.

Now, let Y be a vector field in $\mathcal{F}_\mathcal{V}$. By definition, this means that for every $m' \in \mathcal{V}$, there exists a neighborhood $\mathcal{W}_{m'}$ of m' in $\mathcal{V}$ and a vector field $Y^{m'}$ in $\mathcal{F}$ that coincide with Y on $\mathcal{W}_{m'}$. By assumption on $\mathcal{U}$, the restriction to $\mathcal{U}$ of $Y^{m'}$ reads $\sum_{j=1}^{r} f_j^{m'} X_j$ for some smooth functions $f_1^{m'}, \ldots, f_r^{m'} \in \mathcal{C}^\infty(M)$. Since the open subsets $(\mathcal{W}_{m'})_{m'\in\mathcal{V}}$ cover $\mathcal{V}$, one can extract out of this open cover a sub-open cover $(\mathcal{W}_{m'_i})_{i\in I}$ that admits a partition of unity $(\chi_i)_{i\in I}$ and satisfies that every point in $\mathcal{V}$ has a neighborhood that belongs to finitely many of the open subsets $(\mathcal{W}_{m'_i})_{i\in I}$. By construction, one has on the whole open subset $\mathcal{V}$:

$$Y = \sum_{j=1}^{r}\left(\sum_{i\in I}\chi_i f_j^{m'_i}\right)X_j.$$

Since for every $j = 1, \ldots, r$, the function $\sum_{i \in I} \chi_i f_j^{m_i'}$ is smooth on $\mathcal{V}$, this proves the claim.

2.2 Smooth Singular Foliations: A Definition Using Sheaves

In order to define singular foliations, the use of compactly supported global vector fields is conceptually easy, but some readers may prefer to use sheaves. Sheaves will in any case be essential for the holomorphic case and the real analytical case. Let us therefore use sheaves to give a definition equivalent to Definition 2.1.

In this section, we denote by

$$\mathfrak{X}_\bullet : \mathcal{U} \longrightarrow \mathfrak{X}(\mathcal{U}) \text{ and } \mathcal{C}^\infty_\bullet : \mathcal{U} \longrightarrow \mathcal{C}^\infty(\mathcal{U})$$

the sheaves of vector fields and of smooth functions on the manifold M, respectively.[8]

Definition 2.18 (Definition of a Smooth Singular Foliation: Version 2, with Sheaves) A singular foliation on a smooth manifold M is a subsheaf

$$\mathcal{F}_\bullet : \mathcal{U} \mapsto \mathcal{F}_\mathcal{U}$$

of the sheaf $\mathfrak{X}_\bullet$ of vector fields on M such that

(α) $\mathcal{F}_\bullet$ is involutive,[a]
(β) is a sub-sheaf of $\mathcal{C}^\infty_\bullet$-modules,[b]
(γ) is locally finitely generated.[c]

---

[a] I.e., $[\mathcal{F}(\mathcal{U}), \mathcal{F}(\mathcal{U})] \subset \mathcal{F}(\mathcal{U})$ for all open subsets $\mathcal{U} \subset M$.
[b] I.e., $\mathcal{C}^\infty(\mathcal{U})\mathcal{F}(\mathcal{U}) \subset \mathcal{F}(\mathcal{U})$ for all open subset $\mathcal{U} \subset M$.
[c] For sheaves of modules over functions, the meaning of "locally finitely generated" needs to be made precise. We mean that every point admits an open neighborhood $\mathcal{U}$ on which there exists $X_1, \ldots, X_r \in \mathcal{F}(\mathcal{U})$ such that for every open subset $\mathcal{V} \subset \mathcal{U}$, the restrictions of $X_1, \ldots, X_r$ to $\mathcal{V}$ generate $\mathcal{F}(\mathcal{V})$ as a $\mathcal{C}^\infty(\mathcal{V})$-module.

---

[8] We remind the reader not to confuse, for a given sheaf $\mathcal{F}_\bullet$ on M and a given open subset $\mathcal{U} \subset M$, the space $\mathcal{F}(\mathcal{U})$ or sections of $\mathcal{F}_\bullet$ over $\mathcal{U}$ with the restriction to $\mathcal{U}$ of the sheaf $\mathcal{F}_\bullet$, which should be denoted $\mathcal{F}_\bullet|_\mathcal{U}$.

It is of course embarrassing to have two definitions of singular foliations. Fortunately, as pointed out in Alfonso Garmendia's [29] and Roy Wangs [72, Proposition 2.1.9 and Remark 2.1.13], these two definitions match, as we now see.

Proposition 2.19 (No Difference!) *Let M be a smooth manifold. There is a one to one correspondence between:*

(i) Singular foliations defined as in Definition 2.1.
(ii) Singular foliations defined as in Definition 2.18.

Proof The map $(i) \mapsto (ii)$ consists, given $\mathcal{F}$ as in Definition 2.1, in considering the sheaf $\mathcal{F}_\bullet$ of vector fields that coincide locally with an element in $\mathcal{F}$, see Exercise 2.16. Lemma 2.17 guaranties it is locally finitely generated in the sense of the third item in Definition 2.18.

The map $(ii) \mapsto (i)$ consists in considering global compactly supported sections $\mathcal{F}_c$ of the sheaf $\mathcal{F}_\bullet$ in Definition 2.18. In equation

$$\mathcal{F}_c = \mathcal{F}(M) \cap \mathfrak{X}_c(M).$$

By construction, $\mathcal{F}_c$ is a $\mathcal{C}^\infty(M)$-module stable under Lie bracket. Let us check that it is locally finitely generated in the sense of Definition 2.1. Let $m \in M$ be a point, $\mathcal{U}$ a neighborhood of m, and $X_1, \ldots, X_r \in \mathcal{F}_\mathcal{U}$ be as in item 3 of Definition 2.18. Let $\mathcal{U}' \subset \mathcal{U}$ be a second neighborhood of m such that there exists $\chi \in \mathcal{C}^\infty(M)$ which is equal to 1 on $\mathcal{U}'$ and whose support is a compact subset of $\mathcal{U}$. The vector fields $\chi X_1, \ldots, \chi X_r$ extend to global sections of $\mathcal{F}$ that we still denote by $\chi X_1, \ldots, \chi X_r$. These sections belongs to $\mathcal{F}_c$ since χ is compactly supported. By definition of "locally finitely generated" in the third item in Definition 2.18, for every $Y \in \mathcal{F}_c$, there exists $f_1, \ldots, f_r \in \mathcal{C}^\infty(\mathcal{U})$ such that the restriction of Y to $\mathcal{U}$ and $\sum_{i=1}^{r} f_i X_i$ coincide on $\mathcal{U}$. This implies that the restriction to $\mathcal{U}'$ of Y and of $\sum_{i=1}^{r} \chi f_i \chi X_i$ coincide on $\mathcal{U}'$. Since χf_i is now a smooth function on M, this implies that $\mathcal{U}'$ and the vector fields $\chi X_1, \ldots, \chi X_r$ satisfy the condition of "locally finite generated" as defined in the third item of Definition 2.1.

We have check that both maps above are inverse one to the other. Let $\mathcal{F}$ be a singular foliation as in Definition 2.1. Let $\mathcal{F}_\bullet$ be its associated sheaf as in $(i) \mapsto (ii)$. A global compactly supported section Y of that sheaf is a compactly supported vector field Y on M that locally coincides with a vector field in $\mathcal{F}$. We have to check that Y belongs to $\mathcal{F}$. This goes as follows: every $m \in M$ has a neighborhood $\mathcal{U}_m$ such that there exists $X^m \in \mathcal{F}$ that coincides with X on $\mathcal{U}_m$. Since the support of Y is compact, finitely many of the subsets $\mathcal{U}_m$ cover it. Let us denote by $\mathcal{U}_{m_1}, \ldots, \mathcal{U}_{m_N}$ such a finite family. There exists smooth functions $\chi_1, \ldots, \chi_N$, with support in $\mathcal{U}_{m_1}, \ldots, \mathcal{U}_{m_N}$ respectively, such that $\sum_{i=1}^{N} \chi_i = 1$

on a neighborhood $\mathcal{V}$ of the support of Y. The vector fields Y and $\sum_{i=1}^{N} \chi_i X^{m_i}$ coincide on $\mathcal{V}$. The vector fields Y and $\sum_{i=1}^{N} \chi \chi_i X^{m_i}$, with χ any function which is 1 on the support of Y and 0 outside $\mathcal{V}$, coincide therefore on the whole manifold M. The crucial point is that the sum is now finite, which proves that Y belongs to $\mathcal{F}$. The composition *(i)* $\mapsto$ *(ii)* $\mapsto$ *(i)* is therefore the identity map.

Now, let $\mathcal{F}_\bullet$ be a sheaf as in Definition 2.18. Consider $\mathcal{U} \subset$ on open subset and $Y \in \mathcal{F}(\mathcal{U})$. For every $m \in \mathcal{U}$, let χ be a smooth function on M equal to 1 in a neighborhood of m and whose support is a compact subset of $\mathcal{U}$. Then χY extends to a compactly supported section of $\mathcal{F}_\bullet$ on M, and coincides with Y in a neighborhood of M. The composition *(ii)* $\mapsto$ *(i)* $\mapsto$ *(ii)* is therefore the identity map. This completes the proof.

Remark 2.20 Most of the considerations above, and the proof of Proposition 2.19 in particular, are general phenomenons for what are called *fine sheaves*.

Warning About Notations!
From now on:

1. We will call *foliated manifolds* pairs made of a manifold equipped with a singular foliation, and denote them by $(M, \mathcal{F})$.
2. We will not make any more notation distinction, in the smooth case, between $\mathcal{F}$ and $\mathcal{F}_\bullet$ (i.e., between singular foliations seen as sub-modules of compactly supported vector fields or seen as sheaves, i.e., between singular foliations defined as in Definition 2.1 or as in Definition 2.18). We will mostly use the notation $\mathcal{F}$.

2.3 Singular Foliations on Complex or Real Analytic Manifolds

For complex manifolds, singular foliations have to be defined through sheaves. As a matter of fact, most geometric objects have to be defined through sheaves since there are no or few globally defined functions, vector fields - and so on. For the reader unfamiliar with sheaves, or hostile to them, it simply means that they have to be defined locally. In this section, we fix M a complex or real analytic manifold, and we denote by $\mathcal{O}$ its sheaf of holomorphic or real analytic functions and by $\mathfrak{X}$ the sheaf of vector fields on M. For any open subset $\mathcal{U} \subset M$, we denote by $\mathcal{O}(\mathcal{U})$ the $\mathbb{K}$-algebra of holomorphic or real analytic $\mathbb{K}$-valued functions on $\mathcal{U}$, with $\mathbb{K} = \mathbb{C}$ or $\mathbb{K} = \mathbb{R}$ depending on the context.

Let us start with a question about the limit of what we want to study.

Question 2.21 Let $M = \mathbb{C}$. Consider the sheaf $\mathcal{F}^{0\infty}$ of holomorphic vector fields vanishing together with all their derivatives at 0. Do we want singular foliations to be defined such that this sheaf is one of them? Or do we do not accept such sheaves?

To answer this question, let us describe $\mathcal{F}^{0\infty}$. For $\mathcal{U} \subset \mathbb{C}$ an open subset that does not contain 0, $\mathcal{F}^{0\infty}(\mathcal{U}) = \mathfrak{X}(\mathcal{U})$. For $\mathcal{U}$ a connected open subset containing 0, we have $\mathcal{F}^{0\infty}(\mathcal{U}) = 0$. For a generic open subset $\mathcal{U}$, $\mathcal{F}^{0\infty}(\mathcal{U})$ is the space of vector fields on $\mathcal{U}$ which are equal to zero on the connected component $\mathcal{U}_0$ of 0.

It is therefore tempting to answer "yes" to the question above because (α) for every $\mathcal{U}$, $\mathfrak{X}(\mathcal{U})$ is stable under Lie bracket, (β) $\mathfrak{X}(\mathcal{U})$ is a module over $\mathcal{O}(\mathcal{U})$, and (γ) it is finitely generated: the generator is the holomorphic vector field

$$f(z)\frac{\partial}{\partial z}$$

with f a function which is 0 on $\mathcal{U}_0$ and 1 in $\mathcal{U}\setminus\mathcal{U}_0$ for every open subset $\mathcal{U} \subset \mathbb{C}$. Also, there is a natural candidate for leaves: there are two of them: the point $\{0\}$ and $\mathbb{C}\setminus\{0\}$. Two points are in the same leaf, in this sense, if and only if there exists a vector field in $\mathcal{F}^{0\infty}$ whose time 1 flow maps the first point to the second one.

So, is the answer "yes"? Not so fast! This sheaf is *not* locally finitely generated in the sense of Definition 2.18. For every $\mathcal{U}$ an open subset containing 0, no matter how "small", there is an open subset $\mathcal{V}$ such that $\mathcal{F}^{0\infty}(\mathcal{V})$ is not zero: it suffices to choose $\mathcal{V} \subset \mathcal{U}_0$ an open subset that does not contain 0. The space $\mathcal{F}^{0\infty}(\mathcal{V})$ being not reduced to zero, it can not be obtained by restricting to $\mathcal{V}$ generators of $\mathcal{F}^{0\infty}(\mathcal{U})$, since those are all zero.

So, is the answer "no"? Well, it depends on the context, but we suggest answering "no" to that question, and to define singular foliations as follows:

Definition 2.22 (Holomorphic or Real Analytic Singular Foliations) A singular foliation on a complex (or real analytic) manifold M is a subsheaf $\mathcal{F}$ of the sheaf $\mathfrak{X}$ of holomorphic (or real analytic) vector fields on M which

(α) is involutive,
(β) is stable under multiplication under $\mathcal{O}$,
(γ) is locally finitely generated,[a] i.e., for every point m, there exists a neighborhood $\mathcal{U}$ and vector fields $X_1, \ldots, X_r \in \mathcal{F}(\mathcal{U})$ such that for every open subset $\mathcal{V} \subset \mathcal{U}$, $\mathcal{F}(\mathcal{V})$ is generated as a $\mathcal{O}_\mathcal{V}$-module by the restrictions to $\mathcal{V}$ of $X_1, \ldots, X_r$.

---

[a] In the literature, this property is often referred to as being "of finite type". For consistency of notations throughout the lecture, we prefer to say "locally finitely generated".

Now, complex geometers may be surprised by the previous definition, because the "locally finitely generated" condition is not commonly used in complex geometry, even under its alternative name "of finite type". They would prefer to use *coherent sheaves* [62]. However, it is a classical, but hard, theorem in complex analysis, that locally finitely generated (=of finite type) subsheaves of a coherent sheaf are coherent sheaves and that vector fields form a coherent sheaf.[9] As a consequence, Definition 2.22 can be equivalently restated as in the next proposition.

Proposition 2.23 *A singular foliation on a complex (or real analytic) manifold M is a subsheaf $\mathcal{F}_\bullet$ of the sheaf $\mathfrak{X}$ of vector fields on M which*

(α) *is involutive,*
(β) *is stable under multiplication under $\mathcal{O}$,*
(γ) *is coherent.*

Proposition 2.23 implies that our definition of a singular foliation in the complex case coincides with the one given in Paul Baum and Raoul Bott's [9], Ali Sinan Sertöz's [62], André Belotto da Silva and Daniel Panazzolo's [14], or Tatsuo Suwa's [68].

We just explained why, in the complex setting, one can re-use mutatis mutandis Definition 2.18. However, any sheaf that satisfies items (α) and (β) in Definition 2.18 in fact satisfies a variation of item (γ) as well. Let us be precise: unlike the algebra of germs of smooth functions, the algebra of germs of holomorphic or real analytic functions are Noetherian:

Theorem 2.24 ([69]) *Germs of holomorphic (resp. real analytic) functions near $0 \in \mathbb{C}^n$ (resp. $\mathbb{R}^n$) form a Noetherian ring.*

Remark 2.25 We warn the reader that Theorem 2.24 does *not* imply that any sub-$\mathcal{O}(\mathcal{U})$-module $\mathcal{F}(\mathcal{U}) \subset \mathfrak{X}(\mathcal{U})$ is finitely generated for every $\mathcal{U} \subset M$. Here is a classical counter example: take $M = \mathcal{U} = \mathbb{C}$, and consider $\mathcal{F}$ to be the sheaf of all vector fields vanishing at all the points $\{n \mid n \in \mathbb{N}\}$, except a finite number of them. The space of global sections, i.e., $\mathcal{F}(\mathbb{C})$, is not finitely generated.

Let us explain the consequences of Theorem 2.24 in the holomorphic setting: the real analytic setting is similar. In a chart neighborhood $\mathcal{U}$ of a point $m \in M$, with coordinates $z_1, \ldots, z_d$, holomorphic vector fields decompose as sums

$$\sum_{i=1}^{d} f_i(z_1, \ldots, z_d) \frac{\partial}{\partial z_i}$$

---

[9] It is called "Oka's coherence theorem".

with $f_1, \ldots, f_d \in \mathcal{O}(\mathcal{U})$ being $\mathbb{C}$-valued holomorphic functions on $\mathcal{U}$. In particular, as a module over $\mathcal{O}(\mathcal{U})$, holomorphic vector fields of $\mathcal{U}$ are isomorphic with

$$\mathfrak{X}(\mathcal{U}) \simeq \underbrace{\mathcal{O}(\mathcal{U}) \oplus \cdots \oplus \mathcal{O}(\mathcal{U})}_{d \text{ terms}}$$

(with d the dimension of the manifold). The isomorphism above makes the germs at m of all elements $X \in \mathcal{F}$ a sub-$\mathcal{O}_m$-module of

$$\underbrace{\mathcal{O}_m \oplus \cdots \oplus \mathcal{O}_m}_{d \text{ terms}}$$

with $\mathcal{O}_m$ being the algebra of germs at m of holomorphic functions defined near m. The henceforth obtained sub-module is finitely generated over $\mathcal{O}_m$ by Theorem 2.24. Let $X_1, \ldots, X_r$ be vector fields on $\mathcal{U}_1, \ldots, \mathcal{U}_r$ respectively whose germs at m generate this module. Let $\mathcal{U} = \cap_{i=1}^r \mathcal{U}_i$. For any connected neighborhood $\mathcal{V} \subset \mathcal{U}$ of m in M, and any vector field $X \in \mathcal{F}(\mathcal{V})$, the germ of X at m is in the $\mathcal{O}_m$-module generated by $X_1, \ldots, X_r$. This means that there exists r holomorphic functions $f_1, \ldots, f_r$ defined on some neighborhood $\mathcal{W} \subset \mathcal{V}$ such that $X = \sum_{i=1}^r f_i X_i$. But these functions are defined on a smaller subset $\mathcal{W}$ of $\mathcal{V}$. We cannot therefore conclude that $\mathcal{F}$ is a locally finitely generated module over the sheaf of holomorphic functions. But the previous results are still a phenomenon that does not appear in the smooth setting. In several contexts (as in [44]), it might be enough to work with sheaves $\mathcal{F} \subset \mathfrak{X}$ such that only conditions (α) and (β) hold in Definition 2.22, leaving aside condition (γ). It is always granted "for free" that any point $m \in M$ admits a neighborhood $\mathcal{U}$ and $X_1, \ldots, X_r \in \mathcal{F}(\mathcal{U})$ such that for any $\mathcal{V} \subset \mathcal{U}$ containing m, and any $Y \in \mathcal{F}(\mathcal{V})$, there exists an open subset $\mathcal{W} \subset \mathcal{V}$ containing m and functions $f_1, \ldots, f_r \in \mathcal{O}(\mathcal{W})$ such that $Y = \sum_{i=1}^r f_i X_i$ on $\mathcal{W}$.

2.3.1 Restriction to an Open Subset

When smooth singular foliations are defined as in Definition 2.1, then restrictions to open subsets are easy to define, and subtleties like Lemma 2.9 do not appear.

Let $\mathcal{F}$ be a singular foliation on a smooth, complex or real analytic manifold, seen as a sheaf as in Definition 2.18 or 2.22. For every open $\mathcal{U} \subset M$, the sheaf $\mathcal{F}$ can be restricted to $\mathcal{U}$. The restriction defines a singular foliation on $\mathcal{U}$ that we denote by $\mathcal{F}_\mathcal{U}$.

In the smooth case, under the correspondence introduced by Proposition 2.19, this singular foliation $\mathcal{F}_\mathcal{U}$ corresponds to the subspace of $\mathfrak{X}_c(\mathcal{U})$ denoted by ♣ in Sect. 2.1.

We warn the reader not to confuse $\mathcal{F}_\mathcal{U}$, which is a sheaf on $\mathcal{U}$, and $\mathcal{F}(\mathcal{U})$ which stands for its sections on $\mathcal{U}$, especially in the complex or real analytic settings.

Now, in the complex case, there is a phenomenon called Hartog's theorem or Riemann's extension theorem that says that a holomorphic object defined outside a codimension 2 analytic subset extends to the whole space. It is tempting to apply such results to the complement of the singular locus of a singular foliation (see Sect. 3.4): this is discussed in Remark 3.26.

2.4 Singular Foliation on an Affine Complex Variety

Let us now define singular foliations on affine varieties over the field $\mathbb{C}$. Let us recall that an affine variety (maybe non-irreducible) is a subset $W \subset \mathbb{C}^n$ given by polynomial equations. But we will see them in a more algebraic manner.

Denote by $\mathcal{O}_n$ the algebra of polynomial functions in n-variables, and $\mathcal{I}_W \subset \mathcal{O}_n$ the ideal of functions vanishing on W. We call *functions on* W the quotient ring $\mathcal{O}_W := \frac{\mathcal{O}_n}{\mathcal{I}_W}$. We call *vector fields on* W and denote by $\mathfrak{X}_W$ the $\mathcal{O}_W$-module of derivations of $\mathcal{O}_W$. It is equipped with the commutator as a Lie bracket. Since the algebra $\mathcal{O}_W$ is a Noetherian algebra, and since $\mathfrak{X}_W$ is a $\mathcal{O}_W$-module of finite rank, any sub-$\mathcal{O}_W$-module is finitely generated. The assumption "locally finitely generated" is therefore useless in that context, and we suggest the following definition of a singular foliation on an affine variety.

Definition 2.26 A singular foliation on an affine variety W is a sub-space $\mathcal{F}$ of the $\mathcal{O}_W$-module of $\mathfrak{X}_W$ which

(α) is involutive,
(β) is stable under multiplication under $\mathcal{O}_W$.

Notice that the definition does not make reference to the "ambient space": the definition above only makes use of the algebra of functions on W. If two affine varieties W and W' in $\mathbb{C}^n$ and $\mathbb{C}^{n'}$ respectively satisfy $\mathcal{O}_n/\mathcal{I}_W \simeq \mathcal{O}_{n'}/\mathcal{I}_{W'}$, then they are equipped with the same singular foliations.

For schemes, or quasi-projective varieties, again, the use of sheaves will be necessary.

Exercise 2.27 Write the definition of a singular foliation on an arbitrary scheme.

The above discussion leads to a purely algebraic definition of what a singular foliation is. Let $\mathcal{O}$ be a commutative unital algebra (which may be thought of as being an algebra of "functions").

Definition 2.28 A sub-space $\mathcal{F}$ of $\mathrm{Der}(\mathcal{O})$ is said to be an *algebraic singular foliation* if:

(α) $\mathcal{F}$ is a stable under the Lie bracket of $\mathrm{Der}(\mathcal{O})$,
(β) $\mathcal{F}$ is a sub-$\mathrm{Der}(\mathcal{O})$-module of $\mathrm{Der}(\mathcal{O})$,
(γ) and it is finitely generated as an $\mathcal{O}$-module.

If the algebra $\mathcal{O}$ is *(i)* finitely generated and *(ii)* Noetherian, then every involutive $\mathcal{O}$-submodule of derivations of $\mathcal{O}$ is an algebraic singular foliation: the condition (γ) is therefore useless in that case. In particular:

Proposition 2.29 *An algebraic singular foliation on a smooth*[10] *affine variety in* $W \subset \mathbb{C}^N$ *is also a holomorphic singular foliation on*[11] W.

2.4.1 Germification

Here is an example of an algebraic singular foliation associated to any smooth, real analytic or complex singular foliation $\mathcal{F}$ and a manifold M (with sheaf of functions $\mathcal{O}$). Choose a point $m \in M$.

Definition 2.30 A *function germ at a point* $m \in M$ is an equivalence class of pairs $(\mathcal{U}, f)$ with $\mathcal{U} \subset M$ an open subset containing m, and $f \in \mathcal{O}(\mathcal{U})$, under the equivalence relation $(\mathcal{U}, f) \sim (\mathcal{V}, g)$ if $f = g$ on an open subset of $\mathcal{U} \cap \mathcal{V}$ containing m. The quotient comes equipped with a natural algebra structure denoted by $\mathcal{O}_m$.

Notice that $\mathcal{O}_m$ is a local ring. Any vector field $X \in \mathfrak{X}(M)$, defined in a neighborhood $\mathcal{F}$ of m, induces a derivation X_m of $\mathcal{O}_m$. For $\mathcal{F}$ a singular foliation on M, denote by $\mathcal{F}_m$ the space of derivations of $\mathcal{O}_m$ obtained by considering all such derivations for all vector fields in $\mathcal{F}(\mathcal{U})$ for all open neighborhoods $\mathcal{U}$ of m. Axioms (α), (β), and (γ) in Definition 2.18 or 2.22 imply that axioms (α), (β), and (γ) in Definition 2.28 are satisfied. The following definition then makes sense.

Definition 2.31 Let $\mathcal{F}$ be an algebraic singular foliation. For a given point $m \in M$, we call *germ of $\mathcal{F}$ at m* the algebraic singular foliation $\mathcal{F}_m$ over $\mathcal{O}_m$.

We could also work with formal functions (=formal power series) instead of germs, see [26, 41] and the discussion following Theorem 3.20 in chapter "Canonical Geometric and Algebraic Structures Hidden Behind a Singular Foliation".

2.5 Globally Finitely Generated Singular Foliations

We have been through quite an extensive discussion about the limits and sense of the "locally finitely generated" condition. But quite a few singular foliations are in fact *globally* finitely generated.

---

[10] Smooth here means "that has no singular point"!
[11] W being now seen as a complex manifold.

Notice that the notion was already introduced in Definition 2.7 in the smooth setting using compactly supported vector fields, but we can now define them within the setting of sheaves.

Globally finitely generated can be defined in the same manner in the smooth, complex, or real analytic contexts altogether. We leave it to the reader to check that the next definition matches Definition 2.7[12] in the smooth case.

Definition 2.32 (A Common Definition) A singular foliation $\mathcal{F}$ on a manifold M is said to be *finitely generated* if there exists vector fields[a] $X_1, \ldots, X_r$ defined on the whole manifold M such that for every open subset $\mathcal{U} \subset M$, $\mathcal{F}(\mathcal{U})$ is the $\mathcal{O}(\mathcal{U})$ module generated by the restrictions to $\mathcal{U}$ of $X_1, \ldots, X_r$.

[a]Called *generators of $\mathcal{F}$*.

Remark 2.33 Let $\mathcal{F}$ be a singular foliation on M. Every point $m \in M$ has a neighborhood on which it is finitely generated.

Here is an important result.

Lemma 2.34 *Let $X_1, \ldots, X_r$ be generators of a finitely generated singular foliation $\mathcal{F}$ on a manifold M, then there exists a family of functions $c_{ij}^k \in \mathcal{O}$, indexed by $i, j, k = 1, \ldots, r$, such that for all $i, j = 1, \ldots, r$:*

$$[X_i, X_j] = \sum_{k=1}^{r} c_{ij}^k X_k.$$

Such functions c_{ij}^k are called *Christoffel symbols of $\mathcal{F}$ with respect to $X_1, \cdots, X_r$*. Since there are, in general, relations between the generators $X_1, \ldots, X_k$, the Christoffel symbols c_{ij}^k are <u>not unique</u>.

Exercise 2.35 Let $X_1, \ldots, X_r$ be generators of a finitely generated singular foliation $\mathcal{F}$, and $\left(c_{ij}^k\right)_{i,j,k=1}^r$ a choice of Christoffel symbols of $\mathcal{F}$ with respect to $X_1, \ldots, X_r$.

---

[12] Notice that the vector fields $X_1, \ldots, X_r$ are global sections of the sheaf $\mathcal{F}_\bullet$ if one uses Definition 2.32 but may not be in $\mathcal{F}_c \subset \mathfrak{X}_c(M)$ if one uses Definition 2.7. This makes no practical difference.

1. Show that
$$\left(\frac{c_{ij}^k - c_{ji}^k}{2}\right)^r_{i,j,k=1}$$
is again a choice of Christoffel symbols of $\mathcal{F}$ with respect to $X_1, \ldots, X_r$.
2. Show that, without any loss of generality, Christoffel symbols of $\mathcal{F}$ with respect to $X_1, \ldots, X_r$ can be assumed to satisfy $c_{ji}^k = -c_{ij}^k$ for all possible indices.

Exercise 2.36 The *"non-finitely-many-generators"* singular foliation—an example due to Iakovos Androulidakis and Marco Zambon. On $M = \mathbb{R}^2$, call $\mathcal{F}$ the space of all vector fields $X \in \mathfrak{X}(\mathbb{R}^2)$ that vanish at order n at the point of coordinates $(n, 0)$. I.e., vector fields of the form:
$$X = f(x,y)\frac{\partial}{\partial x} + g(x,y)\frac{\partial}{\partial y}$$
such that for all $a, b, n \in \mathbb{N}_0$ with $a + b \leq n$:
$$\frac{\partial^{a+b} f}{\partial x^a \partial y^b}(0, n) = \frac{\partial^{a+b} g}{\partial x^a \partial y^b}(0, n) = 0.$$

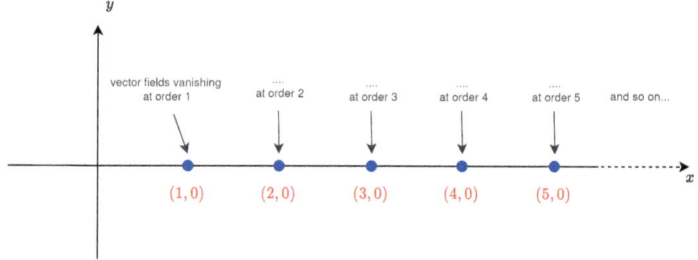

A representation of the "non-finitely-many-generators" singular foliation

Show that

1. $\mathcal{F}$ is an integrable distribution.
2. $\mathcal{F}$ is locally finitely generated.
3. $\mathcal{F}$ is not globally finitely generated.
4. $\mathcal{F}$ is not the image through the anchor map of a Lie algebroid on $\mathbb{R}^2$.

In the smooth setting, if for a singular foliation the number of local generators is bounded, then it is finitely generated.

Proposition 2.37 *Let M be a smooth manifold of dimension n and $\mathcal{F}$ a singular foliation admitting a global bound k of the number of local generators, i.e., M is covered by open sets $U = \{U_i\}_{i \in I}$ such that $\mathcal{F}(U_i)$ is generated by k or less vector fields.*

Then $\mathcal{F}$ can be generated by a family of $(n+1) \cdot k$ vector fields. In particular, it is a finitely generated singular foliation.[13]

Proof Since the cover dimension of a smooth manifold is equal to its dimension as a manifold, the cover U admits a refinement such that every point is covered by at most $n+1$ points. This is the classical paving principle. Now we can apply Theorem 3 from [59], stating that there exists a finite open covering $V_1, \ldots, V_{n+1}$ of M such that $V_j = \bigsqcup_\alpha V_j^\alpha$, where each V_j^α is included in some U_i and V_j^α, V_j^β have disjoint closures for fixed j and $\alpha \neq \beta$. This latter fact means that there exist functions ϕ_j^α such that $\phi_j^\alpha|_{V_j^\alpha} = 1$ and $\phi_j^\alpha|_{V_j^\beta} = 0$ for $\alpha \neq \beta$. We pick for each (j, α) a collection of vector fields $Y_j^{\alpha,1}, \ldots, Y_j^{\alpha,k} \in \mathcal{F}$ whose restrictions generate $\mathcal{F}_{V_j^\alpha}$. We set

$$X_j^s = \sum_\alpha = \phi_j^\alpha Y_j^{\alpha,s}$$

for $j \in \{1, \ldots, n+1\}$ and $s \in \{1, \ldots, k\}$. By construction, this family of vector fields generates the foliation $\mathcal{F}$.

Remark 2.38 One recognizes in the proof of Proposition 2.37 the logic of the argument used to show that any finite-dimensional vector bundle over a smooth manifold is the quotient of (or direct summand in) a trivial vector bundle.

3 Some Basic Notions About Singular Foliations (Symmetries and Regular Part)

For $\mathcal{F}$ a singular foliation on M and $\mathcal{U} \subset M$ an open subset, we denote by $\mathcal{F}_\mathcal{U}$ the induced singular foliation on $\mathcal{U}$.

3.1 Symmetries and Inner Symmetries

In the lines below, we place ourselves in the smooth case. In the real analytic or complex case, the whole story is similar, but the word "diffeomorphism","smooth", and "interval" are to be replaced by "biholomorphism", "holomorphic", and "open ball in $\mathbb{C}$" in the text below.

---

[13] See Definition 2.32.

Let $(M, \mathcal{F})$ and $(M', \mathcal{F}')$ be foliated manifolds in the smooth category. We call *isomorphism of singular foliations* a diffeomorphism $\phi \colon M \simeq M'$ such that $X \in \mathcal{F}$ if and only if $\phi_*(X) \in \mathcal{F}'$. When $M' = M$ and $\mathcal{F}' = \mathcal{F}$, we shall speak of a *symmetry of $\mathcal{F}$*. Sometimes, we will speak of a *local symmetry of a foliated manifold* $(M, \mathcal{F})$: it is a triple $(\mathcal{U}, \mathcal{U}', \phi)$ with $\mathcal{U}, \mathcal{U}'$ open subsets of M and $\phi \colon \mathcal{U} \longrightarrow \mathcal{U}'$ a diffeomorphism which maps $\mathcal{F}_\mathcal{U}$ to $\mathcal{F}'_{\mathcal{U}'}$.

Symmetries of $(M, \mathcal{F})$ form a group, and local symmetries form what is called a pseudo-group.

Definition 3.1 We denote by $\mathrm{Sym}(\mathcal{F})$ the group of symmetries of a singular foliation.

We now intend to define inner symmetries. This goes through the correct definition of time-dependent vector fields. Again, we work on the smooth category.[14]

First, let us define time-dependent vector fields valued in a singular foliation.

Definition 3.2 Let I be an interval of $\mathbb{R}$. Let $(M, \mathcal{F})$ be a foliated manifold. We say that a family $(X_t)_{t \in I}$ of vector fields is a *smooth time-dependent vector field*[15] in $\mathcal{F}$ if, in a neighborhood $\mathcal{U}$ of every point $m \in M$, there exists local generators $X_1, \ldots, X_r$ of $\mathcal{F}$ and smooth functions $f_1, \ldots, f_r \in C^\infty(M \times I)$ such that for all $m \in \mathcal{U}, t \in I$:

$$X_{t\,|m} = \sum_{i=1}^{r} f_i(m, t)\, X_{i\,|m}.$$

Remark 3.3 It deserves to be noticed that a smooth time-dependent vector field valued in $\mathcal{F}$ may *not* be a smooth time-dependent vector field on M such that $X_t \in \mathcal{F}$ for every value of t. Here is a counter example (provided to us by Georges Skandalis). Take the singular foliation on $M = \mathbb{R}$ generated by the vector field

$$X = e^{-\frac{1}{x^2}} \frac{\partial}{\partial x}$$

---

[14] In the real-analytic or holomorphic contexts, there is a subtlety about time-dependent vector fields that appear, e.g., in Definition 3.2: even in the real analytic or holomorphic cases, it suffices that this time-dependency be <u>smooth</u> in the parameter that we denote by t below.

[15] We sometime also say "vector fields depending smoothly on t".

Consider the family of vector fields on M given by

$$X_t = \begin{cases} \frac{t}{x^2+t^2} X & \text{for } t \neq 0 \\ 0 & \text{for } t = 0. \end{cases}$$

Since $\frac{t}{x^2+t^2} e^{-\frac{1}{x^2}}$ extends to a smooth function on $\mathbb{R}^2$ that vanishes with all its derivatives at the point $(0, 0)$, the previously defined family is a smooth time-dependent family of vector fields on $M = \mathbb{R}$. Also, for all $t \in \mathbb{R}$, $X_t \in \mathcal{F}$. But since the function $\frac{t}{x^2+t^2}$ does not extend to a smooth function at $(0, 0)$, the family X_t is *not* a smooth time-dependent vector field in $\mathcal{F}$ in the sense of Definition 3.2.

Exercise 3.4 Show that in Definition 3.2, "there exists local generators $X_1, \ldots, X_r$ of $\mathcal{F}$ and smooth functions $f_1, \ldots, f_r \in C^\infty(M \times I)$" could be replaced by "for every local generators $X_1, \ldots, X_r$ of $\mathcal{F}$, there exists smooth functions $f_1, \ldots, f_r \in C^\infty(M \times I)$".

Exercise 3.5 This exercise supposes the notion of anchored bundle, see Sect. 1.1 in chapter "Canonical Geometric and Algebraic Structures Hidden Behind a Singular Foliation". Let $A \to M$ be a vector bundle. We say that a time-dependent section $(a_t)_{t \in [0,1]}$ of A is smooth if the map $(m, t) \mapsto a_t |_m$ is a smooth map from $M \times [0, 1]$ to A. Show that $(X_t)_{t \in I}$ is a smooth time-dependent vector field in $\mathcal{F}$ if and only if for every anchored bundle (A, ρ) over $\mathcal{F}$, there exists a smooth time-dependent[16] section $(a_t)_{t \in I}$ of $A \to M$ such that $\rho(a_t) = X_t$ for every $t \in I$.

Exercise 3.6 Let $(M, \mathcal{F})$ be a foliated manifold. Let I be an open interval of $\mathbb{R}$.
Let $\hat{\mathcal{F}}$ be the singular foliation on $M \times I$ which is the direct product of $(M, \mathcal{F})$ with $(I, 0)$.
Let $\tilde{\mathcal{F}}$ be the singular foliation on $M \times I$ which is the direct product of $(M, \mathcal{F})$ with $(I, \mathfrak{X}(I))$.
Consider a family $(X_t)_{t \in I}$ of vector fields such that $X_t \in \mathcal{F}$ for all $t \in I$.

1. Show that the following are equivalent:
   (i) $(X_t)_{t \in I}$ is a smooth time-dependent vector field in $\mathcal{F}$,
   (ii) the vector field $X_\bullet$ on $M \times I$ whose value at (m, t) is $(X_t |_m, 0)$ belongs to $\hat{\mathcal{F}}$,
   (iii) the vector field Y on $M \times I$ given by

   $$Y = X_\bullet + \frac{\partial}{\partial t}$$

   belongs to $\tilde{\mathcal{F}}$.

---

[16] A time-dependent section $(a_t)_{t \in I}$ is said to be smooth if $(m, t) \mapsto a_t(m)$ is a smooth map from $M \times I$ to A.

2. Show that the time τ flow ϕ_τ^Y of Y and the time τ-flow $\phi_\tau^{X\bullet}$ of $(X_t)_{t\in I}$ are related by

$$\phi_\tau^Y(m,t) = \left(\phi_\tau^{X\bullet}(m), t+\tau\right)$$

for all t, τ, m for which these flows exist.
3. Conclude that "any inner symmetry (defined below) of $\mathcal{F}$ is obtained by restricting to well-chosen submanifold that cleanly intersects $\tilde{\mathcal{F}}$ the flow of a vector field in $\tilde{\mathcal{F}}$".

> **Definition 3.7 (Inner Symmetries)** Let $(M, \mathcal{F})$ be a foliated manifold. A diffeomorphism of $(M, \mathcal{F})$ is said to be an *inner-symmetry* if it is the time-1 flow of a smooth time-dependent vector field $(X_t)_{t\in[0,1]}$ in $\mathcal{F}$ whose time-1 flow exists. We denote by Inner($\mathcal{F}$) the set of inner symmetries of $\mathcal{F}$.
>
> If the previous diffeomorphism only on some open subset $\mathcal{U} \subset M$, we speak of a *local inner symmetry*.

Here is the first important lemma.

Lemma 3.8 *Inner symmetries of a singular foliation form a group.*

Proof If a smooth time-dependent vector fields $(X_t)_{t\in[0,1]}$ in $\mathcal{F}$ that yields the inner-symmetry ϕ, then $Y_t = -X_t$ is also a smooth time-dependent vector field and it yields ϕ^{-1}. Also, upon replacing X_t by $f'(t)X_{f(t)}$ with $f : [0,1] \to [0,1]$ any smooth function with $f = 0$ near 0 and $f = 1$ near 1, one yields a smooth time-dependent vector field that vanishes near $t = 0$ and $t = 1$ together with all its derivatives with respect to t, whose time-1 flow is still ϕ. This allows gluing: for any two inner symmetries ϕ, ψ given by such smooth time-dependent vector fields $(X_t)_{t\in[0,1]}$ and $(Y_t)_{t\in[0,1]}$ respectively, the vector field

$$Z_t = X_{2t} \text{ if } t \in [0, 1/2] \text{ and } Z_t = Y_{2t-1} \text{ if } t \in [1/2, 1].$$

is a smooth time-dependent vector field in $\mathcal{F}$ that yields $\psi \circ \phi$.

It is a highly non-trivial result that **inner symmetries of $(M, \mathcal{F})$ are in fact symmetries of $(M, \mathcal{F})$**: this will be the point of one of the main theorem in these lectures, namely Theorem 7.12, and more precisely Corollary 7.13. Let us admit this theorem and the corollary for the remaining of this section.

Exercise 3.9 Let $(M, \mathcal{F})$ be a foliated manifold.

1. Show that for any symmetry ψ and any inner symmetry ϕ of $\mathcal{F}$, the composition $\psi \circ \phi \circ \psi^{-1}$ is an inner symmetry.

In the process, describe the smooth time-dependent vector field Y_t whose time-1 flow is $\psi \circ \phi \circ \psi^{-1}$ out of ψ and out of a smooth time-dependent vector field X_t whose time-1 flow is ϕ.

Note: In view of Corollary 7.13 (that states that inner symmetries form a normal subgroup of the group of symmetries), this exercise means that inner symmetries form a *normal* subgroups of the group of symmetries.

For a symmetry ϕ of a singular foliation $(M, \mathcal{F})$, the notion of fixed point is straightforward: it is just a point $m \in M$ such that $\phi(m) = m$. There is a notion of "fixed point" which is much subtle for inner-symmetry, and that we now introduce.

Definition 3.10 Let $(M, \mathcal{F})$ be a foliated manifold. A point $m \in M$ is said to be a *very-fixed point* of a (maybe local) inner symmetry ϕ if ϕ is the time 1-flow of some smooth time-dependent vector field $(X_t)_{t \in [0,1]}$ in $\mathcal{F}$ such that the functions $f_1, \ldots, f_r$ that appear in Definition 3.2 vanish at the point m for all $t \in [0, 1]$.

Equivalently, m is a very-fixed point of ϕ if there exists local generators $X_1, \ldots, X_r$ of $\mathcal{F}$ on some neighborhood $\mathcal{U}$ of m and smooth functions $f_1, \ldots, f_r \in \mathcal{C}^\infty(M \times I)$ satisfying $f_i(m, t) = 0$ for all $t \in [0, 1]$ such that ϕ is the time-1 flow of the vector field given $\forall x \in \mathcal{U}, t \in [0, 1]$ by

$$X_t|_x = \sum_{i=1}^r f_i(x, t) \, X_i|_x.$$

Exercise 3.11 Let $(M, \mathcal{F})$ be a foliated manifold.

1. Show that if m is a very fixed point for an inner symmetry ϕ, then it is also a fixed point, i.e., $\phi(m) = m$.
2. We prove that the converse is not true. Let $\mathcal{F}$ be the singular foliation of vector fields vanishing at 0 on $\mathbb{R}^2$. Show that the rotation of center 0 and of any angle $\theta \neq 0$ is an inner symmetry. Show that 0 is a fixed point of that inner symmetry, but is not a very-fixed point.

Exercise 3.12 This exercise is a continuation of Exercise 3.9. Show that if m is a very-fixed point of an inner symmetry ϕ, and is ψ is a symmetry of $\mathcal{F}$, then $\psi^{-1}(m)$ is a very-fixed point of the inner symmetry $\psi^{-1} \circ \phi \circ \psi$.

Exercise 3.13 This exercise is a continuation of Exercise 3.5. Let (A, ρ) be an anchored bundle for $\mathcal{F}$ (see Sect. 1.1 in chapter "Canonical Geometric and Algebraic Structures Hidden Behind a Singular Foliation"). Let m be a very fixed point for an inner symmetry obtained as the time-1-flow of a vector field $(X_t)_{t \in [0,1]}$ of the form required in Definition 3.10.

1. Show that the smooth time-depending section $(a_t)_{t \in [0,1]}$ such that $\rho(a_t) = X_t$ for all $t \in [0, 1]$, whose existence is granted by Exercise 3.5, can be chosen to satisfy $a_t |_m = 0$ for all $t \in [0, 1]$.
2. Show that any smooth time-depending section $(a_t)_{t \in [0,1]}$ such that $\rho(a_t) = X_t$ satisfies that $a_t |_m$ is valued in the strong kernel[17] of ρ at m.

Remark 3.14 For a smooth time-dependent vector field X_t in $\mathcal{F}$ of the form described in Definition 3.10, notice that for all $t \in [0, 1]$, the class $[X_t]$ of X_t in the isotropy Lie algebra $\mathfrak{g}_m(\mathcal{F})$ of $\mathfrak{F}$ at m is zero (see Sect. 3 in chapter "Canonical Geometric and Algebraic Structures Hidden Behind a Singular Foliation", for a definition of this Lie algebra).

3.2 The Rank at a Point of a Singular Foliation

Given a singular foliation on a smooth, real analytic or complex manifold M, there are two notions that must not be confused: the rank at that point and the dimension of the tangent space at that point.

The *rank* of an $\mathcal{O}$-module $\mathcal{A}$ is the minimal number of its generators. It is denoted by $rk_{\mathcal{O}}(\mathcal{A})$ and takes values in $\mathbb{N} \cup \{+\infty\}$.

Let m be a point in a (smooth, complex, or real analytic) manifold. We say that a sequence $(\mathcal{U}_i)_{i \geq 0}$ of open neighborhoods of m *converges to* m if for any open neighborhood $\mathcal{V}$ of m, there exists i_0 such that for all $i \geq i_0$, we have $\mathcal{U}_i \subset \mathcal{V}$. The third condition (i.e., "locally finitely generated") in Definition 2.18) implies the following result.

> **Proposition 3.15 (The Rank at a Point Is Well-Defined)** *Let $\mathcal{F}$ be a singular foliation on a smooth, complex, or real analytic manifold M. Let $m \in M$ be a point and $(\mathcal{U}_i)_{i \in \mathbb{N}}$ be a sequence of neighborhoods that converges to m. The sequence*
>
> $$i \mapsto rk_{\mathcal{O}_{\mathcal{U}_i}}(\mathcal{F}_{\mathcal{U}_i}).$$
>
> *is finite and constant after a certain rank, and this constant does not depend on the choice of a sequence of open neighborhoods converging to m.*
>
> *It is therefore an integer that depends only on m and $\mathcal{F}$. It is called the* rank of $\mathcal{F}$ *at m, and denoted by* $\mathrm{rk}_m(\mathcal{F})$.

---

[17] See Sect. 3 in chapter "Canonical Geometric and Algebraic Structures Hidden Behind a Singular Foliation" for a definition.

Exercise 3.16 We work in the smooth case only. Let $\mathcal{F}$ be a singular foliation of rank less than or equal to r at every point of the manifold M. Prove that it is finitely generated. *Hint:* use Proposition 2.37.

3.3 The Tangent Space of a Singular Foliation, and Its Dimension

Let $\mathcal{F}$ be a singular foliation on a complex, real analytic or smooth manifold M.

We call *tangent space of $\mathcal{F}$ at $m \in M$* the subspace of $T_m M$, denoted by $T_m \mathcal{F}$, obtained by evaluating at m all vector fields in $\mathcal{F}$, defined in some open neighborhood $\mathcal{U}$ of m in M.

Remark 3.17 If a smooth singular foliation $\mathcal{F}$ is defined through compactly supported vector fields as in Definition 2.1, then $T_m \mathcal{F}$ is defined by:

$$T_m \mathcal{F} := \{X_{|m} \mid X \in \mathcal{F}\}.$$

If a smooth, real analytic or holomorphic singular foliation is defined as a sub-sheaf $\mathcal{F}_\bullet$ of the sheaf $\mathfrak{X}$ of vector fields as in Definition 2.18 or 2.22, then $T_m \mathcal{F}$ is defined[18] by:

$$T_m \mathcal{F} := \cup_{\mathcal{U} \in \mathfrak{V}_m} \{X_{|m} \mid X \in \mathcal{F}_\mathcal{U}\}$$

where $\mathfrak{V}_m$ stands for the set of all open neighborhoods of m in M. However, since singular foliation are locally finitely generated, it follows from the axiom that there exists an open neighborhood $\mathcal{U}$ of m such that:

$$T_m \mathcal{F} := \{X_{|m} \mid X \in \mathcal{F}(\mathcal{U})\}.$$

It suffices to take a neighborhood $\mathcal{U}$ as in the third item in Definition 2.18.

Lemma 3.18 *For every point $m \in M$ in a manifold M equipped with a singular foliation $\mathcal{F}$, the dimension of the tangent space at m is less or equal than the rank of $\mathcal{F}$ at m. In equation:*

$$\dim(T_m \mathcal{F}) \leq \mathrm{rk}_m(\mathcal{F})$$

Proof This follows from the discussion in Remark 3.17.

---

[18] Of course, in the smooth setting, both definitions of $T_m \mathcal{F}$ coincide.

Lemma 3.19 *Let $(M, \mathcal{F})$ be a foliated manifold. The map*

$$\dim^{T\mathcal{F}} : M \to \mathbb{N}$$
$$m \mapsto \dim(T_m\mathcal{F})$$

is lower semi-continuous.

Proof Let $m \in M$ be a point and $r = \dim(T_m\mathcal{F})$. By definition, there exists $X_1, \ldots, X_r \in \mathcal{F}$ such that $X_1|_m, \ldots, X_r|_m$ form a basis of $T_m\mathcal{F}$. There exists a neighborhood $\mathcal{U}$ of m in M such that for every $n \in \mathcal{U}$, $X_1|_n, \ldots, X_r|_n$ are independent, which implies that $T_n\mathcal{F}$ is a vector space of dimension greater or equal to r. This proves that

$$\{m' \in M \mid \dim(T_{m'}\mathcal{F}) \geq r\}$$

is an open subset, which is the content of the lemma. □

3.4 The Regular Part of a Singular Foliation

Let $\mathcal{F}$ be a singular foliation on a manifold M.

> **Definition 3.20 (Regular Point)** A *regular point* of a singular foliation $\mathcal{F}$ on a smooth, complex or real analytic manifold M is a point m_0 in a neighborhood of which the map $\dim^{T\mathcal{F}} : m \mapsto \dim : (T_m\mathcal{F})$ (defined in Lemma 3.19) is constant.
> A point which is not regular is called a *singular point*.

Example 3.21 Any point m where the function $\dim^{T\mathcal{F}}$ reaches its maximal value is a regular point. This is an obvious consequence of lower semi-continuity (see Lemma 3.19).

Remark 3.22 For a smooth singular foliation, there are points which are regular, but are not as in Example 3.21, i.e., are not points where the map $\dim^{\mathcal{F}}$ takes its maximal value.

An example is given as follows. Let us fix $\chi(x)$ a smooth function which is > 0 on $\mathbb{R}_+^*$ and equal to 0 on $\mathbb{R}_-$. A singular foliation on $M = \mathbb{R}$ is given by

$$\mathcal{F} := \left\{ F(x)\chi(x)\frac{\partial}{\partial x} \,\bigg|\, F(x) \in \mathcal{C}^\infty(\mathbb{R}) \right\}.$$

All points of $M = \mathbb{R}$ are regular except 0. On $\mathbb{R}_-^*$, we have $\dim^{T\mathcal{F}} = 0$ while on $\mathbb{R}_+^*$, we have $\dim^{T\mathcal{F}} = 1$.

What Is a Singular Foliation?

In the complex or real analytic settings, if M is connected, it is however true that a point m is regular if and only if $\dim(T_m\mathcal{F})$ reaches its maximal value.

By lower semi-continuity of $\dim^{T\mathcal{F}}$, the subset of all regular points of a singular foliation $\mathcal{F}$ on a smooth, complex or real analytic manifold M is an open subset. By lower semi-continuity again, it is also a dense subset. We call it the *regular part of $\mathcal{F}$* and denoted by M_{reg} (at least when there is no ambiguity on the singular foliation that we consider).

Proposition 3.23 (Regular Part) *The regular part M_{reg} of a singular foliation $\mathcal{F}$ is a dense open subset of M.*

Proof Let $\mathcal{U}$ be an arbitrary neighborhood of an arbitrary point m. Since the restriction to $\mathcal{U}$ of the function $\dim^{T\mathcal{F}}$ takes only finitely many values, there exists $m' \in \mathcal{U}$ where it is maximum. Since it is lower semi-continuous, there is a neighborhood $\mathcal{V} \subset \mathcal{U}$ of m' where this function is constant, i.e., a regular point of $\mathcal{F}$.

Exercise 3.24 Show that each one of the following statements is wrong:

1. The set of singular points of a singular foliation is a submanifold of M.
2. For every $k \in \mathbb{N}$, the set $\{m \in M | \dim(T_m\mathcal{F}) = k\}$ is a manifold[19] of M.
3. Singular points are of measure 0.

Hint: For the last question, one can choose $M = \mathbb{R}$ and $\mathcal{F}$ the singular foliation generated by $\chi(x)\frac{\partial}{\partial x}$ where $\chi(x)$ is a smooth function that vanishes on a fat Cantor subset of $[0, 1]$ (or any subset of $\mathbb{R}$ of empty interior and non-zero measure).

By a *Frobenius regular foliation*, we mean the data, on a manifold N of a distribution $D \subset TN$ which is of constant rank and integrable, i.e., $[\Gamma(D), \Gamma(D)] \subseteq \Gamma(D)$. By the classical Frobenius theorem (cf. e.g.,[1]), every point has a neighborhood which admits local coordinates on which D is generated by $\frac{\partial}{\partial x_1}, \ldots, \frac{\partial}{\partial x_k}$.

Proposition 3.25 *The regular part of a singular foliation is equipped with a "good old" Frobenius regular foliation. Let $(M, \mathcal{F})$ be a foliated manifold and M_{reg} be its regular part. There exists a Frobenius regular foliation D_{reg} on M_{reg} such that the sheaves $\mathcal{F}_{M_{\text{reg}}}$ and $\Gamma(D_{reg})$ coincide.*

---

[19] Foliations that satisfy this property have particular features, studied by David Miyamoto's PhD or [52].

Proof On the regular part, $m \mapsto T_m \mathcal{F}$ is a smooth distribution D of constant rank, whose sections $\Gamma(D)$ are isomorphic to the restriction $\mathcal{F}_{M_{\mathrm{reg}}}$ of the sheaf $\mathcal{F}$ to M_{reg}. Its sections are therefore closed under Lie bracket. Notice that it is also an immediate consequence of the splitting theorems of Sect. 7.3.

In particular, the notions of "singular foliation such that all points are regular" and of "Frobenius regular foliations" coincide.

Remark 3.26 Let M be a complex manifold and $\Sigma \subset M$ a subvariety of codimension ≥ 2. By Hartog's principle, a vector field on $M \setminus \Sigma$ extends to the whole manifold M. Let $D \subset TM$ be a regular integrable distribution on $M \setminus \Sigma$. Its sheaf of sections extends to a sub-sheaf $\mathcal{F}$ of vector fields on $\mathfrak{X}_\bullet$. This sheaf[20] satisfies automatically axioms (α) and (β) of Definition 2.22. It is not clear to us whether this sheaf automatically satisfies the condition (γ) (but this could be the effect of our limited knowledge). For singular foliations which are of codimension 1 on their regular part, it is the case in view of Theorem 4.6.2 in [61]. When Σ has codimension ≥ 3, a theorem of Frisch-Guenot and Siu (cf. e.g., [19] Theorem 2) implies that any coherent sheaf on $M \setminus \Sigma$ extends to a coherent sheaf over M, in particular axiom (γ) is also satisfied.

An Alternative Definition of Regular Points!
Complex geometers may work with an alternative definition of what a regular point is. The issue is that the definition we chose of regular points of a singular foliation conflicts with the definition of regular points of a coherent sheaf. For instance, for Paul Baum and Raoul Bott's [9] or Ali Sinan Sertöz's [62], given $\mathcal{F}$ a holomorphic singular foliation on a complex manifold M, a point $m \in M$ is regular if it admits an open neighborhood $\mathcal{U} \subset M$ such that the restriction $\mathcal{F}_{|\mathcal{U}}$ is a Debord singular foliation (see Sect. 4.2.4). A regular point in our sense is regular in this sense, but the converse is not true in general.

4 Examples of Singular Foliations

The purpose of this section is to give an ordered list of examples of singular foliations.

We start with an exercise (which is in fact the content of Sect. 2.5).

---

[20] Used as a definition of complex singular foliations, for instance in the on-line book "Local dynamics of singular holomorphic foliations" edited by Abate.

Exercise 4.1 Let M be a smooth manifold, and $X_1, \ldots, X_r \in \mathfrak{X}(M)$ be vector fields such that there exists functions $(c_{ij}^k)_{i,j,k\in\{1,\ldots,r\}} \in C^\infty(M)$ with

$$[X_i, X_j] = \sum_{k=1}^{r} c_{ij}^k X_k.$$

1. Show that the space $\mathcal{F}$ of vector fields of the form $\sum_{i=1}^{r} f_i X_i$, with $f_1, \ldots, f_r$ compactly supported functions on M, is a singular foliation[21] on $\mathcal{F}$.
2. State an equivalent result in the real analytic and complex settings, using sheaves.

We also invite the reader to do the next exercise, in order to find counter-examples.

Exercise 4.2 Consider the algebra of smooth functions on $\mathbb{R}^d$.

1. Show that smooth functions on $\mathbb{R}^d$ vanishing at 0 together with their k first partial derivatives, i.e.,

$$\mathcal{I}^k := \left\{ F \in C^\infty(\mathbb{R}^d) \,\middle|\, \frac{\partial^{i_1+\cdots+i_d} F}{\partial x_1^{i_1} \cdots \partial x_d^{i_d}}(0, \ldots, 0) = 0 \right.$$

$$\left. \text{for all } i_1, \ldots, i_d \in \mathbb{N}_0 \text{ with } i_1 + \cdots + i_d \leq k \right\}$$

is an ideal of $C^\infty(\mathbb{R}^d)$ which is finitely generated. Describe explicit generators.

2. Show that the ideal

$$\mathcal{I}^\infty := \left\{ F \in C^\infty(\mathbb{R}^d) \,\middle|\, \frac{\partial^{i_1+\cdots+i_d} F}{\partial x_1^{i_1} \cdots \partial x_d^{i_d}}(0, \ldots, 0) = 0 \text{ for all } i_1, \ldots, i_d \in \mathbb{N}_0 \right\}$$

is not finitely generated. *Hint:* Not easy: one solution is to show that if an ideal $\mathcal{I}$ is finitely generated, then the flow ϕ_X^t of any complete vector field X such that $X[\mathcal{I}] \subset \mathcal{I}$ satisfies $(\phi_t^X)^*(\mathcal{I}) = \mathcal{I}$. Then show that this property is not satisfied[22] for $X := \frac{\partial}{\partial x_1}$.

---

[21] In the sense of Definition 2.1.
[22] To simplify, one can assume that $X = \frac{\partial}{\partial x_1}$ from the very beginning.

4.1 Regular Foliations

Although it seems grammatically problematic, regular foliations are examples of singular foliations. This point was in fact established in Sect. 3.4. This is valid in smooth, complex or real analytic settings.

4.2 Singular Foliations and Lie Algebroids

Let M be a manifold with sheaf of functions $\mathcal{O}_M$ (smooth, real-analytic or holomorphic depending on the context).

Convention In this section, $\mathcal{O}_\mathcal{U}$ (rather than $\mathcal{O}(\mathcal{U})$) will stand for functions over an open subset $\mathcal{U}$. This convention is chosen to be consistent with the convention for sections over $\mathcal{U}$ of a vector bundle $A \to M$, namely $\Gamma_\mathcal{U}(A)$.

Recall that a *Lie algebroid over M* [48] is a triple $(A, \rho, [\cdot, \cdot])$, with A a vector bundle over M, $\rho \colon A \to TM$ a vector bundle morphism over the identity of M called *anchor map*, and $[\cdot, \cdot]$ a Lie bracket on the sheaf of sections of A such that the so-called *Leibniz identity* holds for all $a, b \in \Gamma(A)$, $f \in \mathcal{O}_M$:

$$[a, fb] = f[a, b] + \rho(a)[f]b.$$

Lie algebroids have been extensively reviewed and studied, see e.g., [13, 48, 50]. We will show that any Lie algebroid induces a singular foliation. The following Lemma holds true in the smooth, real-analytic or holomorphic categories.

Lemma 4.3 *For any Lie algebroid $(A, \rho, [\cdot, \cdot])$ over M, and any open subset $\mathcal{U} \subset M$ the anchor map $\Gamma_\mathcal{U}(A) \to \mathfrak{X}(\mathcal{U})$ is a Lie algebra morphism.*

Proof Recall that $\Gamma_\mathcal{U}$ stands for the $\mathcal{O}_V$-module of sections of A over an arbitrary open subset $V \subset M$. The Jacobi identity on $\Gamma_V(A)$ implies that for any $a, b, c \in \Gamma_V(A)$ and any $f \in \mathcal{O}_V$:

$$(\mathrm{ad}_a \circ \mathrm{ad}_b - \mathrm{ad}_b \circ \mathrm{ad}_a)(fc) = \mathrm{ad}_{[a,b]}(fc). \qquad (8)$$

Now, there are three kinds of terms that will appear in the previous equation (8) if one uses the Leibniz identity to allow the function f to "get out":

1. Those for which f will be "differentiated" twice: these terms appear on the left-hand side of the equation (8) as $\rho(a) \circ \rho(b)[f] - \rho(b) \circ \rho(a)[f]c$.
2. Those for which f will be "differentiated" only once, there are two types of such terms
   (2a) $\rho([a, b])[f]c$ on the right-hand side of (8)

(2b) $\rho(a)[f][b,c] + \rho(b)[f][a,c] - \rho(a)[f][b,c] - \rho(b)[f][a,c]$ on the left-hand side of (8). These terms cancel out.
3. Those where f is not "differentiated" at all, those terms are
   (3a) $f(\mathrm{ad}_a \circ \mathrm{ad}_b - \mathrm{ad}_b \circ \mathrm{ad}_a)(c)$ on the left-hand side of (8)
   (3b) $f\,\mathrm{ad}_{[a,b]}(c)$ on the right-hand side of (8)

The Jacobi identity being satisfied for triples of sections of A, the terms (3a) and (3b) cancel each other out. Also, the terms (2b) add up to zero. Hence, the terms 1 and (2a) are the only ones remaining. They transform Eq. (8) into the relation:

$$((\rho([a,b]) - [\rho(a), \rho(b)])[f])\,c = 0.$$

Now, let us assume that $\mathcal{V}$ is chosen such that a nowhere vanishing section $c \in \Gamma_\mathcal{V}(A)$ exists. We then have $(\rho([a,b]) - [\rho(a), \rho(b)])[f] = 0$. Assume also that $\mathcal{V}$ is chosen such that any covector is the differential of at least a function $f \in \mathcal{O}_\mathcal{V}$ (which is always true in the smooth case, but is only true for $\mathcal{V}$ "small" enough in the real analytic or complex cases). This implies that ρ is a Lie algebra morphism from $\Gamma_\mathcal{V}(A)$ to $\mathfrak{X}(\mathcal{V})$. Since every point $m \in M$ admits such a neighborhood $\mathcal{V}$, however, this implies that ρ is a Lie algebra morphism when restricted to any open subset $\mathcal{U} \subset M$.

At this point, it is more convenient to distinguish the smooth case from the complex and real analytic ones.

4.2.1 The Smooth Case

Let us consider that singular foliations on a smooth manifold M are defined as in Definition 2.18, through compactly supported vector fields. Let $(A \to M, [\cdot, \cdot], \rho)$ be a smooth Lie algebroid over M. Let $\mathcal{F} := \rho(\Gamma_c(A))$, where Γ_c stands for compactly supported sections. Lemma 4.3 implies that $\mathcal{F}$ is closed under Lie bracket. The remaining axioms are obviously satisfied. Hence, the following result holds true.

> **Proposition 4.4 (Image Through Anchor Map of Lie algebroids—Smooth Case)**
> *The image through the anchor map of compactly supported sections of a Lie algebroid over M form a singular foliation on M.*

4.2.2 The Complex or Real-Analytic Case

Proposition 4.4 can not be extended immediately from the smooth context to the complex or real analytic contexts altogether.[23] We denote by $\mathcal{O}$ the sheaf of real-analytic or

---

[23] The presentation here also essentially works on affine varieties or schemes.

holomorphic functions. Again, for $A \to M$ a vector bundle, we denote by $\Gamma_{\mathcal{U}}(A)$ the sections of A over an open subset $\mathcal{U}$. Of course, $\Gamma_{\mathcal{U}}(A)$ is a $\mathcal{O}_{\mathcal{U}}$-module, and, assigning to an open subset the sections over it

$$\mathcal{U} \mapsto \Gamma_{\mathcal{U}}(A),$$

one defines a sheaf of $\mathcal{O}$-modules over M. The technical difficulty that appears at this point is that singular foliations are supposed to be sheaves, but

$$\mathcal{U} \mapsto \rho(\Gamma_{\mathcal{U}}(A))$$

is not a sheaf on M (and therefore not a sub-sheaf of the sheaf $\mathfrak{X}$ of vector fields on M). It is only a pre-sheaf.[24] But the difficulty can be circumvented: to turn it into a sheaf, one has to "sheafify" it, i.e., to map an open subset $\mathcal{U} \subset M$ to the sub-$\mathcal{O}_{\mathcal{U}}$-module of vector fields $X \in \mathfrak{X}(\mathcal{U})$ on $\mathcal{U}$ such that every $m \in \mathcal{U}$ admits a neighborhood $\mathcal{V}$ on which there exists $a \in \Gamma_{\mathcal{V}}(A)$ with $\rho(a) = X$ (on $\mathcal{V}$). This defines a sheaf of $\mathcal{O}$-modules $\underline{\rho(\Gamma(A))}$ that we call the *image of the Lie algebroid* $(A, \rho, [\cdot, \cdot])$ *through its anchor map*.

> **Proposition 4.5 (Image Through Anchor Map of Lie Algebroids: Complex Case)** Let $(A, \rho, [\cdot, \cdot])$ be a Lie algebroid over a complex or real analytic manifold M. The sheaf $\underline{\rho(\Gamma(A))}$ (=image of the Lie algebroid $(A, \rho, [\cdot, \cdot])$ through its anchor map) is a singular foliation on M.

Proof This statement is an immediate consequence of Lemma 4.3.

Exercise 4.6 Let X be a vector field on a manifold M. Show that $\mathcal{F}_X = \{fX | f \in \mathcal{C}_c^\infty(M)\}$ is a singular foliation on M that comes from a Lie algebroid of rank 1.

Exercise 4.7 (See [40], Section 3.1.4) Let $\mathcal{F}$ be a singular foliation and $\varphi \in \mathcal{C}^\infty(M)$ a function. Check that

$$\varphi \mathcal{F} := \{\varphi X, X \in \mathcal{F}\}$$

is a singular foliation again. Show that if $\mathcal{F}$ is the image through the anchor map of a Lie algebroid, so is $\varphi \mathcal{F}$.

---

[24] For instance, if A has no globally defined non-constant functions, e.g. A is trivial and M is compact.

4.2.3 Lie Algebra Actions

Let $\mathfrak{g}$ be a Lie algebra. We call a Lie algebra morphism $\mathfrak{g} \to \mathfrak{X}(M)$ a *Lie algebra action* of $\mathfrak{g}$ on M. We denote it by $x \mapsto \underline{x}$. In particular, any Lie group action of a Lie group G on a manifold M induces a Lie algebra action of its Lie algebra $\mathfrak{g}$. The $\mathcal{C}_c^\infty(M)$-module[25] generated by $\{\underline{x}, x \in \mathfrak{g}\}$ is a singular foliation. We have therefore proven the first part of the following result:

Proposition 4.8 (Lie Group Actions) *The infinitesimal (Lie algebra) action of a Lie group action G on a manifold M induces a singular foliation on M. It is the image through the anchor map of a Lie algebroid called transformation Lie algebroid.*

Proof The first part of the statement is already proven. For the second part, consider the trivial vector bundle $\mathfrak{g} \times M \to M$ equipped with the anchor ρ and bracket $[\cdot, \cdot]$ defined on constant sections, which are identified with elements of $\mathfrak{g}$, by

$$\rho(x) = \underline{x} \text{ and } [x, y] = [x, y]_\mathfrak{g}$$

where $x, y \in \mathfrak{g}$ are seen as constant sections of A, and $[x, y]_\mathfrak{g}$ is the bracket of $\mathfrak{g}$. This describes a Lie algebroid structure [48].

4.2.4 Projective or "Debord" Singular Foliations

Here is an important class of singular foliations that come from a Lie algebroid. We will state the results in the smooth case, and leave the generalization to the reader for the complex or real analytic settings.

Definition 4.9 ([17] Generators and No Relations) We say that a singular foliation on a smooth manifold M is *Debord* if $\mathcal{F}$ is a projective $\mathcal{C}^\infty(M)$-module.[a]

---

[a] In the complex or real analytic settings, one has to require $\mathcal{F}$ to be a projective sheaf with respect to the sheaf of functions. Since the rank of $\mathcal{F}$ is (locally) finite, it is equivalent to say that every point has a neighborhood $\mathcal{U}$ where $\mathcal{F}_\mathcal{U}$ is a free $\mathcal{O}_\mathcal{U}$-module.

---

[25] Recall that $\mathcal{C}_c^\infty(M)$ stands for compactly supported smooth functions.

In a concrete manner, Debord foliations are those which admit, in a neighborhood $\mathcal{U}$ of every point, generators $X_1, \ldots, X_r$ between which there is no relation. I.e., if

$$\sum_{i=1}^{r} f_i X_i = 0,$$

then all of the functions $f_1, \ldots, f_r \in \mathcal{C}^\infty(\mathcal{U})$ are zero.

Remark 4.10 Equivalently, we could use the language of Definition 2.18 to define Debord foliations in the smooth case.

Exercise 4.11 Show that the restriction of a Debord singular foliation to an open subset is still a Debord singular foliation.

Remark 4.12 In addition of honoring the important discoveries of Claire Debord about them in [17], the name "Debord foliation" is encouraged by a very practical perspective. Saying "projective foliations" would be ambiguous, since it could mean "foliations on a projective variety".

Debord foliations are dealt with in this section, because they always arise from a Lie algebroid. By the smooth Serre-Swan theorem [58], there exists a vector bundle $A \to M$ and a $\mathcal{C}^\infty(M)$-module isomorphism

$$\Gamma_c(A) \simeq \mathcal{F}.$$

Composing this $\mathcal{C}_c^\infty(M)$-module isomorphism with the inclusion

$$\Gamma_c(A) \simeq \mathcal{F} \hookrightarrow \mathfrak{X}_c(M),$$

we obtain an inclusion $\Gamma_c(A) \hookrightarrow \mathfrak{X}_c(M)$. Since it is a morphism of $\mathcal{C}_c^\infty(M)$-modules, it has to be given by a vector bundle morphism:

$$\rho \colon A \to TM,$$

that we call *anchor map*. The anchor ρ is injective at the level of sections. This does <u>not</u> imply that the anchor ρ needs to be injective at all points of M, but it certainly has to be injective on a dense open subset of M. Lastly, the isomorphism $\Gamma_c(A) \simeq \mathcal{F}$ extends to non-compactly supported sections, and equips $\Gamma(A)$ with a Lie bracket, for which ρ is easily seen to be an anchor map. This proves the following statement:

> **Proposition 4.13 (Debord Algebroids)** *A singular foliation on a smooth, real analytic or complex manifold M is Debord if and only if it is the image of a Lie algebroid whose anchor map is injective on a dense open subset of M.*

Exercise 4.14 Show that the singular foliation on $\mathbb{R}^2$ generated by $x\frac{\partial}{\partial x}$ and $y\frac{\partial}{\partial y}$ is Debord.

Exercise 4.15 Show that the singular foliation in Exercise 4.6 is Debord, at least if there is no open subset where the vector field X is identically zero. Show that the anchor is not injective at a point m where $X_{|m} = 0$.

Exercise 4.16 Show that compactly supported vector fields on a manifold M vanishing on a codimension 1 submanifold form a Debord singular foliation. This theme will be developed further in Sect. 4.5 below.

Example 4.17 An interesting example of a Debord singular foliation will be given in Sect. 4.8, see Example 4.51.

4.3 Vector Fields Vanishing at a Point at Prescribed Order

We can also construct singular foliations by playing with the order of vanishing of vector fields at certain points. We decided to place ourselves within the context of smooth differential geometry in the discussion, but the holomorphic or real-analytic contexts would work as well in more or less a similar manner, and the conclusions are written in a way that merge all three contexts. Let us start with an exercise.

Exercise 4.18 Let $\mathcal{F}_1$ be the space of all compactly supported smooth vector fields on $\mathbb{R}^n$ vanishing at 0. Show that $\mathcal{F}_1$ is a finitely generated singular foliation (see Definition 2.32) generated by the finite family of vector fields

$$\left\{ x_i \frac{\partial}{\partial x_j} \,\middle|\, i, j \in \{1, \ldots, n\} \right\}.$$

Hint: use the so-called "Hadamard's lemma", i.e., the fact that any compactly supported smooth function F on $\mathbb{R}^n$ vanishing at 0 decomposes as

$$F = \sum_{i=1}^n x_i F_i$$

for some compactly supported smooth functions $F_1, \ldots, F_n \in \mathcal{C}^\infty(\mathbb{R}^n)$.

This exercise can be easily generalized. We say that a vector field X on $\mathbb{R}^n$ vanishes to order 2 at the origin if its coefficients

$$X = \sum_{i=1}^{n} F_i(x_1, \ldots, x_n) \frac{\partial}{\partial x_i}$$

satisfy that

$$F_i(0, \ldots, 0) = 0 \text{ and } \frac{\partial F_i}{\partial x_j}(0, \ldots, 0) = 0 \text{ for all } i, j = 1, \ldots, n.$$

It is a classical result that a compactly supported smooth function F on $\mathbb{R}^n$ vanishes at $0 \in \mathbb{R}^n$ together with its differential if and only if it decomposes as

$$F = \sum_{i,j=1}^{n} x_i x_j F_{i,j}$$

for some compactly supported smooth functions $F_{i,j} \in \mathcal{C}^\infty(\mathbb{R}^n)$. Let $\mathcal{F}_2$ be the space of such vector fields. We leave it as an exercise to the reader to check that

1. Show that $\mathcal{F}_2$ is generated, as a $\mathcal{C}_c^\infty(\mathbb{R}^n)$-module, by the family

$$\left\{ x_i x_j \frac{\partial}{\partial k} \,\middle|\, 1 \leq i \leq j \leq n \text{ and } k = 1, \ldots, n \right\}.$$

2. Show that $\mathcal{F}_2$ is stable under Lie bracket.
3. Conclude that $\mathcal{F}_2$ is a finitely generated (see Definition 2.32) singular foliation on $\mathbb{R}^n$.

Let us generalize the previous two exercises. For every $k \in \mathbb{N}$, let $\mathcal{F}_k$ be the space of compactly supported smooth vector fields on $\mathbb{R}^n$ that vanish at 0 together with their partial derivatives of order i for $i \leq k-1$.

1. It is straightforward that $\mathcal{F}_k$ is a $\mathcal{C}^\infty(\mathbb{R}^n)$-module stable under Lie bracket.
2. There is an identification $\mathcal{F}_k = \mathcal{I}_0^k \mathfrak{X}_c(M)$ where $\mathcal{I}_0$ is the ideal of smooth functions on $\mathbb{R}^n$ vanishing at the origin.
3. An explicit family of generators of $\mathcal{F}_k$ over $\mathcal{C}^\infty(\mathbb{R}^n)$ is therefore given by:

$$\left\{ x_{i_1} \cdots x_{i_k} \frac{\partial}{\partial x_j} \,\middle|\, 1 \leq i_1 \leq \cdots \leq i_k \leq n \text{ and } j = 1, \ldots, n \right\}.$$

4. $\mathcal{F}_k$ is therefore a finitely generated singular foliation in the sense of Definition 2.32.

For every $k \geq 1$, the singular foliation $\mathcal{F}_k$ can also be seen as a complex, real analytic or algebraic singular foliations on $\mathbb{K}^n$ with $\mathbb{K} = \mathbb{R}$ or $\mathbb{C}$ depending on the context. In all these realms, it is true that $\mathcal{F}_k = \mathcal{I}_0^k \mathfrak{X}_\bullet$ where $\mathcal{I}_0$ is the ideal of relevant (sheaf of) functions vanishing at the origin. Of course, this discussion can be enlarged to any point in a smooth or complex manifold. In conclusion:

Proposition 4.19 (Vector Fields Vanishing at Given Orders at Given Points) *Let M be a smooth, real analytic or complex manifold. For every point $m \in M$, and every choice of an integer $k \geq 1$, the space of vector fields on M vanishing together with their $k - 1$ first derivatives at m form a singular foliation on M.*

Exercise 4.20 Find all possible relations between the generators of $\mathcal{F}_k$, for any $k \geq 1$.

Exercise 4.21 This exercise supposes the notion of leaves. Find the leaves of each one of the singular foliations $\mathcal{F}_k$ for any $k \geq 1$.

Exercise 4.22 Can k be replaced by $+\infty$ in Proposition 4.19? *Hint:* Look at the second item in Exercise 4.2)

Proposition 4.19 generalizes to a family of distinct points $m_1, \ldots, m_d$ in an arbitrary manifold M, and *orders* $k_1, \ldots, k_d \in \mathbb{N}$. In the smooth case, one can then consider all compactly supported vector fields on M whose order of vanishing at the point m_i is greater or equal to k_i for all $i = 1, \ldots, d$. One can even choose a sequence $(m_i)_{i \in \mathbb{N}}$ of points and $(k_i)_{i \in \mathbb{N}}$ orders, provided that it has no accumulation point. In particular, the Androulidakis-Zambon's "non-finitely-many-generators" singular foliation of Exercise 2.36 is of that type.

4.3.1 More Sophisticated Examples

This example (inspired by Grabowska and Grabowski [32]), appeared in [41], Example 1.11. We present it as a real analytic singular foliation on $\mathbb{R}^n$ (we could see of course also see it as a complex singular foliation on $\mathbb{C}^n$ or a smooth one).

On $M = \mathbb{R}^n$, we attribute to the canonical coordinates $(x_1, \ldots, x_n)$ the strictly positive weights $(i_1, \ldots, i_n)$. Equipped with this weight, the ring $\mathcal{A}$ of real analytic functions on M becomes a graded algebra.

$$\mathcal{A} = \oplus_{i=0}^\infty \mathcal{A}_i.$$

It is also a filtered algebra, with respect to the filtration:

$$\mathcal{A}^{\geq k} = \oplus_{i=k}^{\infty} \mathcal{A}_i.$$

Example 4.23 Assume $i_1 = 1, i_2 = 2$ and so on. The weight of $x_1^3 x_3^2 x_5$ is $1 \times 3 + 3 \times 2 + 1 \times 5 = 14$, so that $x_1^3 x_3^2 x_5 \in \mathcal{A}_{14}$.

Let k be a non-negative integer. The space of real analytic vector fields X such that

$$X\left[\mathcal{A}^{\geq n}\right] \subset \mathcal{A}^{\geq n+k} \text{ for all } n \in \mathbb{N}$$

is a module, that we denote by $\mathcal{F}_k$, over real analytic functions. It is stable under Lie bracket. It is generated by the family

$$\left\{ x_1^{j_1} \ldots x_n^{j_n} \frac{\partial}{\partial x_a} \,\middle|\, i_1 j_1 + i_2 j_2 + \cdots + i_n j_n \geq j_a + k \right\}.$$

If $(j_1, \ldots, j_n)$ satisfies the above condition, then so does $(j'_1, \ldots, j'_n)$ as long as $j'_i \geq j_i$ for all indices $i = 1, \ldots, n$. This implies that the generating family of $\mathcal{F}_k$ can be chosen to be finite. Therefore, $\mathcal{F}_k$ is finitely generated, and is a real analytic singular foliation.

4.4 Singular Foliations Attached to a Sub-variety (I): The Algebraic Case

We now work within the context of complex algebraic geometry. Let $\mathcal{O}$ be the algebra of polynomial functions on an affine variety M. Recall that $\mathcal{O}$ is a quotient of the form:

$$\mathcal{O} = \frac{\mathbb{C}[x_1, \ldots, x_n]}{\mathcal{I}_M} \tag{9}$$

with $\mathcal{I}_M$ a prime ideal of $\mathbb{C}[x_1, \ldots, x_n]$.

Example 4.24 The reader not familiar with algebraic geometry can assume $M = \mathbb{C}^n$ so that $\mathcal{O} = \mathbb{C}[x_1, \ldots, x_n]$ is the algebra of polynomials in n variables.

Recall that, by definition, the $\mathcal{O}$-module $\mathfrak{X}(M)$ of vector fields on M is the $\mathcal{O}$-module $\mathfrak{X}(M)$ of derivations of $\mathcal{O}$.

Example 4.25 For $M = \mathbb{C}^n$, vector fields are simply expressions of the form

$$\sum_{i=1}^{n} P_i(x_1, \ldots, x_n) \frac{\partial}{\partial x_i},$$

and are uniquely determined by the polynomial functions $(P_i(x_1,\ldots,x_n))_{i=1,\ldots,n}$. As an $\mathcal{O}$-module therefore, $\mathfrak{X}(M) = \mathcal{O}^n$.

We recall the following Lemma.

Lemma 4.26 *The $\mathcal{O}$-module $\mathfrak{X}(M)$ of vector fields on an affine variety is of finite rank. In particular, it is Noetherian.*

Proof Any vector field X on M is determined by its values on the functions $\bar{x}_1,\ldots,\bar{x}_n$ (the horizontal refers to the quotient in Eq. (9)). In particular, vector fields are a finitely generated $\mathcal{O}$-module. Now, since $\mathcal{O}$ is Noetherian, so is any finite rank $\mathcal{O}$-module, which concludes the proof.

Let $W \subset M$ be an affine sub-variety, i.e., the zero locus[26] of some prime ideal $\mathcal{I}_W \subset \mathcal{O}$. Since $\mathcal{O}$ is Noetherian, this ideal has finitely many generators $\varphi_\bullet = (\varphi_1,\ldots,\varphi_k)$.

Example 4.27 On $M = \mathbb{C}^n$, an affine subvariety W is a subset given by $\phi_1 = \cdots = \phi_k = 0$ where $\phi_1,\ldots,\phi_k$ generate a prime ideal.

In algebraic geometry, geometrical properties have to be translated in a purely algebraic language. For instance, we say that a vector field X *vanishes on* W if $X[\mathcal{O}] \subset \mathcal{I}_W$ and is *tangent to* W if $X[\mathcal{I}_W] \subset \mathcal{I}_W$. A vector field vanishes on W if and only if it belongs to $\mathcal{I}_W \mathfrak{X}(M)$.

X vanishes on W	$\Leftrightarrow$ $X(w) = 0\ \forall w \in W$,
X is tangent to W	$\Leftrightarrow$ $X \in T_w W$ for every regular point of W.

Here is our main statement:

Proposition 4.28 (Two Foliations Associated to an Affine Variety) *Let $W \subset \mathbb{C}^n$ be an affine variety. Both vector fields on $\mathbb{C}^n$ tangent to W and vector fields on $\mathbb{C}^n$ vanishing on W are algebraic singular foliations.[a]*

---

[a]I.e., are finitely generated sub-$\mathcal{O}$-modules of vector fields on M (=derivations of $\mathcal{O} = \mathbb{C}[x_1,\ldots,x_n]$) stable under Lie bracket.

---

[26] = the subset of points where all elements in $\mathcal{I}_W$ vanish.

Proposition 4.28 is a direct consequence of the following more general result.

Proposition 4.29 *For any ideal $\mathcal{I} \subset \mathcal{O}$, the following families are algebraic singular foliations.*

1. *The $\mathcal{O}$-module $\mathfrak{X}_{\mathcal{I}}$ of all vector fields $X \in \mathfrak{X}(M)$ such that $X[\mathcal{I}] \subset \mathcal{I}$.*
2. *$\mathcal{I}\mathfrak{X}(M)$.*

Proof By Lemma 4.26, it suffices to check that the previous sets are $\mathcal{O}$-modules stable under Lie bracket, which is straightforward.

Exercise 4.30 Let ideal $\mathcal{I}_W \subset \mathcal{O} = \mathbb{C}[z_1, \ldots, z_N]$ be the ideal of functions vanishing on an affine variety W.

1. Show that the space of all vector fields X on $\mathbb{C}^N$ such that $X[\mathcal{I}_W] \subset \mathcal{I}_W^2$ is an algebraic singular foliation.
2. Is this algebraic singular foliation really different from the algebraic singular foliation of vector fields vanishing on W?

There are of course more examples. Let M' be a second affine variety with algebra of functions $\mathcal{O}'$. Let $\phi \colon M \to M'$ be an affine map, i.e., a map such that $\phi^* \colon \mathcal{O}' \to \mathcal{O}$ is an algebra morphism. A vector field $X \in \mathfrak{X}(M)$ such that $X[\varphi^*(F)] = 0$ for all $F \in \mathcal{O}'$ is called *tangent to the fibers of ϕ*.

Example 4.31 For $M = \mathbb{C}^n$ and $M' = \mathbb{C}^{n'}$, ϕ is given by a n'-tuple of polynomial functions, i.e.,

$$\phi \colon (z_1, \ldots, z_n) \mapsto (\phi_1(z_1, \ldots, z_n), \ldots, \phi_{n'}(z_1, \ldots, z_n))$$

and vector fields tangent to fibers are those vector fields X such that $X[\phi_1] = \cdots = X[\phi_{n'}] = 0$.

Vector fields tangent to the fibers of ϕ form an $\mathcal{O}$-module stable under Lie bracket. Here is an obvious consequence of Lemma 4.26.

Proposition 4.32 *For any map of affine varieties $\phi \colon M \to M'$, vector fields on M tangent to fibers of ϕ form an algebraic singular foliation on M.*

Let us finish this section with the following remark.

Remark 4.33 When M is a smooth affine variety, i.e., when M has no singular points (e.g., $M = \mathbb{C}^n$), then M is also a complex manifold. All algebraic singular foliations constructed above, and more generally any algebraic singular foliation $\mathcal{F}$ on M, may be seen as a complex singular foliation: it suffices to consider the sheaf of all vector fields which are linear combinations, with coefficients in holomorphic functions, of vector fields in $\mathcal{F}$. In short, it suffices to take the tensor product of $\mathcal{F}$ with holomorphic functions.

4.5 Singular Foliations Attached to a Submanifold (II): The Smooth or Complex Cases

This section makes sense in the smooth, real analytic or complex contexts indifferently.

Proposition 4.34 (Vector Fields Tangent to L of Vanishing Along L) *Let L be a submanifold of M, and $k \in \mathbb{N}$ an integer. Both*

1. *the space of vector fields tangent to L,*
2. *and the space of vector fields vanishing at order k at all points in L*

are singular foliations.

Proof The proof consists in

1. Checking that the space $\mathfrak{X}_L(M)$ of all vector fields on M tangent to the sub-manifold L,
   (a) is a module over functions,
   (b) is stable under Lie bracket,
   (c) and that, in any local coordinates $(x_1, \ldots, x_a, y_1, \ldots, y_b)$ where L is given by $0 = y_1 = \cdots = y_b$, it is generated by

$$\left\{ \frac{\partial}{\partial x_i},\, y_j \frac{\partial}{\partial y_k} \,\Big|\, 1 \leq i \leq a \text{ and } 1 \leq j, k \leq b \right\}$$

2. Then in checking that the second space is algebraically described by $\mathcal{I}_L^k \mathfrak{X}(M)$, where $\mathcal{I}_L$ stands for the ideal of functions vanishing on L. Since the ideal $\mathcal{I}_L$ is locally finitely generated, this completes the proof.

Exercise 4.35 Let $L_1, L_2 \subset M$ be submanifolds of M that intersect transversally, i.e., such that:

$$T_x L_1 + T_x L_2 = T_x M \qquad \forall x \in L_1 \cap L_2.$$

Consider the space of all vector fields on M tangent to both L_1 and L_2. Show that it is a singular foliation.

Exercise 4.36 We owe to [28]–[10] the idea of the following exercise.[27] Let M be a smooth manifold, and Σ a submanifold defined as the zero locus of a function $\phi \colon M \to \mathbb{R}$. We assume the differential of ϕ to be non-zero at each point of Σ.

1. Show that the space of all vector fields X such that $X[\phi] \in \mathcal{I}_\Sigma^2$, with $\mathcal{I}_\Sigma$ the ideal of smooth functions vanishing on Σ, is a singular foliation on M.
2. Show that it contains vector fields vanishing at order ≥ 2 along Σ.
3. Show that its leaves are the connected components of Σ and $M\backslash\Sigma$.
4. Extend the previous results to any sub-manifold N defined as the zero locus of functions $\phi_1, \ldots, \phi_k$ which are independent in a neighborhood of any point of N.

In [28] and [10], it is explained that such a singular foliation plays the role of "vector fields tangent up to order 2 to the submanifold Σ".

4.6 Hamiltonian Vector Fields and Singular Foliations

Let M be a smooth or holomorphic Poisson structure. We denote by $\mathcal{O}_\bullet$ the corresponding sheaf of functions. Recall [43] that a Poisson structure is a skew-symmetric biderivation $\{\cdot, \cdot\}$ on the sheaf $\mathcal{O}_\bullet$ of functions that satisfies the Jacobi identity. More explicitly, for every open subset $\mathcal{U} \subset M$, one is given a biderivation

$$\begin{aligned} \mathcal{O}_\mathcal{U} \times \mathcal{O}_\mathcal{U} &\longrightarrow \mathcal{O}_\mathcal{U} \\ (f, g) &\mapsto \{f, g\} \end{aligned},$$

compatible with restrictions, and which satisfies the Jacobi identity for all $f, g, h \in \mathcal{O}_\mathcal{U}$:

$$\{f, \{g, h\}\} = \{\{f, g\}, h\} + \{g, \{f, h\}\}.$$

Since $\{\cdot, \cdot\}$ is a biderivation, for any function $h \in \mathcal{O}_\mathcal{U}$, the map

$$f \mapsto \{f, h\}$$

is a derivation of the sheaf $\mathcal{O}_\bullet$. It is therefore given by a vector field X_h which is called the *Hamiltonian vector field of the function h*.

---

[27] Related to Exercise 4.30.

Proposition 4.37 (A Poisson Structure Induces a Singular Foliation) *For any Poisson structure on a manifold M, the $\mathcal{O}$-module generated by Hamiltonian vector fields form a singular foliation on M.*

It is called the **symplectic**[a] *foliation of* $\{\cdot,\cdot\}$.

---

[a] We are not able in the lecture to explain this name: it is justified by the non-trivial observation that the leaves of this singular foliation are naturally equipped with a symplectic structure, see Chapter IV in [43].

Proof Let $\mathcal{F}$ be the sheaf generated by vector fields of the form fX_h with f, h local functions on M. It follows from the Jacobi identity that for any local functions $h_1, h_2 \in \mathcal{O}_\mathcal{U}$ $[X_{h_1}, X_{h_2}] = -X_{\{h_1,h_2\}}$ so that for any local functions $f_1, f_2 \in \mathcal{O}_\mathcal{U}$:

$$[f_1 X_{h_1}, f_2 X_{h_2}] = -f_1 f_2 X_{\{h_1,h_2\}} + f_1 \{f_2, h_1\} X_{h_2} - f_2 \{f_1, h_2\} X_{h_1}. \tag{10}$$

This proves that $\mathcal{F}$ is closed under Lie bracket. Let us prove that it is locally finitely generated. Let $m \in M$ be a point and $(x_1, \ldots, x_d)$ be a local chart on a coordinate neighborhood $\mathcal{V}$. Then for any function $h(x_1, \ldots, x_n) \in \mathcal{O}_\mathcal{V}$, one has:

$$X_h = \sum_{i=1}^n \frac{\partial f}{\partial x_i} X_{x_i}.$$

This is *not* obvious, but it follows from the axioms, see Chapter I in [43]. In particular, the family of vector fields

$$X_{x_1}, \ldots, X_{x_d}$$

generates $\mathcal{F}$ on $\mathcal{V}$. In particular, any point has a neighborhood on which the number of generators is bounded by the dimension of the manifold.

It is a classical result that for any Poisson structure $\{\cdot,\cdot\}$ on a manifold M there exists a vector bundle morphism:

$$\pi^\#: T^*M \longrightarrow TM$$

which is skew-symmetric (and therefore comes from a section $\pi \in \Gamma(\wedge^2 TM)$) such that for any two functions f and $g \in \mathcal{O}_\mathcal{U}$:

$$\{f, g\} = \langle \pi^\#(df), dg \rangle$$

where $\langle \cdot, \cdot \rangle$ stands for the duality pairing between TM and T^*M. It is also a classical result[28] that T^*M has a Lie algebroid structure, whose bracket $[\cdot, \cdot]$ is characterized by the two following properties:

1. its anchor map is $\pi^\#$,
2. on sections of T^*M, i.e., exact 1-forms, it is related to the Poisson structure by[29]
   $[df, dg] = -d\{f, g\}$ for all open $\mathcal{U} \subset M$ and $f, g \in \mathcal{O}_\mathcal{U}$.

We call it the *cotangent Lie algebroid*, see [13]. The following result is obvious, at least if one accepts the existence of the cotangent Lie algebroid.

Lemma 4.38 *The symplectic foliation in Proposition 4.37 is the image of the anchor map of the cotangent Lie algebroid. In particular, $T_m \mathcal{F} = \mathrm{Im}(\pi_m^\#)$ for all $m \in M$.*

Now, there are more singular foliations that are attached to Poisson structures. Here are some of them. First, for any Poisson subalgebra, i.e., any sheaf of sub-algebra $\mathcal{B}_\bullet \subset \mathcal{O}_\bullet$ such that $\{\mathcal{B}_\mathcal{U}, \mathcal{B}_\mathcal{U}\} \subset \mathcal{B}_\mathcal{U}$ for every open $\mathcal{U} \subset M$, the sheaf of $\mathcal{O}_\bullet$-sub-modules $\mathcal{F}_\mathcal{B} \subset \mathfrak{X}_\bullet$ generated the Hamiltonian vector fields of functions in $\mathcal{B}_\bullet$ is stable under Lie bracket in view of Eq. (10). Therefore, as soon as $\mathcal{F}_\mathcal{B}$ is finitely generated, it becomes a singular foliation on M. It happens in particular in the following two contexts.

1. Let M' be a manifold equipped with a Poisson structure $\{\cdot, \cdot\}'$. We say that a map $\phi \colon M \to M'$ is a *Poisson map* if the pull-back map $\phi^* \colon \mathcal{O}'_\mathcal{U} \to \mathcal{O}_{\phi^{-1}(\mathcal{U})}$ is a Lie algebra morphism. Let $\mathcal{B}_\bullet := \phi^* \mathcal{O}'_\bullet$ be the subsheaf of all functions pulled back from functions on M'. This is clearly a Poisson sub-algebra. Moreover, $\mathcal{F}_\mathcal{B}$ is then locally finitely generated. Indeed, for any $m \in M$, it is generated by the Hamiltonian vector fields

   $$X_{\phi^* x'_1}, \ldots, X_{\phi^* x'_{d'}}$$

   with $x'_1, \ldots, x'_{d'}$ being local coordinates in a neighborhood of $\phi(m)$ in M'.
2. We say that a submanifold $N \subset M$ is *coisotropic* if the ideal $\mathcal{I}_N$ of functions vanishing on N is a Poisson subalgebra of $\mathcal{O}$. This is equivalent to require that the Hamiltonian vector field X_h is tangent to N for every $h \in \mathcal{I}_N$, or to assume that for every $p \in N$,

---

[28] See the classical [12] (written in French) or the more recent [13] for an excellent introduction to the subject.

[29] The minus sign can be turned into a "+", it is a matter of convention.

the vector bundle morphism $\pi^{\#}$ maps the annihilator $T_pN^{\perp}$ of $T_pN \subset T_pM$ inside T_pN. Again, $\mathcal{F}_{\mathcal{I}_N}$ is finitely generated. Near any point not in N, $\mathcal{F}_{\mathcal{I}_N}$ coincides with the symplectic singular foliation, while near every point in N, it is generated by the vector fields

$$\{X_{y_i}, y_j X_{x_c} \mid i, j = 1, \ldots, k, \text{ and } c = k+1, \ldots, n\}$$

where $y_1, \ldots, y_k, x_{k+1}, \ldots, x_n$ are local coordinates into which N is given by $y_1 = \cdots = y_k = 0$.

3. The singular foliation $\mathcal{F}_{\mathcal{I}_N}$ restricts to a singular foliation on N, which is generated by the restrictions to N of the Hamiltonian vector fields $X_{y_1}, \ldots, X_{y_k}$. Equivalently, it is the singular foliation[30] on N generated by restrictions to N of Hamiltonian vector fields of functions vanishing on N.

Exercise 4.39 Show that the Hamiltonian vector fields of a Liouville integrable system [43] generate a singular foliation.

Show that this still holds for a non-commutative integrable system as defined in [23].

4.7 Linear Singular Foliations

A faithful finite-dimensional representation of a Lie algebra may be seen as singular foliation: it suffices to consider the singular foliation associated to its transformation Lie algebroid. Let us be more precise.

Notice that for every vector space V of finite dimension, there is a Lie algebra morphism $X \mapsto \hat{X}$ mapping a linear endomorphism of $X \in \text{End}(V)$ to the vector field $\hat{X}$ on V such that $\hat{X}[\alpha] = X^*(\alpha)$ for any $\alpha \in V^*$ (seen as a function on V).

Remark 4.40 Upon choosing a basis $(e_1, \ldots, e_d)$ of V, and the corresponding coordinates $(x_1, \ldots, x_d)$, this morphism maps a matrix $(a_{i,j})_{i=1}^d$ to the vector field $\sum_{i,j=1}^d a_{i,j} x_i \frac{\partial}{\partial x_j}$.

Let $\mathfrak{g}$ be a Lie algebra, and V be a finite-dimensional representation of $\mathfrak{g}$, described by a Lie algebra morphism $\eta \colon \mathfrak{g} \to \text{End}(V)$. Consider the $\mathcal{O}_V$-module[31] $\mathcal{F}^\eta$ generated by the vector fields $\{\widehat{\eta(x)}, x \in \mathfrak{g}\}$.

---

[30] The importance of this singular foliation comes from the following fact: when the quotient of N by the leaves of this singular foliation is a manifold, it is automatically a Poisson manifold, and this procedure is called Poisson reduction, see Chapter V in [43].

[31] With $\mathcal{O}_V$ being smooth, holomorphic, or polynomial functions depending on whether the base field is $\mathbb{R}$ or $\mathbb{C}$, and depending on the preferences of the reader.

Proposition 4.41 *Let (V, η) be a representation of a Lie algebra $\mathfrak{g}$. Then $\mathcal{F}^\eta$ is a singular foliation on V.*

The exercise supposes that the notion of leaves is already familiar to the reader. It also assumes the notion of "isotropy Lie algebra at a point". It explains how the initial representation can be deduced from the induced singular foliation in the faithful case.

Exercise 4.42 Let $(\mathfrak{g}, V, \eta, \mathcal{F}^\eta)$ be as in Proposition 4.41.

1. Show that the leaves of $\mathcal{F}^\eta$ are the orbits for the Lie group action $G \to GL(V)$ integrating η.
2. This question supposes that the notion of isotropy Lie algebra at a point is known. Show that the isotropy Lie algebra of $\mathcal{F}^\eta$ at $0 \in V$ is $\frac{\mathfrak{g}}{\ker(\eta)}$.
3. Is the following statement correct: "Two faithful representations (V, η) and V', η') are isomorphic if and only if their induced singular foliations $\mathcal{F}^\eta$ and $\mathcal{F}^{\eta'}$ are diffeomorphic"?
4. Compare the isotropy Lie algebra of $\mathcal{F}^\eta$ at a point $v \in V$ with the stabilizer of v.

Example 4.43 The singular foliation by concentric spheres, i.e., the singular foliation on $\mathbb{R}^n$ generated by the vector fields

$$\left\{ x_i \frac{\partial}{\partial x_j} - x_j \frac{\partial}{\partial x_i} \;\middle|\; 1 \leq i < j \leq n \right\}$$

comes from the action of $\mathfrak{so}(n)$ on $\mathbb{R}^n$. Its leaves are by concentric spheres.

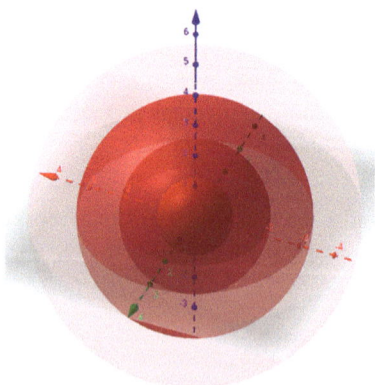

Concentric spheres in three dimensions

4.8 Hörmander Singular Foliations

Assume that we are given, on a manifold M, a family $(\mathcal{G}_i)_{i \in \mathbb{N}}$ of subsheaves of the sheaf $\mathfrak{X}(M)$ such that for all[32] $i, j \in \mathbb{N}$:

$$[\mathcal{G}_i, \mathcal{G}_j] \subset \mathcal{G}_{i+j}.$$

If, moreover, each of the spaces $\mathcal{G}_i$ is a locally finitely generated $C^\infty(M)$-module and if there exists $N \in \mathbb{N}$ such that $\mathcal{G}_N = \mathcal{G}_i$ for all $i \geq N$, then we say that the family $(\mathcal{G}_i)_{i \geq 0}$ is a *filtered subsheaf* of $\mathfrak{X}(M)$. In that case, $\mathcal{G}_0$ and $\mathcal{G}_N$ are singular foliations on M.

Let us associate to it a singular foliation on $\mathbb{R} \times M$. Denote by t the parameter on $\mathbb{R}$. The following Lemma is straightforward.

Lemma 4.44 *Consider a filtered subsheaf* $(\mathcal{G}_i)_{i=0}^N$. *The subsheaf* $\mathcal{G}[\mathbb{R}] \subset \mathfrak{X}_c(\mathbb{R} \times M)$ *of vector fields of the form*[33]

$$\left\{ \sum_{i=0}^N g_i t^i X_i \,\middle|\, g_1, \ldots, g_N \in C^\infty(\mathbb{R} \times M), X_1 \in \mathcal{G}_1, \ldots, X_N \in \mathcal{G}_N \right\}$$

is a singular foliation on $\mathbb{R} \times M$.

We call $\mathcal{G}[\mathbb{R}] \subset \mathfrak{X}(\mathbb{R} \times M)$ the *filtered singular foliation* of $(\mathcal{G}_i)_{i=0}^N$.

Example 4.45 Let α is a contact 1-form on a manifold M. Then the family:

$$\mathcal{G}_i = \begin{cases} 0 & \text{for } i = 0 \\ \Gamma(\mathrm{Ker}(\alpha)) & \text{for } i = 1 \\ \mathfrak{X}(M) & \text{for } i \geq 2 \end{cases}$$

is a filtered subsheaf of $\mathfrak{X}(M)$. Notice that, in this case, the singular foliation

$$\mathcal{G}[\mathbb{R}] := \left\{ ftX + gt^2 Y \,\middle|\, X \in \Gamma(\mathrm{Ker}(\alpha)), Y \in \mathfrak{X}(M), f, g \in C^\infty(\mathbb{R} \times M) \right\}$$

is not of constant rank, although each one of the $\mathcal{G}_i$'s is the section space of a vector bundle of constant rank. The leaves of this singular foliations are the points $\{(0, m)\}$ with $m \in M$ and the submanifolds $\{t\} \times M$ with $t \in \mathbb{R}^*$.

---

[32] Our $\mathbb{N}$ contains zero in the present section.
[33] Below, $X_i \in \mathcal{G}_i \subset \mathfrak{X}(M)$ is considered as a vector field on $\mathbb{R} \times M$ whose value at (t, m) is $(0, X_i|_m)$.

Example 4.46 Consider a family $X_1, \ldots, X_r$ of vector fields. Consider the recursively defined[34] family of sub-sheaves of vector fields:

$$\mathcal{G}_0 = \{0\}, \ \mathcal{G}_1 := \langle X_1, \ldots, X_r \rangle \text{ and } \mathcal{G}_{i+1} = \left\langle \sum_{k=1}^{i} [\mathcal{G}_k, \mathcal{G}_{i+1-k}] + \mathcal{G}_i \right\rangle.$$

If the $\mathcal{C}^\infty(M)$ module generated by $\mathcal{G}_i$ is constant after a certain rank N, then $\mathcal{G}_N$ is a singular foliation on M. The leaves of the corresponding filtered singular foliation are the points $(0, m)$ with $m \in M$ and the submanifolds $\{t\} \times L_m$ with L_m a leaf of the singular foliation $\mathcal{G}_N$ and $t \in \mathbb{R}$.

Exercise 4.47 Show that the singular foliation constructed in Example 4.46 can be alternatively defined by using the sequence

$$\mathcal{G}_1 := \langle X_1, \ldots, X_r \rangle \text{ and } \mathcal{G}_{i+1} = [\mathcal{G}_1, \mathcal{G}_i] + \mathcal{G}_i.$$

Definition 4.48 (Hörmander's Condition) Let M be a manifold. A singular foliation on $\mathbb{R} \times M$ is said to be a Hörmander singular foliation if it is a singular foliation associated to a filtered sub-sheaves $(\mathcal{G}_i)_{i \in \mathbb{N}}$ of $\mathfrak{X}(M)$ such that $\mathcal{G}_i = \mathfrak{X}(M)$ after a certain rank.

For any Hörmander singular foliation such that $\mathcal{G}_0 = 0$, the leaves are the points $\{(0, m)\}$ with $m \in M$ and the submanifolds $\{t\} \times M$ for $t \in \mathbb{R}^*$. In general, leaves are the sets $\{0\} \times L$ with L a leaf of the singular foliation generated by $\mathcal{G}_0$ and the submanifolds $\{t\} \times M$ for $t \in \mathbb{R}^*$.

Exercise 4.49 Show that all the isotropy Lie algebras (see Sect. 3 in chapter "Canonical Geometric and Algebraic Structures Hidden Behind a Singular Foliation") of a Hörmander's singular foliation such that $\mathcal{G}_0 = 0$ are nilpotent Lie algebras.

The name comes from Hörmander's condition in operator theory, whose content we briefly explain. Consider a differential operator on $\mathcal{C}^\infty(M)$ which is a sum of squares plus one "linear" term, i.e., is of the form $D = \sum_{i=1}^{r} X_i^2 + X_{r+1}$ with $X_1, \ldots, X_r, X_{r+1} \in \mathfrak{X}(M)$. Then it has been shown by Hörmander that D is hypo-elliptic if the family $X_1, \ldots, X_r, X_{r+1}$ and its successive Lie brackets (i.e., the outcome of the construction

---

[34] The brackets $\langle \cdot \rangle$ below stand for "the module over functions generated by".

in Example 4.46) generate all vector fields on M, see [7, 34]. This condition is known as Hörmander's condition, hence the name of the previously described singular foliations.

Example 4.50 The following example is given in [7]. Let $M = \mathbb{R}^2$ with parameters x, y and consider $\partial_x, x\partial_y$ as being $\mathcal{G}_1$ and $\mathcal{G}_i = \mathfrak{X}(\mathbb{R}^2)$ for $i \geq 2$. The associated Hörmander singular foliation is the singular foliation on $\mathbb{R} \times \mathbb{R}^2$ generated by the three vector fields

$$t\partial_x, tx\partial_y, t^2\partial_y.$$

Its isotropy Lie algebra (see Sect. 3 in chapter "Canonical Geometric and Algebraic Structures Hidden Behind a Singular Foliation") at the points $t = x = 0$ is the Heisenberg Lie algebra. It is an Abelian Lie algebra of dimension 2 at all other singular points (i.e., the set $\{t = 0, x \neq 0\}$).

Example 4.51 The following example appears in Erik van Erp and Robert Yuncken's [71]. Let M be a manifold of dimension d. Assume that the tangent bundle TM of a manifold M comes equipped with an increasing sequence of sub-bundles

$$0 = E_0 \subsetneq \cdots \subsetneq E_i \subsetneq E_{i+1} \subsetneq \cdots \subsetneq E_k = TM$$

such that $\mathcal{E}_i = \Gamma(E_i)$ is a filtered subsheaf of $\mathfrak{X}(M)$. Then its associated singular foliation $\mathcal{E}[t]$ on $\mathbb{R} \times M$ is Debord. This can be seen as follows. Let

$$0 = r_0 < r_1 < \cdots < r_i < r_{i+1} < \cdots < r_k = d$$

be the ranks of the subbundles $(E_i)_{i=1}^k$. Let $e_1, \ldots, e_d$ be a local trivialization on $\mathcal{U} \subset M$ of the tangent bundle $TM \to M$ such that for every $i = 1, \ldots, k$ the family $e_1, \cdots, e_{r_i}$ is a trivialization of E_i on $\mathcal{U}$. Then the family of vector fields on $\mathbb{R} \times \mathcal{U} \subset \mathbb{R} \times M$

$$\left(te_1, \cdots, te_{r_1}, t^2 e_{r_1+1}, \cdots, t^2 e_{r_2}, t^3 e_{r_2+1}, \cdots, t^{k-1} e_{r_{k-1}}, t^k e_{r_{k-1}+1}, \cdots, t^k e_d\right)$$

(where t is the parameter on $\mathbb{R}$) generates $\mathcal{E}[t]$. Since $e_1, \ldots, e_d$ is a trivialization of TM, there is of course no non-trivial relations between these generators, hence the result.

Notice that the isotropy Lie algebras (see Sect. 3 in chapter "Canonical Geometric and Algebraic Structures Hidden Behind a Singular Foliation") of $\mathcal{E}[t]$ at $t = 0$ are nilpotent Lie algebras of dimension d.

4.9 Miscellaneous Examples

1. An important class of examples of singular foliations are the so-called *Riemannian singular foliations*. Those are smooth singular foliations on a Riemannian manifold such that any geodesic orthogonal to a leaf is orthogonal to all leaf it crosses. Leaves given by actions of compact groups (for instance, concentric spheres as in Example 4.43) are singular Riemannian foliations. In [56], Hadi Nahari and Thomas Strobl gave an interpretation of those in terms of Poisson structures. See also Oleksii Kotov and Thomas Strobl's [39].
2. We said very little about dimension 1 and codimension 1 singular foliations on varieties, sometimes of small dimension. There is a wide literature on the matter, see e.g., the book [61].

Let us now describe a type of singular foliation that appeared in [26]. Given

1. G a connected[35] Lie group with Lie algebra $\mathfrak{g}$,
2. $P \xrightarrow{\pi} L$ be a principal G-bundle,
3. and S be a manifold on which the Lie group G acts,

a singular foliation can be constructed as follows. To start with, G acts freely and properly on $P \times S$, so that the quotient space

$$M = \frac{P \times S}{G} \tag{11}$$

is a manifold whose elements are denoted by $\overline{(p, s)}$ with $p \in P$ and $s \in S$, while the horizontal bar stands for the class modulo the diagonal G-action. The map $\Pi \colon \overline{(p, s)} \mapsto \pi(p)$ is a surjective submersion, that turns the manifold M into a fiber bundle over L, with fibers diffeomorphic to S (i.e., each fiber of Π is diffeomorphic to S, and every point ℓ in L has a neighborhood $\mathcal{U}$ such that $P^{-1}(\mathcal{U}) \simeq \mathcal{U} \times S$, the projection Π being then given by the projection on the first component). Now, let us equip M with a singular foliation. To start with:

1. the sheaf $\mathfrak{X}(P)$ of vector fields on P is a singular foliation on P,
2. and the infinitesimal $\mathfrak{g}$-action induces a singular foliation $\mathcal{F}_S$ on S (as in Sect. 4.2),
3. hence $P \times S$ is equipped with the direct product singular foliation (as in Sect. 5.1) that we denote by $\mathcal{G}$.

---

[35] The idea of [26] is that any singular foliation is of this type near a leaf, at least at formal level, but G has to be taken infinite dimensional.

Since the projection Π satisfies condition *(ii)* in Proposition 5.12, there exists an unique singular foliation $\mathcal{F}$ on the quotient space M, whose pull back singular foliation $\Pi^{-1}(\mathcal{F})$ (see Sect. 5.2) coincides with $\mathcal{G}$. Moreover, every point ℓ in L has a neighborhood $\mathcal{U}$ such that the restriction of $\mathcal{F}$ to $\Pi^{-1}(\mathcal{U}) \simeq \mathcal{U} \times S$ is the direct product of the singular foliation of all vector field on $\mathcal{U}$ with the singular foliation $\mathcal{F}_S$ on S.

Exercise 4.52 Let $E \to M$ be a vector bundle or rank N, and $P \to M$ be its frame bundle. Let G be the Lie group of $N \times N$ invertible matrices. Show that the previous construction applied to the singular foliation of vector fields on $S := \mathbb{R}^N$ that are zero at the origin gives the singular foliation of vector fields on E tangent to the zero section.

Exercise 4.53 This exercise requires the notion of leaves, and of transverse singular foliation to a leaf. For the singular foliation $\mathcal{F}$ on M as in Eq. (11) in the discussion above, establish the following points.

1. The classes $\overline{(p_1, s_1)}, \overline{(p_2, s_2)}$ of two points $(p_1, s_1), (p_2, s_2) \in P \times S$ in M belong to the same leaf of $\mathcal{F}$ if and only if s_1, s_2 belong to the same G-orbit.
2. Any leaf of $\mathcal{F}$ is a fiber bundle over L.
3. For every fixed point x of the G-action, $\Pi(P \times \{x\})$ is a leaf of $\mathcal{F}$ diffeomorphic to L.
4. Show that for a leaf in the previous item, a representative of the transverse singular foliation is given by a neighborhood of x in S.

We list as exercises several instances of singular foliations that do not enter any of the previous categories.

Exercise 4.54 Let ω be a closed 2-form. Show that

$$\{X \in \mathfrak{X}(M) \mid i_X \omega = 0\}$$

is a singular foliation on M, provided that it is locally finitely generated.

Exercise 4.55 Does the conclusion of Exercise 4.54 still holds for ω a closed n-form?

Exercise 4.56 Yahya Turki [70] introduced the following notion: we say that a bivector field $\pi \in \Gamma(\wedge^2 TM)$ is *foliated* if $\pi^\sharp(\Omega^1(M))$ is closed under the Lie bracket, i.e., if is a singular foliation.

1. Show that for any twisted Poisson structure (π, Ω) (also, called "Poisson structures with background" or "WSW-structures", see [37]–[35] for a definition) on a manifold M, π is a foliated bi-vector field.
2. Show that in the neighborhood of a regular point of π, there exists a closed 3-form Ω such that the pair (π, Ω) is a twisted Poisson structure.

3. Give an example of a foliated bivector field π that admits a singular point m such that there is no closed 3-form Ω, defined on a neighborhood of m, such that (π, Ω) is a twisted Poisson structure.

(*Hint:* this is done in [70].)

5 New Constructions from Old Ones

In the present section, we work indifferently in the context of smooth, complex or real analytic geometry. Most arguments presented here, however, make no sense in algebraic geometry, and have to be adapted. Conversely, some of them only make sense in algebraic geometry. We will be more precise in due time.

Here is a first exercise to train on these notions.

Exercise 5.1 Let $\mathcal{F}$ be a smooth singular foliation on M and $\varphi \in \mathcal{O}_M$ be a function. Show that

$$\varphi \mathcal{F} := \{\varphi X, X \in \mathcal{F}\}$$

is a singular foliation again. State and show the corresponding result in the real analytic, complex and algebraic settings.

5.1 Direct Products of Singular Foliations

For X_1, X_2 vector fields on M_1, M_2 respectively, we shall denote by (X_1, X_2) the vector field on $M_1 \times M_2$ whose value at $(m_1, m_2) \in M_1 \times M_2$ is $(X_1|_{m_1}, X_2|_{m_2}) \simeq T_{(m_1,m_2)} M_1 \times M_2$.

For $(M_1, \mathcal{F}_1)$ and $(M_2, \mathcal{F}_2)$ foliated manifolds, the product manifold $M_1 \times M_2$ can be equipped with the direct product of both foliations.

Definition 5.2 (Direct Product of Singular Foliations) The *direct product* of two foliated manifolds $(M_1, \mathcal{F}_1)$ and $(M_2, \mathcal{F}_2$ is the singular foliation $\mathcal{F}_1 \times \mathcal{F}_2$ on $M_1 \times M_2$ such that, for every open subset $\mathcal{U}_1 \subset M_1, \mathcal{U}_2 \subset M_2$, $\mathcal{F}_1 \times \mathcal{F}_2$ is the $\mathcal{O}_{\mathcal{U}_1 \times \mathcal{U}_2}$-module generated by vector fields of the form (X_1, X_2) with $X_1 \in \mathcal{F}_1$ and $X_2 \in \mathcal{F}_2$.
It is denoted by $(M_1 \times M_2, \mathcal{F}_1 \times \mathcal{F}_2)$.

Exercise 5.3 Show that the direct product of finitely generated singular foliations is a finitely generated singular foliation. Compare their ranks.

5.2 Pull-Back (Through Surjective Submersions)

Let us give the easiest version of the pull-back of a singular foliation: the pull-back through surjective submersion. We will come back to this notion later on, using a more general definition due to Androulidakis and Skandalis [3].

We work indifferently in the smooth, holomorphic or real analytic settings. We have two manifolds P and M, with respective sheaves of sections $\mathcal{O}^P$ and $\mathcal{O}^M$. Given $\psi : P \to M$ a map in the relevant category.

Definition 5.4 A vector field $X \in \mathfrak{X}(P)$ said to be ψ-*related* to a vector field $\tilde{X}$ on M if for all $p \in P$,

$$(T\psi)_p(X_{|p}) = \tilde{X}_{|\psi(p)} \qquad (12)$$

Equivalently, for any $f \in \mathcal{O}^M$, $X[f \circ \psi] = \tilde{X}[f] \circ \psi$, or, equivalently, such that the following diagram commutes:

$$\begin{array}{ccc} \mathcal{O}^P & \xrightarrow{\tilde{X}} & \mathcal{O}^P \\ \psi^* \downarrow & & \downarrow \psi^* \\ \mathcal{O}^M & \xrightarrow{X} & \mathcal{O}^M \end{array}$$

A vector field on M is said to be a *vertical vector field* if it takes values in $\ker(d_p \psi) \subset T_p P M$ for all $p \in M$. Equivalently, vertical vector fields are vector fields ψ-related to $0 \in \mathfrak{X}(M)$.

Definition 5.5 (Pull-Back of Singular Foliations) Let $\mathcal{F}$ be a singular foliation on a manifold M and let $\psi : P \to M$ be a surjective submersion. We call *pull-pack of $\mathcal{F}$ by ψ* and denote by $\psi^{-1}(\mathcal{F})$ the singular foliation on P generated, as an $\mathcal{O}^P$-module, by vector fields ψ-related to a vector field in $\mathcal{F}$.

The definition needs to be justified. First, one has to check that this definition indeed gives a sheaf of $\mathcal{O}^P$-modules on $\mathcal{X}(P)$. We then have to check that all three conditions $(\alpha), (\beta), (\gamma)$ in Definition 2.18 or 2.22 are satisfied.

(α) is an obvious consequence of the fact that if Y_1, Y_2 are ψ-related to X_1, X_2 respectively, then $[Y_1, Y_2]$ is ψ-related to $[X_1, X_2]$.

(β) holds by definition.

(γ) holds since, as for every submersion, every point $p \in P$ admits a neighborhood on which isomorphic to the product of two open balls $\mathcal{U}_1, \mathcal{U}_2$ such that ψ is the projection onto $\mathcal{U}_1$, which is seen as an open subset of M. Under this isomorphism, $\psi^{-1}(\mathcal{F})_\mathcal{U}$ is the direct product of $\mathcal{F}_{\mathcal{U}_1}$ with $\mathfrak{X}_{\mathcal{U}_2}$. For $\mathcal{U}_1, \mathcal{U}_2$ small enough, it is therefore finitely generated (see Exercise 5.3).

Remark 5.6 In particular, all vertical vector fields, i.e., vector fields tangent to the fibers of ψ, are contained in $\psi^{-1}(\mathcal{F})$.

Remark 5.7 In the smooth case, one can equivalently define singular foliations using compactly supported vector fields as in Definition 2.1. Assume $\mathcal{F}_c$ is a singular foliation on M as in Definition 2.1. Defining pull-back causes then a technical difficulty: if the fibers of $\psi : P \to M$ are not connected, then the pull-back singular foliation $\psi^{-1}(\mathcal{F}_c)$ can not be defined as being the $\mathcal{C}^\infty(P)$-module generated by compactly supported vector fields on P which are ψ-related to a vector field in $\mathcal{F}_c$. Indeed, if the fibers of ψ are not compact, there is no compactly supported vector field on P which is ψ-related to a non-zero vector field on M. The singular foliation $\psi^{-1}(\mathcal{F})$ has to be then defined as the $\mathcal{C}_c^\infty(P)$-module generated by vector fields on P which are ψ-compatible to a vector field in $\mathcal{F}$. In equation, if one denotes by $\mathfrak{X}(P)_\psi$ the Lie algebra of vector fields on P which are ψ-compatible to a vector field on M, and $\psi_* : \mathfrak{X}(P)_\psi \to \mathfrak{X}(M)$ the natural Lie algebra morphism, we have:

$$\psi^{-1}(\mathcal{F}_c) := \mathcal{C}_c^\infty(P)\,(\psi_*)^{-1}(\mathcal{F}_c).$$

Exercise 5.8 Let us work in the setting of smooth differential geometry. A *horizontal distribution* on the surjective submersion $\psi : P \to N$, is a regular distribution $p \mapsto \mathcal{H}_p$ on P such that[36]

$$\mathcal{H}_p \oplus \ker(T_p\psi) = T_pP \text{ for all } p \in P.$$

We call *horizontal lift* of $X \in \mathfrak{X}(M)$ and denote by $\mathcal{H}(X)$ the unique section of $\mathcal{H}$ such that $T\psi(\mathcal{H}(X)|_p) = X|_{\psi(p)}$ for all $p \in P$. Show that $\psi^{-1}(\mathcal{F})$ is generated, as a sheaf of $\mathcal{C}^\infty(P)$-modules, by horizontal lifts of vector fields in $\mathcal{F}$ and vertical vector fields (= vector fields tangent to the fibers of ψ).

---

[36] Those are also called Ehresmann connection. They exist for any fiber bundles.

Exercise 5.9 Let G be a discrete group acting freely and properly on a smooth manifold M by smooth diffeomorphism. Recall that this implies that M/G is a manifold and that the natural projection $\Pi \colon M \to M/G$ is a surjective local diffeomorphism. Let $\mathcal{F}$ be a singular foliation on M.

1. Show that if G acts on symmetries of $\mathcal{F}$, then there exists a singular foliation $\mathcal{F}^G$ on M/G whose pull-back to M is $\mathcal{F}$.
2. Is the converse true?

Exercise 5.10 Show that the p-vertical vector fields are contained in $p^{-1}(\mathcal{F}_B)$.

Exercise 5.11 Let $(M, \mathcal{F})$ be a foliated manifold. Let $\psi \colon P \to M$ be a surjective submersion

1. Show that the leaves of $\psi^{-1}(\mathcal{F})$ are the connected components of the inverse images through ψ of the leaves of $\mathcal{F}$.
2. Show that the isotropy Lie algebra of $\psi^{-1}(\mathcal{F}_B)$ at a point m is canonically isomorphic to the isotropy Lie algebra of $\mathcal{F}_B$ at $\psi(m)$.
3. Show that the transverse singular foliation of $\psi^{-1}(\mathcal{F}))$ of a leaf through a point $p \in P$ is canonically isomorphic to the transverse singular foliation of $\mathcal{F}$ at the leaf through $\psi(p)$.

We conclude this section with a statement which we insist is not an obvious statement, for it will use the assumption "locally finitely generated". It is wrong for general involutive distributions. For instance, for the "infinite comb" of Exercise 2.5, the projection $(x, y) \mapsto y$ onto the horizontal axis satisfies Conditions *(ii)* and *(iii)* below but does not satisfy condition *(i)*.

Proposition 5.12 *Let $\psi \colon P \to M$ be a surjective submersion with connected fibers, and $\mathcal{F}_P$ a singular foliation on P. Then the following are equivalent:*

(i) *There exists a singular foliation $\mathcal{F}_M$ on M such that $\mathcal{F}_P = \psi^{-1}(\mathcal{F}_M)$.*
(ii) *Each fiber of ψ is contained in a leaf of L.*
(iii) *For every $m \in M$, we have $\ker(T_m \psi) \subset T_m \mathcal{F}_P$.*

5.3 Restriction of a Singular Foliation to a Transverse Submanifold

Let $\mathcal{F}$ be a singular foliation on a smooth manifold M, and let $S \subset M$ be a sub-manifold. We would like to restrict the singular foliation $\mathcal{F}$ to S.

The next exercise presents a naive idea – which works, but has to be made more precise.

Exercise 5.13 Let $\mathcal{F}$ be a singular foliation on a singular foliation M, seen as a subspace of $\mathfrak{X}_c(M)$ as in Definition 2.1. Let $S \subset M$ be a closed embedded submanifold. Consider $\mathfrak{i}_S^* \mathcal{F}_{naive} \subset \mathfrak{X}_c(S)$ to be the sub-space of all vector fields on S obtained by restricting to S vector fields in $\mathcal{F}$ that happen to be tangent to S. Show that $\mathfrak{i}_S^* \mathcal{F}_{naive}$

1. is a sub-Lie algebra of $\mathfrak{X}(S)$,
2. is a sub-$\mathcal{C}^\infty(S)$-module of $\mathfrak{X}(S)$,

It is therefore a Lie-Rinehart subalgebra of vector fields on S (i.e., it satisfies (α) and (β) in Definition 2.1). However,

3. Show that if the submanifold S is embedded but not closed in M, the restriction to S of a vector field compactly supported on M may even not be compactly supported on S. (*Hint*: take $M = \mathbb{R}^2$, $\mathcal{F} = \mathfrak{X}_c(M)$ and Σ the spiral obtained as an integral curve of $X = x\partial_y - y\partial_x - x^2\partial_x - y^2\partial_y$. Then multiply X by a compactly supported function which is 1 at $(0, 0)$.).
4. Show that if S is only immersed and not embedded, $\mathfrak{i}_S^* \mathcal{F}_{naive}$ many even not be a $\mathcal{C}^\infty(S)$-module. *Hint*: the previous counter-example will do as well...

In order to have an induced singular foliation on S, we have to be more sophisticated. Let $S \subset M$ be an immersed submanifold of M: we now work in the smooth, real analytic or complex settings altogether, and we consider that $\mathcal{F}$ is a sheaf as in Definition 2.18. We denote by $\mathfrak{i}\colon S \hookrightarrow M$ the canonical inclusion. We define a sheaf $\mathfrak{i}_S^* \mathcal{F} \subset \mathfrak{X}(S)$ as follows. To every $\mathcal{U} \subset S$, we associate the space of all vector fields $Y \in \mathfrak{X}(S)_\mathcal{U}$ such that for every $s \in \mathcal{U}$, there exists $X \in \mathcal{F}_\mathcal{W}$ (for some open subset $\mathcal{W} \subset M$ containing $\mathfrak{i}(s)$) such that

$$T_{s'}\mathfrak{i}(Y|_{s'}) = X|_{\mathfrak{i}(s')}$$

for every s' is a neighborhood of s in S. Check that the previously defined object:

5. is a sub-sheaf of the sheaf of vector fields on S,
6. is closed under Lie bracket,
7. and is a module over the relevant sheaf of functions on S.

Consider the foliation of $\mathbb{R}^2$ by horizontal lines, i.e., $\mathcal{F}$ is generated by $\frac{\partial}{\partial x}$. Let f be a function which has support $[0, \infty)$. Then the graph of f, namely,

$$S = \{(x, f(x) | x \in \mathbb{R}\}$$

is an embedded submanifold of M.

8. Show that, in this case, $i_S^*\mathcal{F}$ is not locally finitely generated. (*Hint:* $i_S^*\mathcal{F}$ is exactly the space of vector fields which are supported in $(-\infty, 0] \subset \mathbb{R}$.)

However, there is a situation where the sheaf $i_S^*\mathcal{F}$ defined in Exercise 5.13 is a locally finitely generated module, and is therefore an induced singular foliation on S.

Definition 5.14 We say that a submanifold S of a foliated manifold $(M, \mathcal{F})$ *intersects cleanly* $\mathcal{F}$ if $T_s S + T_s \mathcal{F} = T_s M$ for all $s \in S$. We also say that S is *transverse to* $\mathcal{F}$, or is a transverse sub-manifold.

The condition about a clean intersection is enough to guarantee that $i_S^*\mathcal{F}$ is locally finitely generated.

Proposition 5.15 (Submanifolds Intersecting $\mathcal{F}$ Cleanly) *Let $S \subset M$ be a submanifold that intersects cleanly[a] a smooth singular foliation $\mathcal{F}$. Then $i_S^*\mathcal{F}$ is a singular foliation on S.*

It is called the restriction of the singular foliation to S.

---

[a] Also called transverse. See Definition 5.14.

The following exercises describe this structure more precisely.

Exercise 5.16 Let S be a submanifold that cleanly intersects $(M, \mathcal{F})$, and let $i_S^*\mathcal{F}$ be its induced singular foliation.

1. Show that the rank of $i_S^*\mathcal{F}$ at a point s is $\text{rk}_s \mathcal{F} - \text{codim}(S)$.
2. Show that $T_s i_S^*\mathcal{F} = T_s \mathcal{F} \cap T_s S$ for all $s \in S$.
3. (Supposes that the notion of leaf is known, see Sect. 7.) Show that the leaf of $i_S^*\mathcal{F}$ through a point $s \in S$ is the connected component containing s of S with the leaf through s of $\mathcal{F}$.
4. (Supposes that the notion of isotropy Lie algebra is known, see Sect. 3 in chapter "Canonical Geometric and Algebraic Structures Hidden Behind a Singular Foliation".) Show that the isotropy Lie algebra of $\mathcal{F}$ and $i_S^*\mathcal{F}$ coincide at any point $s \in S$.

Exercise 5.17 The goal of this exercise is to show that there is a neighborhood of a transverse submanifold S in a foliated manifold $(M, \mathcal{F})$ on which $\mathcal{F}$ coincides with a neighborhood of the zero section in the normal bundle $N_S := TL/TS \xrightarrow{p} S$, equipped with the pull-pack singular foliation $p^* i_S^* \mathcal{F}$. We work in the smooth setting (it is not true in the holomorphic or real analytic settings).

1. Show the "tubular neighborhood theorem", i.e., that there is a neighborhood $\mathcal{U}$ of S in L diffeomorphic to a neighborhood $\mathfrak{U}$ of the zero section in the normal bundle $N_S := TL/TS \xrightarrow{p} S$, through a diffeomorphism which is the identity on S.
2. Show that the tubular neighborhood $\mathcal{U}$ in the previous item can be chosen such that vector fields tangent to the fibers of $p \colon \mathcal{U} \to S$ are included in $\mathcal{F}$.
3. Conclude that the restriction to $\mathcal{U}$ of the singular foliation $\mathcal{F}$ is isomorphic to the pull-back singular foliation $p^{-1}(i_S^* \mathcal{F})$ (*Hint:* $\mathcal{U}$ see Sect. 5.2, and Proposition 5.12.)

Exercise 5.18 This exercise requires the notion of anchored bundle over a singular foliation, see Sect. 1.1 in chapter "Canonical Geometric and Algebraic Structures Hidden Behind a Singular Foliation". Let (A, ρ) be an anchored bundle over a singular foliation $\mathcal{F}$. Let $S \subset M$ be a submanifold that intersects $\mathcal{F}$ cleanly. Show that $\rho^{-1}(TS) \subset i_S^* A$ is an anchored bundle over $\mathcal{F}_S$, when equipped with the restriction of the anchor map. (Here, $i_S^* A$ stands for the restriction of the vector bundle A to S.)

Exercise 5.19 This section requires the notion of symmetry of a singular foliation (see Sect. 3.1). Let $\Phi \colon M \to M$ be a symmetry of a singular foliation $\mathcal{F}$ on a manifold M. Let $S \subset M$ be a submanifold.

1. Show that S intersects $\mathcal{F}$ cleanly if and only if $\Phi(S)$ does.
2. Show that, in that case, the restriction of Φ to S is an isomorphism of foliated manifolds $(S, i_S^* \mathcal{F}) \longrightarrow (\Phi(S), i_{\Phi(S)}^* \mathcal{F})$.

5.4 Pull-Back of Singular Foliations (Beyond Immersions and Submersions)

We have already defined pull-back through surjective submersions, but also the restriction to some submanifolds. Let us unify these constructions, following an idea of Androulidakis and Skandalis [3].

In this section, we restrict ourselves to the case of smooth manifolds: the complex or real analytic cases are similar.

Let L, M be manifolds together with a smooth map $\phi \colon L \to B$. Let $\phi^* TM$ be the pull-back through ϕ of the tangent bundle TM:

$$\phi^* TM := \{(\ell, u) \in L \times TM \mid u \in T_{\phi(\ell)} M\}.$$

There are two natural maps:

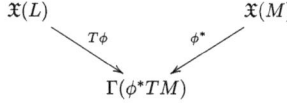

defined as follows.

1. Any vector field X on M gives a section ϕ^*X of ϕ^*TM defined by

$$\ell \mapsto (\ell, X_{\phi(\ell)}).$$

called the pull-back of X.

2. There is a natural vector bundle morphism defined for all $u \in T_\ell L$ by

$$T\phi : TL \to \phi^*TM$$
$$u \mapsto (\ell, T_\ell\phi(u))$$

At the level of sections, it induces a map $\mathfrak{X}(L) \to \Gamma(\phi^*TM)$.

Let $\mathcal{F}_M$ be a singular foliation on M, seen as in Definition 2.1 as a subspace of compactly supported vector fields. We denote by $\phi^*\mathcal{F}_M$ the $\mathcal{C}_c^\infty(L)$-submodule[37] of $\Gamma_c(\phi^*TM)$ generated by $\{\phi^*X \mid X \in \mathcal{F}_M\} \subset \Gamma(\phi^*TM)$. We now present the construction of a singular foliation on L as an exercise.

Exercise 5.20 Consider the submodule of $\mathfrak{X}_c(L)$ defined by[38]

$$\phi^{-1}(\mathcal{F}_M) := \{X \in \mathfrak{X}_c(L) \mid T\phi(X) \in \phi^*\mathcal{F}_M\}.$$

1. Show that $\phi^{-1}(\mathcal{F}_M)$ is involutive, i.e., closed under Lie bracket.
2. We say that $\phi : L \to M$ is *transverse to* $\mathcal{F}_M$ if for all $\ell \in L$, we have $T_{\phi(\ell)}\mathcal{F}_M + T_\ell\phi(T_\ell L) = T_{\phi(\ell)}M$. Show that for ϕ an immersion, this definition matches the transversality condition given in Eq. (2). Show that submersions are transverse to any singular foliation on M.
3. Show that if ϕ is transverse to $\mathcal{F}_M$, then $\phi^{-1}(\mathcal{F}_M)$ is a singular foliation on M.

---

[37] Recall that the index $_c$ means "compactly supported".
[38] Equivalently, a vector field $X \in \mathfrak{X}_c(L)$ belongs to $\phi^{-1}(\mathcal{F}_M)$ if and only if there exists smooth functions $g_1, \ldots, g_k \in \mathcal{C}_c^\infty(L)$ and $Y_1, \ldots, Y_k$ in $\mathcal{F}_M$ such that $T_\ell\phi(X|_\ell) = \sum_{i=1}^k g_i(\ell) Y_i|_{\phi(\ell)}$ for all $\ell \in L$.

This exercise justifies the following definition.

Definition 5.21 (Pull-Back w.r.t. a Transverse Map) Let $(M, \mathcal{F}_M)$ be a foliated manifold, and L be a manifold and $\phi : L \to M$ a map transverse to $\mathcal{F}_M$. We call the singular foliation $\phi^{-1}(\mathcal{F}_M)$ the *pull-back of $\mathcal{F}_M$ through ϕ*.

Exercise 5.22 Explain why this notion "unifies" (= i.e., admits as particular cases) pull-back with respect to surjective submersions seen in Sect. 5.2 and restrictions to transverse submanifolds seen in Sect. 5.3.

5.5 The Suspension of a Singular Foliation

We work here in the setting of smooth differential geometry. Most results could be adapted to real analytic or complex settings.

In this section, we denote by $\mathfrak{X}(N)$ the singular foliation of all vector fields on a manifold N.

5.5.1 Suspension in Dimension 1

We call *suspension* of the manifold M with respect to a diffeomorphism $\phi \colon M \to M$ the quotient of $M \times \mathbb{R}$ by the action of the additive group $\mathbb{Z}$ by:

$$k \cdot (m, t) := (\phi^k(m), t + k) \tag{13}$$

for all $k \in \mathbb{Z}, m \in M, t \in \mathbb{R}$. Since the action of $\mathbb{Z}$ is free and proper, the quotient is a manifold—that we call *suspension of M by ϕ* and denote by $M_\phi := \frac{M \times \mathbb{R}}{\mathbb{Z}_\phi}$. We also denote by $\pi \colon M \times \mathbb{R} \to \frac{M \times \mathbb{R}}{\mathbb{Z}_\phi}$ the natural projection.

Let us assume now that M comes equipped with a singular foliation $\mathcal{F}$ (seen as a sheaf as in Definition 2.18) and that $\phi \colon M \to M$ is a symmetry of $\mathcal{F}$, i.e., that $\phi_*(\mathcal{F}) = \mathcal{F}$. Then $M \times \mathbb{R}$ comes with the direct product singular foliation $(M \times \mathbb{R}, \mathcal{F} \times \mathfrak{X}(\mathbb{R}))$ (where $\mathfrak{X}(\mathbb{R})$ stands for the sheaf of all vector fields on $\mathbb{R}$). For all $k \in \mathbb{Z}$:

$$(m, t) \mapsto (\phi^k(m), t + k)$$

is a symmetry of the direct product singular foliation $(M, \mathcal{F}) \times (\mathbb{R}, \mathfrak{X}(\mathbb{R}))$. By Exercise 5.9, the following results hold true.

Proposition 5.23 *Let $\phi : M \to M$ be a symmetry for a singular foliation $\mathcal{F}$. There exists an unique singular foliation on the suspension $M_\phi := \frac{M \times \mathbb{R}}{\mathbb{Z}_\phi}$ of M by ϕ whose pull-back on $M \times \mathbb{R}$ is the direct product singular foliation $(M \times \mathbb{R}, \mathcal{F} \times \mathfrak{X}(\mathbb{R}))$.*

We call the singular foliation in Proposition 5.23 the *suspension of $\mathcal{F}$ by the symmetry ϕ* and denote it by $\mathcal{F}_\phi$. Before describing this singular foliation in more details, let us recall a classical result of differential geometry about suspensions of diffeomorphisms:

Lemma 5.24 *If a diffeomorphism ϕ of a manifold M is the time 1 flow of a complete vector field $X \in \mathfrak{X}(M)$, then the suspension $M_\phi := \frac{M \times \mathbb{R}}{\mathbb{Z}_\phi}$ of M by ϕ is diffeomorphic to the direct product[39] $M \times S^1$.*

Proof The vector field on $M \times \mathbb{R}$ whose value at (m, t) is $tX_{|m}$ has a flow at time t that intertwines the $\mathbb{Z}$-action as in Eq. (13) with the $\mathbb{Z}$-action:

$$k \cdot (m, t) = (m, t + k).$$

Since the quotient of $M \times \mathbb{R}$ through this action is $M \times S^1$, this completes easily the proof.

When X belongs to $\mathcal{F}$, then the vector field that appears in the proof of Lemma 5.24 belongs to the suspension of $\mathcal{F}$. Hence (by the highly non-trivial Corollary 7.13 that will be proven later on), the next statement holds true.

Proposition 5.25 (Inner Symmetries Have Trivial Suspensions) *If a symmetry ϕ of a singular foliation $\mathcal{F}$ on M is the time 1 flow of a complete vector field[a] in $\mathcal{F}$, then its suspension $(M_\phi, \mathcal{F}_\phi)$ is isomorphic to the direct product singular foliation $(M \times S^1, \mathcal{F} \times \mathfrak{X}(S^1))$.*

---
[a] With a little more work, this result can be extended to the case where ϕ is an inner symmetry of $\mathcal{F}$.

Example 5.26 An important example of such a singular foliation is the so-called *self-eating snake*, which is defined by:

1. $M = \mathbb{R}^2$ with coordinates (x, y),

---
[39] I.e., the suspension $\frac{M \times \mathbb{R}}{\mathbb{Z}_{id_M}}$ associated to the identity map of M.

2. $\mathcal{F}$ is the singular foliation defined by the vector field $x\partial_y - y\partial_x$ (whose leaves are concentric circles),
3. $\phi \colon M \to M$ is the division by 2, namely $\phi(x, y) = (x/2, y/2)$.

The leaves of the suspension of $(M, \mathcal{F})$ by ϕ are then as follows. To start with, $M_\phi = M \times S^1$, and the singular foliation is as follows:

1. the circle $\{0\} \times S^1$ is a leaf,
2. all the other leaves are diffeomorphic to a cylinder. These cylinders wrap around this circle.

5.5.2 Suspension in Dimension ≥ 2

The suspension can be defined in a more general context. Let Σ be any connected manifold, $\sigma \in \Sigma$ a point. Let $\pi_1(\Sigma, \sigma)$ be the fundamental group[40] based at σ. Let $\tilde{\Sigma}_\sigma$ be the universal cover of Σ, again computed[41] with respect to σ. The fibers of the natural projection

$$\tilde{\Sigma}_\sigma \to \Sigma$$
$$[\gamma] \mapsto \gamma(1)$$

are equipped with a natural $\pi_1(\Sigma, \sigma)$-action by

$$[g] \cdot [\gamma] := [g \star \gamma]$$

where $\star$ refers to concatenation of paths. This action turns $\tilde{\Sigma}_\sigma$ it into a $\pi_1(\Sigma, \sigma)$-principal bundle.

Let M be a manifold. For any group morphism:

$$\varphi \colon \pi_1(\Sigma, \sigma) \longrightarrow \mathrm{Diff}(M),$$

we call *suspension of φ* the quotient manifold:

$$\frac{M \times \tilde{\Sigma}_\sigma}{\pi_1(L, \ell)_\varphi}$$

---

[40] I.e., $\pi_1(\Sigma, \sigma)$ is the group of homotopy classes of loops based at σ.
[41] I.e., $\tilde{\Sigma}_\sigma$ is the set of all homotopy classes $[\gamma]$ of paths $\gamma \colon [0, 1] \to \Sigma$ with $\gamma(0) = \sigma$.

where the quotient is with respect to the diagonal action

$$[g] \cdot (m, [\gamma]) = (\psi([g])(m), [g \star \gamma]) \tag{14}$$

for all $[g] \in \pi_1(\Sigma, \sigma), m \in M, [\gamma] \in \tilde{\Sigma}_\sigma$. Let $\mathcal{F}$ be a singular foliation on M, and assume that φ is valued in symmetries of $\mathcal{F}$. Consider the direct product singular foliation

$$(M, \mathcal{F}) \times (\Sigma, \mathfrak{X}(\tilde{\Sigma}_\sigma)). \tag{15}$$

The group action defined in Eq. (14) is valued in symmetries of the direct product singular foliation (15). In particular, the singular foliation (15) descends to the quotient to define a singular foliation $\frac{\mathcal{F} \times \mathfrak{X}(\tilde{\Sigma}_\sigma)}{\pi_1(L, \ell)_\varphi}$ on $\frac{M \times \tilde{\Sigma}_\sigma}{\pi_1(L, \ell)_\varphi}$. We call *suspension of the singular foliation* $(M, \mathcal{F})$ *with respect to* $\varphi \colon \pi_1(\Sigma, \sigma) \to \mathrm{Sym}(M, \mathcal{F})$ the pair

$$\left(\frac{M \times \tilde{\Sigma}_\sigma}{\pi_1(\Sigma, \sigma)_\varphi}, \frac{\mathcal{F} \times \mathfrak{X}(\tilde{\Sigma}_\sigma)}{\pi_1(\Sigma, \sigma)_\varphi} \right)$$

Example 5.27 For $\Sigma = S^1$ one recovers the previous construction.

Remark 5.28 It is tempting to generalize Proposition 5.25 to an arbitrary Σ by stating that if φ is valued in inner symmetries of $\mathcal{F}$, then the suspension should be isomorphic to the direct product of $(M, \mathcal{F}) \times (\Sigma, \mathfrak{X}(\Sigma))$. But such a statement is wrong, see the discussion about torus in [26].

Exercise 5.29 This exercise requires the notion of leaves. Show that if $m \in M$ is a point where $T_m \mathcal{F} = \{0\}$, and if $\varphi([g])(m) = m$ for every $[g] \in \pi_1(\Sigma, \sigma)$, then the suspension of $(M, \mathcal{F})$ with respect to φ has a leaf L diffeomorphic to Σ, whose inverse image is $\{m\} \times \tilde{\Sigma}_\sigma$.

Exercise 5.30 This exercise requires the notion of flat leaves of Exercise 8.16. Show that the leaf $L \simeq \Sigma$ in Exercise 5.29 is flat.

Hint: Show that the quotient of the direct product singular foliation $(M, 0) \times (\tilde{\Sigma}_\sigma, \mathfrak{X}(\tilde{\Sigma}_\sigma))$ under the $\pi_1(\Sigma, \sigma)$-action is a regular foliation admitting $L \simeq \Sigma$ as a leaf.

5.6 New Constructions From Old Ones in Algebraic Geometry

In this section, we work with algebraic singular foliations as in Definition 2.28. Let us repeat the context. In this section, $\mathcal{O}$ is an Abelian algebra and $\mathrm{Der}(\mathcal{O})$ stands for the $\mathcal{O}$-

module of derivations of $\mathcal{O}$ (which is a Lie algebra[42] when equipped with the commutator). We define algebraic singular foliation over $\mathcal{F}$ with respect to $\mathcal{O}$ to be sub-$\mathcal{O}$-modules of Der($\mathcal{O}$) which are finitely generated[43] and stable under the Lie bracket of derivations. The purpose of the section is to explain how purely algebraic methods, allow defining new algebraic singular foliations out of this one.

We invite the reader to think that $\mathcal{O}$ is the algebra of polynomial functions on $\mathbb{K}^d$, or is the algebra of functions on some affine variety[44] W, i.e., a quotient of the algebra of polynomial functions on $\mathbb{C}^d$ by a prime ideal $\mathcal{I}$. Recall that since $\mathcal{O}$ is Noetherian, any sub-module of the module of derivations is finitely generated. In this particular case, our constructions have a geometric meaning that we will detail.

Throughout this section, we choose an algebraic singular foliation $\mathcal{F}$ over the algebra $\mathcal{O}$.

5.6.1 Restriction

An ideal $\mathcal{I}$ is said to be a *foliated ideal* if

$$\mathcal{F}[\mathcal{I}] \subset \mathcal{I}.$$

The quotient space $\mathcal{F}/\mathcal{I}\mathrm{Der}(\mathcal{O}) \cap \mathcal{F}$ then inherits a natural algebraic singular foliation structure over $\mathcal{O}/\mathcal{I}$. We call the latter algebraic singular foliation the *restriction w.r.t the ideal $\mathcal{I}$*.

Example 5.31 When $\mathcal{O}$ is the algebra of functions on an affine variety W, and $\mathcal{I}$ is the ideal of functions vanishing on an affine subvariety $W' \subset W$, then $\mathcal{I}$ is a foliated ideal if and only if all vector fields in $\mathcal{F}$ are tangent to W, and the previous construction corresponds to the restriction of the singular foliation $\mathcal{F}$ to W'.

5.6.2 Algebra Extension

Assume that the algebra $\mathcal{O}$ has no zero divisor, and let $\mathbb{O}$ be its field of fractions. Any derivation of $\mathcal{O}$ extends to a derivation of $\mathbb{O}$, so that we have a natural inclusion[45] $\mathbb{O} \otimes_{\mathcal{O}} \mathrm{Der}(\mathcal{O}) \subset \mathrm{Der}(\mathbb{O})$. For any subalgebra $\tilde{\mathcal{O}}$ with $\mathcal{O} \subset \tilde{\mathcal{O}} \subset \mathbb{O}$ such that every derivation $X \in \mathcal{F}$ (extended to a derivation of $\mathbb{O}$) preserves $\tilde{\mathcal{O}}$, there is natural algebraic singular foliation over $\tilde{\mathcal{O}}$ given by $\tilde{\mathcal{O}} \otimes_{\mathcal{O}} \mathcal{F} \subset \mathrm{Der}(\tilde{\mathcal{O}})$.

---

[42] It is even a Lie-Rinehart algebra.

[43] The assumption "finitely generated" is never used in this section and could be erased. However, we decided to keep it, since it is inherited to the new sub-modules that we will construct.

[44] It is here implicit that our affine varieties are over the field $\mathbb{C}$.

[45] This inclusion is even an equality for $\mathcal{O}$ a polynomial algebra over $\mathbb{K}$.

Example 5.32 We will not try to give a complete geometric description of this construction in the context of affine varieties. However, let us mention that the blow-up of a singular foliation at a point of Sect. 5.7 is a construction of that type on any affine chart.

5.6.3 Localization

Let us recall the definition of localization.

Definition 5.33 A subset $S \subset \mathcal{O}$ is called *multiplicatively closed* if $1 \in S$ and if S is stable under multiplication. For a multiplicative set $S \subset \mathcal{O}$, the *localization* of $\mathcal{O}$ at S is the algebra defined as follows:

1. If $\mathcal{O}$ has no zero divisor, then it is the subalgebra $S^{-1}\mathcal{O}$ of its field $\mathbb{O}$ of fractions given by:

$$S^{-1}\mathcal{O} := \left\{ \frac{f}{s} \,\middle|\, f \in \mathcal{O}, s \in S \right\} \qquad (16)$$

2. If $\mathcal{O}$ has zero divisor, the previous definition can be enlarged by considering the quotient of $\mathcal{O} \times S$ by the equivalence relation $(f, s) \sim (g, t)$ if there is an element $u \in S$ such that $u(ft - gs) = 0$. Addition and multiplication are then defining by checking that the following operations go to the quotient:

$$(f, s) + (g, t) := (ft + gs, st) \quad \text{and} \quad (f, s) \times (g, t) := (fg, st). \qquad (17)$$

Remark 5.34 The algebra $\mathcal{O}$ is a subalgebra of $S^{-1}\mathcal{O}$ via the homomorphism $\mathcal{O} \hookrightarrow S^{-1}\mathcal{O}$, $f \mapsto \frac{f}{1}$.

Example 5.35 For an affine variety W, localization can be interpreted as meaning that we restrict ourselves to the Zarisky open subset $\mathcal{U}$ of W. Then S is the multiplicative subset of all polynomial functions P whose zeros are outside $\mathcal{U}$ (i.e., such that $P(w) \neq 0$ for all $w \in \mathcal{U}$). This interpretation explains the name.

Any derivation $X \in \text{Der}(\mathcal{O})$ extends to a derivation of the localization $S^{-1}\mathcal{O}$. When $\mathcal{O}$ has no zero divisor, the extension is defined by:

$$X : S^{-1}\mathcal{O} \longrightarrow S^{-1}\mathcal{O}$$
$$\frac{f}{s} \longmapsto \frac{X[f]s - fX[s]}{s^2}.$$

When zero divisors exist, then one has to check that the following map goes to the quotient:

$$X: (f, s) \mapsto (X[f]s - fX[s], s^2)$$

with respect to the equivalence relation defined above.

In both cases, the $S^{-1}\mathcal{O}$-module generated by extensions to $S^{-1}\mathcal{O}$ of derivations in an algebraic singular foliation $\mathcal{F}$ over $S^{-1}\mathcal{O}$ that we denote by $S^{-1}\mathcal{F}$ and call *localization of $\mathcal{F}$ at S*. When no zero divisor exists on $\mathcal{O}$, elements in $S^{-1}\mathcal{F}$ can be thought of as quotients $\frac{X}{s}$ with $X \in \mathcal{F}$ and $s \in S$. The Lie bracket restricted to $S^{-1}\mathcal{F}$ is given as follows:

$$\forall X, Y \in \mathcal{F}, \forall (s, t) \in S^2, \quad \left[\frac{1}{s}X, \frac{1}{t}Y\right] = \frac{1}{st}[X, Y] + \frac{Y[s]}{s^2 t}X - \frac{X[t]}{st^2}Y. \tag{18}$$

Example 5.36 Geometrically, localization corresponds to restriction to a Zariski open subset.

5.7 Blowup of a Singular Foliation

In this section, we work in the realm of complex algebraic geometry over $\mathbb{C}$. Most construction extend to the smooth setting: indeed, this is the context in which Debord and Skandalis [18] introduced the notion of blow-up of a singular foliation.

5.7.1 Blow-Up at a Point

Recall that for any $d \in \mathbb{N}$, the set $\mathbb{P}^d$ of all straight lines through the origin of $\mathbb{C}^{d+1}$ is a complex manifold of dimension d over $\mathbb{C}$, called the *d-dimensional projective space*. Formally, it is defined as the equivalence classes of relation on the quotient $\mathbb{C}^{d+1} \setminus \{(0, \ldots, 0)\}$ under the equivalence relation:

$$u = (u_0, u_1, \ldots, u_d) \sim v = (v_0, v_1, \ldots, v_d) \iff \exists \lambda \in \mathbb{C} \setminus \{0\} \text{ such that } u = \lambda v.$$

Equivalently, it can be defined as the quotient manifold

$$\mathbb{P}^d := \mathbb{C}^{d+1} \setminus \{(0, \ldots, 0)\}/\mathbb{C} \setminus \{0\},$$

where the group $\mathbb{C} \setminus \{0\}$ acts by diagonal multiplication on $\mathbb{C}^{d+1}$. In particular, elements in $\mathbb{P}^d$ shall be denoted as $d+1$-tuples of elements not all equal to zero and defined up to a non-zero constant, and denoted by $[x_1 : \ldots : x_{d+1}]$.

What Is a Singular Foliation?

Lemma 5.37 *The projective space $\mathbb{P}^d$ is a complex manifold of dimension d. It is given by the $d+1$ following charts:*

$$\psi_i : (x_1, \ldots, \hat{x}_i, \ldots, x_{d+1}) \mapsto [x_1 : \ldots : x_{i-1} : \underbrace{1}_{i^{th}\ term} : x_{i+1} : \ldots : x_{d+1}].$$

The idea of the blow-up at the origin consists in replacing $\mathbb{C}^{d+1}$, by pairs made of straight lines through the origin (=elements of $\mathbb{P}^d$) and a point on that straight line.

Definition 5.38 The blow-up $Bl_0(\mathbb{C}^{d+1})$ of $\mathbb{C}^{d+1}$ at the origin consists of all pairs $(D, z) \in \mathbb{P}^d \times \mathbb{C}^{d+1}$ such that $z \in D$.

Given coordinates $[x_0 : \ldots : x_d]$ and $(z_0, \ldots, z_d)$ on $\mathbb{P}^d$ and $\mathbb{C}^{d+1}$ respectively, we can describe $Bl_0(\mathbb{C}^{d+1})$ in terms of coordinates:

$$Bl_0(\mathbb{C}^{d+1}) = \{(x, z) \in \mathbb{P}^d \times \mathbb{C}^{d+1} \mid z_i x_j = z_j x_i,\ i, j = 0, \ldots, d\}.$$

These equations make sense, because multiplying all the x_i's by a non-zero factor leave them invariant.

Lemma 5.39 $Bl_0(\mathbb{C}^{d+1})$ *is a complex manifold of dimension $d+1$. It is given by the $d+1$ following charts:*

$$\phi_i : (x_1, \ldots, x_d) \mapsto ([x_1 : \ldots : x_{i-1} : \underbrace{1}_{i^{th}\ term} : x_{i+1} : \ldots : x_d],$$

$$(x_i x_1, \ldots x_i x_{i-1}, \underbrace{x_i}_{i^{th}\ term}, x_i x_{i+1}, \ldots, x_i x_d)).$$

In these charts, the natural projection $\sigma : Bl_0(\mathbb{C}^{d+1}) \longrightarrow \mathbb{C}^{d+1}$ mapping the pair (D, z) to z is given by the projection onto the second factor.

For $z \neq 0$ the pre-image $\sigma^{-1}(z)$ is pair (D, z) with D being the unique line $D \in \mathbb{P}^d$ passing through $z \in \mathbb{C}^{d+1}$. But $\sigma^{-1}(0) \simeq \mathbb{P}^d$. Last:

$$\sigma : Bl_0(\mathbb{C}^{d+1}) \setminus \sigma^{-1}(\{0\}) \longrightarrow \mathbb{C}^{d+1} \setminus \{0\} \tag{19}$$

is a biholomorphism. In particular, any vector field on $\mathbb{C}^{d+1}$ can be transported to a vector field on $Bl_0(\mathbb{C}^d) \setminus \sigma^{-1}(\{0\})$. It is natural to ask whether this vector field can be extended to $Bl_0(\mathbb{C}^{d+1})$ or not. In general the answer is "no", but it is "yes" if the vector field vanishes at 0, as we now see.

Proposition 5.40 *For a holomorphic vector field X of $\mathbb{C}^{d+1}$, the following two points are equivalent:*

(i) *X vanishes at 0*
(ii) *there exists a vector field $\tilde{X}$ on $\mathrm{Bl}_0(\mathbb{C}^{d+1})$ σ-related[46] to X.*

If it exists, then the vector field in item (ii) is unique.

Proof On the i-th of the $d+1$ charts of Lemma 5.39, σ reads:

$$(x_1, \ldots, x_{d+1}) \mapsto (x_i x_1, \ldots x_i x_{i-1}, x_i, x_i x_{i+1}, \ldots, x_i x_{d+1}).$$

The pull-back of the coordinate functions $(z_1, \ldots, z_{d+1})$ of $\mathbb{C}^{d+1}$ are therefore given by

$$\sigma^*(z_j) = \begin{cases} x_i x_j & j \neq i \\ x_i & j = i \end{cases}$$

This implies that the unique vector field X_j on that chart such that

$$\sigma_*(X_j) = \frac{\partial}{\partial z_j}$$

is given in the coordinates $(x_1, \ldots, x_d)$ by

$$X_j = \begin{cases} \frac{1}{x_i} \frac{\partial}{\partial x_j} & j \neq i \\ \frac{\partial}{\partial x_i} - \sum_{j \neq i} \frac{x_j}{x_i} \frac{\partial}{\partial x_j} & j = i \end{cases}$$

In turn, this implies that for every vector field $X = \sum_{i=1}^{d+1} P_i(z_1, \ldots, z_d) \frac{\partial}{\partial z_j}$, the unique vector field on the i-th chart such that $\sigma_*(Z) = X$ is

$$Z = \sum_{j \neq i} \left(\frac{P_j(x_1 x_i, \ldots, x_i, \ldots, x_i x_d)}{x_i} - \frac{x_j P_i(x_1 x_i, \ldots, x_i, \ldots, x_i x_d)}{x_i} \right) \frac{\partial}{\partial x_j}$$

$$+ \frac{P_i(x_1 x_i, \ldots, x_i, \ldots, x_i x_d)}{x_i} \frac{\partial}{\partial x_i}$$

This vector field is well-defined on the whole charts if and only if the functions $P_1, \ldots, P_{d+1}$ vanish at the origin. This proves the claim.

---

[46] I.e., $T_{(D,z)}\sigma(\tilde{X}_{(D,z)}) = X_{|z}$. Equivalently $\sigma_*\tilde{X} = X$.

What Is a Singular Foliation?

> **Proposition 5.41 (Blow-Up of a Singular Foliation at the Origin)** Let $\mathcal{F}$ be a holomorphic or algebraic singular foliation on $\mathbb{C}^{d+1}$. Assume all vector fields on $\mathbb{C}^{d+1}$ vanish at 0. Then there exists a unique singular foliation $\widetilde{\mathcal{F}}$ on $\mathrm{Bl}_0(\mathbb{C}^d)$ such that (19) is an isomorphism of foliated manifolds.
> We call $\widetilde{\mathcal{F}}$ the blow-up of $\mathcal{F}$ at the origin.

Exercise 5.42 In this exercise, we call $\sigma^{-1}(0)$ the *exceptional divisor* of the blow-up: its points are canonically identified with straight lines through the origin. Let $\mathcal{F}$ be a singular foliation on $\mathbb{C}^{d+1}$ made of vector fields vanishing at 0, and let $\widetilde{\mathcal{F}}$ be its blow up.

1. Let $X \in \mathfrak{X}(\mathbb{C}^d)$ be a vector field vanishing at 0 and $\tilde{X} \in \mathfrak{X}(\mathrm{Bl}_0(\mathbb{C}^d))$ such that $\sigma_*(\tilde{X}) = X$. Show that $\tilde{X}$ vanishes at every point of the exceptional divisor if and only if

$$X = \lambda \sum_{i=1}^{d+1} z_i \frac{\partial}{\partial z_i} + \text{quadratic terms}$$

for some $\lambda \in \mathbb{C}$ where "quadratic terms" means vector fields vanishing at least quadratically at zero.

2. Show that some point D in the exceptional divisor is a point-leaf[47] if and only if D (seen now as a straight line) is an eigenvector for all the linearizations of all vector fields in $\mathcal{F}$.

5.7.2 Blow-Up Along a Smooth Submanifold

This construction of the blow-up of a singular foliation at a point can be extended considerably to a blow up along a submanifold to which the singular foliation is tangent. Since we presented the previous construction in the complex setting, we decided to present this construction in the smooth setting: it can of course be extended to the complex or real setting, but this requires using a procedure more algebraic than the one described below, that uses the tubular neighborhood theorem (which holds in the smooth case only).

Let N be a submanifold of M. Denote by $\mathcal{N}_{N/M}$ the normal bundle $TM_{|N}/TN$ of a submanifold $N \subset M$. The fiber bundle $\mathbb{P}(\mathcal{N}_{N/M}) \to N$ can be interpreted as the projective space of directions normal to N in M. The blow-up of M along N is a manifold obtained by gluing $M \backslash N$ and[48] $\mathbb{P}(\mathcal{N}_{N/M})$. The construction goes as follows. According to the tubular neighborhood theorem, there exists a diffeomorphism Φ from a neighborhood $\mathcal{U}$ of the

---

[47] I.e, a leaf reduced to a point - equivalently a point where the tangent space of the foliation is zero.
[48] For a vector bundle $E \to N$, we denote by $\mathbb{P}(E)$ the projective bundle over N namely, the complex manifold obtained by taking the projective space of all the fibers of E.

zero section in $\mathcal{N}_{N/M}$ to a neighborhood $\mathcal{V}$ of N in M: it can be obtained, for instance, by considering $\mathcal{N}_{N/M}$ as a sub-bundle of $TM_{|N}$, then using the geodesic flow: this requires to choose a metric on M. This diffeomorphism Φ is the identity when restricted to N.

Now, let $\mathrm{Bl}_N(\mathcal{N}_{N/M})$ be the fiber bundle over N obtained by applying to each fiber of $\mathcal{N}_{N/M}$ the construction of the blow-up at a point: an element in $\mathrm{Bl}_N(\mathcal{N}_{N/M})$ is a pair (P, u) with P a straight line through 0 in some fiber of $\mathcal{N}_{N/M}$ and u a point in that straight line. We denote by τ the projection $\mathrm{Bl}_N(\mathcal{N}_{N/M}) \to \mathcal{N}_{N/M}$. By construction, τ is a diffeomorphism from $\tau^{-1}(\mathcal{N}_{N/M} \backslash N)$ to $\mathcal{N}_{N/M} \backslash N$. The idea now consists in gluing

1. the manifold $M \backslash N$
2. with the manifold $\mathrm{Bl}_N(\mathcal{N}_{N/M})$

by identifying

1. the open subset $\mathcal{V} \backslash N$
2. with the open subset $\tau^{-1}(\mathcal{U} \backslash N)$

with the help of the diffeomorphism

$$\Phi \circ \tau : \tau^{-1}(\mathcal{U} \backslash N) \longrightarrow \mathcal{V} \backslash N$$

The result is a manifold $\mathrm{Bl}_N(M)$, called *blow-up of M along N*, equipped with a natural projection σ onto M, which is a diffeomorphism from $\sigma^{-1}(M \backslash N)$ to $M \backslash N$, and whose fiber over a point $n \in N$ is the projective space of the normal bundle $\mathcal{N}_{N/M}$ at n. The set $\sigma^{-1}(N) \simeq \mathbb{P}(\mathcal{N}_{N/M}) \subset \mathrm{Bl}_N(M)$ is a hypersurface called the *exceptional divisor* of the blow-up $\sigma : \mathrm{Bl}_N(M) \longrightarrow M$.

Example 5.43 Let $N = \mathbb{R}^m$ be a linear subspace of $\mathbb{R}^d$ be a linear subspace. Then $\mathrm{Bl}_N(M) \simeq \mathrm{Bl}_0(\mathbb{R}^m) \times \mathbb{R}^{d-m}$.

The next statement has been established by Debord and Skandalis [18].

Proposition 5.44 (Blow-Up of a Singular Foliation, General Case) *Let $\mathcal{F}$ be a singular foliation on M and $L \subset M$ a leaf.*
There exists a unique singular foliation $\tilde{\mathcal{F}}$ on the blow-up $\mathrm{Bl}_L(M)$ of M along L such that σ is an isomorphism from $\mathrm{Bl}_L(M) \backslash \sigma^{-1}(L)$ to $M \backslash L$.
It is called the the blow-up of $\mathcal{F}$ along L.

In fact, we not need to take L to be a leaf: the construction would work for any submanifold to which all vector fields in $\mathcal{F}$ are tangent. The proof is based on a lemma: a vector field $X \in \mathfrak{X}(M)$ reads $X = \sigma(\tilde{X})$ for $\tilde{X}$ a vector field in $\mathrm{Bl}_L(M)$ if and only if X is tangent to L.

5.7.3 Monoidal Transformation

The constructions of blow-ups above can be enlarged to any monoidal transformation with respect to any foliated ideal.

Let us work in the algebraic setting.[49] Let $M := \mathbb{C}^n$ and $\mathcal{O} := \mathbb{C}[z_1, \ldots, z_d]$. Consider an ideal $\mathcal{I}$, and choose $\varphi_1, \ldots, \varphi_k$ generators of this ideal. The monoidal transformation of M with respect to $\mathcal{I}$ is the sub-variety $M_\mathcal{I}$ of $\mathbb{P}^{k-1} \times \mathbb{C}^d$ defined by the equations

$$x_i \phi_j(z_1, \ldots, z_d) = x_j \phi_i(z_1, \ldots, z_d) \text{ for } 1 \leq i < j \leq k, \tag{20}$$

where $z_1, \ldots, z_d$ are the coordinates on $\mathbb{C}^d$ and $[x_1 : \cdots : x_k]$ are the homogeneous coordinates on the projective space $\mathbb{P}^{k-1}$, then by choosing the component of this subvariety that projects onto M. Alternatively, $M_\mathcal{I}$ can be seen as the Zariski closure of the graph of the map

$$M \backslash V_\mathcal{I} \to \mathbb{P}^{k-1}$$
$$(z_1, \ldots, z_d) \mapsto [\phi_1(z_1, \ldots, z_d) : \cdots : \phi_k(z_1, \ldots, z_d)].$$

Here $V_\mathcal{I} \subset M$ is the zero locus of $\mathcal{I}$, i.e., the sub-variety of all points where all functions in $\mathcal{I}$ are zero. It is a classical result that $M_\mathcal{I}$ is a quasi-projective variety, and that the natural projection on the first component:

$$\phi : M_\mathcal{I} \to M \tag{21}$$

is a proper map, which is a biholomorphism[50] when restricted to

$$\phi^{-1}(M \backslash V_\mathcal{I}) \to M \backslash V_\mathcal{I}. \tag{22}$$

Now, consider $\mathcal{F}$ an algebraic singular foliation on $\mathbb{C}^d$.

If $\mathcal{I}$ is a foliated ideal, i.e., if[51] $\mathcal{F}[\mathcal{I}] \subset \mathcal{I}$, then $M_\mathcal{I}$ comes induced with a natural singular foliation $\mathcal{F}_\mathcal{I}$ such that the map defined in (22) is an isomorphism of foliated manifolds. In particular, it means that for every $X \in \mathcal{F}$, the vector field on $\phi^{-1}(M \backslash V_\mathcal{I})$

---

[49] Sertöz does the construction on any complex manifold, see [62].
[50] In fact, it is even biregular.
[51] Geometrically, this should be interpreted as meaning that a leaf of $\mathcal{F}$ is either contained in $V_\mathcal{I}$ or has empty intersection in $V_\mathcal{I}$, i.e that $V_\mathcal{I}$ is a union of leaves of $\mathcal{F}$.

defined by transporting X through Eq. (22) extends to a vector field $\tilde{X}$ to the whole variety $M_\mathcal{I}$. We will not make a more precise statement, simply because doing it would require to make sense of singular foliations on schemes, or at least quasi-projective varieties - something that we have not done in the present manuscript. We prefer to refer the reader to the literature on the subject, namely [62], or the more recent [46, 47].

Exercise 5.45 Let us explain how to extend a vector field X on $\mathbb{C}^d$ such that $X[\mathcal{I}] \subset \mathcal{I}$ to a vector field on $M_\mathcal{I}$. Let $\varphi_1, \ldots, \varphi_k$ be generators of $\mathcal{I}$.

1. Show that there exists $(\lambda_{ij})_{i,j=1,\ldots,k} \in \mathcal{O}$ such that

$$X[\varphi_i] = \sum_{j=1}^{k} \lambda_{ij} \varphi_j.$$

2. Show that the vector field defined on $\mathbb{P}^{k-1} \times \mathbb{C}^d$ by[52]

$$\left(\sum_{a,b=1}^{k} \lambda_{ba} x_a \frac{\partial}{\partial x_b}, X \right)$$

is tangent to the subset defined by Eq. (20).
3. Conclude.

5.8 Nash-Blowup of a Singular Foliation

We now introduce another construction of a "blow-up" of a singular foliation, which is due to Omar Mohsen [54] who used it, with his collaborators, to prove several hard results about PDEs [7]. In fact, this blowup belongs to the large class of Nash blowups, and we will therefore call it the Nash blow-up of a singular foliation. A particular feature of this construction is that it produces a <u>Debord</u> singular foliation, which comes therefore from a Lie algebroid. To be more precise, it might be that the Nash blowup of a foliated manifold is not a smooth manifold, i.e., it may have singular points. But (1) whenever the Nash blowup is a smooth manifold, it becomes a foliated manifold with a Debord singular foliation, and (2) the conclusion is still valid in general, but we have to make sense of the notion of projective foliation on a singular quasi-projective variety, see [46]. The original article of Mohsen insists more on the topological groupoid that lies on the top of this Lie algebroid, showing that it has several desirable topological properties that we do not discuss here, see [54].

---

[52] Notice the inversion of the indices a and b.

> **Warning!**
> The Nash blowup method uses several concepts that are only introduced much later in the text. We recommend the reader unfamiliar with these notions to look first at Sect. 2 in chapter "Canonical Geometric and Algebraic Structures Hidden Behind a Singular Foliation". The "level" of this section is, in general, higher than the level of the neighboring sections.
>
> As usual, we try to deal with the smooth, real analytic and complex cases altogether.

5.8.1 Generalities About Grassmann Bundles

We start by recalling the notion of k-Grassmannian. We work in the complex setting, but we will mention at some point what happens in the real one. For any $N, k \in \mathbb{N}$ such that $0 \leq k \leq N$, the set $\text{Grass}_k(\mathbb{C}^N)$ of all k-dimensional vector subspaces of $\mathbb{C}^N$ is a complex compact manifold of dimension $k(N-k)$, called the *Grassmanian of k-planes in $\mathbb{C}^N$* (see e.g., [11, 57, 63]). For $k=1$, one recovers the notion of projective space.

Let us spell out the topology and the manifold structure of $\text{Grass}_k(\mathbb{C}^N)$. The groups $\text{GL}(\mathbb{C})$ and $\text{U}(N)$ of invertible $\mathbb{C}$-linear maps, and of unitary linear transformations acts transitively on $\text{Grass}_k(\mathbb{C}^N)$ by $g \cdot V = g(V)$ for all $V \in \text{Grass}_k(\mathbb{C}^N)$ and $g \in \text{GL}(\mathbb{C})$ or $\text{U}(N)$. This action is well-defined since g is invertible. Therefore,

$$\text{Grass}_k(\mathbb{C}^N) \stackrel{set}{\simeq} \frac{\text{U}(N)}{\text{U}(k) \times \text{U}(N-k)}$$

$$\stackrel{set}{\simeq} \frac{\text{GL}(N)}{\text{GL}(k, N)}$$

Above the subgroups $\text{U}(k) \times \text{U}(N-k)$ and $\text{GL}(k, N)$ are the respective stabilizers of $\mathbb{C}^k \times \{0\}^{N-k} \subset \mathbb{C}^N$ in their corresponding groups. Since both groups in the first line above are compact Lie groups, the first description equips $\text{Grass}_k(\mathbb{C}^N)$ with a structure of compact manifold. Since both groups in the second line are complex Lie groups, the second description equips the quotient with a structure of complex manifold. If one replaces $\mathbb{C}$ by $\mathbb{R}$, then this second structure does not make any sense, but the compact smooth manifold structure still exists.

Remark 5.46 The manifold $\text{Grass}_k(\mathbb{C}^N)$ has natural holomorphic coordinates, called *canonical affine coordinates*. One such chart $\phi := \mathbb{C}^{k(N-k)} \to \text{Grass}_k(\mathbb{C}^N)$ maps a $k \times (N-k)$-matrix S to the subspace generated by the column vectors:

$$\begin{pmatrix} \text{id} \\ S \end{pmatrix}$$

The image of this chart is the set of k-dimensional subspace that do not intersect $\{0\}^k \times \{\mathbb{C}\}^{N-k}$. The others natural charts are obtained by composition with the $U(N)$-action, i.e., are of the form $g \cdot \phi$ for some $g \in U(N)$.

Convention 5.47 *For our current purpose, it will be convenient to consider the Grassmannian of all sub-spaces of <u>co</u>-dimension r in $\mathbb{C}^N$ rather than the set of subspaces of a given dimension. It is convenient to denote this manifold by* $\mathrm{Grass}_{-r}(\mathbb{C}^N)$ *(notice the use of a minus sign). In other words, we set* $\mathrm{Grass}_{-r}(\mathbb{C}^N) := \mathrm{Grass}_{N-r}(\mathbb{C}^N)$.

Now, given a vector bundle over a manifold, one can the Grassmannian at each fiber. The next definition makes sense in the smooth or complex cases without adaptation.

Definition 5.48 Let $A \to M$ be a vector bundle of rank d over a manifold M. Let $k \leq \mathrm{rk}(A) = d$. The disjoint union:

$$\mathrm{Grass}_k(A) := \coprod_{x \in M} \mathrm{Grass}_k(A_x)$$

comes equipped with a natural manifold structure. Also

$$\Pi \colon \mathrm{Grass}_k(A) \longrightarrow M \qquad (23)$$

is a fibration with fiber $\simeq \mathrm{Grass}_k(\mathbb{K}^d)$ ($\mathbb{K} = \mathbb{R}$ or $\mathbb{C}$). It is called *k-th Grassmann bundle of A*. The same holds upon replacing k by $-k$.

The following exercise uses the notion of linear vector fields on an vector bundle (see Sect. 2 in chapter "Canonical Geometric and Algebraic Structures Hidden Behind a Singular Foliation" for details). Again, it is valid as it is in both smooth or complex cases.

Exercise 5.49 Let $A \to M$ be a vector bundle over M. Show that a linear vector field on A induces a vector field on $\Pi \colon \mathrm{Grass}_{-r}(A) \to M$ that is Π-projectable[53] on M.

Hint: the flow of a linear vector fields on A is a pseudo-group of diffeomorphism of A, hence it induces a pseudo-group of vector bundle isomorphisms, which then induces a pseudo-group of fiber bundle diffeomorphisms on $\mathrm{Grass}_{-r}(A)$.

---

[53] I.e. Π-related (see Definition 5.4) to a vector field in $\mathfrak{X}(M)$.

5.8.2 Nash-Blowup I: The Space

Throughout this subsection $(M, \mathcal{F})$ is a foliated manifold with M connected, and (A, ρ) is an anchored bundle over $\mathcal{F}$.

We denote by $M_{\text{reg}} \subset M$ the regular part of $(M, \mathcal{F})$ (see Sect. 3.4). By construction, it is the open dense subset such that the maps $x \mapsto \text{im}(\rho_x)$ and $x \mapsto \text{ker}(\rho_x)$ are locally constant, i.e., the set of regular points of $\mathcal{F}$.

Here we have to make a distinction between the smooth or complex settings.

- In the complex case, the codimension r of $\text{im}(\rho_x) = T_x \mathcal{F} \subseteq T_x M$ is the same for all points $x \in M_{\text{reg}}$ since M is connected. This is also true in the real analytic setting.
- In the smooth case, we will have from now on to <u>assume</u> that it is the case.[54]

Notice that for every point $x \in M_{\text{reg}}$ in the regular part,

1. $\text{im}(\rho_x)$ is an element of the Grassmannian $\text{Grass}_r(T_x M)$, and
2. $\text{ker}(\rho_x)$ is an element of $\text{Grass}_{-r}(A_x)$.

Consider now:

1. the Grassmann bundle $\text{Grass}_r(TM)$, and
2. the Grassmann bundle $\text{Grass}_{-r}(A)$.

As before, denote by Π, in both cases, their natural projections onto M. Consider the two natural sections of Π defined on the regular part M_{reg} by:

1. $\sigma_{\text{im}} \colon M_{\text{reg}} \longrightarrow \text{Grass}_r(TM)$, $x \longmapsto \text{im}(\rho_x)$, and
2. $\sigma_{\text{ker}} \colon M_{\text{reg}} \longrightarrow \text{Grass}_{-r}(A)$, $x \longmapsto \text{ker}(\rho_x)$.

Then we define

1. $\text{Bl}_{tgt}(M, \mathcal{F}) := \overline{\sigma_{\text{im}}(M_{\text{reg}})}$ to be the closure of the image of the section σ_{im} in $\text{Grass}_r(TM)$, and
2. $\text{Bl}_A(M, \mathcal{F}) := \overline{\sigma_{\text{ker}}(M_{\text{reg}})}$ to be the closure of the image of the section σ_{ker} in $\text{Grass}_{-r}(A)$.

We shall denote by π the restriction of Π to both $\text{Bl}_{\text{im}}(M, \mathcal{F})$ and $\text{Bl}_{\text{ker}}(M, \mathcal{F})$.

---

[54] Equivalently, we have to assume that all the regular leaves have the same dimension. It is not the case, for instance, for the singular foliation on $M = \mathbb{R}$ generated by the vector field $\chi(t)\partial_t$ with $\chi(t) = e^{-1/t^2}$ for $t \geq 0$ and $\chi(t) = 0$ for $t \leq 0$, since then $M_{\text{reg}} = \mathbb{R}_-^* \cup \mathbb{R}_+^*$ but regular leaves have dimension 0 on $\mathbb{R}_-^*$ and one on $\mathbb{R}_+^*$. See Sect. 3.4 for an overview of these notions.

Comparison with Existing Literature

We claim the definition of $\mathrm{Bl}_A(M, \mathcal{F})$ above matches the definition of the blowup of a singular foliation given by Omar Mohsen in [54], and used in [7]. This requires a careful line by line check, but the difference is mainly a difference of presentation. It also matches Ali Sinan Sertöz's construction [62] of a Nash blowup coherent of a coherent sheaf when applied to the coherent sheaf $\mathcal{F}$ (see Proposition 2.23) up to a difference of context: smooth setting in Omar Mohsen, complex setting in Sertöz.[a]

Also, in [47], a sequence depending on $n \in \mathbb{N}$ of "blow-ups" is constructed, which for $n = 0$ gives back $\mathrm{Bl}_{\mathrm{tgt}}(M, \mathcal{F})$, and for $n = 1$ gives $\mathrm{Bl}_A(M, \mathcal{F})$.

---

[a] To be precise, Ali Sinan Sertöz 's construction consists in applying the method described above to a vector bundle morphism $B \xrightarrow{d} A$ whose co-image is $\mathcal{F}$: the existence of such a vector bundle morphism is the very definition of coherent sheaves.

Exercise 5.50 Spell out using adapted coordinates the definition of $\mathrm{Bl}_A(M, \mathcal{F})$. Check that if there exists coordinates in which the anchored bundle has polynomial expressions (i.e., the anchor map of constant sections is a polynomial vector field), then $\mathrm{Bl}_A(M, \mathcal{F})$ is an affine variety in every chart given by adapted coordinates. (*Hint:* A solution to this problem is given in [47]–[46], where it is proven that it is even a monoidal transformation.)

Remark 5.51 Intuitively, for $x \in M$, $\pi^{-1}(x) = \mathrm{Bl}_A(M, \mathcal{F}) \cap \Pi^{-1}(x)$ is the set of all possible limits of the subspaces $\ker \rho_y$ when $y \in M_{\mathrm{reg}}$ converges to x. More precisely, for any $x \in M$, there is a an open neighborhood $\mathcal{U} \subset M$ of x such that $\mathrm{Grass}_{-r}(A) \simeq \mathcal{U} \times \mathrm{Grass}_{-r}(\mathbb{K}^{\mathrm{rk}(A)})$. By construction,

$$\pi^{-1}(x) = \left\{ \text{Codim. } r \text{ subspaces } V \subset A_x \;\middle|\; \exists (x_n) \in M_{\mathrm{reg}}, \right.$$
$$\left. \text{such that, } \ker \rho_{x_n} \xrightarrow[n \to +\infty]{} V \text{ as } x_n \xrightarrow[n \to +\infty]{} x \right\}.$$

One can make a similar construction with $\mathrm{Bl}_{tgt}(M, \mathcal{F})$:

$$\pi^{-1}(x) = \left\{ \text{Dim. } r \text{ subspaces } W \subset T_x M \;\middle|\; \exists (x_n) \in M_{\mathrm{reg}}, \right.$$
$$\left. \text{such that, } \mathrm{im}\rho_{x_n} \xrightarrow[n \to +\infty]{} W \text{ as } x_n \xrightarrow[n \to +\infty]{} x \right\}.$$

Remark 5.52 Since $\mathrm{im}(\rho_x)$ coincides with the vector subspace of $T_x M$ that we denoted by $T_x \mathcal{F}$, which also coincides with the tangent space of the leaf through x, the set

$\mathrm{Bl}_{tgt}(M, \mathcal{F})$ does not depend on the choice of an anchored bundle. It is also the case for $\mathrm{Bl}_A(M, \mathcal{F})$ as we now see.

Lemma 5.53 *For any two anchored bundle (A, ρ) and (A', ρ'), there is an unique homeomorphism[55] making the following diagram commutative*

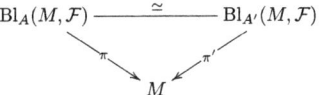

Proof [56] This homeomorphism, if it exists, is unique, by density of the regular part. We saw in Proposition 1.7, Sect. 1.1 in chapter "Canonical Geometric and Algebraic Structures Hidden Behind a Singular Foliation", that there exists anchored bundle morphisms (see Sect. 1.1 in chapter "Canonical Geometric and Algebraic Structures Hidden Behind a Singular Foliation")

$$(A, \rho) \underset{\Psi}{\overset{\Phi}{\rightleftarrows}} (A', \rho') . \tag{24}$$

Now, $(A \oplus A', \rho + \rho')$ is also an anchored bundle over $\mathcal{F}$, and there is a commutative diagram as follows:

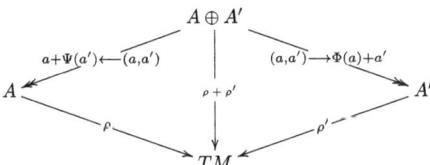

Since the vector bundle morphisms $A \oplus A' \to A$ and $A \oplus A' \to A'$ above are surjective anchored bundle morphisms, it suffices to show that Lemma 5.53 holds under the additional assumption that the anchored bundles be related by a surjective anchored bundle morphism.

Consider two anchored bundles (A, ρ) and, say, (C, ρ_C) over $\mathcal{F}$ such that there exists a surjective anchored bundle morphism $\phi : C \to A$. Let $K \to M$ be the kernel of ϕ. Consider the subset

$$\mathrm{Grass}_{-r}(C, K) \subset \mathrm{Grass}_{-r}(C)$$

---

[55] A close look at the proof shows that it is in fact more than a simple homeomorphism: the pull-back of a function on $\mathrm{Bl}_{A'}(M, \mathcal{F})$ which is, locally, the restriction of a locally defined smooth, real analytic or holomorphic function on $\mathrm{Grass}_{-r}(A)$ is a function on $\mathrm{Bl}_A(M, \mathcal{F})$ of the same type.

[56] We acknowledge for a discussion with Cédric Rigaud for this proof.

of all vector sub-spaces of a fiber C_m which satisfy two conditions: their codimension is r and they contain K_m. Since K is a sub vector bundle of C, $\text{Grass}_{-r}(C, K)$ is a closed submanifold. Since K is in the kernel of ϕ, this submanifold is moreover canonically diffeomorphisc[57] to $\text{Grass}_{-r}(A)$. We denote by ϕ this diffeomorphism. Since, the anchor of C is zero on K, the section $\sigma^C_{\text{ker}} \colon M_{\text{reg}} \longrightarrow \text{Grass}_{-r}(C)$ used to construct $\text{Bl}_C(M, \mathcal{F})$ is valued in $\text{Grass}_{-r}(C, K)$. Moreover, it is related with the section $\sigma^A_{\text{ker}} \colon M_{\text{reg}} \longrightarrow \text{Grass}_{-r}(A)$ used to construct $\text{Bl}_A(M, \mathcal{F})$ by

$$\phi \circ \sigma^C_{\text{ker}} = \sigma^A_{\text{ker}}.$$

The diffeomorphism ϕ therefore also intertwines the closures of the graphs of σ^A_{ker} and σ^C_{ker}, hence induces an homeomorphism $\text{Bl}_A(M, \mathcal{F}) \simeq \text{Bl}_C(M, \mathcal{F})$ that has all desired properties.

Lemma 5.53 implies[58] that $\text{Bl}_A(M, \mathcal{F})$ is a global object, defined over the whole manifold M. Since anchored bundles exist near every point, $\text{Bl}_A(M, \mathcal{F})$ can be defined in a neighborhood of every point. Since it does not depend on the choice of an anchored bundle, these locally defined sets glue in a natural manner to define a bundle $\text{Bl}(M, \mathcal{F})$ over the whole complex manifold M.

The next exercise[59] describes some important features of that construction.

Exercise 5.54 Let $\mathcal{F}$ be a singular foliation on a connected manifold M. In the smooth case, we assume that all regular leaves have the same dimension r. Assume that an anchored bundle (A, ρ) exists. Prove that $\pi \colon \text{Bl}_A(M, \mathcal{F}) \to M$ satisfies the following properties:

1. π is proper and surjective. In particular, for each point $x \in M$, the fiber $\pi^{-1}(x)$ is a non-empty compact set. (*Hint:* Use the fact that the projection Π admits compact fibers.)
2. For every $x \in M$ and $V \in \pi^{-1}(x)$, one has $\text{Sker}(\rho, x) \subseteq V \subseteq \text{ker}(\rho_x)$.

---

[57] Replace by biholomorphic in the complex case in the present discussion, and similarly for the real analytic case.

[58] For singular foliations $\mathcal{F}$ which are not finitely generated - which is the case of most of them, especially in the complex case, see Sect. 2.5.

[59] We refer the reader to Sect. 3 in chapter "Canonical Geometric and Algebraic Structures Hidden Behind a Singular Foliation", for definitions of the isotropy Lie algebra and the definition of strong kernel at a point. We also invite the reader to use Remark 5.51. The questions asked in this exercise are solved in [47], although in a slightly different context.

3. For every $x \in M$ and $V \in \pi^{-1}(x)$, the image $[V]$ of V in the isotropy Lie algebra at x, i.e.,

$$\mathfrak{g}_x(\mathcal{F}) = \frac{\ker(\rho_x)}{\operatorname{Sker}(\rho, x)}$$

is a sub-space codimension $r - \dim(L_x)$, where $\dim(L_x)$ is the dimension the leaf through x. Also, show that $V \to [V]$ is an injective map from $\pi^{-1}(x)$ to $\operatorname{Grass}_{-r+\dim(L_x)}(\mathfrak{g}_x(\mathcal{F}))$.

4. For every $x \in M$ and $V \in \pi^{-1}(x)$, $[V]$ is a Lie sub-algebra of $\mathfrak{g}_x(\mathcal{F})$ of codimension $r - \dim(L_x)$, where $\dim(L_x)$ is the dimension of the leaf through x. *Hint:* equip the anchored bundle with an almost Lie algebroid bracket and show that the almost Lie algebroid bracket restricts to a bilinear map $\wedge^2 V \to V$.

5. For every $x \in M_{\text{reg}}$, $\pi^{-1}(x) = \ker(\rho_x)$ is reduced to a point in $\operatorname{Grass}_{-r}(A)$. Also, $\pi^{-1}(M_{\text{reg}})$ is a smooth manifold and the restriction $\pi \colon \pi^{-1}(M_{\text{reg}}) \longrightarrow M_{\text{reg}}$ is invertible.[60]

The conclusion of the previous exercise is the following. Let $x \in M$. Within the Grassmannian $\operatorname{Grass}_{r-\dim(L_x)}(\mathfrak{g}_x(\mathcal{F}))$ of all sub-spaces of codimension $r - \dim(L_x)$ in the isotropy Lie algebra $\mathfrak{g}_x(\mathcal{F})$ at x lies a subset, denoted by

$$\operatorname{Grass}^{Lie}_{-r+\dim(L_x)}(\mathfrak{g}_x(\mathcal{F}))$$

made of all sub-Lie algebras of codimension $r - \dim(L_x)$. This subset is compact. What the previous exercise gives is an injective inclusion of $\pi^{-1}(x) \subset \operatorname{Grass}_r(A_{|x})$ as a compact subset inside $\operatorname{Grass}^{Lie}_{-r+\dim(L_x)}(\mathfrak{g}_x(\mathcal{F}))$.

Remark 5.55 A remarkable result by Omar Mohsen [54] is that $\pi^{-1}(x) \subset \operatorname{Grass}^{Lie}_{-r+\dim(L_x)}(\mathfrak{g}_x(\mathcal{F}))$ lies in fact inside the set of sub-Lie algebras that integrate to a <u>closed</u> sub-Lie group of the simply connected Lie group[61] integrating $\mathfrak{g}_x(\mathcal{F})$. This is highly non-trivial: such Lie sub-algebras do not form a compact subset of $\operatorname{Grass}^{Lie}_{-r+\dim(L_x)}(\mathfrak{g}_x(\mathcal{F}))$.

---

[60] Invertible here means: diffeomorphism, in the smooth case, biholomorphism, in the complex case.
[61] In fact, [54] established an even stronger result: it integrates to a closed subgroup of the isotropy of Androulidakis-Skandalis holonomy groupoid, which is a quotient of the universal one.

Overall, we have therefore constructed an inclusion

$$\mathrm{Bl}_A(M, \mathcal{F}) \hookrightarrow \coprod_{x \in M} \mathrm{Grass}^{Lie}_{-r+\dim(L_x)}(\mathfrak{g}_x(\mathcal{F})) \qquad (25)$$

The image of this inclusion does not depend on the choice of an anchored bundle. Lemma 5.53 and the previous discussion lead to the following conclusion:

Proposition 5.56 (Bl$(M, \mathcal{F})$ **is well-defined**) Bl$_A(M, \mathcal{F})$ *does not depend on the choice of an anchored bundle (A, ρ) over $\mathcal{F}$, or precisely:*

1. *the image of the injective map (25) does not depend on (A, ρ),*
2. *and for any other anchored bundle (A', ρ') over $\mathcal{F}$, there is canonical homeomorphism* Bl$_A(M, \mathcal{F}) \simeq$ Bl$_{A'}(M, \mathcal{F})$.

This independence allows the next definition.

Definition 5.57 (Nash Blow-Up of a Singular Foliation) In view of Proposition 5.56, it makes sense to denote by Bl$(M, \mathcal{F})$ (with no reference to the chosen anchored bundle) the blowup of a foliated manifold $(M, \mathcal{F})$ whose regular leaves all have the same dimension. We call it[a] the *Nash blowup of $(M, \mathcal{F})$*.

---

[a]Following [16]–[62].

The following definition now makes sense.

Definition 5.58 Let $(M, \mathcal{F})$ be a singular foliation and r the dimension of the regular leaves. For every point $x \in M$, we call *limit Lie subalgebras* the subalgebras of codimension $r - \dim(L_x)$ of the holonomy Lie algebra $\mathfrak{g}_x(\mathcal{F})$ that appear in Exercise 5.54, i.e., which are the projections on $\mathfrak{g}_x(\mathcal{F})$ of the limits of kernels of the anchor maps.

Let us recapitulate what we have established on $\mathrm{Bl}(M, \mathcal{F}) \subset \mathrm{Grass}_{-r}(A)$ and its natural projection on M. Firstly, it satisfies two properties resolutions have to satisfy, namely

1. the map π is proper - in particular, it is surjective,
2. and $\pi^{-1}(M_{\mathrm{reg}})$ is a manifold which is isomorphic[62] to M_{reg}

Secondly, despite its extremely pleasant two properties, the closed subset $\mathrm{Bl}(M, \mathcal{F}) \subset \mathrm{Grass}_{-r}(A)$ has a major problem: **it is not a sub-manifold in general**.

Remark 5.59 In [47], it is shown that $\mathrm{Bl}(M, \mathcal{F})$ is obtained by a locally monoidal transformation, whose center lies within the closed subset of points in M that admit no neighborhood on which $\mathcal{F}$ is Debord (a subset of M_{sing}). Also, smoothness is addressed.

5.8.3 Some Natural Vector Bundles

Proposition 5.56 has given the space on which the Nash blowup will be defined. We now have to equip it with a singular foliation. Before doing so, we have to describe two exact sequences of vector bundles over $\mathrm{Gr}_{-r}(A)$, for $A \to M$ a vector bundle and r an integer $\leq \mathrm{rk}(A)$.

1. Let $\Pi^! A$ be the pull-back of $A \to M$ to $\mathrm{Grass}_{-r}(A)$.
2. We call *tautological subbundle* the vector subbundle of $\Pi^! A$ whose fiber at a point $V_x \in \mathrm{Grass}_{-r}(A)$ is precisely the vector space V_x, seen a subspace of $(\Pi^! A)_{V_x} = A_x$. We denote it by τ_A^{-r} (the $-r$ being a reminder that it is of corank r in $\Pi^! A$).
3. We denote by A_{Bl}^r and call *tautological quotient bundle* the bundle over $\mathrm{Grass}_{-r}(A)$ obtained by taking the quotient of the first bundle by the second one, i.e., $A_{Bl}^r := \Pi^! A / \tau_A^{-r}$. (The "$r$" reminds that it has rank r).

These three vector bundles fit into the exact sequence

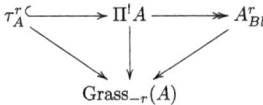

Remark 5.60 There is of course a dual canonical isomorphism and inclusion of vector bundles:

---

[62] I.e., diffeomorphic, diffeomorphic through real analytic maps, or biholomorphic depending on the context.

$$(A^r_{Bl})^* \xrightarrow{\sim} (\tau^r_A)^{\perp} \hookrightarrow \Pi^! A^*$$
$$\searrow \quad \downarrow \quad \swarrow$$
$$\text{Grass}_{-r}(A) \tag{26}$$

where the symbol $\perp$ stands for the annihilator.

5.8.4 Nash-Blowup II: The Debord Foliation

We now explain how to lift the singular foliation $\mathcal{F}$ to a singular foliation on $\text{Bl}(M, \mathcal{F})$. Let (A, ρ) be an anchored bundle over $\mathcal{F}$. Lemma 2.13, in Sect. 2 in chapter "Canonical Geometric and Algebraic Structures Hidden Behind a Singular Foliation" implies that for every $X \in \mathcal{F}$ there exists a linear vector field ξ_X on A (by choosing an almost Lie algebroid bracket on (A, ρ)) fulfilling the following two properties

1. it is p-projectable[63] to X
2. and its flow $\phi_t^{\xi_X}$ is an isomorphism of anchored bundles whenever defined

$$\begin{array}{ccc} A & \xrightarrow{\phi_t^{\xi_X}} & A \\ \rho \downarrow & & \downarrow \rho \\ TM & \xrightarrow{T\phi_t^X} & TM. \end{array} \tag{27}$$

Using Exercise 5.49, we see that ξ_X induces a vector field on $\text{Grass}_{-r}(A)$ which is Π-projectableI.e. Π-related (see Definition 5.4) to X. on X.

We call $\hat{\xi}_X$ a *lift of X on* $\text{Grass}_{-r}(A)$.

Lemma 5.61 (Lift of Vector Fields of $\mathcal{F}$) *Consider the lift $\hat{\xi}_X$ of some $X \in \mathcal{F}$ to the Grassmann bundle* $\text{Grass}_{-r}(A)$ *of some anchored bundle* (A, ρ).

1. *The flow of $\hat{\xi}_X$ of X, whenever it is defined, preserves the subset[64] $\text{Bl}(M, \mathcal{F}) \subset \text{Grass}_{-r}(A)$, i.e.,*

$$\begin{array}{ccc} \text{Bl}(M, \mathcal{F}) & \xrightarrow{\phi_t^{\xi_X}} & \text{Bl}(M, \mathcal{F}) \\ \pi \downarrow & & \downarrow \pi \\ M & \xrightarrow{\phi_t^X} & M \end{array} \tag{28}$$

---

[63] I.e. p-related with X.
[64] Recall that we erased the index "A" from the notation $\text{Bl}_A(M, \mathcal{F})$, since it does not depend on the anchored bundle (A, ρ) by Proposition 5.56, but, when an anchored bundle (A, ρ) is given, we still see it as a subset of $\text{Grass}_{-r}(A)$.

2. In particular, $\hat{\xi}_X$ is tangent to $\mathrm{Bl}(M, \mathcal{F})$ in the neighborhood of any point where $\mathrm{Bl}(M, \mathcal{F})$ is a sub-manifold.

Proof The second item in the lemma is a direct consequence of the first one. Let us prove the first one. For any x is the regular part of $\mathcal{F}$, it follows from (27)

$$\phi_t^{\xi_X}|_x (\ker \rho_x) = \ker \rho_{\phi_t^X(x)}.$$

Equivalently, in terms of the section $\sigma_{\ker} : M_{\mathrm{reg}} \to \mathrm{Grass}_{-r}(A)$, it means that

$$\phi_t^{\hat{\xi}_X} \circ \sigma_{\ker} = \sigma_{\ker} \circ \phi_t^X.$$

In particular, since $\phi_t^X(M_{\mathrm{reg}}) \subseteq M_{\mathrm{reg}}$, this implies that $\phi_t^{\hat{\xi}_X}$ preserves the closure $\overline{\sigma_{\ker}(M_{\mathrm{reg}})}$, i.e., it preserves $\mathrm{Bl}(M, \mathcal{F})$. Let us spell out this argument: for any element $V_x \in \mathrm{Bl}(M, \mathcal{F})$, there exists a sequence $(x_n)_{n \in \mathbb{N}}$ in M_{reg} such that $x_n \xrightarrow[n \to +\infty]{} x$ and such that $\ker \rho_{x_n} \xrightarrow[n \to +\infty]{} V_x$. In view of Eq. (27), one has

$$\phi_t^{\xi_X}|_{x_n} (\ker \rho_{x_n}) = \ker \rho_{\phi_t^X(x_n)}, \quad \text{for every } n \in \mathbb{N}_0.$$

The limit $\phi_t^{\hat{\xi}_X}(V_x)$ of the sequence

$$n \mapsto \phi_t^{\hat{\xi}_X} (\ker \rho_{x_n}) = \phi_t^{\xi_X}|_{x_n} (\ker \rho_{x_n})$$

therefore belongs to $\mathrm{Bl}(M, \mathcal{F})$ by construction. This completes the proof.

From now on, we assume for the sake of simplicity that $\mathrm{Bl}(M, \mathcal{F})$ is a smooth submanifold. However, we claim that this assumption is not required for most statement below, provided that a correct notion of singular foliation on a singular subset is given: we refer to [46] for a detailed study of the matter.

Proposition 5.62 *Let $(M, \mathcal{F})$ be a foliated manifold such that*

(a) all regular leaves have the same dimension[65] and
(b) $\mathrm{Bl}(M, \mathcal{F})$ is a sub-manifold[66] of $\mathrm{Grass}_{-r}(A)$ for some anchored bundle (A, ρ).

---

[65] Again, this is always true if M is connected in the complex and real analytic settings.
[66] In [46], a framework is given to make sense of the three statements of the theorem even when this second assumption is not satisfied.

Then:

1. *Every vector field $X \in \mathcal{F}$, there exists a unique vector field $\pi^!(X)$ on $\mathrm{Bl}(M, \mathcal{F})$ which is π-related to X;*
2. *The vector fields $\{\pi^!(X), X \in \mathcal{F}\}$ generate a singular foliation. The latter shall be denoted by $\pi^!(\mathcal{F})$.*
3. *The restriction of π to $\pi^{-1}(M_{reg})$ is an isomorphism of singular foliation from $(\pi^{-1}(M_{reg}), \pi^!(\mathcal{F}))$ to $(M_{reg}, \mathcal{F})$.*

Proof Uniqueness in item 1 is a direct consequence of the fact that π is an isomorphism when restricted to the dense open subset $\pi^{-1}(M_{reg})$. Existence is an immediate consequence of the second item Lemma 5.61: the restriction of $\hat{\xi}_X$ to $\mathrm{Bl}(M, \mathcal{F})$ is the required vector field. Now, since π is an isomorphism when restricted to the dense open subset $\pi^{-1}(M_{reg})$, it is obvious that:

$$[\pi^!(X), \pi^!(Y)] = \pi^!([X, Y]) \quad \text{for all } X, Y \in \mathcal{F}$$

The sheaf considered in item 2 is therefore stable under Lie bracket: it is also obviously locally finitely generated. Item 3 is obvious.

This statement allows to make sense of the following definition. In the complex case, it does correspond to Nash blowup of coherent sheaves[67] in [62], hence the chosen name. We will explain later on how it relates with Mohsen's groupoid in the smooth setting.

Definition 5.63 (The Nash-Blowup) The singular foliation $(\mathrm{Bl}(M, \mathcal{F}), \pi^!(\mathcal{F}))$ shall be referred to as the *Nash blowup of $(M, \mathcal{F})$*.

Here is now the most surprising point about the Nash blowup: the singular foliation that we obtain is Debord, i.e., it is isomorphic, as a sheaf, to the sheaf of sections of a vector bundle, which is of course A^r_{Bl} restricted to $\mathrm{Bl}(M, \mathcal{F})$. Equivalently, it is given by a Lie algebroid whose anchor map is injective on a dense open subset (see Sect. 4). This is established in items 1 and 2 of the next theorem.

---

[67] Holomorphic singular foliations are instances of coherent sheaves, see Proposition 2.23.

Theorem 5.64 (The Nash-Blow Up Is a Debord Foliation) *Let $(M, \mathcal{F})$ be a foliated manifold such that*

(a) *all regular leaves have the same dimension[a] and*
(b) $\mathrm{Bl}(M, \mathcal{F})$ *is a manifold[b]*

Then

1. *The Nash blowup $\pi^!(\mathcal{F})$ is a Debord singular foliation on $\mathrm{Bl}(M, \mathcal{F})$.*
2. *Its associated Lie algebroid is the restriction to $\mathrm{Bl}(M, \mathcal{F})$ of the canonical quotient bundle A^r_{Bl}. In equation:*

$$\pi^!(\mathcal{F}) \simeq \Gamma(A^r_{Bl}|_{\mathrm{Bl}(M,\mathcal{F})}).$$

3. *In the smooth setting, it is the Lie algebroid of Mohsen's Lie groupoid[c] of $\mathcal{F}$. In the complex setting, as a vector bundle, it matches the vector bundle described in [62], section IV.*

---

[a] Again, the first assumption is always valid in the complex and real analytic settings if M is connected.

[b] Again, [46, 47] give a framework that allows to make sense of the conclusion of the theorem even when this second assumption is not satisfied.

[c] Item 3 in the theorem does not make much sense yet, since Mohsen's Lie groupoid of a singular foliation has not been defined. It is a topological groupoid defined in [54], which is Lie when $\mathrm{Bl}(M, \mathcal{F})$ is a manifold.

In view of the second item, we can make sense of the following definition:

Definition 5.65 We call the Lie algebroid in item 2 of Theorem 5.64 the Nash blow up Lie algebroid, and we denote it by A_{Bl}.

We will only prove items 1 and 2 of this theorem in the smooth setting only, and leave it to the reader to adapt to real analytic or complex settings. We refer to [46, 47] for item 3. We start with a lemma.

Lemma 5.66 *There exists a vector bundle morphism*

$$\hat{\rho} \colon \Pi^! A \longrightarrow T(\mathrm{Grass}_{-r}(A))$$

over $\text{Grass}_{-r}(A)$ such that

(a) the following diagram commutes:

$$\begin{array}{ccc} \Pi^! A & \xrightarrow{\hat{\rho}} & T\text{Grass}_{-r}(A)) \\ \downarrow & & \downarrow T\Pi \\ A & \xrightarrow{\rho} & TM \end{array}$$

(b) and for any $a \in \Gamma(A)$, $\hat{\rho}(\Pi^! a) \in \mathfrak{X}(\text{Grass}_{-r}(A))$ is a lift of $X = \rho(a) \in \mathcal{F}$ (induced by a linear vector field $\xi_X \in \mathfrak{X}(A)$ as in Lemma 5.61).

Proof Let $V_x \in \text{Grass}_{-r}(A)$ be an element in the fiber of $x \in M$. Consider $a_1, \ldots, a_n$ a local trivialization of A in a neighborhood $\mathcal{U}$ of x. The value at $V_x \in \text{Grass}_{-r}(A)$ of the lifts $\hat{\xi}_{a_1}, \ldots, \hat{\xi}_{a_i} := \hat{\xi}_{\rho(a_i)}, \cdots \hat{\xi}_{a_n}$ constructed as in Lemma 5.61 define a vector bundle morphisms $\hat{\rho}_\mathcal{U}$ which satisfies both properties above by construction. Using a partition of unity $(U_i, \chi_i, \rho_i)_{i \in I}$, this construction can be made global: it suffices to check that the gluing $\sum_i \Pi^* \chi_i \hat{\rho}_i$ still satisfies the required properties if each $\hat{\rho}_i$ defined as before does.

We can now prove the two first items of the theorem.

Proof By the second item in Lemma 5.66, for every $X \in \mathcal{F}$, and for every $a \in \Gamma(A)$ such that $\rho(a) = X$, the vector field $\hat{\rho}(\Pi^! a)$ comes from a linear vector field as in Lemma 5.61 (here $\Pi^! a$ is the pull-back section of a). As a consequence, by construction

1. the vector field $\pi^! X$
2. the restriction of $\hat{\rho}(\Pi^! a)$ to $\text{Bl}(M, \mathcal{F})$

coincide on $\text{Bl}(M, \mathcal{F})$. In other words,

$$\hat{\rho}_{|\text{Bl}(M,\mathcal{F})}(\Pi^! a) = \pi^! X.$$

This implies that the pair

$$\left(\Pi^! A|_{\text{Bl}(M,\mathcal{F})}, \hat{\rho}_{|\text{Bl}(M,\mathcal{F})}\right)$$

is an anchored bundle over $\pi^!(\mathcal{F})$. Let us check that the tautological sub vector bundle τ_A^{-r} lies in the kernel of the anchor map $\hat{\rho}$, restricted to $\text{Bl}(M, \mathcal{F})$. This follows from the fact that $\ker \hat{\rho}_{V_x} \subseteq \ker \rho_x$ for all $x \in M$ by its very definition, and that both spaces coincide on M_{reg} since $\dim(\text{im} \hat{\rho}_{V_x}) = \dim(\text{im} \rho_x)$ for all $x \in M_{\text{reg}}$.

What Is a Singular Foliation?

Hence, the anchor map $\hat{\rho}_{|\text{Bl}(M,\mathcal{F})}$ goes to the quotient to define a map as follows:

$$0 \longrightarrow T_A^{-r} \longrightarrow \Pi^! A \longrightarrow A_{Bl}^r \longrightarrow 0$$
$$\downarrow \swarrow$$
$$T(\text{Bl}(M,\mathcal{F})) \qquad\qquad (29)$$

This map is injective on M_{reg} with image $T\mathcal{F}$. Therefore, $\pi^!(\mathcal{F})$ is a projective singular foliation and is the image of a Lie algebroid bracket on $A^r_{Bl|\text{Bl}(M,\mathcal{F})}$. This finishes the proof of the two first items.

We finish this section by presenting several exercises, describing examples and properties.

Exercise 5.67 Show that if $(M, \mathcal{F})$ is a Debord singular foliation (in particular a regular foliation) then $\text{Bl}(M, \mathcal{F}) \simeq M$ and $\pi^!(\mathcal{F}) \simeq \mathcal{F}$.

Exercise 5.68 Show that

1. $\pi^!\mathcal{F}$ is a regular foliation of rank r if and only if $\tau_A^r = \ker(\hat{\rho})$, i.e., $V_x = \ker(\hat{\rho}_{V_x})$ for all $x \in M$. In particular, if $\text{Sker}(\hat{\rho}, x) \neq 0$ for all $x \in M$ and τ_A^r is a line bundle, then $\pi^!\mathcal{F}$ is regular.
2. $\text{Sker}(\rho, x) = V_x$ for all $V_x \in \pi^{-1}(x)$ with $x \in M$, if only if $\mathcal{F}$ is a regular foliation.

Exercise 5.69 Show that for $\mathcal{F}$ the singular foliation on $\mathbb{R}^n$ of all vector fields vanishing at the origin, the Nash blowup coincides with the blowup at 0 as in Sect. 5.7.

Exercise 5.70 Let M be a manifold and a $N \subset M$ submanifold. Consider $\mathcal{F}_N \subset \mathfrak{X}(M)$ the singular foliation of vector fields that are tangent to N. Show that the Nash blowup coincides with the blowup along the submanifold N in this case.

Exercise 5.71 Let $(M = \mathbb{R}^3, \mathcal{F})$ the singular foliation given by the transformation Lie algebroid of the action of $\mathfrak{so}(3)$ on $\mathbb{R}^3$. It is generated by the vector fields

$$X = y\frac{\partial}{\partial x} - x\frac{\partial}{\partial y}, \quad Y = z\frac{\partial}{\partial x} - x\frac{\partial}{\partial z}, \quad Z = z\frac{\partial}{\partial y} - y\frac{\partial}{\partial z}.$$

1. Show that $\text{Bl}(M, \mathcal{F})$ is the usual blowup of $\mathbb{R}^3$ at 0. In particular, $\text{Bl}(M, \mathcal{F})$ is smooth.
2. Show that $\pi^!(\mathcal{F})$ is generated in the x-chart by

$$\pi^!(X) = xy\frac{\partial}{\partial x} - (y^2+1)\frac{\partial}{\partial y} - yz\frac{\partial}{\partial z} \text{ and } \pi^!(Y) = xz\frac{\partial}{\partial x} - yz\frac{\partial}{\partial y} - (z^2+1)\frac{\partial}{\partial z}$$

while $\pi^!(Z)$ still has the same expression. Compute the generators on the other charts.
3. Deduce that $\pi^!(\mathcal{F})$ is a regular foliation on $\text{Bl}(M, \mathcal{F})$ of rank 2.

The next exercises will describe more precisely the leaves of the Nash blowup. We claim that we never really need the simplifying assumption that $\mathrm{Bl}(M, \mathcal{F})$ is a submanifold, but we invite the reader to make it for the sake of simplicity.

Exercise 5.72 Let $(\mathrm{Bl}(M, \mathcal{F}), \pi^!(\mathcal{F}))$ be the Nash blowup of a singular foliation $\mathcal{F}$, and let $\pi : \mathrm{Bl}(M, \mathcal{F}) \to M$ be the projection.

1. Show that $\pi : (\mathrm{Bl}(M, \mathcal{F}), \pi^!(\mathcal{F})) \longrightarrow (M, \mathcal{F})$ maps leaves into leaves. *Hint: Use* $\Pi \circ \phi_{t_1}^{\hat{X}_k} \circ \cdots \circ \phi_{t_k}^{\hat{X}_k} = \phi_{t_1}^{X_1} \circ \cdots \circ \phi_{t_k}^{X_k} \circ \Pi$.
2. Let us fix $x \in M$, and consider $\pi^{-1}(x)$ to be a subset of $\mathrm{Grass}^{Lie}_{-r+\dim(L_x)}(\mathfrak{g}_x(\mathcal{F}))$ as in Exercise 5.54. Let $G_x(\mathcal{F})$ be the simply connected Lie group integrating $\mathfrak{g}_x(\mathcal{F})$. It naturally acts on

$$\mathrm{Grass}^{Lie}_{-r+\dim(L_x)}(\mathfrak{g}_x(\mathcal{F})).$$

There is therefore a Lie algebra morphism

$$\phi : \mathfrak{g}_x(\mathcal{F}) \longrightarrow \mathfrak{X}\left(\mathrm{Grass}^{Lie}_{-r+\dim(L_x)}(\mathfrak{g}_x(\mathcal{F}))\right),$$

which in turn yields an action of $G_x(\mathcal{F})$ on $\mathrm{Grass}^{Lie}_{-r+\dim(L_x)}(\mathfrak{g}_x(\mathcal{F}))$. Show the following points.

(a) The fiber $\pi^{-1}(x)$ of $(\mathrm{Bl}(M, \mathcal{F}) \to M$ is stable under the $G_x(\mathcal{F})$-action.
(b) We assume that $T_x\mathcal{F} = \{0\}$. Show that the leaf of $\pi^!(\mathcal{F})$ through some point in $\pi^{-1}(x)$ is precisely the $G_x(\mathcal{F})$-orbit.
(c) Show that a leaf of $\pi^!(\mathcal{F})$ is reduced to a point[68] $V \in \pi^{-1}(x)$ if and only if $T_x\mathcal{F} = \{0\}$ and V is a Lie ideal of the isotropy Lie algebra $\mathfrak{g}_x(\mathcal{F})$.
(d) Show that a point[69] $V \in \pi^{-1}(x)$ is a regular point if and only if it is equal to its own stabilizer.

Exercise 5.73 Assume that $A \to M$ is a Lie algebroid over M. Show that $A \to M$ acts naturally on $\mathrm{Bl}(M, \mathcal{F}) \to M$.

5.9 Push-Forward

Let $\phi : P \to M$ be a smooth, complex or real analytic map, depending on the context. We will assume that ϕ is a surjective submersion.

---

[68] Seen as a codimension r Lie algebra of $\mathfrak{g}_x(\mathcal{F})$.
[69] Seen as a codimension $r - \dim(L_x)$ Lie algebra of $\mathfrak{g}_x(\mathcal{F})$.

The *push-forward* $T_p\phi : T_p P \to T_{\phi(p)} M$ does not extend in general to vector fields: for X a vector field on P and $n = \phi(p) = \phi(p')$ with $p \neq p' \in P$, then X_p and $X_{p'}$ are both pushed forward to tangent vectors at $n \in M$, but in general $T_p\phi(X_p) \neq T_{p'}\phi(X_{p'})$. When this happens, we denote this vector field by $\phi_*(X)$ and we call it the *push-forward of X through ϕ*.

Let us introduce a notation: for $\phi \colon P \to M$ a surjective submersion, we denote by $\mathfrak{X}(P)_\phi$ the space of vector fields X on P which are ϕ-related to a vector field on M, that we denote by $\phi_*(X)$.

Assume that we are now given a singular foliation $\mathcal{F}^P$ on P. Then $\mathcal{F}^P \cap \mathfrak{X}(P)_\phi$ is both a $C^\infty(M)$-module and stable under Lie bracket, and so is

$$\phi_*(\mathcal{F}^P \cap \mathfrak{X}(P)_\phi) \subset \mathfrak{X}(M).$$

When the latter is finitely generated, it is a singular foliation that we call *push-forward singular foliation* and denote by $\phi_*(\mathcal{F}^P)$.

Exercise 5.74 Let $\mathcal{F}^M$ be a singular foliation on M, and $\phi \colon P \to M$ a surjective submersion. Show that the push-forward of the pull-back $\phi^{-1}(\mathcal{F}^M)$ (see Sect. 5.2) is the singular foliation $\mathcal{F}^M$.

Exercise 5.75 Here are examples where $\phi_*(\mathcal{F}^P \cap \mathfrak{X}(M)_\phi)$ is not finitely generated.

Let $\mathcal{F}^P$ be Androulidakis-Zambon's "non-finitely-many-generators" singular foliation of Exercise 2.36. In this case, we have $P := \mathbb{R}^2$. Let $M := \mathbb{R}$, and $\phi \colon (x, y) \mapsto y$ the projection onto the horizontal axis.

1. Show that for every ϕ-projectable[70] vector field X on $\mathbb{R}^2$ whose derivatives vanish at order n at the point of coordinates $(n, 0)$, its projection $\phi_*(X) \in \mathfrak{X}(\mathbb{R})$ is a vector field that vanishes at order n at 0.
2. Show that $\phi_*(\mathcal{F}^P \cap \mathfrak{X}(P)_p)$ coincides with the space of vector fields on $\mathbb{R}$ vanishing at 0 with all their derivatives.
3. Conclude (*Hint*: use the second item in Exercise 4.2).

6   Morphisms of Singular Foliations

Isomorphisms of singular foliations are easily defined; they are diffeomorphisms (biholomorphisms in the complex case) of the underlying varieties that intertwine their respective singular foliations: see Sect. 3.1. But defining general morphisms of singular foliations is more involved: we even dare to say that finding a definition that makes consensus is still an open problem.

---

[70] I.e., ϕ-related to a vector field on $\mathbb{R}$, see Definition 5.4.

There is a case, however, for which an easy definition of morphism exists: surjective submersions.

Definition 6.1 Let $(P, \mathcal{G})$ and $(M, \mathcal{F})$ be foliated manifolds. A submersion $\phi \colon P \to M$ is said to be a morphism of singular foliation if $\mathcal{G} \subset \varphi^{-1}(\mathcal{F})$.

This definition satisfies the following two interesting properties.

1. If two points are in the same leaf of $\mathcal{G}$, then their images are in the same leaf of $\mathcal{F}$.
2. For every $p \in P$, the inclusion $T_p\phi\left(T_p\mathcal{G}\right) \subset T_{\phi(p)}\mathcal{F}$ holds.
3. If $P = M$ and Φ is the identity map, then morphisms are simply inclusions of singular foliations. More precisely, the identity map is a morphism $(M, \mathcal{G})$ and $(M, \mathcal{F})$ if and only if $\mathcal{G} \subset \mathcal{F}$.

However, it is strange that with this definition, an immersion $S \hookrightarrow M$ could not be a morphism of singular foliation. For instance, for sub-manifolds intersecting $S \subset M$ cleanly[71] a singular foliation $\mathcal{F}_M$ on M, we would like the inclusion map to be a morphism of singular foliation from $(S, i_S^* \mathcal{F}_M)$ to $(M, \mathcal{F}_M)$. More generally, for the Androulidakis-Skandalis pull-back $(L, p^{-1}(\mathcal{F}_M))$ of $(M, \mathcal{F}_M)$ by a map $p \colon L \to M$ transverse to $\mathcal{F}$ as in Sect. 5.4, we would like p to be a morphism. Recall from that section that we say that a smooth map $p \colon L \to M$ is transverse to $\mathcal{F}_M$ if for all $\ell \in L$

$$T_{\phi(\ell)}\mathcal{F}_M + T_\ell\phi(T_\ell P) = T_{\phi(\ell)}M.$$

This is enough to define the pull-back $\varphi^{-1}(\mathcal{F}_M)$. The latter is a singular foliation on L, see Sect. 5.4. This clears the way to our next definition, which is more general than Definition 6.1, since any submersion $P \to M$ is transverse to any singular foliation on M.

Definition 6.2 (Morphisms of Singular Foliations: The Transverse Case) Let $(L, \mathcal{F}_L)$ and $(M, \mathcal{F}_M)$ be foliated manifolds. A map $\phi \colon L \to M$ is said to be a morphism of singular foliation if

1. ϕ is a transverse to $\mathcal{F}_M$,
2. $\mathcal{F}_L \subset \phi^{-1}(\mathcal{F}_M)$.

The above notion is constructed exactly such that every map $L \to M$ transverse to $\mathcal{F}_M$ is a morphism of singular foliation $(L, \phi^{-1}(\mathcal{F}_M))$ to $(M, \mathcal{F}_M)$. Again, this definition

---

[71] See Sect. 5.3.

satisfies several interesting properties.

1. If two points are in the same leaf of $\mathcal{F}_L$, then their images through ϕ are in the same leaf of $\mathcal{F}_M$.
2. For every $\ell \in L$, the inclusion $T_\ell \phi\, (T_\ell \mathcal{F}_L) \subset T_{p(\ell)} \mathcal{F}_M$ holds.

Exercise 6.3 Show that the inclusion of a transverse submanifold S in a foliated manifold $(M, \mathcal{F})$ is a morphism in the previous sense for every sub-singular foliation included in the restriction $i_S^* \mathcal{F}$ (see Sect. 5.3).

Exercise 6.4 Show that for the direct product singular foliation $\mathcal{F}_1 \times \mathcal{F}_2$ on $M = M_1 \times M_2$, the projections are morphisms. Moreover, show that the direct product singular foliation is the largest singular foliation on M with this property.

Exercise 6.5 This exercise requires the notion of Lie algebroid morphism [49]. Assume that the base map of a Lie algebroid morphism is a submersion: is it a morphism of singular foliations? Does a Poisson submersion induce morphism of their respective symplectic foliations?

While the above definition is sufficient to treat many important cases, it has weaknesses. For instance, it is strange that for L a leaf in a singular foliation $\mathcal{F}$ on M (see Sect. 7 for a definition of leaves), the inclusion $i \colon L \hookrightarrow M$ is never a morphism, unless L is an open subset of M. Hence a natural question:

Question 6.6 Is Definition 6.2 the definitive answer to the question of defining "morphisms of singular foliations"?

For the coming lines, we are indebted to a discussion with Hadi Nahari and his advisor, Thomas Strobl. Let $(L, \mathcal{F}_L)$ and $(M, \mathcal{F}_M)$ be singular foliations and $\phi \colon L \to M$ be a smooth map.

1. The most naive condition to impose on ϕ in order to say that ϕ is a morphism would be to ask that $T_m \phi\, (T_m \mathcal{F}_1) \subset T_{\phi(m)} \mathcal{F}_2$ for any $m \in M_1$. However, with this notion, the map $t \mapsto t^3$ would be a morphism of singular foliations from $(\mathbb{R}, \mathfrak{X}_c(\mathbb{R}))$ to $(\mathbb{R}, t \cdot \mathfrak{X}_c(\mathbb{R}))$ (= compactly supported vector fields on $\mathbb{R}$ that vanish at 0). This latter map, however, does not preserve leaves and hence should not be called morphism. This is therefore *not* a good definition.
2. A more clever idea is to use the vector bundle $\phi^* TM \to L$ as in Sect. 5.4. Recall that both $T\phi(\mathcal{F}_L)$ and the $C_c^\infty(L)$-module generated by $p^* X$ with $X \in \mathcal{F}_M$ belong to the compactly supported sections of that vector bundle. We denote the second one by $C_c^\infty(M) p^* \mathcal{F}_M$. A possible definition of morphism of singular foliations would be to say that ϕ is a morphism if $T\phi(\mathcal{F}_L) \subset C_c^\infty(M) p^* \mathcal{F}_M$.

Assume for a moment that we define morphisms using the second item above. Then the condition of the first item would also be satisfied. Also, a morphism as in Definition 6.2 would then still be a morphism in the new sense. Last, with such a definition, it would still be true that two points in the same leaf $\mathcal{F}_L$ are mapped to the same leaf of $\mathcal{F}_M$.

This notion of morphism is however too weak for certain purposes.[72] An alternative definition can be found in Garmendia and Villatoro's [30], where morphisms of foliated manifolds are defined as sheaf comorphisms compatible with the Lie bracket. Hence, we prefer not to take the risk of making this definition into a formal one, and to leave room for more discussion on the matter. We only insist that, whatever definition one chooses, maps as in Definition 6.2 should be morphisms.

7 Leaves of a Singular Foliations

We show in this section that to any singular foliation is attached a smooth partitionifold, given by its leaves.

7.1 What Is a Leaf?

Let $\mathcal{F}$ be a singular foliation on a manifold M. For the present discussion, we will place ourselves in the context of smooth differential geometry, and consider $\mathcal{F}$, as in Definition 2.1, as an involutive $\mathcal{C}^\infty(M)$-submodule of compactly supported smooth vector fields on M. The discussion can be easily adapted to the complex and real analytic settings.[73]

> *Question 7.1 (What Is a Leaf?)* What are the leaves of $\mathcal{F}$? And do they exist?

There are two natural notions of leaves, two different notions that deserve to be called "leaves".

1. The first idea is that leaves are "reachable points". That is, we will define an equivalence relation on M by pairing two points in M such that one can be reached one to the other by following the flow of vector fields in $\mathcal{F}$.

---

[72] For instance, it does not induce a morphism to the level of isotropy Lie algebras (cf. Sect. 3 in chapter "Canonical Geometric and Algebraic Structures Hidden Behind a Singular Foliation") or fundamental groupoids (cf. Sect. 5.5 in chapter "Canonical Geometric and Algebraic Structures Hidden Behind a Singular Foliation").

[73] But can *not* be adapted to the algebraic setting.

2. But we may also use the tangent space of $\mathcal{F}$. A leaf should be a submanifold (by definition!) whose tangent space at a point m is the tangent distribution $T_m\mathcal{F}$ at that point.

Here is a formal definition.

Definition 7.2 Let $\mathcal{F} \subset \mathfrak{X}_c(M)$ be a singular foliation. We say that a point $y \in M$ is *reachable from a point* $x \in M$ if there exists:

1. a finite sequence $x_0, \ldots, x_N$ of points in M with $x_0 = x$ and $x_N = y$
2. time-dependent vector fields[74] $(X_t^{(i)})_{t \in \mathbb{R}} \in \mathcal{F}$ for $i = 0, \ldots, N-1$, with X_i being defined in a neighborhood of x_i and x_{i+1},

such that for all indices $i = 0, \ldots, N-1$, the integral curve starting at x_i at time $t = 0$ of $X_t^{(i)}$ reaches x_{i+1} at time $t = 1$.

Exercise 7.3 Show that could suppress "time-dependent" in the second item Definition 7.2, and that the R-leaves would stay the same.

If one uses the sheaf Definition 2.18, the definition is easily adapted: one has to assume that the time 1-flow of $X_t^{(i)}$ is defined near x_i for all $i = 1, \ldots, N-1$. The following statement is obvious.

Proposition 7.4 *The relation on M defined by $x \sim y$ if y is reachable from x is an equivalence relation.*

We call *reachable leaves* or *R-leaves* for short the equivalence classes of the previous relation.
But there is a second natural definition of what a leaf should be.

Definition 7.5 A *tangent-leaf*, or *T-leaf* for short, is a connected submanifold $L \subset M$ such that for every $\ell \in L$,

$$T_\ell L = T_\ell \mathcal{F},$$

and which is maximal among connected sub-submanifolds that satisfy the same property.[75]

---

[74] I.e. a linear combination of the type $\sum_{i=1}^r f_i X_i$ with $X_i \in \mathcal{F}$ and $f_i \in C^\infty(M \times \mathbb{R})$. These vector fields being zero for all t outside some compact subset of M, their flows at time t are defined on the whole manifold M.

[75] I.e., it cannot be strictly included in a submanifold that satisfies the same property.

While defining singular foliation in Definition 2.1, we assumed "locally finitely generated". R-leaves and T-leaves could be defined for any sub-module $\mathcal{F} \subset \mathfrak{X}_c(M)$ stable under Lie bracket, even if they do not satisfy the "locally finitely generated" assumption. The next exercise shows that without this assumption, R-leaves and T-leaves are different concepts.

Exercise 7.6 *"The infinite comb (revisited) (after Exercise 1.36 and 2.5)"* Let $M := \mathbb{R}^2$ be the Cartesian plane with coordinates (x, y). Let $\mathcal{I}_- \subset \mathcal{C}^\infty(\mathbb{R}^2)$ be the ideal of functions vanishing identically on $\mathbb{R}_- \times \mathbb{R}$.

Consider all vector fields of the form

$$\mathcal{F}_{comb} = \left\{ f(x,y)\frac{\partial}{\partial x} + g(x,y)\frac{\partial}{\partial y} \, \middle| \, g \in \mathcal{C}^\infty(\mathbb{R}^2), \, f \in \mathcal{I}_- \right\}$$

1. Show that $\mathcal{F}_{comb}$:
   (a) is stable under multiplication by $\mathcal{C}^\infty(\mathbb{R}^2)$,
   (b) is involutive, i.e., is closed under the Lie bracket of vector fields:

$$[\mathcal{F}_{comb}, \mathcal{F}_{comb}] \subset \mathcal{F}_{comb}.$$

2. Draw what vector fields in $\mathcal{F}_{comb}$ look like.
3. Determine $T_{x,y}\mathcal{F}_{comb}$ for all (x, y) (*Hint*: it depends on the sign of x).
4. Show that any point in $M = \mathbb{R}^2$ is reachable from any point in $M = \mathbb{R}^2$. How many R-leaves exists?
5. Does $\mathcal{F}_{comb}$ admit T-leaves?

Definition 7.7 (Definition of Leaves) A *leaf* of an involutive distribution $\mathcal{F} \subset \mathfrak{X}_c(M)$ is a submanifold $L \subset M$ which:

(i) is a T-leaf,
(ii) and a R-leaf.

Here is the main result of this section, which is attributed to Hermann [33].

Theorem 7.8 (Hermann: Singular Foliations Do Admit Leaves!) *Every singular foliation on a smooth manifold M partitions M into leaves. Also, R-leaves are also T-leaves.*

Here is an immediate consequence of this theorem. Any leaf of $\mathcal{F}$ is an immersed submanifold, since so are T-leaves by definition. Leaves partition M, since R-leaves partition M par definition. Leaves form therefore a partitionifold. It is moreover a smooth partitionifold, since the tangent space of the leaf through a point is $T_m \mathcal{F}$ by the definition of T-leaves. Hence, every singular foliation induces a smooth partitionifold on M that we will denote by $L_\bullet$. By definition, for every $m \in M$, L_m is the set of reachable points from m, and it satisfies that $T_m \mathcal{F} = T_m L_m$.

Here is an even more precise statement than Theorem 7.8. It is the one that we will indeed prove, and it immediately implies Theorem 7.8.

Theorem 7.9 (Second Version) *Let $\mathcal{F}$ be a singular foliation on a smooth manifold M. Every R-leaf L is a (maybe immersed) submanifold of M, whose tangent space $T_\ell L$ coincides with $T_\ell \mathcal{F}$ at every $\ell \in R$, i.e. is a T-leaf. Lastly, leaves form a smooth partitionifold of M.*

The rest of the present section is mainly dedicated to the proof of this statement. In Sect. 7.2, we will prove that the flow of a vector field in $\mathcal{F}$ is a symmetry of $\mathcal{F}$, provided that it exists. In Sect. 7.3, we will prove an important splitting theorem explaining the local structure of a singular foliation. Only then, we will be able to address Theorem 7.9 in Sect. 7.4.

7.2 A Singular Foliation Is a Symmetry of Itself

The first step to prove Theorems 7.8 and 7.9 is to prove that vector fields in a singular foliation have flow which are infinitesimal symmetries of themselves. The arguments presented in this section are elementary, but quite complicated. Much better conceptual arguments proving the same results will be given using the notion of anchored bundle and almost Lie algebroids.

This is actually a particular instance of the following more general statement, see e.g. [31].

Proposition 7.10 (The Flow of an Infinitesimal Symmetry Is a Symmetry) *Choose $t \in \mathbb{R}$. Let $Y \in \mathfrak{X}(M)$ be a vector field whose time t-flow $\varphi_t^Y : M \to M$ exists. If[a] $[Y, \mathcal{F}] \subset \mathcal{F}$, then φ_t^Y is a symmetry of $\mathcal{F}$.*

---

[a] That is, if Y is an "infinitesimal symmetry" of $\mathcal{F}$.

In fact, we are going to prove a more general result.

Proposition 7.11 *Let Y be a vector field such that $[Y, \mathcal{F}] \subset \mathcal{F}$. For every open neighborhood $\mathcal{U}$ on which $\mathcal{F}$ is generated by vector fields $X_1, \ldots, X_r$, and any $\mathcal{V} \subset \mathcal{U}$ an open subset such that $\phi_t^Y(x)$ exists and takes values in $\mathcal{U}$ for all $x \in \mathcal{V}$ and $|t| \leq \epsilon$, there exists a matrix $\mathbf{A}(t, x)$, whose coefficients are functions on $\mathcal{V}$ depending on t such that for all $i = 1, \ldots, r$:*

$$(\phi_t^Y)_* \begin{pmatrix} X_1 \\ \vdots \\ X_r \end{pmatrix} = \left(\mathbf{A}(t, x) \right) \begin{pmatrix} X_1 \\ \vdots \\ X_r \end{pmatrix}$$

Moreover, we can assume that

$$\mathbf{A}(s, \phi_t^Y(x)) \circ \mathbf{A}(t, x) = \mathbf{A}(t + s, x) \tag{30}$$

for all $s, t \in \mathbb{R}, x \in \mathcal{V}$ for which $\phi_t^Y(x) \in \mathcal{V}$, and $|t|, |t + s| \leq \epsilon$.

Proof Consider an open neighborhood $\mathcal{U}$ of a point $m \in M$ on which $\mathcal{F}$ is generated by $X_1, \ldots, X_n$. Let us chose $\epsilon > 0$ and a smaller neighborhood $\mathcal{V} \subset \mathcal{U}$ such that if $|t| \leq \epsilon$, $\phi_t^Y(\mathcal{V}) \subset \mathcal{U}$. By definition of a symmetry of a singular foliation, there exist smooth functions $b_i^j \in C^\infty(\mathcal{U})$, such that $[Y, X_i] = \sum_{j=1}^r b_i^j X_j$. Let us write this expression as a matrix:

$$\mathrm{ad}_Y \begin{pmatrix} X_1 \\ \vdots \\ X_r \end{pmatrix} = \left(\mathbf{ad}_Y(x) \right) \begin{pmatrix} X_1 \\ \vdots \\ X_r \end{pmatrix} \tag{31}$$

with $\mathbf{ad}_Y$ being a shorthand for the matrix of functions on $\mathcal{U}$ whose i-th line and j-th column is b_j^i.

For any diffeomorphism $\phi: \mathcal{V} \to \phi(\mathcal{V})$, the push-forward map $\phi_*: \mathfrak{X}(\phi(\mathcal{V})) \simeq \mathfrak{X}(\mathcal{V})$ is defined by $\phi_*(X)|_m = T_{\phi(m)}\phi^{-1}(X_{\phi(m)})$. It satisfies for all $F \in C^\infty(\phi(\mathcal{V}))$ and $X \in \mathfrak{X}(\phi(\mathcal{V}))$ the relation:

$$\phi_*(FX) = \phi^* F \ \phi_*(X) \tag{32}$$

Also, if $\phi = \phi_t^Y$ is the flow of Y at time t:

$$\frac{\partial}{\partial t}(\phi_t^Y)_* X = (\phi_t^Y)_*[Y, X] = [Y, (\phi_t^Y)_* X] \tag{33}$$

We want to show that there exist time-dependent functions $A_i^j(t, x)$ on $\mathcal{V}$ such that

$$(\phi_t^Y)_*(X_i) = \sum_{j=1}^{r} A_i^j(t) X_j \tag{34}$$

where $(\phi_t^Y)_*(X_i)$ is to be understood as the image through the push-forward map of the restriction of X_i to $\phi_t^Y(\mathcal{V})$. We also want the matrix of functions $(A_i^j(t, x))$ to be invertible for all t, x.

Again, let us write the expression we wish to obtain in a matrix form. Below, both sides are column vectors of vector fields on $\mathcal{V}$:

$$(\phi_t^Y)_* \begin{pmatrix} X_1 \\ \vdots \\ X_r \end{pmatrix} = \begin{pmatrix} \mathbf{A}(t, x) \end{pmatrix} \begin{pmatrix} X_1 \\ \vdots \\ X_r \end{pmatrix} \tag{35}$$

with $\mathbf{A}(t, x)$ being a shorthand for the matrix of functions on $\mathcal{V}$ whose i-th line and j-th column is $A_j^i(t, x)$. Consider the initial value problem with parameters $x \in \mathcal{V}$:

$$\frac{\partial A_i^j(t, x)}{\partial t} = \sum_{k=1}^{r} b_i^k(\phi_t^X(x)) A_k^j(t, x)$$

with initial conditions $A_i^j(0, x) = \delta_{i,j}$. Or, equivalently, consider the initial value problem on the vector space $r \times r$ matrices:

$$\frac{\partial}{\partial t}\begin{pmatrix} \mathbf{A}(t, x) \end{pmatrix} = \begin{pmatrix} \mathbf{ad}_Y\left(\phi_t^Y(x)\right) \end{pmatrix}\begin{pmatrix} \mathbf{A}(t, x) \end{pmatrix} \tag{36}$$

with initial condition $\mathbf{A}(0, x) = \mathrm{id}$. The initial value problem have solutions for all $x \in \mathcal{V}$ and $|t| \leq \epsilon$, upon changing $\mathcal{V}$ for a smaller neighborhood that we still call $\mathcal{V}$ if necessary. Those solutions depend smoothly on the parameters $x \in \mathcal{V}$. Also, the matrix $\mathbf{A}(t, x)$ is invertible for all $|t| \leq \epsilon$ and $x \in \mathcal{V}$. Last, as any differential equation, it satisfies (30).

We claim that Eq. (35) holds. To show it, let us introduce the column vector whose components are vector fields on $\mathcal{V}$:

$$R(t, x) = \begin{pmatrix} \mathbf{A}(t, x) \end{pmatrix}^{-1} \circ (\phi_t^Y)_* \begin{pmatrix} X_1 \\ \vdots \\ X_r \end{pmatrix}$$

An easy computation gives (we now abbreviate the matrix notations, also $(X_\bullet)$ stands for the column vector $X_1, \ldots, X_r$):

$$\frac{\partial R(t,x)}{\partial t} = -\mathbf{A}^{-1} \circ \frac{\partial \mathbf{A}}{\partial t} \circ \mathbf{A}^{-1} \circ (\phi_t^Y)_*(X_\bullet) + \mathbf{A}^{-1} \circ \underbrace{(\phi_t^Y)_* \circ \mathrm{ad}_Y(X_\bullet)}_{\text{by Eq. (33)}}$$

$$= -\mathbf{A}^{-1} \circ \frac{\partial \mathbf{A}}{\partial t} \circ \mathbf{A}^{-1} \circ (\phi_t^Y)_*(X_\bullet) + \mathbf{A}^{-1} \circ (\phi_t^Y)_* \circ \underbrace{\mathrm{ad}_Y(x)\,(X_\bullet)}_{\text{by Eq. (31)}}$$

$$= -\mathbf{A}^{-1} \circ \frac{\partial \mathbf{A}}{\partial t} \circ \mathbf{A}^{-1} \circ (\phi_t^Y)_*(X_\bullet) + \mathbf{A}^{-1} \circ \underbrace{\mathrm{ad}_Y(\phi_t^Y(x)) \circ (\phi_t^Y)_*(X_\bullet)}_{\text{by Eq. (32)}}$$

$$= \mathbf{A}^{-1} \circ \underbrace{\left(-\frac{\partial \mathbf{A}}{\partial t} \circ \mathbf{A}^{-1} + \mathrm{ad}_Y(\phi_t^Y(x))\right)}_{=0 \text{ by Eq. (36)}} \circ (\phi_t^Y)_*(X_\bullet)$$

$$= 0.$$

Since $R(0,x) = (X_\bullet)$, we have $R(t,x) = (X_\bullet)$ for all $t \leq \epsilon$ and the (35) follows. This implies that the push-forward of any vector field in $\mathcal{F}$ under the flow of Y is a vector field in $\mathcal{F}$ at least for t small enough. Composing such push-forward maps, we obtain that it is still true for all t such that the flow of Y is well-defined.

Let us restate Proposition 7.11 differently. We call *infinitesimal symmetry of $\mathcal{F}$* a vector field Y such that $[Y, \mathcal{F}] \subset \mathcal{F}$ (in contrast with *symmetry of $\mathcal{F}$* which are diffeomorphisms such that $\phi_*(\mathcal{F}) = \mathcal{F}$).

Proposition 7.12 (Symmetries and Infinitesimal Symmetries) *When the flow of an infinitesimal symmetry of $\mathcal{F}$ exists, it is a symmetry of $\mathcal{F}$.*

Again, a much better proof will be given using the notion of anchored bundle and almost Lie algebroids. Proposition 7.12 has several immediate and very important corollaries.

Corollary 7.13 *Let $X \in \mathcal{F}$ be a vector field whose time t-flow $\varphi_t^X : M \to M$ exists. Then φ_t^X is a symmetry of $\mathcal{F}$.*

Remark 7.14 It deserves to be noticed that the conclusion of the corollary is *not* true for the infinite comb (see Exercises 1.36–2.5–7.6). Its proof indeed made an intense use of the assumptions "locally finitely generated".

Here is a second corollary, which is not totally trivial. Inner symmetries were defined in Sect. 3.1.

Corollary 7.15 *Let $(M, \mathcal{F})$ be a foliated manifold. Any inner symmetry[76] of $\mathcal{F}$ is a symmetry of $\mathcal{F}$.*

Moreover, the group $\mathrm{Inner}(\mathcal{F})$ of inner symmetries is a normal subgroup in the group $\mathrm{Sym}(\mathcal{F})$ of symmetries of $\mathcal{F}$.

Proof This is not totally obvious, since Corollary 7.13 only deals with vector fields in $\mathcal{F}$ that do not depend on the time. However, it can be deduced from Corollary 7.13 as follows.

Let I be an open interval of $\mathbb{R}$ containing $[0, 1]$. We denote the real parameter by t. Let $(X_t)_{t \in I}$ be a smooth time-dependent vector field[77] in $\mathcal{F}$, whose time-1 flow is an inner symmetry that we denote by ϕ and let $\tilde{\mathcal{F}}$ be the singular foliation on $M \times I$ which is the direct product of $(M, \mathcal{F})$ with $(I, \mathfrak{X}(I))$.

By the equivalence between item (i) and (iii) in the first question of Exercise 3.6, the vector field

$$Y := X_t + \frac{\partial}{\partial t}$$

belongs to $\tilde{\mathcal{F}}$. By Proposition 7.12, its time 1 flow ϕ_1^Y is a symmetry of $\tilde{\mathcal{F}}$, at least in the open subset where it is well-defined. Since ϕ_1^Y maps $M \times \{0\}$ to $M \times \{1\}$, the symmetry ϕ_1^Y induces a singular foliation isomorphism from the restriction of $\tilde{\mathcal{F}}$ to $M \times \{0\}$ to the restriction of $\tilde{\mathcal{F}}$ to $M \times \{0\}$. Since these restrictions coincide with $\mathcal{F}$ (upon identifying $M \times \{0\}$ and $M \times \{1\}$ with M), $\phi_1^Y|_{M \times \{0\}}$ induces a symmetry of $(M, \mathcal{F})$. Now, the time-1 flow ϕ_1^Y of Y and the time 1-flow ϕ of $(X_t)_{t \in I}$ are related by

$$\phi_1^Y(m, 0) = (\phi(m), 1) \text{ for all } m \in M.$$

This completes the proof of the claim.

Remark 7.16 The results of this section can be extended to the real analytic or complex settings.

7.3 The Local Splitting Theorem

The second step in the proof of Theorems 7.8 and 7.9 is an equivalent of Weinstein's splitting theorem in Poisson geometry. It was apparently rediscovered many times, see e.g., Paul Baum and Raoul Bott's Theorem 0.30 in [9], or Proposition 1.12 in [3].

---

[76] See Definition 3.7.
[77] See Definition 3.2.

Let us state this theorem first.

7.3.1 The Statements

The results of this section are valid in the smooth, complex (upon replacing $\mathbb{R}$ by $\mathbb{C}$ in the statements below), or real analytic cases. They are *not* true in algebraic geometry.

Theorem 7.17 (Local Splitting, Version 1) *Consider $\mathcal{F}$ a singular foliation on a manifold M of dimension d.*

Any $m \in M$ a point admits a neighborhood on which $\mathcal{F}$ is isomorphic to the direct product of

1. *the singular foliation of all vector fields on an open ball in $\mathbb{R}^l$, with $l = \dim(T_m \mathcal{F})$,*
2. *with a singular foliation $\mathcal{T}$ on an open ball in $\mathbb{R}^{d-l}$, contained in the space of vector fields vanishing at the center of open ball.*

Moreover, the rank of $\mathcal{T}$ at the center of the open ball is $r = \mathrm{rk}_m(\mathcal{F}) - l$.

Alternatively, it can be practical to state this result in local coordinates.

Theorem 7.18 (Local Splitting, Version 2) *Let $\mathcal{F}$ be a singular foliation on a smooth, complex or real analytic singular foliation M. Let $m \in M$ be a point and let $l = \dim(T_m \mathcal{F})$. Every point m admits a chart $\mathcal{U}_m$ with local coordinates $(x_1, \ldots, x_l, y_1, \ldots, y_{d-l})$, centered at m, on which the restriction of $\mathcal{F}$ admits the following generators:*

(a) the l vector fields $\frac{\partial}{\partial x_1}, \ldots, \frac{\partial}{\partial x_l}$,
(b) and k vector fields of the form

$$f_1(y_1, \ldots, y_{d-l}) \frac{\partial}{\partial y_1} + \cdots + f_{n-l}(y_1, \ldots, y_{d-l}) \frac{\partial}{\partial y_{d-l}}$$

with $f_1(0, \ldots, 0) = \cdots = f_{n-l}(0, \ldots, 0) = 0$.

Moreover, one can assume that $k + l$ is equal to $\mathrm{rk}_m(\mathcal{F})$.

Here is a third version of the local splitting theorem (inspired by [3]). Notice that there is no equivalent statement for the Weinstein splitting theorem in Poisson geometry.

Theorem 7.19 (Local Splitting, Version 3) *Let $\mathcal{F}$ be a singular foliation on a smooth, complex or real analytic singular foliation M of dimension d. For every $m \in M$, there exists*

1. *an open neighborhood $\mathcal{U}$ of m in M*
2. *a singular foliation $\mathcal{T}$ of rank $\mathrm{rk}_m(\mathcal{F}) - l$ on an open neighborhood $\mathcal{V}$ of 0 in $\mathbb{K}^{d-l}$, with $l = \dim(T_m \mathcal{F})$, admitting $\{0\}$ as a leaf,*
3. *a surjective submersion $\phi \colon \mathcal{U} \to \mathcal{V}$,*

such that the restriction of $\mathcal{F}$ to $\mathcal{U}$ coincides with the pull-back singular foliation $\phi^{-1}(\mathcal{T})$.

Before proving these theorems, let us recall the following lemma:

Lemma 7.20 *If a vector field X is not zero at some point $m \in M$, then there exists a local chart $\mathcal{U}$ with coordinates $(x, y_1, \ldots, y_{d-1})$, centered at m, such that, on $\mathcal{U}$, we have $X = \frac{\partial}{\partial x}$.*

Proof of Theorems 7.17, 7.18, 7.19 We leave it to the reader to verify that all three versions of the local splitting theorem are equivalent. We will prove Theorem 7.18. Our proof is by recursion: since the statement is local by nature, it suffices to consider the following recursion assumption

$\mathcal{H}_l = $ "The statement is proved at $m = 0$ for any singular foliation $\mathcal{F}$ on an open ball in a finite dimensional vector space such that $\dim(T_0 \mathcal{F}) \le l$".

For $l = 0$, $\mathcal{H}_0$ is automatically true and there is nothing to prove. Assume now $\mathcal{H}_l$ is valid, and let us prove $\mathcal{H}_{l+1}$.

Let $X^1, \ldots, X^r$ be generators of a singular foliation $\mathcal{F}$ defined in an open neighborhood of $0 \in \mathbb{K}^d$. Without any loss of generality, one can assume $X^r|_m \ne 0$. By the Hadamard lemma 7.20, there exists local coordinates $(x, y_1, \ldots, y_{d-1})$ centered at 0, such that, on $\mathcal{U}$, in which $X^r = \frac{\partial}{\partial x}$. In these coordinates, the remaining generators read as:

$$X^j = \sum_{i=1}^{d-1} F_i^j(x, y_1, \ldots, y_{d-1}) \frac{\partial}{\partial y_i} + g^j(x, y_1, \ldots, y_{d-1}) \frac{\partial}{\partial x}.$$

Since $X^r = \frac{\partial}{\partial x}$ belongs to $\mathcal{F}$, there is a second family of generators of $\mathcal{F}$ given by X^r together with the $r-1$ vector fields:

$$\widehat{X}^j := X^j - g^j(x, y_1, \ldots, y_{d-1})\frac{\partial}{\partial x} = \sum_{i=1}^{d-1} F_i^j(x, y_1, \ldots, y_{d-1})\frac{\partial}{\partial y_i}.$$

Let $\mathcal{G}$ be the module generated by $\widehat{X}^1, \ldots, \widehat{X}^{r-1}$. This module has the following description: $\mathcal{G}$ is the intersection of $\mathcal{F}$ with vector fields on the fiber of the map

$$\Pi \colon (x, y_1, \ldots, y_{d-1}) \mapsto x.$$

In equation:

$$\mathcal{G} = \mathcal{F} \cap \{\Pi - vertical\}.$$

Since both $\mathcal{F}$ and Π-vertical vector fields are closed under Lie bracket, it defines, in particular, a singular foliation of rank $r-1$ on some neighborhood of 0.

Now, $[X^r, \mathcal{G}] \subset \mathcal{F}$, since $X^r \in \mathcal{F}$ and $\mathcal{G} \subset \mathcal{F}$. Also, $\mathcal{G}$ being vertical with respect to Π while X^r is Π-related to the vector field $\frac{\partial}{\partial x}$ on $\mathbb{R}$, the Lie bracket $[X^r, \mathcal{G}]$ is valued in Π-vertical vector fields, so that

$$[X^r, \mathcal{G}] \subset \mathcal{F} \cap \{\Pi - vertical\} = \mathcal{G}.$$

Said differently, X^r is an infinitesimal symmetry of $\mathcal{G}$. By Theorem 7.11, its flow is a symmetry of $\mathcal{G}$. Concretely, it means that for all $(x, y_1, \ldots, y_{d-1})$ and all $t \in \mathbb{R}$ such that $(x+t, y_1, \ldots, y_{d-1})$ is still within the considered open subset,

$$\left(\phi_t^{X^r}\right)_* \begin{pmatrix} \widehat{X}_1(x, y) \\ \vdots \\ \widehat{X}_{r-1}(x, y) \end{pmatrix} = \begin{pmatrix} A(t, x, y) \end{pmatrix} \begin{pmatrix} \widehat{X}_1(x, y) \\ \vdots \\ \widehat{X}_{r-1}(x, y) \end{pmatrix},$$

where $A(t, x, y)$ is an invertible matrix that satisfies:

$$A\left(s, \phi_t^{X^r}(x, y)\right) \circ A(t, x, y) = A(t+s, x, y).$$

Since the flow at time t of X^r reads

$$\phi_t^{X^r} \colon (x, y_1, \ldots, y_{d-1}) \longrightarrow (x+t, y_1, \ldots, y_{d-1}),$$

What Is a Singular Foliation?

it means that there exists an invertible matrix $\mathbf{A}(t, x, y)$ such that:

$$\begin{pmatrix} \widehat{X}_1(x+t, y) \\ \vdots \\ \widehat{X}_{r-1}(x+t, y) \end{pmatrix} = \begin{pmatrix} \mathbf{A}(t, x, y) \end{pmatrix} \begin{pmatrix} \widehat{X}_1(x, y) \\ \vdots \\ \widehat{X}_{r-1}(x, y) \end{pmatrix},$$

where the invertible matrix $\mathbf{A}(t, x, y)$ satisfies:

$$\mathbf{A}\left(s, \phi_t^{X^r}(x, y)\right) \circ \mathbf{A}(t, x, y) = \mathbf{A}(t+s, x, y).$$

In particular, the vector fields

$$\begin{pmatrix} Z_1(x, y) \\ \vdots \\ Z_{r-1}(x, y) \end{pmatrix} = \begin{pmatrix} \mathbf{A}(x, 0, y) \end{pmatrix}^{-1} \begin{pmatrix} \widehat{X}_1(0, y) \\ \vdots \\ \widehat{X}_{r-1}(0, y) \end{pmatrix}.$$

are well-defined in a neighborhood of 0 and satisfy the following two properties:

1. they are local generators of $\mathcal{G}$ (since the matrix $\mathbf{A}(t, x, y)$ is invertible for t, x, y small enough),
2. they are invariant under the flow of X^r, since

$$(\phi_t^{X'})_* \begin{pmatrix} Z_1(x, y) \\ \vdots \\ Z_{r-1}(x, y) \end{pmatrix} = (\phi_t^{X'})_* \left(\begin{pmatrix} \mathbf{A}(x, 0, y) \end{pmatrix}^{-1} \begin{pmatrix} \widehat{X}_1(0, y) \\ \vdots \\ \widehat{X}_{r-1}(0, y) \end{pmatrix} \right).$$

$$= \begin{pmatrix} \mathbf{A}(x+t, 0, y) \end{pmatrix}^{-1} \begin{pmatrix} \mathbf{A}(t, 0, y) \end{pmatrix} \begin{pmatrix} \widehat{X}_1(0, y) \\ \vdots \\ \widehat{X}_{r-1}(0, y) \end{pmatrix}$$

$$= \begin{pmatrix} \mathbf{A}(x, 0, y) \end{pmatrix}^{-1} \begin{pmatrix} \widehat{X}_1(0, y) \\ \vdots \\ \widehat{X}_{r-1}(0, y) \end{pmatrix} = \begin{pmatrix} Z_1(x, y) \\ \vdots \\ Z_{r-1}(x, y) \end{pmatrix}.$$

In coordinates, it means that they are of the form:

$$Z^i = \sum_{j=1}^{d-1} f_j^i(y_1, \ldots, y_{d-1}) \frac{\partial}{\partial y_j}.$$

They therefore define a singular foliation $\mathcal{G}$ of rank $r - 1$ on $\mathbb{K}^{d-1}$. By construction, the dimension of $T_0\mathcal{G}$ is $l - 1$. We can then apply the recursion hypothesis, and we obtain the existence of coordinates $(x_1, \ldots, x_l, y'_1, \ldots, y'_{d-l-1})$ on which $\mathcal{G}$ is of the form described in Theorem 7.18. These variables, together with $x_{l+1} := x$ form a system of coordinates on which $\mathcal{F}$ is also of the form described by Theorem 7.18.

7.4 Proof of Theorems 7.8 and 7.9

We will use the following property of immersed submanifolds:

Proposition 7.21 *If a connected subset $L \subset M$ satisfies that every $m \in L$ has a neighborhood $\mathcal{U}$ such that the connected component of m in $L \cap \mathcal{U}$ is a submanifold of dimension k, then it is a (maybe immersed) submanifold of dimension k.*

Now we can prove Theorem 7.9 as follows. Choose a R-leaf L. An immediate consequence of the local splitting theorem is that every point $m \in L$ admits a neighborhood $\mathcal{U} \subset M$ admitting the following property: For the restriction $i^*_\mathcal{U}\mathcal{F}$ the set of reachable points $L^\mathcal{U}_m$ is the submanifold $y_1 = \cdots = y_{d-\ell} = 0$ in some local coordinates $(x_1, \ldots, x_d, y_1, \cdots = y_{d-\ell})$ on which $m = (0, \ldots, 0)$. Said otherwise, the connected component of m in $L \cap \mathcal{U}$ is a submanifold. It therefore satisfies the assumptions of Proposition 7.21 and is an immersed submanifold. It is therefore also a T-leaf. This concludes the proof of Theorem 7.9. Theorem 7.8 is an immediate consequence.

Remark 7.22 Notice that the functions $x_1, \ldots, x_{d-l}$ that appear in the local splitting theorem define a diffeomorphism $\Phi^\mathcal{U}_m$ from the submanifold $L^\mathcal{U}_m$ to an open neighborhood of $\mathbb{K}^{d-l}$. The families $(L^\mathcal{U}_m, \Phi^\mathcal{U}_m)_{m \in M}$ form an atlas for the leaf L_m.

8 Near a Leaf: The Transverse Singular Foliation and $\mathcal{F}$-connections

In Sect. 7, we saw that a singular foliation indeed decomposes the underlying manifold into submanifolds called leaves. In this subsection, we will discuss the "shape" of singular foliations near a fixed leaf. We refer to [26] to a general classification, which is way beyond the purpose of the present section, and uses the notion of Yang-Mills bundle as in [24, 25, 38].

8.1 Traveling Along a Leaf

Our first result means that "if you travel along a leaf, the landscape you will see is always the same", i.e., it means that any two points in the same leaf of a singular foliation $\mathcal{F}$ have arbitrarily small open neighborhoods on which the restrictions of $\mathcal{F}$ are isomorphic.

> **Proposition 8.1 (Along a Leaf, the Landscape Is Always the Same)** *Let L be a leaf of a singular foliation.*
>
> 1. *In the smooth case, for any two points ℓ_0, ℓ_1 of the leaf L of a singular foliation $\mathcal{F}$, there exists an inner symmetry of $\mathcal{F}$ mapping ℓ_0 to ℓ_1.*
> 2. *In the holomorphic or real analytic cases, the same results hold, but the inner symmetry is in general only defined in a neighborhood of ℓ_0.*

We start with a lemma. Recall that we say that two points $\ell_0, \ell_1 \in L$ are $\mathcal{F}$-reachable one from the other if there exists vector fields $X_1, \ldots, X_s \in \mathcal{F}$ and $t_1, \ldots, t_s \in \mathbb{R}$ such that R-let

$$\Phi^{X_1}_{t_1} \ldots \Phi^{X_s}_{t_s}(\ell_0) = \ell_1. \tag{37}$$

In the smooth setting, we assume $X_1, \ldots, X_s$ to be complete. In the complex or real-analytic settings, we can only impose that the composition of flows $\Phi^{X_1}_{t_1} \ldots \Phi^{X_s}_{t_s}$ that appears in Eq. (37) is well-defined in a neighborhood of ℓ_0. The previous lemma was in fact already proven in the course of Sect. 7, but we reprove it so that the chapter can be read independently.

Lemma 8.2 *Any two points in the same leaf of $\mathcal{F}$ are $\mathcal{F}$-reachable one from the other.*

Proof The notion of "being $\mathcal{F}$-reachable one from the other" defines an equivalence relation on L that we call the $\mathcal{F}$-reachability relation. Since the tangent space of the leaf $T_\ell L$ at a point $\ell \in \mathcal{F}$ coincides with $T_\ell \mathcal{F}$, every $u \in T_\ell L$ is of the form $X_{|\ell}$ for some $X \in \mathcal{F}$ that one can without any loss of generality assume to be of compact support, hence complete. Applying this reasoning to a basis $e_1, \ldots, e_k$ of $T_\ell L$, we find complete vector fields $Y_1, \ldots, Y_k \in \mathcal{F}$ through $e_1, \ldots, e_k$. Now the differential of the map

$$\mathbb{R}^k \to L$$
$$(t_1, \ldots, t_k) \mapsto \Phi^{Y_1}_{t_1} \ldots \Phi^{Y_k}_{t_k}(\ell_0)$$

is invertible at $t_1 = \cdots = t_k = 0$, since $Y_1|_\ell = e_1, \ldots, Y_k|_\ell = e_k$ are in the image. This implies that the image of this map contains a neighborhood of ℓ, so that the set of points of L which are $\mathcal{F}$-reachable from any point $\ell \in L$ contains a neighborhood of ℓ. In particular, each equivalence class of the $\mathcal{F}$-reachability relation is an open subset of L. Since the leaf L is a connected manifold by definition, this implies that there is only one equivalence class, namely L itself.

Proof of Proposition 8.1 The result is now in an immediate consequence of Proposition 7.12, which states that each flow in Eq. (37) is a symmetry of $\mathcal{F}$ (local in the complex or real analytic settings and global in the smooth setting).

Proposition 8.1 has several natural consequences.

Definition 8.3 Let L be a leaf. A pointed submanifold (S, ℓ) of M with $\ell \in L \cap S$ is said to be a $\mathcal{F}$-*cut* of the leaf L if

(i) S is transverse to L at ℓ, i.e., $T_\ell L \oplus T_\ell S = T_\ell M$, and
(ii) S intersects $\mathcal{F}$ cleanly, i.e., $T_s \mathcal{F} + T_s S = T_s M$ for every $s \in S$.

Remark 8.4 Notice that any pointed submanifold (S, ℓ) that satisfies condition (i) admits a neighborhood $S' \subset S$ of ℓ such that condition (ii) is also satisfied. Notice also that any neighborhood S' of a $\mathcal{F}$-cut S is a $\mathcal{F}$-cut again.

Consider a $\mathcal{F}$-cut (S, ℓ). Then consider the restriction $\mathcal{T}_S := i_S^* \mathcal{F}$ as in Sect. 5.3, i.e., the restriction to S of vector fields in $\mathcal{F}$ that are tangent to S. The following lemma holds true.

Lemma 8.5 *For every $\mathcal{F}$-cut (S, ℓ) of a leaf L of a singular foliation $\mathcal{F}$, $\mathcal{T}_S := i_S^* \mathcal{F}$ is a singular foliation on S. Also, the point $\{\ell\}$ is a leaf of $\mathcal{T}_S$.*

Proof The first statement is a consequence of Proposition 5.15, since item (ii) in the definition of a $\mathcal{F}$-cut means that S intersects $\mathcal{F}$ cleanly (see Definition 5.14). The second statement follows from the fact that $T_\ell \mathcal{T}_S = T_\ell S \cap T_\ell \mathcal{F}$ (see Question 2 in Exercise 3.21 in chapter "Canonical Geometric and Algebraic Structures Hidden Behind a Singular Foliation"). Since L in the leaf, $T_\ell \mathcal{F} = T_\ell L$. Item (i) in the definition of a S-cut then implies that $T_\ell \mathcal{T}_S = T_\ell S \cap T_\ell L = \{0\}$. Hence, the leaf through L of $\mathcal{T}_S$ reduces to $\{\ell\}$.

> **Theorem 8.6 (Any Two $\mathcal{F}$-Cuts to L Have Isomorphic Germs)** *Let L be a leaf of a smooth, real analytic or complex singular foliation. For any two $\mathcal{F}$-cuts (S_0, ℓ_0) and (S_1, ℓ_1), there exists neighborhoods S_0' of $\ell_0 \in S_0$ and S_1' of $\ell_1 \in S_1$ and an isomorphism of singular foliations:*
>
> $$(S_0', \ell_0, \mathcal{T}_{S_0'}) \xrightarrow{\simeq} (S_1', \ell_1, \mathcal{T}_{S_1'})$$
>
> *where $\mathcal{T}_{S_0'}, \mathcal{T}_{S_1'}$ are the induced singular foliations on S_0' and S_1'.*

Proof We prove it in two steps:

Step 1 We prove that there exists a $\mathcal{F}$-cut $(\tilde{S}_0, \ell_1)$ through the point ℓ_1 whose restricted singular foliation $\mathcal{T}_{\tilde{S}_0}$ is isomorphic[78] to $(S_0, \ell_0, \mathcal{T}_{S_0})$.

Step 2 We then prove Theorem 8.6 for the case $\ell_0 = \ell_1$.

By the first step, one can assume without any loss of generality that $\ell_0 = \ell_1$ in Theorem 8.6. The second step then provides a proof of the result.

Step 1 Let Φ be an inner symmetry as in Proposition 8.1 such that $\Phi(\ell_0) = \ell_1$. Since Φ is a symmetry of $\mathcal{F}$, and since it restricts to a diffeomorphism of L such that $\Phi(\ell_0) = \ell_1$, $(\Phi(S_0), \ell_1)$ is a $\mathcal{F}$-cut. Also, the restriction of Φ to a diffeomorphism $S_0 \to \Phi(S_0)$ is an isomorphism between their respective restricted singular foliations. This proves the first point.

Step 2 Let $\ell \in L$ be a point. After Theorem 7.18, we can without loss of generality assume ℓ to be the origin of $U \subset \mathbb{R}^n$ with coordinates $(x_1, \ldots, x_l, y_1, \ldots, y_{n-l})$ such that the foliation is generated by $\frac{\partial}{\partial x_1}, \ldots, \frac{\partial}{\partial x_l}$ and $Y_1, \ldots, Y_{r-l}$ where Y_i only depend on the y variables and only contain $\frac{\partial}{\partial y_j}$ components. In particular $L = \{y = 0\}$. By construction, $S_0 = \{x = 0\}$ is a $\mathcal{F}$-cut, so it suffices to show that any other $\mathcal{F}$-cut S_1 through the origin is (locally) equivalent to it. The space $T_\ell S_1$ being transverse to $T_\ell L$ translates to $T_\ell S_1 \to T_\ell S_0$ being surjective, i.e., shrinking S_0 to $S_0' \ni \ell$, there is a local section $\sigma : S_0' \to S_1$. i.e., near ℓ we have $S_1 = \{(f(y), y)\}$ for some function $f : \mathbb{R}^{n-l} \to \mathbb{R}^l$ (with $f(0) = 0$). The family of diffeomorphisms $\Phi_t(x, y) = (x + t \cdot f(y), y)$ map S_0 to S_1 and are the flow of a vector field in $\mathcal{F}$ (fixing ℓ), hence induces an isomorphism of transverse foliations. This completes the proof of the Theorem.

---

[78] In the holomorphic or real analytic settings, one has to replace S_0 by a neighborhood of ℓ_0 in S_0. This does not affect the argument.

We call *germ at* $0 \in \mathbb{K}^s$ of a singular foliation on $\mathbb{K}^s$ an equivalence class of pairs $(\mathcal{U}, \mathcal{T}_\mathcal{U})$ where

1. $\mathcal{U}$ is a neighborhood of 0 in $\mathbb{K}^s$, and
2. $\mathcal{T}_\mathcal{U}$ is a singular foliation on $\mathcal{U}$,

under the equivalence relation that consists in identifying $(\mathcal{U}, \mathcal{T}_\mathcal{U})$ and $(\mathcal{V}, \mathcal{T}_\mathcal{V})$ if there exists a neighborhood $\mathcal{W}$ of 0 contained in $\mathcal{U} \cap \mathcal{V}$ on which the restrictions of $\mathcal{T}_\mathcal{U}$ and $\mathcal{T}_\mathcal{V}$ coincide. Lastly, we call *isomorphism classes of germs at* $0 \in \mathbb{K}^s$ of singular foliations the equivalence classes for the equivalence relation on germs at 0 of singular foliations on $\mathbb{K}^s$ that identifies two germs if they have representatives $(\mathcal{U}, \mathcal{F}_\mathcal{U})$ and $(\mathcal{V}, \mathcal{F}_\mathcal{V})$ which are isomorphic as foliated manifolds, through an isomorphism that maps 0 to 0.

Remark 8.7 Alternatively, an isomorphism class of germs of singular foliations at $0 \in \mathbb{K}^s$ is an equivalence class for the equivalence relation on pairs as in items 1. and 2. above under the equivalence relation that identifies $(\mathcal{U}, \mathcal{T}_\mathcal{U})$ and $(\mathcal{V}, \mathcal{T}_\mathcal{V})$ if and only if there exists open neighborhoods $\mathcal{U}' \subset \mathcal{U}$ and $\mathcal{V}' \subset \mathcal{V}$ of 0 and an isomorphism of singular foliation

$$\Phi \colon (\mathcal{U}', i^*_{\mathcal{U}'} \mathcal{T}_\mathcal{U}) \longrightarrow (\mathcal{V}', i^*_{\mathcal{V}'} \mathcal{T}_\mathcal{V})$$

that maps 0 to 0.

Let L be a leaf of dimension k of a singular foliation $\mathcal{F}$ on a manifold M of dimension d. A pair $(\mathcal{U}, \mathcal{T}_\mathcal{U})$, with $\mathcal{U}$ a neighborhood of 0 in $\mathbb{K}^{d-k}$, is called a *representative of the transverse singular foliation of* L if there exists a $\mathcal{F}$-cut (S, ℓ) whose restricted singular foliation is isomorphic to $(\mathcal{U}, \mathcal{T}_\mathcal{U})$ through an isomorphism that maps $\ell \in S$ to $0 \in \mathbb{K}^{d-k}$. Theorem 8.6 implies that any two representatives of the transverse singular foliation of L are in the same class for the equivalence relation defining isomorphism classes of germs at 0 of singular foliations. The next definition therefore makes sense.

Definition 8.8 ("The" Transverse Singular Foliation of a Leaf) Let L be a dimension k leaf of a singular foliation $\mathcal{F}$ on a manifold of dimension d. We call *transverse singular foliation of* L the class in isomorphism classes of germs at $0 \in \mathbb{K}^{d-k}$ of representatives of the transverse singular foliation of L.

8.2 Tubular Neighborhoods and $\mathcal{F}$-connections

We now introduce a type of Ehresmann connection in a neighborhood of a leaf that appeared in [42], Section 2.2., and is generalized and used in [41] and [26] to classify neighborhoods of leaves. In this section, we work in the smooth setting. In the real analytic or holomorphic setting, the objects introduced here (tubular neighborhood, $\mathcal{F}$-connections) make sense and can be defined mutatis mutandis, but the issue is that they may not exist, because their construction goes through partitions of unity.

Let L be an embedded submanifold of a manifold M. There always exist a pair $(\mathcal{U}_L, p)$ where:

1. $\mathcal{U}_L$ is an open neighborhood of L in M,
2. $p \colon \mathcal{U}_L \longrightarrow L$ is a surjective submersion (whose restriction to L is the identity)

Moreover, one can assume $\mathcal{U}_L$ is diffeomorphic to a neighborhood of the zero section in the normal bundle $N_L := TM_{|L}/TL$ through a diffeomorphism that intertwines p and the natural projection of N_L on its base. Here, we will never need such an isomorphism, and we will simply call *tubular neighborhoods* pairs $(\mathcal{U}_L, p)$ as above.

In the smooth context, tubular neighborhood always exist, and any two tubular neighborhoods $(\mathcal{U}_L, p)$ and $(\mathcal{U}'_L, p')$ have restrictions near L which are isomorphic through an isomorphism that intertwines p and p'. In the real analytic or complex setting, tubular neighborhoods do not exist in general.

In the smooth setting moreover, any tubular neighborhood admits an *Ehresmann distribution*, i.e., there exists a smooth distribution $H \subset T\mathcal{U}_L$ of constant rank $k = \dim(L)$ such that for every point $m \in \mathcal{U}_L$, we have

$$\mathrm{Ker}(T_m p) \oplus H_m = T_m \mathcal{U}_L.$$

Moreover, there exists an Ehresmann distribution H such that $H_\ell = T_\ell L$ for every $\ell \in L$. Given a vector field $X \in \mathfrak{X}(L)$, one defines a smooth vector field $H(X)$ on $\mathcal{U}_L$ by imposing that $H(X)$ be section of H such that $T_m p(H_m(X)) = X_{|m}$ for all $m \in \mathcal{U}_L$. Equivalently, $H(X)$ is the unique vector field on $\mathcal{U}_L$ which is valued in the distribution H at every point and is p-related to X. One call *horizontal lift* of an Ehresmann distribution H the henceforth induced map:

$$H \colon \mathfrak{X}(L) \longrightarrow \mathfrak{X}(\mathcal{U}_L) \tag{38}$$
$$X \mapsto H(X)$$

The horizontal lift satisfies several properties that we list below:

1. for every $X \in \mathfrak{X}(L)$ and $f \in \mathcal{C}^\infty(L)$, we have $H(fX) = p^* f \, H(X)$,
2. for every $X \in \mathfrak{X}(L)$, we have $p_*(H(X)) = X$,

3. for every $X, Y \in \mathfrak{X}(L)$, $\kappa_H(X, Y) := H([X, Y]) - [H(X), H(Y)]$ is a vector field tangent to the fiber of p.
4. $\kappa_H(X, Y)$ is a skew-symmetric and $C^\infty(L)$-bilinear map. It can therefore be seen as a 2-form on L valued in p-vertical vector fields. It is called the *curvature* of the Ehresmann distribution.

Now, in the context of singular foliations, when L is a leaf, one will impose one more constraint on the Ehresmann connection. In Definition 8.9 below, items 1. and 2. mean that $(\mathcal{U}_L, p)$ is a tubular neighborhood.

Definition 8.9 ($\mathcal{F}$-connection) Let L be an embedded leaf of a singular foliation $\mathcal{F}$. We say that a triple $(\mathcal{U}_L, p, H)$ where:

1. $\mathcal{U}_L$ is an open neighborhood of L in M,
2. $p\colon \mathcal{U}_L \longrightarrow L$ is a surjective submersion (whose restriction to L is the identity), and
3. H is an Ehresmann distribution with respect to p,

is a $\mathcal{F}$-connection if sections of H are included in $\mathcal{F}$.

Remark 8.10 [26] explains why $\mathcal{F}$-connections are a particular case of the so-called Yang-Mills connections [24, 25, 38], at least at formal level.

Exercise 8.11 Could the phrase "sections of H are included in $\mathcal{F}$" in Definition 8.9 above could be equivalently replaced by

1. "$H \subset T\mathcal{F}$",
2. or by "H_m is tangent to the leaf through m for every $m \in M$".

Hint: Consider the singular foliation on $\mathbb{R}^2$ generated by $\frac{\partial}{\partial x}$ and $y^2 \frac{\partial}{\partial y}$, the leaf $L = \{(x, 0), x \in \mathbb{R}\}$, the projection $p\colon (x, y) \mapsto (x, 0)$, and $H = \left\langle \frac{\partial}{\partial x} + y \frac{\partial}{\partial y} \right\rangle$.

Remark 8.12 If on a tubular neighborhood $(\mathcal{U}_L, p)$ there exists an Ehresmann connection which is an $\mathcal{F}$-connection, then for every $\ell \in L$, the fiber $p^{-1}(\ell)$ of $p\colon \mathcal{U}_L \longrightarrow L$ is a submanifold that cleanly intersects $\mathcal{F}$. Each fiber is therefore a $\mathcal{F}$-cut of the leaf L. In particular, for every $\ell \in M$, $p^{-1}(\ell)$ admits a restricted singular foliation that we will denote by $\mathcal{T}_\ell$ when needed.

Remark 8.13 For any $\mathcal{F}$-connection $(\mathcal{U}_L, p, H)$, we have $H_m \subset T_m\mathcal{F}$ for all $m \in M$. This condition is however not sufficient to guarantee that H defines an $\mathcal{F}$-connection, see Exercise 8.11.

The phrase "sections of H are included in $\mathcal{F}$" in Definition 8.9 above could be equivalently replaced by the following condition "the horizontal lift H (see Eq. (38)) is valued in $\mathcal{F}$".

Lemma 8.14 *A $\mathcal{F}$-connection (U, p, H) for an embedded leaf L is equivalent to the data[79] given by:*

1. *a neighborhood $\mathcal{U}$ of L in M and a projection p as above,*
2. *and a $C^\infty(L)$-linear section of the natural projection $\mathcal{F}_\mathcal{U}^{proj} \to \mathfrak{X}(L)$, where $\mathcal{F}_\mathcal{U}^{proj}$ stands for vector fields in $\mathcal{F}_\mathcal{U}$ which are p-related with a vector field in $\mathcal{F}$.*

Proof Given a $\mathcal{F}$-connection as in Definition 8.9, the morphism s of item 2 is the horizontal lift defined as in Eq. (38). Conversely, given a section s as in item 2, the horizontal distribution H is the distribution generated by the vector fields $s(\mathfrak{X}(L))$.

Using this characterization one can verify that, at least in the smooth setting, $\mathcal{F}$-connections exist:

> **Proposition 8.15 ($\mathcal{F}$-connections Exist)** *Any embedded leaf of a smooth singular foliation admits an $\mathcal{F}$-connection.*

Proof Fix a tubular neighborhood $(\mathcal{U}, p)$ of an embedded l-dimensional leaf L of the singular foliation $\mathcal{F}$. The local splitting Theorem 7.17 obviously implies that $\mathcal{F}$-connections exist locally, i.e., every point $\ell \in L$ admits a neighborhood $\mathcal{V}$ in M which an $\mathcal{F}$-connection exists: it suffices to consider (we use notations of Theorem 7.18) the vector fields

$$H := \left\langle \frac{\partial}{\partial x_1}, \ldots, \frac{\partial}{\partial x_l} \right\rangle,$$

which define a distribution H whose sections are in $\mathcal{F}$ and is in direct sum with $\mathrm{Ker}(Tp)$ in a neighborhood of $y = 0$. As a consequence, upon shrinking $\mathcal{U}$ if necessary, we can cover

---

[79] In [26], which works in the transversally formal setting, this lemma becomes the definition of an $\mathcal{F}$-connection.

a neighborhood of a leaf L by small open subsets $(\mathcal{U}_i)_{i \in I}$ such that $p^{-1}(\mathcal{U}_i) \cap \mathcal{U}$ admits a $\mathcal{F}$-connections for every index i. Consider $s_i : \mathfrak{X}(\mathcal{U}_i) \to \mathcal{F}^{proj}_{p^{-1}(\mathcal{U}_i)}$ their horizontal lifts. One then glues these local connections by the use of a partition of unity $(\chi_i)_{i \in I}$ for the open cover $(\mathcal{U}_i)_{i \in I}$ of L, and defines $s(X) = \sum_i p^*\chi_i \cdot s_i(X|_{U_i})$ for every vector field $X \in \mathfrak{X}(L)$. This completes the construction.

For L a leaf a singular foliation, here is a natural question: does there exist a regular foliation $\mathcal{R}$ contained in $\mathcal{F}$ and admitting L as a leaf? The answer is that such a regular foliation exists near a given embedded leaf L if and only there exists a flat $\mathcal{F}$-connection, i.e., a $\mathcal{F}$-connection for which the curvature (defined above) is zero (i.e., if H is an integrable distribution).

Exercise 8.16 Let L be an embedded leaf of a smooth singular foliation $\mathcal{F}$. Show that the following items are equivalent:

(i) There exists a neighborhood U of L and a regular foliation $\mathcal{R} \subset \mathcal{F}_U$ of rank $dim(L)$.
(ii) L admits a flat $\mathcal{F}$-connection.

We say that a leaf is *flat* if one of the equivalent conditions in Exercise 8.16 is satisfied.

8.3 Fibered Products Along a Leaf

We now present a construction of "pull-back" of a singular foliation near a leaf, when an $\mathcal{F}$-connection is given. It does not depend on the choice of an $\mathcal{F}$-connection, but the existence of an $\mathcal{F}$-connection is however needed to guarantee its existence of the pullback. Let L be an embedded leaf of a singular foliation $\mathcal{F}$. Assume that one is given:

1. An $\mathcal{F}$-connection $(\mathcal{U}_L, p, H)$
2. Any manifold Σ and any map $\varphi \colon \Sigma \longrightarrow L$. Notice that we do not require φ to be a submersion, nor an immersion: it just has to be in the relevant category of maps (smooth, real analytic or holomorphic).

Since p is a surjective submersion, the fibered product

$$\varphi^!\mathcal{U}_L := \mathcal{U}_L \times_{p,L,\varphi} \Sigma := \{(m, \sigma) \in M \times \Sigma \mid p(m) = \varphi(\sigma)\}$$

is a submanifold of $\mathcal{U}_L \times \Sigma$. Now, $\mathcal{U}_L \times \Sigma$ comes with the direct product singular foliation $\mathcal{F} \times \mathfrak{X}(\Sigma)$, i.e., the direct product of the foliation $\mathcal{F}$ (restricted to $\mathcal{U}_L$) and of the foliation $\mathfrak{X}(\Sigma)$ of all vector fields on Σ.

Lemma 8.17 *The submanifold $\varphi^! \mathcal{U}_L := \mathcal{U}_L \times_{p,L,\varphi} \Sigma$ intersects cleanly the singular foliation $\mathcal{F} \times \mathfrak{X}(\Sigma)$.*

Proof The proof relies on the existence of an Ehresmann connection H. Since H is in direct sum with $\mathrm{Ker}(Tp)$, it is easily checked that any vector of $T(\mathcal{U}_L \times \Sigma)$ reads as a sum of an element in $u + v + w$ with $u \in H$, $v \in \mathrm{Ker}(Tp)$ and $w \in T\Sigma$. Now by definition, v is tangent to the submanifold $T(\mathcal{U}_L \times_{p,L,\varphi} \Sigma)$, w belongs to $T(\mathfrak{X}(\Sigma)) \subset T(\mathcal{F} \times \mathfrak{X}(\Sigma))$. Moreover, u belongs to $T(\mathcal{F}) \subset T(\mathcal{F} \times \mathfrak{X}(\Sigma))$ since H is an $\mathcal{F}$-connection. The conditions in Definition 5.14 are therefore satisfied.

We denote by $\varphi^! \mathcal{F}$ the restriction of the direct product singular foliation to $\varphi^! \mathcal{U}_L := \mathcal{U}_L \times_{p,L,\varphi} \Sigma$.

Remark 8.18 The existence of a distribution H whose sections are in $\mathcal{F}$ was used to claim that it is a singular foliation, $\varphi^! \mathcal{F}$ was defined without any reference to H. It therefore does not depend on the latter.

We have to make two lists of comments. The first one is a list of generalities about fibered products, and the second one relates the items of the first list to the properties of the foliated manifold $(\varphi^! \mathcal{U}_L, \varphi^! \mathcal{F})$.

1. The natural projection $\varphi^! p : \varphi^! \mathcal{U}_L \to \Sigma$ is a surjective submersion, whose fiber over $\sigma \in \Sigma$ is diffeomorphic to the fiber of p over $\varphi(\sigma)$.
2. The submanifold $L \times_{p,L,\varphi} \Sigma$ is canonically diffeomorphic to Σ, making Σ a submanifold of $\mathcal{U}_L \times_{p,L,\varphi} \Sigma$. This inclusion is a right inverse of the above projection.
3. Any Ehresmann connection H on $(\mathcal{U}_L, p)$ induces an Ehresmann connection $\varphi^! H$ on $\mathcal{U}_L \times_{p,L,\varphi} \Sigma$. By construction,

$$\varphi^! H_{(m,\sigma)} := \{(h, v) \in H_m \times T_\sigma \Sigma \mid T_m p(h) = T_\sigma \varphi(u)\}.$$

We leave it to the reader to check the following list of points:

1. The fibers of $\varphi^! p$ and of p intersect cleanly $\varphi^! \mathcal{F}$ and $\mathcal{F}$ respectively, and the diffeomorphism described in the first item above is an isomorphism of their respective restrictions.
2. The manifold $\Sigma \in \phi^! \Sigma$ is a leaf of $\varphi^! \mathcal{F}$.
3. If the Ehresmann connection H is an $\mathcal{F}$-connection for $\mathcal{F}$, then $\varphi^! H$ is a $\mathcal{F}$-connection for $\varphi^! \mathcal{F}$.

As a conclusion, $(\varphi^! \mathcal{U}_L, \varphi^! p, \varphi^! H)$ is an $\mathcal{F}$-connection for the leaf Σ of $\varphi^! \mathcal{F}$. We call it the *fibered product* of $(\mathcal{U}_L, p, H)$ with respect to $\varphi \colon \Sigma \to L$.

8.4 Parallel Transportation and $\mathcal{F}$-connections

For this section, in addition to the own works of the authors [41, 42], we acknowledge ideas coming from the PhD [24, 25] of Simon Raphael Fischer, see also [26].

Let $(\mathcal{U}_L, p, H)$ be an $\mathcal{F}$-connection for an embedded leaf L of a singular foliation $\mathcal{F}$. Let $H \colon \mathfrak{X}(L) \to \mathcal{F}_{proj}$ be the horizontal lift as in Eq. (38). Recall that the notion of smooth time-dependent vector field valued in $\mathcal{F}$ has some subtleties: they are dealt with in Definition 3.2. Let I be an open interval of $\mathbb{R}$.

Lemma 8.19 *For any smooth time dependent vector field $(X_t)_{t \in I}$ on the leaf L, $H(X_t)$ is a smooth time dependent vector field in $\mathcal{F}$.*

Proof For every $m \in \mathcal{U}_L$, let $X_1, \ldots, X_k \in \mathfrak{X}(L)$ be a local trivialization of TL on a neighborhood $\mathcal{W}$ of $p(m)$. There exists functions $f_1(\ell, t), \ldots, f_k(\ell, t)$ on $\mathcal{W} \times I$ such that

$$X_t|_m = \sum_{i=1}^{k} f_i(m, t) \, X_i|_m.$$

This implies that

$$H(X_t)|_m = \sum_{i=1}^{k} f_i(p(m), t) H(X_i)$$

is a smooth time-dependent vector field on $\mathfrak{X}(L)$.

Here is an immediate consequence of this lemma, together with the fact that $H(X)$ is p-related with X.

Proposition 8.20 *For every time dependent vector field $(X_t)_{t \in I}$ on L whose time t_0 flow is well-defined, the flow $\Phi_{t_0}^{H(X_t)}$ at time t_0 of the horizontal lift $H(X_t)$*

1. *is well-defined in a neighborhood $\mathcal{U}'_L$ of L in $\mathcal{U}_L$,*
2. *is a symmetry of $\mathcal{F}$,*

3. commutes with p, more precisely

$$\begin{array}{ccc} \mathcal{U}'_L & \xrightarrow{\Phi^{H(X_t)}_{t_0}} & \mathcal{U}_L \\ {\scriptstyle p}\downarrow & & \downarrow{\scriptstyle p} \\ L & \xrightarrow{\Phi^{X_t}_{t_0}} & L \end{array}$$

Recall from Sect. 8.3 that for every $\ell \in L$, $p^{-1}(\ell)$ is an $\mathcal{F}$-cut for the leaf L, so that the restriction of $\mathcal{F}$ to $p^{-1}(\ell)$ is a singular foliation that we denote by $\mathcal{T}_\ell$. Recall from Theorem 8.6 that the foliated manifolds $(p^{-1}(\ell_0), \mathcal{T}_{\ell_0})$ and $(p^{-1}(\ell_1), \mathcal{T}_{\ell_1})$ have restrictions to neighborhoods of ℓ_0 and ℓ_1 respectively which are isomorphic, through an isomorphism of foliated manifolds that maps ℓ_0 to ℓ_1.

We now explain how parallel transportation with respect to an $\mathcal{F}$-connection realizes such an isomorphism of foliated manifolds. Let us briefly recall the notion of parallel transportation. Given an Ehresmann connection H on a tubular neighborhood $(\mathcal{U}_L, p)$ of an embedded manifold L, one says that a path $\tilde{\gamma} : [0, 1] \to \mathcal{U}_L$ in $\mathcal{U}_L$ is *parallel* if for all $t \in [0, 1]$:

$$\frac{d\gamma}{dt}(t) \in H_{\gamma(t)}.$$

Given a path $\gamma : [0, 1] \to L$ on L such that $\gamma(0) = \ell_0$, and given a point $m \in p^{-1}(\ell_0)$, there exists at most one parallel path $\gamma^H_m : [0, 1] \to \mathcal{U}_L$ such that

$$\tilde{\gamma}_m(0) = m \text{ and } p\left(\gamma^H_m(t)\right) = \gamma(t) \text{ for all } t \in [0, 1].$$

We call this path the *horizontal lift* of $\gamma(t)$ starting from m. If, moreover, $H_\ell = T_\ell L$ for every $\ell \in L$, then the path $\gamma^H_m(t)$ is well defined at time 1 for every m in a neighborhood $\mathcal{U}_{p^{-1}(\ell_0)}$ of ℓ_0 in $p^{-1}(\ell_0)$. We call *parallel transportation* over $\gamma(t)$ the map:

$$P^H_\gamma : \mathcal{U}_{p^{-1}(\ell_0)} \to p^{-1}(\ell_1)$$
$$m \mapsto \gamma^H_m(1)$$

The following results are classical, cf. e.g., [36]:

1. The parallel transportation P^H_γ is a diffeomorphism onto its image.
2. Its inverse is the parallel transportation over the path $t \mapsto \gamma(1 - t)$.
3. The parallel transportation does not depend on a parametrization of $\gamma(t)$. To be more precise $P^H_\gamma = P^H_{\gamma \circ \psi}$ for every diffeomorphism $\psi : [0, 1] \to [0, 1]$ such that $\psi(0) = 0$ and $\psi(1) = 1$.

4. The previous item allows making sense of the following statement: given two paths $\gamma_1, \gamma_2 \colon [0, 1] \to L$ such that $\gamma_1(1) = \gamma_2(0)$, one has:

$$P_{\gamma_2}^H \circ P_{\gamma_1}^H = P_{\gamma_1 * \gamma_2}^H$$

where $\gamma_1 * \gamma_2 \colon [0, 1] \to L$ is any path that merges γ_1 and γ_2, for instance

$$\gamma_1 * \gamma_2(t) = \begin{cases} \gamma_1(2t) & \text{for } t \in [0, 1/2] \\ \gamma_2(2t - 1) & \text{for } t \in [1/2, 1] \end{cases}.$$

Recall that an Ehresmann connection H is said to be *complete*[80] if its associated lift s_H (defined as in Eq. (38)) maps complete vector fields on L to complete vector fields on $\mathcal{U}_L$. In this case, it can be shown that P_γ^H is defined on the whole fiber $p^{-1}(\ell_0)$, and is a diffeomorphism onto the fiber $p^{-1}(\ell_1)$. *We will from now on assume that the H-connection is complete in order to simplify the proofs and the statements.*

For an arbitrary $\mathcal{F}$-connection we would get the corresponding statement on neighborhoods of L, but we will leave the reader to make the generalization.

Now, assume the complete Ehresmann connection is a $\mathcal{F}$-connection on some tubular neighborhood of an embedded leaf.

Lemma 8.21 *Let $\gamma \colon [0, 1] \to L$ be a smooth path from ℓ_0 to ℓ_1. The parallel transportation $P_\gamma^H \colon p^{-1}(\ell_0) \to p^{-1}(\ell_1)$ is an isomorphism of singular foliations from $(p^{-1}(\ell_0), \mathcal{T}_{\ell_0})$ to $(p^{-1}(\ell_1), \mathcal{T}_{\ell_1})$. We denote it by $P_{\ell_0, \gamma, \ell_1}^H$*

Proof Let $I = [0, 1]$. For any path $\gamma \colon I \to L$, there exists a compactly supported time dependent vector field $(X_t)_{t \in [0,1]}$ on L such that

$$X_t|_{\gamma(t)} = \frac{d\gamma(t)}{dt}$$

for all $t \in I$. The integral curve of $(X_t)_{t \in I}$ starting from ℓ_0 coincides with the path $\gamma \colon I \to L$ by construction. Since X_t and H are complete, Proposition 8.20 applies with $\mathcal{U}_L' = \mathcal{U}_L$, and yields the time 1-flow $\Phi_1^{H(X_t)}$ in $H(X_t)$. By construction, the restriction to the fiber $p^{-1}(\ell_0)$ of $\Phi_1^{H(X_t)}$ coincides with $P_{\ell_0, \gamma, \ell_1}$. Since $\Phi_1^{H(X_t)}$ is a symmetry of $\mathcal{F}$, Exercise 5.19 applies and yields the desired isomorphism of the restricted singular foliations.

---

[80] For instance, the central circle of the restriction to a relatively compact neighborhood of it of the so-called self-eating snake does not admit a complete $\mathcal{F}$-connection, see Example 5.26. One can avoid completeness by working at the formal level (as in [26]), or at the level of germs.

Let us recall some vocabulary from Sect. 3.1. Given two isomorphisms of singular foliations $\Phi, \Psi : (M_1, \mathcal{F}_1)$ and $(M_2, \mathcal{F}_2)$, the following are equivalent:

(i) $\Psi^{-1} \circ \Phi$ is an inner symmetry of $(M_1, \mathcal{F}_1)$.
(ii) $\Phi \circ \Psi^{-1}$ is an inner symmetry of $(M_2, \mathcal{F}_2)$,

We say that Φ, Ψ *differ by an inner symmetry* of one of these equivalent conditions are satisfied. In the next lemma, again, the assumption "complete" could be deleted, at the expense of restricting ourselves to a smaller neighborhood of L

Lemma 8.22 *Let H, H' be two complete $\mathcal{F}$-connections on the same tubular neighborhood $(\mathcal{U}_L, p)$. Let $\gamma : [0, 1] \to L$ be a smooth path from ℓ_0 to ℓ_1. The singular foliations isomorphisms $P^H_{\ell_0, \gamma, \ell_1}$ and $P^{H'}_{\ell_0, \gamma, \ell_1}$ obtained by parallel transportation over $\gamma(t)$ with respect to H and H' differ by an inner symmetry.*

Proof Observe that $H(X) - H'(X)$ is for every $X \in \mathfrak{X}(L)$ valued in the singular foliation of vector fields in $\mathcal{F}$, which are p-vertical, which we denote by $\mathcal{T}_\bullet$. Let $(X_t)_{t \in I}$ be as in the proof of Lemma 8.21. Then $Y_t := H(X_t) - H'(X_t)$ is a smooth time dependent vector field in $\mathcal{T}_\bullet$. We will use the following classical formula (see, e.g., [60]) for time-dependent vector fields A_t, B_t, where $\phi_{t,s}$ denotes the flow at time t with starting from time s:

$$\phi^{A,B}_{t,t_0} = \phi^A_{t,0} \circ \phi^{(\phi^A_{0,t})_* B_t}_{t,t_0} \circ \phi^A_{0,t_0}$$

Evaluating it at $t_0 = 0$, $A = s(X)$ and $B = Y$ yields:

$$\phi^{s(X)+Y}_{t,0} = \phi^{s(X)}_{t,0} \circ \phi^{(\phi^{s(X)}_{0,t})_* Y_t}_{t,0}.$$

Since the flow $\phi^{s(X)}$ preserves both $\mathcal{F}$ and verticality, $Z = (\phi^{s(X)}_{0,1})_* Y_1|_{p^{-1}(x)} \in \mathcal{T}_x$, i.e., $\phi^{s(X)+Y}_{1,0} = \phi^{s(X)}_{1,0} \circ \psi$ for an inner symmetry ψ. A similar technique has been used to prove Proposition 2.3 in [5]. □

Now we can finally show the following result. The setting is the one of Lemmas 8.21–8.22.

Lemma 8.23 *Let H be a complete $\mathcal{F}$-connection. The isomorphisms of singular foliations obtained by parallel transportation with respect to H over homotopic paths differ by an inner symmetry.*[81]

---

[81] See Definition 3.7.

Let us restate Lemma 8.23 more precisely: it says that for γ_0 and γ_1 two homotopic paths in L from ℓ_0 and ℓ_1, the parallel transportations $P^H_{\ell_0,\gamma_0,\ell_1}$ and $P^H_{\ell_0,\gamma_1,\ell_1}$, which are isomorphisms of singular foliations by Lemma 8.21, differ by an inner symmetry.

Proof In this proof we essentially follow part of the proof of [36] of the (generalized) Ambrose-Singer theorem, adapted to our setting. Given two homotopic paths, $\gamma, \tilde{\gamma}$ we can reparametrize them such that they are constant near the boundary, so that composing them yields a smooth null homotopic loop. Hence, it suffices to show the statement for a (smoothly) null homotopic loop γ. Let H be a homotopy (with fixed endpoint x) such that $H_1 = \gamma$ and $H_0 \equiv x$. We can consider $f_t = P^s(H_t)$ and want to show that $\frac{\partial f_t}{\partial t} \circ f_t^{-1}$ is a smooth time-dependent[82] vector field Z_t on $(p^{-1}(x), \mathcal{T}_x)$, then f_t is its flow and the claim follows. In order to show the claim, one can pull back the whole situation along H and obtain a tubular neighborhood with foliation and connection over $[0, 1]^2$. In [36, 9.11, Claim 2] it is shown that for $X = \frac{\partial}{\partial x}, Y = \frac{\partial}{\partial y}$ on the unit square we have:

$$Z_t = \frac{\partial f_t}{\partial t} \circ f_t^{-1} = \int_0^1 -(\phi_\tau^{s(X)})^*[s(X), s(Y)] + (\phi_\tau^{s(X)})^*[s(X), (\phi_t^{s(Y)})^*(\phi_{-\tau}^{s(X)})^*s(Y)]$$

$$-(\phi_\tau^{s(X)})^*(\phi_t^{s(Y)})^*(\phi_{-\tau}^{s(X)})^*[s(X), s(Y)] d\tau$$

Applying flows, commutators and integrals to (time-dependent) elements in the foliation yields elements in the foliation, so Z_t is a smooth time-dependent vector field in the transverse foliation, i.e., its flow f_t is an inner symmetry. □

Let $(\mathcal{U}_L, p, H)$ be an $\mathcal{F}$-connection with the Ehresmann connection H being complete. Altogether, Lemmas 8.21–8.23 above imply that for every $\ell \in L$, there exists a group morphism from $\pi_1(L, \ell)$ (the fundamental group of the leaf L) to the group of outer symmetries of the singular foliation $(\pi^{-1}(\ell), \mathcal{T}_\ell)$, i.e., the quotient of the group of symmetries of the singular foliation $(\pi^{-1}(\ell), \mathcal{T}_\ell)$ by the group of inner symmetries of $(\pi^{-1}(\ell), \mathcal{T}_\ell)$, see Sect. 3.1. In equation:

$$\Xi \colon \pi_1(L, \ell) \longrightarrow \mathrm{Out}(\pi^{-1}(\ell), \mathcal{T}_\ell). \tag{39}$$

Lemma 8.22 implies that Ξ does not depend on the choice of an Ehresmann connection H, provided it is complete. Moreover, if one drops this completeness assumption, the Lemmas above can be extended, but the group morphism above is only defined at the level of germs.[83]

---

[82] Defined as Definition 3.2.

[83] Let us be more precise on this point. Let $(\mathcal{V}, \mathcal{T})$ be a representative of the transverse singular foliation of $\mathcal{F}$. By its definition (Definition 8.8), there exists an isomorphism of singular foliations from a neighborhood of 0 in $(\mathcal{V}, \mathcal{T})$ to a neighborhood of ℓ in $(\pi^{-1}(\ell), \mathcal{T}_\ell)$ mapping 0 to ℓ.

This group morphism appeared first in C.L.G. and L.R.'s [42] (last line of Definition[84] 2.29). It extends an older construction by Dazord [15]. This morphism[85] is called *outer holonomy* by Simon Raphael Fischer and C.L.G. in [26]. It also appears for some important particular singular foliations associated to submanifolds in the works of Francis [27, 28] and of Bischoff et al. [10]. There are analogous statement by Rui Loja Fernandes [21], and by Rui Loja Fernandes and Yvan Struchiner [22], for Lie algebroids.

An Easy Mistake!
For a regular foliation, the outer holonomy is just the usual holonomy. It is therefore tempting to believe that if it is a trivial group morphism, then there is a neighborhood of L in M where the singular foliation is trivial, i.e., is isomorphic to a neighborhood of $(L, \{\ell\})$ in the direct product of $(L, \mathfrak{X}_c(L))$ with $(\pi^{-1}(\ell), \mathcal{T}_\ell)$. For regular foliation, hence for regular leaves, this is true.

But for a leaf of a singular foliation, this is completely wrong: let E be a nontrivial orientable vector bundle, and consider the singular foliation of all vector fields on E tangent to the zero section. Then the outer holonomy is trivial (because orientation preserving symmetries are inner symmetries) but the singular foliation can not be a direct product otherwise E itself would be a trivial vector bundle.

However, for transversally quadratic singular foliation, it is true that outer holonomy determines $\mathcal{F}$ in a formal neighborhood of L, see [26] for details.

The outer holonomy depends only on the choice of the isomorphism Ψ. Its class modulo conjugation by an outer isomorphism of singular foliation on the arrival space is however canonical. In particular, its kernel K_ℓ does only depend on the choice of the point ℓ. Finally, this notion of holonomy has been generalized in [42] as a sequence of group morphisms from $\pi_n(L)$, or all $n \geq 2$. This construction relies on the notion of universal Lie-∞ algebroid of a singular foliation, which we will encounter later in this text.

There is a similar but however slightly different approach of this "holonomy of a leaf" which consists of seeing it as a groupoid morphism.

More precisely, there are two transitive groupoids over L.

Using this local isomorphism, the group morphism Ξ becomes a group morphism valued in outer symmetries of the germs of singular foliations at 0 represented by $(\mathcal{V}, \mathcal{T})$.

[84] $\mathrm{Out}(\pi^{-1}(\ell), \mathcal{T}_\ell)$ is denoted by $\mathrm{Diff}(p^{-1}(\ell)/\mathcal{T}_\ell)$ in the referenced work.

[85] Fischer and Laurent-Gengoux [26] work in the slightly different context of formal singular foliations, the construction is however similar.

1. The fundamental groupoid $\Pi_1(L)$ is the set of all homotopy classes of paths with fixed end points [48]. It is a transitive groupoid over L, that we denote by $\Pi_1(L) \rightrightarrows L$. By construction, arrows between ℓ_0, ℓ_1 are the homotopy classes of paths from ℓ_0 to ℓ_1. Composition and inversion of paths amount to a Lie groupoid structure.
2. Let $(\mathcal{U}_L, p, H)$ be an $\mathcal{F}$-connection. Consider the groupoid over L for which the arrows with source $\ell_0 \in L$ and target $\ell_1 \in L$ is the set of all isomorphisms of singular foliations from $(p^{-1}(\ell_0), \mathcal{T}_{\ell_0})$ to $(p^{-1}(\ell_1), \mathcal{T}_{\ell_1})$. This groupoid admits a natural quotient if one identifies two arrows that differ by an inner symmetry. We denote by $\mathrm{Out}(\mathcal{U}_L, p) \rightrightarrows L$ this quotient groupoid.

   In the construction of item 2, one could also consider germs of isomorphisms of singular foliations from $(p^{-1}(\ell_0), \mathcal{T}_{\ell_0})$ to $(p^{-1}(\ell_1), \mathcal{T}_{\ell_1})$, defined from a neighborhood of ℓ_0 to a neighborhood of ℓ_1. We then denote the henceforth obtained groupoid by $\mathrm{Out}_L(\mathcal{U}_L, p)$.

We can now define the holonomy Hol as a groupoid morphism.

> **Proposition 8.24 (The Holonomy of a Singular Foliation [15, 26, 42])** *Let L be an embedded leaf of a smooth singular foliation $\mathcal{F}$ on a manifold M. Let $(\mathcal{U}_L, p, H)$ be an $\mathcal{F}$-connection, with H a complete Ehresmann connection. There is a groupoid morphism, called outer holonomy:*
>
> $$\begin{array}{ccc} \Pi_1(L) & \xrightarrow{Hol} & \mathrm{Out}(\mathcal{U}_L, p) \\ \downarrow\downarrow & & \downarrow\downarrow \\ L & = & L \end{array}$$
>
> *This morphism does not depend on the choice of H. Moreover, if no complete $\mathcal{F}$-connection H exists, then Hol still exists, provided that $\mathrm{Out}(\mathcal{U}_L, p)$ is replaced by the groupoid $\mathrm{Out}_L(\mathcal{U}_L, p) \rightrightarrows L$.*

> **Warning!**
> We warn the reader not to confuse the groupoid $\mathrm{Out}(\mathcal{U}_L, p) \rightrightarrows L$ with the groupoid $\mathrm{OutSym}_{\mathcal{F}} \rightrightarrows M$ of Sect. 5.1 in chapter "Canonical Geometric and Algebraic Structures Hidden Behind a Singular Foliation". They seem similar, but there is an important difference. The first groupoid is over a leaf L and the second one is over M. But even the restriction of the second one to a leaf is *not* the same as the first one. In short, $\mathrm{Out}(\mathcal{U}_L, p)$ is obtained by dividing by all inner symmetries of the transversal, while $\mathrm{OutSym}_{\mathcal{F}}$ is obtained by dividing by inner symmetries having a very-fixed point.

Let us conclude this section by recalling that Androulidakis and Zambon [5] have defined a more sophisticated holonomy, using the holonomy groupoid of the leaf instead of the fundamental groupoid of the leaf. The group morphism[86] above can be seen as a quotient of that morphism.

References

1. R. Abraham, J.E. Marsden, T. Ratiu, *Manifolds, Tensor Analysis, and Applications*, vol. 75, 2nd edn. Appl. Math. Sci. (Springer, New York, 1988)
2. A.A. Agrachev, Y.L. Sachkov, *Control Theory from the Geometric Viewpoint*, vol. 87. Encycl. Math. Sci. (Springer, Berlin, 2004)
3. I. Androulidakis, G. Skandalis, The holonomy groupoid of a singular foliation. J. Reine Angew. Math. **626**, 1–37 (2009)
4. I. Androulidakis, G. Skandalis, The analytic index of elliptic pseudodifferential operators on a singular foliation. J. K-Theory **8**(3), 363–385 (2011)
5. I. Androulidakis, M. Zambon, Holonomy transformations for singular foliations. Adv. Math. **256**, 348–397 (2014)
6. I. Androulidakis, M. Zambon, Stefan-Sussmann singular foliations, singular subalgebroids and their associated sheaves. Int. J. Geom. Methods Mod. Phys. **13**, 17 (2016). Id/No 1641001
7. I. Androulidakis, O. Mohsen, R. Yuncken, A pseudodifferential calculus for maximally hypoelliptic operators and the Helffer-Nourrigat conjecture. arXiv:2201.12060 (2022)
8. C. Araujo, J.P. Figueredo, Foliations on complex manifolds. Not. Am. Math. Soc. **69**(7), 1137–1144 (2022)
9. P. Baum, R. Bott, Singularities of holomorphic foliations. J. Differ. Geom. **7**, 279–342 (1972)
10. F. Bischoff, Á. del Pino, A. Witte, Jets of foliations and b^k-algebroids. arXiv:2311.17045 (2023)
11. G.E. Bredon, *Topology and Geometry*. Graduate Texts in Mathematics, vol. 139 (Springer, New York, 1993)
12. A. Coste, P. Dazord, A. Weinstein, *Groupoïdes symplectiques*. Publications du Département de Mathématiques. Nouvelle Série. A, vol. 2, i–ii, 1–62, Publ. Dép. Math. Nouvelle Sér. A, 87-2, Univ. Claude-Bernard, Lyon (1987)
13. M. Crainic, R.L. Fernandes, I. Mărcuț, *Lectures on Poisson Geometry*, vol. 217. Grad. Stud. Math. (American Mathematical Society, Providence, 2021)
14. A.B. da Silva, D. Panazzolo, Generalized flow-box property for singular foliations. Rev. R. Acad. Cienc. Exactas Fís. Nat., Ser. A Mat., RACSAM **113**(4), 3949–3965 (2019)
15. P. Dazord, Feuilletages à singularités. Indag. Math. **47**, 21–39 (1985)
16. T. de Fernex, R. Docampo, Nash blow-ups of jet schemes. Ann. Inst. Fourier **69**(6), 2577–2588 (2019)
17. C. Debord, Holonomy groupoids of singular foliations. J. Differ. Geom. **58**(3), 467–500 (2001)
18. C. Debord, G. Skandalis, Blow-up constructions for Lie groupoids and a Boutet de Monvel type calculus. Münster J. Math. **14**(1), 1–40 (2021)
19. A. Douady, Prolongement de faisceaux analytiques cohérents (Travaux de Trautmann, Frisch-Guenot et Siu). Semin. Bourbaki 1969/70, No. 366, 39–54 (1971)
20. L.D. Drager, J.M. Lee, E. Park, K. Richardson, Smooth distributions are finitely generated. Ann. Global Anal. Geom. **41**(3), 357–369 (2012)

---

[86] They work at the level of germs, and therefore do not need the completeness assumption.

21. R.L. Fernandes, Lie algebroids, holonomy and characteristic classes. Adv. Math. **170**(1), 119–179 (2002)
22. R.L. Fernandes, I. Struchiner, The classifying Lie algebroid of a geometric structure. II: G-structures with connection. São Paulo J. Math. Sci. **15**(2), 524–570 (2021)
23. R.L. Fernandes, C. Laurent-Gengoux, P. Vanhaecke, Global action-angle variables for non-commutative integrable systems. J. Symplectic Geom. **16**(3), 645–699 (2018)
24. S.-R. Fischer, Geometry of curved Yang-Mills-Higgs gauge theories. PhD thesis, Genova and Lyon (2021)
25. S.-R. Fischer, Integrating curved yang-mills gauge theories. arXiv:2210.02924 (2022)
26. S.-R. Fischer, C. Laurent-Gengoux, A classification of neighborhoods around leaves of a singular foliation. arXiv:2401.05966 (2024)
27. M. Francis, Groupoids and algebras of certain singular foliations with finitely many leaves. PhD thesis, Pennsylvania State University (2021)
28. M. Francis, On singular foliations tangent to a given hypersurface. arXiv:2311.03940 (2023)
29. A. Garmendia, The geometry of singular foliations. Master's thesis. PhD thesis, Leuven (2019)
30. A. Garmendia, J. Villatoro, Integration of singular foliations via paths. Int. Math. Res. Not. **2022**(23), 18401–18445 (2022)
31. A. Garmendia, O. Yudilevich, On the inner automorphisms of a singular foliation. Math. Z. **293**(1–2), 725–729 (2019)
32. K. Grabowska, J. Grabowski, Solvable Lie algebras of vector fields and a Lie's conjecture. SIGMA Symmetry Integrability Geom. Methods Appl., 16: Paper No. 065, 14 (2020)
33. R. Hermann, The differential geometry of foliations. II. J. Math. Mech. **11**, 303–315 (1962)
34. L. Hörmander, Hypoelliptic second order differential equations. Acta Math. **119**, 147–171 (1967)
35. C. Klimčík, T. Strobl, WZW-Poisson manifolds. J. Geom. Phys. **43**(4), 341–344 (2002)
36. I. Kolář, P.W. Michor, J. Slovák, *Natural Operations in Differential Geometry* (Springer, Berlin, 1993)
37. Y. Kosmann-Schwarzbach, Quasi, twisted, and all that ... in Poisson geometry and Lie algebroid theory, in *The Breadth of Symplectic and Poisson Geometry. Festschrift in Honor of Alan Weinstein* (Birkhäuser, Boston, 2005), pp. 363–389
38. A. Kotov, T. Strobl, Curving Yang-Mills-Higgs gauge theories. Phys. Rev. D **92**, 085032 (2015)
39. A. Kotov, T. Strobl, Lie algebroids, gauge theories, and compatible geometrical structures. Rev. Math. Phys. **31**(4), 31 (2019). Id/No 1950015
40. C. Laurent-Gengoux, R. Louis, Lie-Rinehart algebras $\simeq$ acyclic lie ∞-algebroids. J. Algebra **594**, 1–53 (2022)
41. C. Laurent-Gengoux, L. Ryvkin, The neighborhood of a singular leaf. J. Éc. Polytech. Math. **8**, 1037–1064 (2021)
42. C. Laurent-Gengoux, L. Ryvkin, The holonomy of a singular leaf. Sel. Math. **28**(2) (2022)
43. C. Laurent-Gengoux, A. Pichereau, P. Vanhaecke, *Poisson Structures*, vol. 347. Grundlehren Math. Wiss. (Springer, Berlin, 2012)
44. C. Laurent-Gengoux, S. Lavau, T. Strobl, The universal Lie ∞-algebroid of a singular foliation. Doc. Math. **25**, 1571–1652 (2020)
45. S. Lavau, A short guide through integration theorems of generalized distributions. Differ. Geom. Appl. **61**, 42–58 (2018)
46. R. Louis, On Nash resolution of (singular) Lie algebroids. arXiv:2404.08840 (2024)
47. R. Louis, A series of Nash resolutions of a singular foliation. arXiv:2301.08706 (2024)
48. K. Mackenzie, *Lie Groupoids and Lie Algebroids in Differential Geometry*. London Mathematical Society Lecture Note Series, vol. 124 (Cambridge University Press, Cambridge, 1987)
49. K.C.H. Mackenzie, *General Theory of Lie Groupoids and Lie Algebroids*, vol. 213. London Mathematical Society Lecture Note Series (Cambridge University Press, Cambridge, 2005)

50. E. Meinrenken, Lie algebroids. arXiv:2401.03034 (2024)
51. D. Miyamoto, The basic de Rham complex of a singular foliation. Int. Math. Res. Not. **2023**(8), 6364–6401 (2023)
52. D. Miyamoto, Singular foliations through diffeology, in *Recent Advances in Diffeologies and Their Applications. AMS-EMS-SMF Special Session*, Université de Grenoble-Alpes, Grenoble, July 18–20, 2022 (American Mathematical Society, Providence, 2024), pp. 139–160
53. I. Moerdijk, J. Mrčun, *Introduction to Foliations and Lie Groupoids*, vol. 91. Camb. Stud. Adv. Math. (Cambridge University Press, Cambridge, 2003)
54. O. Mohsen, Blow-up groupoid of singular foliations. arXiv:2105.05201 (2021)
55. T. Nagano, Linear differential systems with singularities and an application to transitive Lie algebras. J. Math. Soc. Jpn. **18**, 398–404 (1966)
56. H. Nahari, T. Strobl, Singular Riemannian foliations and $\mathcal{I}$-poisson manifolds. Doc. Math. arXiv:2210.17306 (2022, to appear)
57. M. Nakahara. *Geometry, Topology and Physics*. Graduate Texts in Physics (Hilger, Bristol, 1990)
58. J. Nestruev, *Smooth Manifolds and Observables*, vol. 220, 2nd revised and expanded edn. Grad. Texts Math. (Springer, Cham, 2020)
59. P.A. Ostrand, Covering dimension in general spaces. Gen. Topol. Appl. **1**, 209–221 (1971)
60. A. Posilicano, A Lie group structure on the space of time-dependent vector fields. Monatsh. Math. **105**(4), 287–293 (1988)
61. B. Scárdua, *Holomorphic Foliations with Singularities. Key Concepts and Modern Results*. Lat. Am. Math. Ser. (Springer; Instituto de Matemática y Ciencias Afines (IMCA), Cham, 2021)
62. S. Sertöz, Residues of singular holomorphic foliations. Compos. Math. **70**(3), 227–243 (1989)
63. S.B. Sontz, *Principal Bundles: The Classical Case* (Universitext, Springer International Publishing, Berlin, 2015)
64. P. Stefan, Accessibility and foliations with singularities. Bull. Am. Math. Soc. **80**, 1142–1145 (1974)
65. P. Stefan, Integrability of systems of vector fields. J. Lond. Math. Soc. II. Ser. **21**, 544–556 (1980)
66. H.J. Sussmann, Orbits of families of vector fields and integrability of distributions. Trans. Am. Math. Soc. **180**, 171–188 (1973)
67. H.J. Sussmann, Orbits of families of vector fields and integrability of systems with singularities. Bull. Am. Math. Soc. **79**, 197–199 (1973)
68. T. Suwa, Residues of complex analytic foliation singularities. J. Math. Soc. Jpn. **36**, 37–45 (1984)
69. J.-C. Tougeron, Idéaux de fonctions différentiables. I. Ann. Inst. Fourier (Grenoble) **18**(fasc., fasc. 1), 177–240 (1968)
70. Y. Turki, A Lagrangian for Hamiltonian vector fields on singular Poisson manifolds. J. Geom. Phys. **90**, 71–87 (2015)
71. E. van Erp, R. Yuncken, On the tangent groupoid of a filtered manifold. Bull. Lond. Math. Soc. **49**(6), 1000–1012 (2017)
72. R. Wang, On integrable systems & rigidity for PDEs with symmetry. arXiv:1712.00808 (2017)

Canonical Geometric and Algebraic Structures Hidden Behind a Singular Foliation

Camille Laurent-Gengoux, Ruben Louis, and Leonid Ryvkin

Throughout this chapter, we take many sophisticated ideas and break them down into simpler parts to explain the hidden structures of a singular foliation. We begin in Sect. 1 with the concept of "anchored bundles" in the context of a singular foliation, and proceed to present their morphisms and equivalences. This part concentrates exclusively on the $C^\infty(M)$-module structure of singular foliation $\mathcal{F}$. In Sect. 2, we go further by adding a bracket to an anchored bundle. This brings us to a concept known as "almost Lie algebroid" associated to a singular foliation. This part now makes use of the Lie bracket. Subsequently, in Sect. 3, we discuss the notion of "isotropy Lie algebra and holonomy Lie algebroid" of Androulidakis-Skandalis. In Sect. 4, we discuss the concept of "bisubmersions", also introduced by Androulidakis and Skandalis. These ideas help to explain how to define the "holonomy groupoid" of a singular foliation in Sect. 5.1. In Sect. 6, we discuss the notion of geometric resolution of a singular foliation (again, this uses only the structure of module over functions of a singular foliation), while Sect. 7

C. Laurent-Gengoux (✉)
Université de Lorraine, CNRS, IECL, Metz, France
e-mail: camille.laurent-gengoux@univ-lorraine.fr

R. Louis
Department of Mathematics, Jilin University, Changchun, Jilin, China

Institut für Mathematik, Georg-August-Universität Göttingen, Göttingen, Germany
e-mail: ruben.louis@mathematik.uni-goettingen.de

L. Ryvkin
Université Claude Bernard Lyon 1, Villeurbanne, France

University of Göttingen, Göttingen, Germany
e-mail: ryvkin@math.univ-lyon1.fr

© The Author(s), under exclusive license to Springer Nature Switzerland AG 2025
M. Garcia-Fernandez et al. (eds.), *Advances in Poisson Geometry*,
Advanced Courses in Mathematics - CRM Barcelona,
https://doi.org/10.1007/978-3-031-86657-9_4

expends the notion of almost Lie algebroid over a singular foliation to something more general called the "universal Lie ∞-algebroid" (or "universal Q-manifold") of a singular foliation.

1 Anchored Bundles over a Singular Foliation

Throughout this section M is a smooth, real analytic or complex manifold.[1] Also, $\mathcal{O}$ stands for the corresponding sheaf of functions on M.

1.1 Anchored Bundles

As we will see, the smooth setting is considerably simpler, and has much better properties. However, we aim to address all possible all settings, as much as we can. The reader not interested in sheaves can, in the discussion below, simply ignore the sheaf vocabulary, and consider compactly supported sections, smooth functions and vector fields instead.

We choose M a manifold in the relevant category. We denote by $\mathcal{O}_\bullet$ or simply by $\mathcal{O}$ the sheaf of functions and by $\mathfrak{X}_\bullet$ the sheaf of vector fields on a manifold M.

> **Definition 1.1 (Anchored Bundle)** An *anchored vector bundle* is a pair (A, ρ) made of a vector bundle $A \to M$, and a vector bundle morphism called its *anchor map*.
>
> $$\begin{array}{ccc} A & \xrightarrow{\rho} & TM \\ \downarrow & & \downarrow \\ M & =\!=\!= & M \end{array}$$

Readers who are exclusively interested in the smooth setting are welcome to bypass the following lines. As we saw about Lie algebroids in Sect. 4.2 in chapter "What Is a Singular Foliation?",

$$\mathcal{U} \mapsto \rho(\Gamma_\mathcal{U}(A))$$

is a pre-sheaf in the complex and real analytic cases, but it can be sheafified (in the smooth case, it is always a sheaf, so sheafification is useless). We denote by $\rho(\Gamma(A))$ this sheaf and call it the *image* of $\Gamma(A)$ through ρ. By construction, $\rho(\Gamma(A)) \subseteq \mathfrak{X}_\bullet$ is a sub-sheaf of $\mathcal{O}$-module which is locally finitely generated. Furthermore, it is generated, locally, by a maximum of $\mathrm{rk}(A)$ generators.

---

[1] It could also mostly be an affine variety or a Zarisky open subset of $\mathbb{C}^d$, but we will not detail these cases.

Definition 1.2 (Anchored Bundle over $\mathcal{F}$)
Let $\mathcal{F}$ be a singular foliation on M. We say that an anchored bundle (A, ρ)

1. *terminates within* $\mathcal{F}$ if $\rho(\Gamma(A)) \subseteq \mathcal{F}$,
2. *is over* $\mathcal{F}$ if $\rho(\Gamma(A)) = \mathcal{F}$.

Notice that anchored bundle over $\mathcal{F}$ could be defined for any locally finitely generated sub-sheaf of $\mathfrak{X}_\bullet$. We have not used $[\mathcal{F}, \mathcal{F}] \subset \mathcal{F}$ at this point.

Question 1.3 (Behind a Singular Foliation?) *Let $\mathcal{F}$ be a singular foliation on M.*

1. *Does there always exist an anchored bundle (A, ρ) over $\mathcal{F}$?*
2. *If yes, how unique (= canonical) are they?*
3. *If yes, what properties and additional structures do they have?*

Remark 1.4 For Debord foliations (see Sect. 4.2.4 in chapter "What Is a Singular Foliation?"), an anchored bundle exists on the whole manifold M, by Serre-Swan theorem.

Proposition 1.5 (Answer to the First Part of Question 1.3) *Let $\mathcal{F}$ be a singular foliation on M.*

1. *If $\mathcal{F}$ is finitely generated, then there exists an anchored bundle (A, ρ) over $\mathcal{F}$, and A can be chosen to be a trivial vector bundle. In particular, an anchored bundle exists in a neighborhood of any point.*
2. *In the smooth setting, the following points are equivalent.[a]*
   (i) $\mathcal{F}$ is finitely generated.[b]
   (ii) There exists an anchored bundle (A, ρ) over $\mathcal{F}$.

---
[a] Notice that a statement equivalent to this one was already established in Proposition 2.37 in chapter "What Is a Singular Foliation?". We give here an alternative proof.
[b] See Sect. 2.5 in chapter "What Is a Singular Foliation?".

Proof Assume that $\mathcal{F}$ is finitely generated, and $X_1, \ldots, X_r$ are generators. Let A be the trivial vector bundle of rank r, i.e.,

$$A = M \times \mathbb{K}^r \longrightarrow M.$$

Denote the canonical trivialization of A by $a_1 \ldots, a_r$ and define the anchor map by $\rho(a_i) = X_i$ for all $i = 1, \ldots, r$. We have $\rho(\Gamma(A)) = \mathcal{F}$ by construction. This proves the first item of the statement. It also proves the implication $(i) \implies (ii)$. Let us show that $(ii) \implies (i)$. Let (A, ρ) be as in 1. It is a classical theorem in smooth differential geometry that there exists a vector bundle $B \to M$ such that $A \oplus B$ is a trivial vector bundle $E \to M$. Define a vector bundle morphism on that trivial vector bundle by

$$\rho_E : E \xrightarrow{\text{pr}_A} A \xrightarrow{\rho} TM$$

where pr_A is the projection onto A with respect to B. The pair (E, ρ_E) is a trivial vector bundle such that $\rho_E(\Gamma(E)) = \mathcal{F}$. In particular, $\mathcal{F}$ has $\text{rk}(E)$ generators. This concludes the proof. □

1.1.1 Are Two Anchored Bundles over $\mathcal{F}$ Really Different?

Let us define morphisms of anchored bundles—and add an equivalence class of them. *Until the end of the present section, we work in the smooth setting, and deal with the complex or real analytic contexts in remarks or footnotes.*

> **Definition 1.6 (Morphisms and Equivalences)** Let $(A_1 \to M_1, \rho_1)$ and $(A_2 \to M_2, \rho_2)$ be anchored bundles on a smooth manifold M.
>
> 1. We call *morphism of anchored bundles* any vector bundle morphism $\Phi : A_1 \longrightarrow A_2$ over a map $\phi : M_1 \to M_2$ making the following diagram commutative:
>
>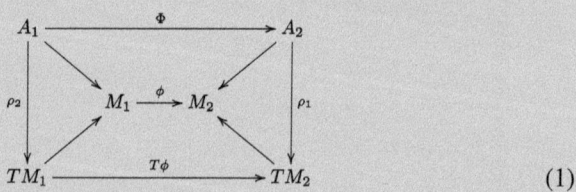
>
> (1)
>
> We speak of an *isomorphism of anchored bundle* when Φ is an isomorphism of vector bundles.
> 2. Two morphisms of anchored bundles Φ, Φ' as in item 1 are said to be *equivalent* if $\rho \circ (\Phi - \Phi') = 0$.

(continued)

Definition 1.6 (continued)

3. An *equivalence of anchored bundles* is a pair of anchored bundle morphisms[a]

$$A_1 \underset{\Psi}{\overset{\Phi}{\rightleftarrows}} A_2 \qquad (2)$$

such that $\Psi \circ \Phi$ and $\Phi \circ \Psi$ are equivalent to the identities of A_1 and A_2.
It is easily checked that both equivalences above are indeed equivalence relations on the class of anchored bundles and their sets of morphisms.

---

[a] In the complex or real analytic settings, an equivalence of vector bundle morphisms shall be a covering $(\mathcal{U}_i)_{i \in I}$ of M and an equivalence (Φ_i, Ψ_i) on each one of the open sets $\mathcal{U}_i$. We also assume Φ_i, Φ_j and Ψ_i, Ψ_j to be equivalent on $\mathcal{U}_i \cap \mathcal{U}_j$.

For a good understanding of the next theorem, recall that an anchored bundle terminates within a singular foliation $\mathcal{F}$ if $\rho(\Gamma(A)) \subset \mathcal{F}$ and is over $\mathcal{F}$ if $\rho(\Gamma(A)) = \mathcal{F}$.

Proposition 1.7 (The Unique (up to Equivalence) Anchored Bundle) *Any two anchored bundles over the same singular foliation are equivalent.*

Proof Let[2] $(A_1 \to M_1, \rho_1)$ and $(A_2 \to M_2, \rho_2)$ be anchored bundles over a singular foliation $\mathcal{F}$. Let $U \subset M$ be an open subset of M and fix a local trivialization $e_1, \ldots, e_r$ of A_1. To define a $\mathcal{O}(U)$-linear map

$$\varphi_U : \Gamma_U(A_1) \longrightarrow \Gamma_U(A_2)$$

such that $\rho_1(a) = \rho_2(\varphi(a))$ for every $a \in \Gamma_U(A_1)$, it suffices to map e_i to any $f_i \in \Gamma_U(A_2)$ such that $\rho_2(f_i) = e_i$ for all $i = 1, \ldots, r$, then extend by linearity. Likewise, we have a map

$$\psi_U : \Gamma_U(A_2) \longrightarrow \Gamma_U(A_1)$$

---

[2] The reader used to an algebraic point of view can prove this statement in one sentence: "$\Gamma(A_2)$ is a projective module over functions".

In the smooth case, we use partition of unity to glue these local maps to a global one.[3] It is straightforward that those define an equivalence. □

Exercise 1.8 Let L be an embedded leaf of a singular foliation L on a smooth manifold M. Let $A_L \to L$ be the holonomy Lie algebroid.[4] Show that there exists a neighborhood $\mathcal{U}$ of L in M, equipped with a submersion $p \colon \mathcal{U} \to L$, on which there exists an anchored bundle of the form (A, ρ) with $A = p^*A_L$ while ρ is an anchor whose restriction to L coincides with the anchor bundle of the Lie algebroid A_L.

1.1.2 Leaves of an Anchored Bundle

Up to this point, we have mainly relied on the fact that $\mathcal{F}$ is a locally finitely generated module over functions. The property of stability under the Lie bracket has not played a role yet. Here is, however, a first result that makes use of leaves.

Proposition 1.9 *Let (A, ρ) be an anchored bundle over a singular foliation $\mathcal{F}$. Any two points in the same leaf have neighborhoods on which the restrictions of (A, ρ) are isomorphic.*

Proof An even stronger statement will in fact be proven, namely Proposition 2.11, which immediately implies this one. □

2 Almost Lie Algebroids: Definition and Existence

The existence and (up to equivalence) uniqueness of an anchored bundle over a singular foliation now clarified. Now comes the third part of Question 1.3: What kind of structure does this bundle possess? Here, we suggest a potential candidate.

> **Definition 2.1 (Almost Lie Algebroids [17])** Let (A, ρ) be an anchored vector bundle over a smooth, real analytic or complex manifold M. We call *almost-Lie algebroid structure* a skew-symmetric bilinear (over $\mathbb{K}$) map
>
> $$[\cdot, \cdot]_A : \Gamma(A) \wedge \Gamma(A) \longrightarrow \Gamma(A)$$

(continued)

---

[3] The gluing cannot be completed in the complex setting, but it is not needed in view of the definition suggested in the footnote of Definition 1.6.
[4] See Sect. 3.6.

Definition 2.1 (continued)
that satisfies the *Leibniz identity*,

$$[x, fy]_A = \rho(x)[f]\, y + f[x, y]_A, \quad \text{for all } x, y \in \Gamma(A),\, f \in \mathcal{O}(M) \qquad (3)$$

and the *anchor condition*:

$$\rho([x, y]_A) = [\rho(x), \rho(y)], \quad \text{for all } x, y \in \Gamma(A). \qquad (4)$$

Remark 2.2 In the definition of an almost Lie algebroid, we do not assume $[\cdot, \cdot]_A$ to satisfy the Jacobi identity, i.e., for all $x, y, z \in \Gamma(A)$, the *Jacobiator*

$$J(x, y, z) = [x, [y, z]_A]_A + [y, [z, x]_A]_A + [z, [x, y]_A]_A$$

does not vanish. When it does, it turns to a Lie algebroid whose image through the anchor map is $\mathcal{F}$. However, it satisfies for any sections $x, y, z \in \Gamma(A)$:

$$\rho\left(J(x, y, z)\right) = 0. \qquad (5)$$

The following Lemma makes almost Lie algebroids a good candidate to answer item 3 in Question 1.3.

Lemma 2.3 *For every almost-Lie algebroid on $(A \twoheadrightarrow M, \rho, [\cdot, \cdot]_A)$, the image of the anchor map $\rho(\Gamma(A)) \subseteq \mathfrak{X}_\bullet$ is a singular foliation on M.*

Proof It is an immediate consequence of the anchor condition. □

We can now answer the third point of Question 1.3. We learned from Marco Zambon the following result (the earliest written form we found is Proposition 2.1.4 of [4]):

Proposition 2.4 (Almost Lie Algebroids) *Every finitely generated foliation on M is the image under the anchor map of an almost-Lie algebroid.*
   In the smooth case, moreover,

1. *Every anchored vector bundle (A, ρ) over M such that $\rho(\Gamma(A)) = \mathcal{F}$ can be endowed with an almost-Lie algebroid bracket.*

(continued)

Proposition 2.4 (continued)
2. *A singular foliation is the image under the anchor map of an almost-Lie algebroid if and only if it is finitely generated.*

Proof Let $X_1, \ldots, X_r$ be generators of $\mathcal{F}$. By Lemma 2.34 and Exercise 2.35 in chapter "What Is a Singular Foliation?", there exist functions[5] c_{ij}^k such that

$$[X_i, X_j] = \sum_{k=1}^n c_{ij}^k X_k \text{ and } c_{ji}^k = -c_{ij}^k.$$

Let A be the trivial vector bundle over M with fibers $\mathbb{K}^r$, and let $e_1, \ldots, e_r$ be the canonical trivialization of this vector bundle. We define the almost Lie algebroid anchor and brackets on generators by

$$\begin{cases} \rho(e_i) = X_i \\ [e_i, e_j]_A = \sum_{k=1}^r c_{ij}^k e_k \end{cases}$$

and extend them using linearity (for the anchor) or Leibniz identity (for the bracket). This is easily checked to be an almost Lie algebroid.

The second part of the statement (i.e., the smooth case) comes from the observation that almost Lie algebroid brackets on a given anchored bundle (A, ρ) can be glued using a partition of unity. More precisely, given an anchored bundle $(A \to M, \rho)$, a partition of unity $(\chi_i)_{i \in I}$ relative to an open cover $(\mathcal{U}_i)_{i \in I}$, and almost Lie algebroid brackets $[\cdot, \cdot]_i$ (relative to ρ) on $\mathcal{U}_i$ for all $i \in I$, the following expression:

$$[\cdot, \cdot] = \sum_{i \in I} \chi_i [\cdot, \cdot]_i$$

is an almost Lie algebroid bracket[6] on $\cup_{i \in I} \mathcal{U}_i$—relative to the anchor ρ. □

We call an almost Lie algebroid that terminates, as an anchored bundle, within a given singular foliation $\mathcal{F}$ on M an *almost Lie algebroid that terminates within* $\mathcal{F}$. Let us turn

---

[5] Called Christoffel symbols in Lemma 2.34, Exercise 2.35 and in the discussion around Definition 2.32 in chapter "What Is a Singular Foliation?". See also Proposition 1.12 in chapter "State of the Art and Open Questions".

[6] This would be totally wrong with Lie algebroids: this comes from the fact that Jacobi identity is quadratic in the bracket, while the Leibniz identity and the anchor condition are linear with respect to the bracket, once the anchor is fixed.

it into a category by defining morphisms. In fact, we will only deal with morphisms over the identity of M, which are much simpler. The subtlety is that we do not assume morphisms of almost Lie algebroid structures to be compatible with the bracket, but only to be compatible with the anchor! This is absolutely counter-intuitive, but makes perfect sense having Lie ∞-algebroids[7] in mind.

Definition 2.5 Let M be a manifold.

1. We call *morphism of almost Lie algebroids* morphisms of anchored bundles—forgetting the almost Lie algebroid bracket.
2. Two such morphisms are equivalent if and only if they are equivalent as anchored bundle morphisms.
3. In particular, an equivalence between almost Lie algebroids is simply an equivalence of their underlying anchored bundles.

This deserves justification: why did we not require that "morphisms" respect the almost Lie algebroid brackets? The answer comes from the following proposition that says that they automatically do, up to an element in the kernel of the anchor.

Proposition 2.6 *Let* $(A_1, [\cdot, \cdot]_{A_1}, \rho_1)$ *and* $(A_2, [\cdot, \cdot]_{A_2}, \rho_2)$ *be almost Lie algebroids that terminate within the same singular foliation* $\mathcal{F}$. *For any morphism* Φ *from the first one to the second one:*

$$[\Phi(a), \Phi(b)]_{A_2} - \Phi([a, b]_{A_1}) \in \ker(\rho_2)$$

Proof By definition of an almost Lie algebroid:

$$\rho_2([\Phi(a), \Phi(b)]_{A_2} - \Phi([a, b])) = [\rho_2 \circ \Phi(a), \rho_2 \circ \Phi(b)]_{A_2} - \rho_2 \circ \Phi([a, b]))$$
$$= [\rho_1(a), \rho_1(b)]_{A_2} - \rho_1([a, b]))$$
$$= 0$$

This proves the claim. □

Let us conclude this section by a theorem that follows from Propositions 2.4 and 2.7.

Proposition 2.7 *Any two almost Lie algebroids over a finitely generated singular foliation* $\mathcal{F}$ *are equivalent. Moreover, any almost Lie algebroids that terminates within* $\mathcal{F}$ *admits*

---

[7] Lie ∞-algebroid is introduced in Sect. 7.4.

a morphism to any almost Lie algebroids over $\mathcal{F}$, and this morphism is unique up to equivalence.

This means that, given a singular foliation $\mathcal{F}$, in the category where

1. objects are almost Lie algebroids that terminates within $\mathcal{F}$ and
2. arrows are equivalence classes of morphisms of almost Lie algebroids,

the terminal[8] objects are almost Lie algebroids over $\mathcal{F}$.

Exercise 2.8 Let φ be a function on a smooth manifold M.

1. Show that the module $\mathcal{F}_{d\phi}$ generated by vector fields of the form:

$$\{X[\phi]Y - Y[\phi]X \mid X, Y \in \mathfrak{X}_c(M)\}$$

form a singular foliation.
2. Show that $(\wedge^2 TM, \rho = i_{d\phi})$ (i.e., the bivectors, equipped with the contraction by the 1-form $d\phi$) is an anchored bundle over $\mathcal{F}_{d\phi}$
3. Is it true[9] that $[P, Q] := \mathcal{L}_{\rho(P)}Q$, with P, Q bivector fields, satisfies $\rho([P, Q]) = [\rho(P), \rho(Q)]$. Is it is an almost Lie algebroid bracket on $(\wedge^2 TM, \rho = i_{d\phi})$?
4. For $M = \mathbb{R}^n$, show that there exists an almost Lie algebroid bracket on $(\wedge^2 TM, \rho = i_{d\phi})$ whose restriction to constant bivector fields is given by $[\cdot, \cdot]$.

Hint: A related example is dealt with in Example 3.13 in [23] and in Section 3.2.1 in [22].

In [1], Androulidakis et al. introduced the Helffer-Nourrigat cone. This is a very important object, that can be described easily out of the notion of an anchored bundle, as we do in the first question of the next exercise (the subsequent questions involve notions that will be seen only later on in the text).

Exercise 2.9 Let $(M, \mathcal{F})$ be a smooth singular foliation such that all regular leaves have the same dimension r. We denote by $M_{\text{reg}} \subset M$ the open dense subset of all regular points of $\mathcal{F}$.

1. Let (A, ρ) be an anchored bundle over $\mathcal{F}$. Let $\rho^* \colon T^*M \to A^*$ be the dual of the anchor map. We call *Helffer-Nourrigat cone, computed with respect to* (A, ρ), the closed subset

---

[8] As we will see, these properties are a sort of "toy-model" for the properties of the universal Lie ∞-algebroid of Sect. 7.
[9] $\mathcal{L}$ stands for the Lie derivative.

of A^* given by

$$\mathrm{NH}_A(\mathcal{F}) := \overline{\mathrm{Im}(\rho^*)|_{M_{\mathrm{reg}}}} = \overline{\coprod_{m \in M_{\mathrm{reg}}} \rho_m^*(T_m^*M)}.$$

The horizontal bar refers to the closure in the usual topology. We denote by $\pi : \mathrm{NH}_{(A,\rho)}(\mathcal{F}) \to M$ the restriction to the Helffer-Nourrigat cone of the projection $A^* \to M$. Show that

(a) Show that for every $m \in M_{\mathrm{reg}}$, the fiber of π over M coincides[10] with $\mathrm{Im}(\rho_m^*) = (\ker(\rho_m))^\perp$.

(b) Show that for every $m \in M$, the fiber of π is non-empty, and is contained the annihilator of the strong kernel[11] of ρ at m.

(c) Show that for every $m \in M$, the fiber of π is a union of sub-vector spaces of A_m^*, all of dimension r.

(d) (*For Poisson geometers*) Show that if (A, ρ) admits a Lie algebroid bracket, the Helffer-Nourrigat cone is a union of symplectic leaves of the Poisson structure on A^* associated to the Lie algebroid bracket.

We will now explain how the Helffer-Nourrigat cone can be seen as a subset of $\coprod_{L \in \mathrm{Leaves\ of\ } \mathcal{F}} A_L^*$. Here A_L is the holonomy Lie algebroid of a leaf L, defined in Sect. 3.6.

2. Use question 1.b to show that for every $m \in M$, the fiber of π over m can be included into $A_L^*|_m$. *Hint*: We recall that $A_L|_m$ is the quotient of A_m by the strong kernel of ρ at m.

3. Show that the image of the inclusion in question 2. is made of the union, for all $m \in M$, of the annihilator of all limit subalgebras[12] at m.

We denote this set by $\mathrm{HN}(\mathcal{F})$, without reference to a particular anchored bundle, since by the previous two questions, it does depend on the choice of an anchored bundle. We call it the *Helffer-Nourrigat cone* of $\mathcal{F}$.

The presentation above seems different from the presentation done in [1] but the difference is only a difference of presentation, see [27]. We acknowledge discussions with Fani Petalidou and Mohsen Masmoudi when writing this exercise.

---

[10] We use the symbol $\perp$ for the annihilator.

[11] See Sect. 3.

[12] See Definition 5.58 in chapter "What Is a Singular Foliation?". The limit Lie algebras being vector subspaces of the isotropy Lie algebra $\mathfrak{g}_m(\mathcal{F})$, which is included into $A_L|_m$, can be seen as vector subspaces of $A_L|_m$. Since the limit Lie algebras are precisely the points of the Nash resolution (which is a sub-set of the Grassmannian of $\mathfrak{g}_m(\mathcal{F})$, see Sect. 5.8 in chapter "What Is a Singular Foliation?"), we can also say the fiber of the Helffer-Nourrigat cone over a point m is the union of all annihilators of the "points" in fiber over m of the Nash resolution.

Exercise 2.10 Show that the Helffer-Nourrigat cone for the singular foliation on $M = \mathbb{R}^n$ made of all vector fields vanishing at the origin is given as follows:

1. if m is not the origin, the fiber of the Helffer-Nourrigat cone over m is $\simeq \mathbb{R}^n$,
2. if m is the origin, the fiber is made of square $n \times n$ matrices of rank ≤ 1.

We thank Cédric Rigaud for this exercise.

2.1 An Alternative Proof of Proposition 7.11 in Chapter "What Is a Singular Foliation?"

We use the notions of anchored bundle and the almost Lie algebroids to give a much simpler proof of a result that was crucial to establish the existence of leaves: Proposition 7.11 in chapter "What Is a Singular Foliation?". This proof is inspired from a proof that appeared in [4] by Henrique Bursztyn, Hudson Lima, and Eckhard Meinrenken for almost-Lie algebroids and internal symmetries. The result that we will prove, and which immediately implies Proposition 7.11 in chapter "What Is a Singular Foliation?", is the following proposition.[13]

Proposition 2.11 *Let $\mathcal{F}$ be a singular foliation on M. Choose $Y \in \mathfrak{X}(M)$ be a vector field such that $[Y, \mathcal{F}] \subset \mathcal{F}$. For any open subset $\mathcal{U} \subset M$ on which the time 1-flow ϕ_1^Y of Y is well-defined, the restrictions of any anchored bundle (A, ρ) over M to $\mathcal{U}$ and $\phi_t^Y(\mathcal{U})$ are isomorphic.*[14]

These isomorphisms can be seen, when A is a trivial bundle, as families indexed by $m \in M$ of invertible matrices as in Proposition 7.11 in chapter "What Is a Singular Foliation?". Proposition 2.11 is therefore a generalization of Proposition 7.11 in chapter "What Is a Singular Foliation?".

The proof is based on the notion of linear vector field. A vector field Y on a vector bundle $E \xrightarrow{p} M$ is said to be *linear* if one of the following equivalent conditions holds:

(i) For any function f on E whose restriction to any fiber of $p: E \to M$ is a polynomial of degree $\leq k$, $Y[f]$ is a polynomial of degree $\leq k$.
(ii) $Y[p^*\mathcal{O}] \subset p^*\mathcal{O}$ and $Y[\Gamma(A^*)] \subset \Gamma(E^*)$, with the understanding that $\Gamma(A^*)$ must be considered as a smooth function on E linear on each fiber of $p: E \to M$.

---

[13] We work in the smooth setting in the present section: it can be adapted to the complex and algebraic settings, but not to the algebraic one.
[14] We insist on "isomorphic" and not "equivalent". Of course, it is part of the statement that the base map of the isomorphism is ϕ_1^X.

(iii) In any local coordinates $(x_1, \ldots, x_n, y_1, \ldots, y_r)$ on E, with $(x_i)_{i=1,\ldots,n}$ being local coordinates on the base manifold M, and $(y_j)_{j=1,\ldots,r}$ linear coordinates on the fibers, the vector field Y is of the form:

$$Y = \sum_{i=1}^{n} A_i(x_1, \ldots, x_n) \frac{\partial}{\partial x_i} + \sum_{i,j=1}^{n} B_{ij}(x_1, \ldots, x_n) y_i \frac{\partial}{\partial y_j}.$$

Linear vector fields on $E \to M$ are stable under Lie bracket. Moreover, for any linear vector fields Y, there is a unique vector field[15] $p_*Y \in \mathfrak{X}(M)$ such that

$$Y[p^*F] = p^*(p_*Y)[F] \text{ for every } F \in \mathcal{O}_M,$$

and the assignment $Y \mapsto p_*Y$ is a Lie algebra morphism that we will call projection.

Lemma 2.12 *Let* $E \xrightarrow{p} M$ *be a vector bundle and* $X \in \mathfrak{X}(M)$ *be a vector field. For any linear map:*

$$\delta_X \colon \Gamma(E^*) \longrightarrow \Gamma(E^*)$$

such that for every function $f \in \mathcal{O}_M$ *and every* $\epsilon \in \Gamma(E^*)$:

$$\delta_X(f\epsilon) = f \delta_X(\epsilon) + X[f] \epsilon \tag{6}$$

there exists a unique linear vector field on E *that projects on* X *and whose restriction to fiberwise linear functions on* E *is* δ_X.

The Lie algebra of linear vector fields on $E \to M$ can be seen as the Lie algebra of the group of vector bundle isomorphism of $E \to M$. Below is a more precise statement, that we leave to the reader:

Lemma 2.13 *Let* Y *be a linear vector field on a vector bundle* $E \xrightarrow{p} M$, *and* p_*Y *the vector field on* M *to which it is p-related. The flow* ϕ_t^Y *at time t of a linear vector field is defined if and only the flow at time t of its projection* p_*Y *on* M *is defined. In that case, it is vector bundle isomorphism*

$$\begin{array}{ccc} E & \xrightarrow{\phi_t^Y} & E \\ \downarrow & & \downarrow \\ M & \xrightarrow{\phi_t^{p_*Y}} & M \end{array}.$$

---

[15] I.e., a unique vector field p-related to Y, if one uses the terminology of Sect. 5.2 in chapter "What Is a Singular Foliation?".

Proof of Proposition 2.11 Let Y be a vector field satisfying $[Y, \mathcal{F}] \subset \mathcal{F}$. Let us first observe that the classical Lie derivative:

$$\delta_Y^{TM} : \Gamma(T^*M) \to \Gamma(T^*M)$$
$$\alpha \mapsto L_Y \alpha$$

satisfies condition (6) and therefore defines a linear vector field $\widehat{Y}^{TM}$ on $TM \xrightarrow{p} M$ which is p-related to Y. This vector field is called the tangent prolongation of Y. Our goal is to construct a linear vector field on A which is ρ-related with $\widehat{Y}^{TM}$.

Let us first assume that $A = M \times \mathbb{R}^r$ is a trivial bundle.

- In this case, there is a matrix-valued smooth function $\mathfrak{Y}$ on M such that $[Y, \rho(e_i)] = \sum_{j=1}^{r} \mathfrak{Y}_{i,j} \rho(e_j)$ for every $i \in \{1, \ldots, r\}$. This allows to define $\gamma_Y^A : \Gamma(E) \to \Gamma(E)$ by

$$\gamma_Y^A \left(\sum_{i=1}^{r} f_i e_i \right) = \sum_{i=1}^{r} Y[f_i]\, e_i + \sum_{i,j=1}^{r} f_i \mathfrak{Y}_{i,j}\, e_j$$

for every r-tuple of functions $f_1, \ldots, f_r$. This construction imposes that γ_Y^A adheres to Eq. (6).

- Next, we construct a "dual" of γ_Y^A, that we denote by $\delta_Y^A : \Gamma(A^*) \to \Gamma(A^*)$, by imposing the following duality relation:

$$\left\langle \delta_Y^A(\alpha) \big| e \right\rangle := \gamma_Y^A[\langle \alpha, e \rangle] - \left\langle \alpha, \gamma_Y^A(e) \right\rangle. \tag{7}$$

for every $e \in \Gamma(A), \alpha \in \Gamma(A^*)$. This definition requires a justification: the right-hand term in Eq. (7) is $C^\infty(M)$-linear in e, so that there exists a section $\delta_Y^A(\alpha)$ that satisfies Eq. (7). Moreover, $\alpha \to \delta_Y^A(\alpha)$ satisfies for any $f \in C^\infty(M)$ the relation
$\delta_Y^A(f\alpha) = Y(f)\alpha + f\delta_Y^A(\alpha)$.

- In view of Lemma 2.12 therefore, δ_Y^A defines a linear vector field $\widehat{Y}^A$ p-related to Y. Let us check that $\widehat{Y}^A$ is ρ-related to $\widehat{Y}^{TM}$, i.e.,

$$\widehat{Y}^A \circ \rho^*(F) = \rho^* \circ \widehat{Y}^{TM}(F)$$

for every $F \in C^\infty(TM)$. For $F \in p^*C^\infty(M)$, the condition holds automatically true. It suffices therefore to check that the condition holds true when applied on a fiberwise linear function, i.e., a 1-form $\eta \in \Omega^1(M)$. In view of the definition of $\widehat{Y}^A$ and $\widehat{Y}^{TM}$ on such function, it therefore suffices to check that

$$\delta_Y^A \circ \rho^*(\eta) = \rho^* \circ L_Y(\eta)$$

Let us check this relation: for every $\eta \in \Gamma(T^*M)$, $a \in \Gamma(A)$, we have by construction

$$\left\langle \delta_Y^A(\rho^*\eta), a \right\rangle = Y\left[\langle \rho^*\eta, a\rangle\right] - \langle \rho^*\eta, \gamma_Y^A(a)\rangle$$

$$= Y[\langle \eta, \rho(a)\rangle] - \langle \eta, \rho(\gamma_Y^A(a))\rangle$$

$$= L_Y \circ \iota_{\rho(a)} \eta - \iota_{\rho(\gamma_Y^A(a))} \eta$$

$$= \iota_{\rho(a)} \circ L_Y \eta + \iota_{[Y,\rho(a)]-\rho(\gamma_Y^A(a))} \eta$$

We therefore have

$$\left\langle \left(\delta_Y^A \circ \rho^* - \rho^* \circ L_Y\right) \eta, a \right\rangle = \left\langle \eta, [Y, \rho(a)] - \rho\left(\gamma_Y^A(a)\right) \right\rangle$$

By definition of γ_Y^A, the previous relation holds true if $a = e_i$ belongs to the canonical trivialization of the trivial bundle $A \to M$. By $C^\infty(M)$-linearity, it holds true for any $a \in \Gamma(A)$. This proves that $\hat{Y}^A$ is ρ-related with $\hat{Y}^{TM}$.

Let us now turn to the case of a non-trivial anchored bundle $(A \to TM, \rho)$. By choosing a second vector bundle E such that $A' = A \oplus E$ is a trivial vector bundle and setting $\rho_{A'} = \rho_A \circ \pi_A$, with π_A the projection onto A, we obtain a new anchored bundle. Since it is trivial, we can apply the previous considerations and obtain $\delta_Y^{A'}$. We now define $\delta_A^Y : \Gamma(E^*) \to \Gamma(E^*)$ by:

$$\delta_Y^A(\alpha)(e) = \delta_Y^{A'}(\pi_A^*\alpha)(i_{A,*}e),$$

where π_A and i_A are the projection and inclusion of $A \subset A'$. This map can also be obtained from $\pi_A \circ \gamma_Y^{A'} \circ i_A$ with the dualization procedure, as in Eq. (7). We can now verify:

$$\delta_Y^A \rho_A^* \beta(e) = \delta_Y^{A'}(\pi_A^* \rho_A^* \beta)(i_{A,*}e) = \delta_Y^{A'}(\rho_{A'}^* \beta)(i_{A,*}e)$$

$$= (\rho_{A'}^* L_Y \beta)(i_{A,*}e) = (i_A^* \rho_{A'}^* L_Y \beta)(e)$$

$$= (\rho_A^* L_Y \beta)(e)$$

i.e., the corresponding linear vector field $\hat{Y}^A$ is ρ-related to $\hat{Y}^{TM}$. In particular, where defined, the flow of $\hat{Y}^A$ induces an isomorphism of anchored bundles. □

3 Isotropy Lie Algebra and Holonomy Lie Algebroids

Let us use the almost Lie algebroids associated to a singular foliation in the previous section to associate a Lie algebra, called isotropy Lie algebra, to any point of a singular foliation. We then relate it with the original definition of Androulidakis and Skandalis.

We work here in the smooth setting. All arguments can easily be adapted to the complex or real analytic ones.

3.1 Kernel and Strong-Kernel of a Morphism of Vector Bundles

Consider a vector bundle morphism over the identity of M:

$$\begin{array}{ccc} B & \xrightarrow{\Phi} & C \\ \downarrow & & \downarrow \\ M & = & M \end{array}$$

Choose a point $m \in M$. There are two subspaces in B_m that deserve to be called "kernels".

1. the usual kernel $\ker(\Phi_m)$, i.e.,

$$\{u \in B_m \mid \Phi_{|m}(u) = 0\},$$

2. and there is the *strong kernel*, i.e., the subspace $\mathrm{Sker}(\Phi, m) \subset B_m$ of all elements through which there is a neighborhood $\mathcal{U}$ of m in M and a local section in the kernel of $\Phi: \Gamma_{\mathcal{U}}(B) \to \Gamma_{\mathcal{U}}(C)$. In equation:

$$\mathrm{Sker}(\Phi, m) := \{u \in B_m \text{ s.t. } \exists U \in \Gamma(B) \text{ with } \Phi(U) = 0 \text{ and } U_{|m} = u\}.$$

Of course, there is an inclusion:

$$\mathrm{Sker}(\Phi, m) \subset \ker(\Phi|_m).$$

Moreover, the dimensions of the distributions have opposite behavior:

1. the map $m \mapsto \dim(\ker(\Phi_{|m}))$ is upper semi-continuous, i.e., if a sequence (x_n) in M has limit x, then

$$\dim(\ker(\Phi_{|x})) \geq \text{ upper limit of } \dim(\ker(\Phi_{|x_n}))$$

2. the map $m \mapsto \dim(\text{Sker}(\Phi, m))$ is lower semi-continuous, i.e., if a sequence (x_n) in M has limit x, then

$$\dim(\text{Sker}(\Phi, x)) \leq \text{lower limit of } \dim(\text{Sker}(\Phi, x_n)).$$

In particular, if kernel and strong kernel coincide at a point, they coincide in a neighborhood of that point. In particular, their dimensions are constant at all points in a neighborhood, so that they form a vector sub-bundle. Let us state this conclusion for future reference.

Proposition 3.1 *Let $\Phi \colon B \to C$ be a vector bundle morphism over the identity of M. For any $m \in M$, the following two assertions are equivalent:*

1. *the kernel and the strong kernel coincide at m.*
2. *There is a neighborhood $\mathcal{U}$ of m in M on which the kernel and the strong kernel coincide at all points.*

In this case, moreover, these coinciding kernels form a sub-vector bundle of the restriction to $\mathcal{U}$ of B.

3.2 The Isotropy Lie Algebra (I: The Space)

Let $\mathcal{F}$ be a singular foliation on a manifold M. Let $\mathcal{U}$ be an open neighborhood of m on which $\mathcal{F}$ is finitely generated. In view of Proposition 2.4, there exists an anchored bundle $(A \to \mathcal{U}, \rho)$ over $\mathcal{F}$. We call *isotropy vector space at m* the quotient space:

$$\mathfrak{g}_m(\mathcal{F}) = \frac{\ker(\rho_m)}{\text{Sker}(\rho, m)}.$$

Notice that the notation $\mathfrak{g}_m(\mathcal{F})$ makes no reference to the chosen anchored bundle. This is justified by the following proposition:

Proposition 3.2 (The Isotropy Vector Space at m Makes Sense) *Let $\mathcal{F}$ be a singular foliation. The isotropy vector spaces associated to any two anchored bundles are canonically isomorphic.*[a]

---

[a] It makes sense, therefore, to denote it by $\mathfrak{g}_m(\mathcal{F})$.

Proof This is an immediate consequence of Theorem 1.7, since a morphism of anchored bundle morphisms over M maps kernel to kernel and Strong kernel to Strong kernel, and since homotopic morphisms induce the same map at the quotient level. □

Here is an important theorem, due to [2].

Theorem 3.3 (Ranks and Dimensions) *Let $(M, \mathcal{F})$ be a singular foliation. For every $m \in M$,*

1. *the rank $\mathrm{rk}_m(\mathcal{F})$ of $\mathcal{F}$ at m (i.e., the minimal number of local generators),*
2. *the dimension $\dim(L_m)$ of the leaf through m,*
3. *and the dimension $\dim(\mathfrak{g}_m(\mathcal{F}))$ of the holonomy vector space[a] at m,*

are related by the relation

$$\mathrm{rk}_m(\mathcal{F}) = \dim(\mathfrak{g}_m(\mathcal{F})) + \dim(L_m).$$

---

[a]That will be soon equipped with a Lie bracket making it Androulidakis-Skandalis isotropy Lie algebra.

Proof For simply notations, we set $r = \mathrm{rk}_m(\mathcal{F})$, $g := \dim(\mathfrak{g}_m(\mathcal{F}))$, and $\ell = \dim(L_m)$. Let $Y_1, \ldots, Y_r$ be a minimal family local generators of $\mathcal{F}$ near m. Without any loss of generality, one can assume that $Y_1|_{r-\ell+1}, \ldots, Y_r|_m$ form a basis of $T_m L_m$ while $Y_1, \ldots, Y_{r-\ell}$ are all zero at the point m.

Since these vectors vanish at m, the classes $[Y_1], \ldots, [Y_{r-\ell}]$ of these vectors in $\mathfrak{g}_m(\mathcal{F})$ are well defined, and, altogether, these classes form a system of generators of $\mathfrak{g}_m(\mathcal{T})$. Hence $g \leq r - \ell$. It remains to show that the classes $[Y_1], \ldots, [Y_{r-\ell}] \in \mathfrak{g}_m(\mathcal{T})$ are linearly independent. Assume that they are not, i.e., (without loss of generality): $[Y_1] = \sum_{i=2}^{r} \alpha_i [Y_i]$ for some $\alpha_i \in \mathbb{K}$. This means that $Y_1 = \sum_{i=2}^{r} \alpha_i Y_i + Y$ for some Z in $\mathcal{I}_m \mathcal{F}$. This element Z reads as a sum $Z = \sum_{i=1}^{r} g_i Y_i$ for some functions g_i that vanish at m, so that we obtain

$$(1 - g_1) Y_1 = \sum_{i=2}^{r-\ell} (\alpha_i + g_i) Y_i + \sum_{i=1}^{\ell} g_i Y_{r-\ell+i}$$

Now, there exists a neighborhood of m where we can invert $(1-g_1)$. On that neighborhood, Y_1 is a linear combination of the other Y's which contradicts the minimality of the family $Y_1, \ldots, Y_r$. This concludes the proof.

3.3 The Isotropy Lie Algebra II: The Bracket

Now, let $\mathcal{F}$ be a singular foliation, m is point, $\mathcal{U}$ an open subset containing m and $(A \to \mathcal{U}, \rho)$ an anchored bundle of $\mathcal{F}$ restricted to $\mathcal{U}$.

According to Proposition 2.4, $(A \to \mathcal{U}, \rho)$ can be equipped with an almost Lie algebroid bracket[16] that we denote by

$$[\cdot, \cdot]_A : \Gamma(A) \times \Gamma(A) \longrightarrow \Gamma(A).$$

Let us fix a point $m \in M$. Consider two elements $a, b \in A_m$. For any two sections $\tilde{a}, \tilde{b}$ of $A \to \mathcal{U}$ through a, b. In view of the Leibniz identity, the value at m of the almost Lie algebroid bracket $[\tilde{a}, \tilde{b}]_A$ depends on the 1-jet at m of the sections $\tilde{a}, \tilde{b}$. However, if $\rho(a) = 0$, then

$$[\tilde{a}, f\tilde{b}]|_m = f(m)[\tilde{a}, \tilde{b}]|_m + \rho(a)[f]b$$
$$= f(m)[\tilde{a}, \tilde{b}]|_m$$

for any local function f. This implies that $[\tilde{a}, \tilde{b}]|_m$ depends only on the value of the section $\tilde{b}$ at m, i.e., depends only on b. As a consequence, if $a, b \in \text{Ker}(\rho_m)$, then $[\tilde{a}, \tilde{b}]|_m$ depends only on a and b so that the bracket $[\cdot, \cdot]_A$ induces a bilinear map

$$[\cdot, \cdot]_{A,m} : \wedge^2 \ker(\rho_m) \longrightarrow A_m$$

given for all $a, b \in \text{Ker}(\rho_m)$ by

$$[a, b]_{A,m} = [\tilde{a}, \tilde{b}]_A(m)$$

for any sections $\tilde{a}, \tilde{b}$ through a and b. Moreover, $[a, b]_{A,m}$ is in fact valued in $\ker(\rho_m)$: this follows easily from the anchor condition. Lastly, the anchor condition implies that the strong kernel at m is an "ideal" of that bracket, i.e., $[\text{Sker}(\rho, m), \ker(\rho_m)]_{A,m} \subset \text{Sker}(\rho, m)$ so that the skew-symmetric bilinear map $[\cdot, \cdot]_{A,m}$ goes to the quotient to a bilinear map

$$[\cdot, \cdot]_m : \wedge^2 \mathfrak{g}_m(\mathcal{F}) \longrightarrow \mathfrak{g}_m(\mathcal{F}) \qquad (8)$$

---

[16] At least in the smooth setting. In complex or real analytic setting, one may have to restrict to a smaller open neighborhood.

Proposition 3.4 (The Lie Bracket) *The bilinear map* (8):

1. *is a Lie bracket on the holonomy vector space* $\mathfrak{g}_m(\mathcal{F})$,
2. *is canonically defined, i.e., does not depend on the choice of an anchored bundle and of an almost Lie algebroid bracket.*

Proof The Jacobi identity follows from the fact the Jacobiator (see Remark 2.2) of three sections of the bracket $[\cdot,\cdot]_A$ lies in the kernel of $\rho \colon \Gamma(A) \to \mathfrak{X}(M)$, see Eq. (5). Now, for any two almost Lie algebroid brackets $[\cdot,\cdot]$ and $[\cdot,\cdot]'$ on A, we have, for any two sections $\tilde{a}, \tilde{b}$

$$\rho\left([\tilde{a},\tilde{b}]'_A - [\tilde{a},\tilde{b}]_A\right) = 0$$

so that $[\tilde{a},\tilde{b}]'_A - [\tilde{a},\tilde{b}]_A$ is valued in the Strong kernel of ρ. This implies that the induced bracket (8) does not depend on the choice of an almost Lie algebroid bracket on a given anchored bundle (A, ρ). More generally, given two anchored bundles (A, ρ) and (A', ρ'), Proposition 1.7 implies the existence of anchored bundle morphisms $\Phi \colon (A, \rho) \longrightarrow (A', \rho')$. For any two almost Lie algebroid structures on A and A', we have by Proposition 2.6

$$\rho'\left(\Phi([\tilde{a},\tilde{b}]_A) - [\Phi(\tilde{a}), \Phi(\tilde{b})]_{A'}\right) = 0$$

so that $\Phi([\tilde{a},\tilde{b}]_A) - [\Phi(\tilde{a}), \Phi(\tilde{b})]_A$ is valued in the Strong kernel of ρ' at m. This implies that Φ induces a Lie algebra morphism

$$(\Phi, m) \colon \frac{\ker(\rho_m)}{\operatorname{Sker}(\rho, m)} \longrightarrow \frac{\ker(\rho'_m)}{\operatorname{Sker}(\rho', m)}.$$

Equivalent morphism would induce the same Lie algebra morphism. The same construction, applied to an "inverse" $\Psi \colon A' \to A$ as in Proposition 1.7, gives an inverse map to that Lie morphism. The result follows □

Definition 3.5 (Isotropy Lie Algebra: Definition) We call *isotropy Lie algebra of* $\mathcal{F}$ *at* $m \in M$ the pair $(\mathfrak{g}_m(\mathcal{F}), [\cdot,\cdot])$ with $\mathfrak{g}_m(\mathcal{F})$ the vector space defined as in Proposition 3.2 $[\cdot,\cdot]$ as in Proposition 3.4.

We end this subsection with the following lemma.

Lemma 3.6 *Let $(M, \mathcal{F})$ be a singular foliation and $x \in M$ such that $T_x\mathcal{F} = \{0\}$. Then, $\mathrm{rk}_x(\mathcal{F}) = \dim \mathfrak{g}_x(\mathcal{F})$ and $\mathcal{F}$ admits on an open neighborhood $\mathcal{U}$ of x an almost Lie algebroid structure on the trivial bundle $A = \mathcal{U} \times \mathfrak{g}_x(\mathcal{F})$ whose anchor is 0 at the point x and whose bracket is given by the Lie bracket of $\mathfrak{g}_x(\mathcal{F})$ at this point.*

Proof If $T_x\mathcal{F} = \{0\}$, then the leaf through x is reduced to a point. By Theorem 3.3 (3), $\mathrm{rk}_x(\mathcal{F}) = \dim \mathfrak{g}_x(\mathcal{F})$. It follows that $\mathcal{F}$ admits a minimal set of generators $X_1, \ldots, X_k$ in a neighborhood $\mathcal{U}$ of m with $k = \dim \mathfrak{g}_x(\mathcal{F})$ elements. that induce a basis $e_1, \ldots, e_k$ of $\mathfrak{g}_x(\mathcal{F})$. Hence, $\mathcal{U} \times \mathfrak{g}_x(\mathcal{F})$ admits an almost Lie algebroid whose image is $\mathcal{F}|_\mathcal{U}$. The bracket is constructed as follows. The functions c_{ij}^k such that

$$[X_i, X_j] = \sum_k c_{ij}^k X_k$$

are not unique but their values $c_{ij}^k(x)$ at x have to coincide with the Christoffel symbol of the Lie algebra $\mathfrak{g}_x(\mathcal{F})$ in the basis $e_1, \ldots, e_k$. The functions c_{ij}^k then define an almost Lie algebroid structure as in Sect. 1 which satisfies the required properties. □

Exercise 3.7 Let m be a point in a foliated manifold $(M, \mathcal{F})$ Show that the following points are equivalent:

(i) m is a regular point,
(ii) $\mathfrak{g}_m(\mathcal{F}) = 0$.

Exercise 3.8 Let $(M, \mathcal{F})$ be a foliated manifold and $m \in M$ a point. Show the following points:

1. Let U be an open subset of M containing m, then $\mathfrak{g}_m(\mathcal{F}) = \mathfrak{g}_m(\mathcal{F}|_U)$.
2. Let $(\tilde{M}, \tilde{\mathcal{F}})$ be another foliated manifold and $\tilde{m} \in \tilde{M}$. Then $\mathfrak{g}_{(m,\tilde{m})}(\mathcal{F} \times \tilde{\mathcal{F}}) = \mathfrak{g}_m(\mathcal{F}) \oplus \mathfrak{g}_{(\tilde{m})}(\tilde{\mathcal{F}})$. (The direct sum should be understood as a Lie algebra direct sum).
3. For $\varphi \colon N \mapsto M$ a surjective submersion, $\mathfrak{g}_n(\varphi^{-1}(\mathcal{F})) = \mathfrak{g}_m(\mathcal{F})$ for every $n \in \varphi^{-1}(m)$.

Exercise 3.9 Several exercises in Sect. 4 in chapter "What Is a Singular Foliation?" consisted into computing explicitly the isotropy Lie algebra at a given singular point. We invite the reader to look for the word "isotropy" is that section.

Exercise 3.10 Let m be a point of a foliated manifold $(M, \mathcal{F})$.

1. Show that the isotropy Lie algebra $\mathfrak{g}_m(\mathcal{F})$ at m is canonically isomorphic to the singular foliation defined by the vectors fields as in item (b) of the splitting Theorem 7.18 in chapter "What Is a Singular Foliation?".
2. Show that the isotropy Lie algebra at any point of L is isomorphic to the isotropy Lie algebra at the origin of any representative of the transverse singular foliation (see Definition 8.8 in chapter "What Is a Singular Foliation?"). *Hint:* This in fact the same question as the previous one, with different wording.
3. Let L be the leaf through m. Show that the isotropy Lie algebras at two points in L are isomorphic (*Hint:* Use Proposition 8.1 in chapter "What Is a Singular Foliation?").
4. Is this isomorphism canonical?
5. *Hard*[17] Show that if the leaf L is simply connected, then this isomorphism is canonical up to an inner Lie algebra automorphism.
6. *Hard.* Construct a group morphism $\pi_1(L, \ell) \to \mathrm{Out}(\mathfrak{g}_m(\mathcal{F}))$, where Out is the Lie group of outer automorphism. *Hint*: this of course related to the outer holonomy constructed in Eq. (39) in chapter "What Is a Singular Foliation?". Can be also seen as a general result on transitive Lie algebroids, then one applies the results of Sect. 3.6.

the isotropy Lie algebras of $\mathcal{F}$ at m and the isotropy Lie algebra of any representative of the transverse singular foliation (as in Theorem 8.6 in chapter "What Is a Singular Foliation?") at ℓ are canonically isomorphic. (*Hint:* use the splitting theorem of Sect. 7.3 in chapter "What Is a Singular Foliation?")

3.4 Androulidakis-Skandalis Construction of the Isotropy Lie Algebra

Let us now present the original definition of the isotropy Lie algebra of a singular foliation by Androulidakis and Skandalis.

Let $\mathcal{F}$ be a singular foliation, defined as Definition[18] 2.1 in chapter "What Is a Singular Foliation?". For every $m \in M$, we consider $\mathcal{F}_m \subset \mathcal{F}$ the sub-Lie algebra of vector fields in $\mathcal{F}$ vanishing at m. Let $\mathcal{I}_m$ be the ideal of functions vanishing at m. There is an inclusion $\mathcal{I}_m \mathcal{F} \subset \mathcal{F}_m$, where $\mathcal{I}_m \mathcal{F}$ stands for the space of vector fields on the form $\sum_{i=1}^s f_i X_i$ with $f_1, \ldots, f_s \in \mathcal{I}_m$ and $X_1, \ldots, X_s \in \mathcal{F}$. Moreover, $\mathcal{I}_m \mathcal{F}$ is a Lie ideal of $\mathcal{F}_m$, since for all $X \in \mathcal{F}, Y \in \mathcal{F}_m$ and $F \in \mathcal{I}_m$:

$$[FX, Y] = \underbrace{F}_{\in \mathcal{I}_m} \underbrace{[X, Y]}_{\in \mathcal{F}} - \underbrace{Y[F]}_{\in \mathcal{I}_m} \underbrace{X}_{\in \mathcal{F}}.$$

---

[17] Except for the reader that knows (1) the holonomy Lie algebroid, see Sect. 3.6, and (2) that there is a similar phenomenon for the kernels of two points on any transitive Lie algebroid.

[18] Again, the interested reader will easily adapt these arguments to the real analytic and complex settings.

Proposition 3.11 (The Isotropy Lie Algebra: Original Construction) *Let $(M, \mathcal{F})$ be a singular foliation. For every $m \in M$, the isotropy Lie algebra at m is canonically isomorphic to the quotient Lie algebra:*

$$\mathfrak{g}_m(\mathcal{F}) \xrightarrow{\simeq} \tfrac{\mathcal{F}_m}{\mathcal{I}_m \mathcal{F}}.$$

Proof Let (A, ρ) be an anchored bundle over $\mathcal{F}$, defined in a neighborhood of m. For any $a \in \ker(\rho|_m)$, consider $\tilde{a}$ is any section of A through a. By construction, $\rho(\tilde{a})$ is in $\mathcal{F}_m$. Since any two such sections $\tilde{a}_1, \tilde{a}_2$ through a satisfy

$$\tilde{a}_1 - \tilde{a}_2 = \sum_{i=1}^{\mathrm{rk}(A)} F_i b_i$$

with $F_i \in \mathcal{I}_m$ and $b_i \in \Gamma(A)$ for $i = 1, \ldots, \mathrm{rk}(A)$, we have

$$\rho(\tilde{a}_1) - \rho(\tilde{a}_2) = \sum_{i=1}^{\mathrm{rk}(A)} F_i \, \rho(b_i).$$

As a consequence, $\rho(\tilde{a}_1) - \rho(\tilde{a}_2) \in \mathcal{I}_m \mathcal{F}$, and the map

$$\ker(\rho_m) \longrightarrow \tfrac{\mathcal{F}_m}{\mathcal{I}_m \mathcal{F}}$$

is therefore well-defined. It is also surjective by construction. It has the strong kernel in its kernel, and therefore goes down to a morphism:

$$\mathfrak{g}_m(\mathcal{F}) = \tfrac{\ker(\rho_m)}{\mathrm{Sker}(\rho, m)} \longrightarrow \tfrac{\mathcal{F}_m}{\mathcal{I}_m \mathcal{F}}.$$

The anchor condition implies that it is a Lie algebra morphism from $\mathfrak{g}_m(\mathcal{F})$ onto $\tfrac{\mathcal{F}_m}{\mathcal{I}_m \mathcal{F}}$. Let us show that the kernel of this map is zero. An element $a \in \mathrm{Ker}(\rho_m)$ has a class in $\mathfrak{g}_m(\mathcal{F})$ mapped to zero by the previous Lie algebra morphism if there exists a section $\tilde{a} \in \Gamma(A)$ through a be such that $\rho(\tilde{a}) \in \mathcal{I}_m \mathcal{F}$. This implies that $\rho(\tilde{a}) = \sum_{i=1}^{k} f_i \rho(\tilde{a}_i)$ for some sections $\tilde{a}_1, \ldots, \tilde{a}_k$ and some functions $f_1, \ldots, f_k$ vanishing at m. In particular, the section $\tilde{a} - \sum_{i=1}^{k} f_i \tilde{a}_i$ belongs to the strong kernel of ρ at m. Since its value at m is a, this completes the proof. $\square$

Exercise 3.12 Let $(M, \mathcal{F})$ be a singular foliation. For $m \in M$, consider the evaluation map $\mathrm{ev}_m \colon \mathcal{F} \to T_m M$, $X \mapsto X(m)$. Show that

1. the image of ev_m is $T_m \mathcal{F}$ and the kernel is $\mathcal{F}_m$. Also, ev_m goes to quotient to a map $\frac{\mathcal{F}}{\mathcal{I}_m \mathcal{F}} \to T_m M$ that we denote by $\underline{\mathrm{ev}}_m$.
2. $\mathfrak{g}_m(\mathcal{F}) \simeq \ker(\underline{\mathrm{ev}}_m) = \frac{\mathcal{F}_m}{\mathcal{I}_m \mathcal{F}}$.
3. $\mathrm{rk}_m(\mathcal{F}) = \dim(\frac{\mathcal{F}}{\mathcal{I}_m \mathcal{F}})$.

Exercise 3.13 Let m be a point where all vector fields in a singular foliation $\mathcal{F}$ vanish. Show that the tangent bundle of M at m is the dual of $\mathcal{I}_m/\mathcal{I}_m^2$, where $\mathcal{I}_m$ is as above. Then show that the Lie algebra $\mathcal{F}$ acts on the vector space $\mathcal{I}_m/\mathcal{I}_m^2$, and that this action is zero for any element in $\mathcal{I}_m \mathcal{F}$. Conclude that $\mathfrak{g}_m(\mathcal{F})$ acts on $T_m^* M$ and therefore on $T_m M$.

3.5 The Linear Isotropy Lie Algebra

3.5.1 The Linear Part of a Vector Field Vanishing at a Point

Let m be a point in a manifold M, and $\mathcal{I}_m$ be the ideal of functions vanishing at $m \in M$. Denote by $\mathfrak{X}_m(M)$ the Lie algebra of vector fields vanishing at m. The purpose of this preliminary section is to show that there exists a natural Lie algebra morphism:

$$\mathfrak{X}_m(M) \longrightarrow \mathrm{gl}(T_m M).$$

There are several equivalent manners to see this Lie algebra homomorphism, that we now detail.

1. One manner is simply to take local coordinates $(x_1, \ldots, x_n)$ in which m has coordinates $(0, \ldots, 0)$. The vector fields

$$\frac{\partial}{\partial x_1}, \ldots, \frac{\partial}{\partial x_n},$$

restricted to $T_m M$ form a basis of that vector space that we shall denote by $\delta_1, \ldots, \delta_n$. We then map a vector field:

$$\sum_{i=1}^n X_i(x_1,\ldots,x_n) \frac{\partial}{\partial x_i}$$

to the linear endomorphism of $T_m M$ whose matrix in the basis $\delta_1, \ldots, \delta_n$ is

$$\left(\frac{\partial X_i}{\partial x_j}(0,\ldots,0) \right)$$

We leave it to the reader to check that this is indeed a Lie algebra morphism.
Although very explicit, this method has a drawback: we have to check it does not depend on the choice of local coordinates. It is therefore better to use the coming two descriptions then show that, in local coordinates, they take the previous form.

2. The second manner is to use the flow ϕ_t^X of a vector field $X \in \mathfrak{X}_m(M)$. Since X vanishes at m, for every $\eta > 0$, there is a neighborhood $\mathcal{U}_m$ of m on which ϕ_t^X is well-defined for all $t \in \,]-\eta, \eta[$. Also, $\phi_t^X(m) = m$, so that the differential of ϕ_t^X at m is a family depending on $t \in \,]-\eta, \eta[$ of invertible linear endomorphisms

$$T_m \phi_t^X : T_m M \longleftrightarrow T_m M$$

We then define a linear endomorphism of $T_m M$ by

$$X \mapsto \left.\frac{\partial}{\partial t}\right|_{t=0} T_m \phi_t^X$$

The previous map is well-defined, but it is not clear that it is a Lie algebra morphism. Also, defining it required the notion of flow, which does not make sense in algebraic geometry.

3. The third manner is to look, for any vector field X vanishing at m, at the adjoint action:

$$Y \mapsto [X, Y]$$

and to check that $[X, Y]|_m$ only depends on $Y|_m$, so that the adjoint action induces a linear endomorphism of $T_m M$. The Jacobi identity implies that this map is a Lie algebra morphism.

4. A fourth manner is to use the canonical identification

$$T_m^* M \simeq \frac{\mathcal{I}_m}{\mathcal{I}_m^2}$$

with $\mathcal{I}_m$ the ideal[19] of functions vanishing on M (this, in the algebraic geometry setting, is in fact the definition of the cotangent space). Consider vector fields as

---

[19] In real analytic or complex geometry setting, "ideal" must be understood as "sheaf of ideals".

derivations of the sheaf of functions: a vector field X vanishes at m if and only if $X[\mathcal{I}_m] \subset \mathcal{I}_m$. By derivation properties, this implies $X[\mathcal{I}_m^2] \subset \mathcal{I}_m^2$, so that X induces a linear endomorphism of of $T_m^*M \simeq \frac{\mathcal{I}_m}{\mathcal{I}_m^2}$. Since the bracket of vector fields is their commutator, when seen as a derivation, it follows that the map above is a Lie algebra morphism:

$$\mathfrak{X}_m(M) \longrightarrow \mathrm{gl}(T_m^*M).$$

The desired Lie algebra morphism is obtained by composing the latter morphism with the canonical dualization Lie algebra isomorphism $\mathrm{gl}(T_m^*M) \simeq \mathrm{gl}(T_mM)$.

5. Last, one can recognize that for the singular foliation $\mathfrak{X}_m(M)$ of all vector fields vanishing at m, the isotropy Lie algebra at m is $\mathrm{gl}(T_mM)$. The map is then just the Lie algebra morphism $\mathfrak{X}_m(M) \longrightarrow \frac{\mathfrak{X}_m(M)}{\mathcal{I}_m \mathfrak{X}_m(M)} = \mathrm{gl}(T_mM)$.

> **Notation 3.14 (Linear Part of a Vector Field)** *We denote by $\mathfrak{Lin}$ the Lie algebra morphism*
>
> $$\{\text{Vector fields vanishing at } m\} \longrightarrow \mathrm{gl}(T_mM).$$
>
> *described in the lines above.*

3.5.2 The Linear Isotropy Lie Algebra of a Foliation Vanishing at a Point

Let us consider a singular foliation $\mathcal{F}$ on a manifold M made of vector fields vanishing at a point m (equivalently, such that $\{m\}$ is a leaf). To our knowledge, Dominique Cerveau [6] is the first to have understood the importance and studied the following Lie algebra.

> **Definition 3.15 (Linear Isotropy Lie Algebra)** Let $\mathcal{F}$ be a singular foliation, and m a point. We call *linear isotropy Lie algebra of* $\mathcal{F}$ and denote by $\mathfrak{g}_m^{lin}(\mathcal{F})$ the image of $\mathcal{F}$ through the linear part morphism $\mathfrak{Lin}$. In equation:
>
> $$\mathfrak{g}_m^{lin}(\mathcal{F}) := \mathfrak{Lin}(\mathcal{F}).$$

Remark 3.16 Upon choosing local coordinates, and therefore a basis of $T_m M$, the linear isotropy Lie algebra of $\mathcal{F}$ at m at the origin is the sub-Lie-algebra of all matrices (a_{ij}) such that there exists $X \in \mathcal{F}$ whose Taylor expansion at the origin reads:

$$X = \sum_{i,j} a_{ij} x_i \frac{\partial}{\partial x_j} + \text{higher order terms}$$

Example 3.17 Let $\mathcal{F}$ be the singular foliation induced by a Lie algebra action of $\mathfrak{g} \subset \mathfrak{gl}(\mathbb{R}^d)$ on $\mathbb{R}^d$. Then the linear holonomy of $\mathcal{F}$ at 0 is $\mathfrak{g}/\mathfrak{k}$ with $\mathfrak{k} \subset \mathfrak{g}$ being the Lie algebra of all elements that act trivially.

Example 3.18 Let $\mathcal{F} \subset \mathcal{I}_0^2 \mathfrak{X}(\mathbb{R}^d)$, i.e., a foliation made of vector fields vanishing quadratically in the origin (See Sect. 4.3 in chapter "What Is a Singular Foliation?"). Then the linear isotropy Lie algebra of $\mathcal{F}$ at 0 is $\{0\}$.

By construction,

$$\mathfrak{Lin} \colon \mathcal{F} \longrightarrow \mathfrak{g}_m^{lin}(\mathcal{F}) \tag{9}$$

is a surjective morphism of Lie algebras. Here is therefore a natural question, that one could ask for any Lie algebra morphism: Does (9) admit a section which is a Lie algebra morphism? If yes, it means, geometrically, that $\mathcal{F}$ contains a sub-singular foliation associated to the Lie algebra action of $\mathfrak{g}_m^{lin}(\mathcal{F})$ on M.

Question 3.19 *Does $\mathfrak{Lin}$ admit sections? I.e, does $\mathcal{F}$ contain, in a neighborhood $\mathcal{U}$ of m, a sub-singular foliation given by a Lie algebra action of $\mathfrak{g}_m^{lin}(\mathcal{F})$ on $\mathcal{U}$?*

In general, the answer to this kind of question tends to be "no, unless the image is semi-simple". And in the infinite dimensional case, the answer tends to be "no, unless the image is compact and semi-simple. If the image is semi-simple, then there are only formal sections". There are several results in that vein, by Conn for Poisson structures and Zung for Lie algebroids. To our knowledge, the singular foliation case is widely open: we will discuss this in Sect. 5 in chapter "State of the Art and Open Questions".

Here is an important result by Dominique Cerveau for the semi-simple case. A more recent proof can also be found in [24], Theorem 2.8.

> **Theorem 3.20 (A Linearization Theorem by Dominique Cerveau)** *If the linear isotropy Lie algebra of $\mathcal{F}$ at m is a semi-simple Lie algebra, then the map:*
>
> $$\mathfrak{Lin} \colon \mathcal{F} \longrightarrow \mathfrak{g}_m^{lin}(\mathcal{F})$$
>
> *admits a formal section[a] which is a Lie algebra morphism.*
>
> ---
> [a] See discussion around Eq. (10).

We will not prove this theorem, but we will at least us say a word about its meaning.[20] Formal functions[21] at a point $m \in M$ form an algebra that we should denote by $\hat{\mathcal{O}}_m$. Formal functions $\hat{\mathcal{O}}_m$ are a module over over germs of smooth, complex, polynomials, or real analytic functions (that we should denote by $\mathcal{O}$). As a consequence, the tensor product

$$\hat{\mathcal{O}}_m \otimes_\mathcal{O} \mathcal{F}$$

is a finitely generated $\hat{\mathcal{O}}_m$ module stable under Lie bracket,[22] and $\mathfrak{Lin}$ extends by linearity to a Lie algebra morphism:

$$\mathfrak{Lin} \colon \hat{\mathcal{O}}_m \otimes_\mathcal{O} \mathcal{F} \longrightarrow \mathfrak{g}_m^{lin}(\mathcal{F}) \qquad (10)$$

The result of Dominique Cerveau states that this Lie algebra morphism admits a section which is a Lie algebra morphism.

---

[20] Also, a more general result by Dominique Cerveau will be stated later, see Proposition 3.25.

[21] In the smooth setting, it is the quotient of $\mathcal{C}^\infty(M)$ by the ideal of functions vanishing with all their derivatives. In the other settings, it is a formal completion, i.e., the ring of formal power series in d variables near m.

[22] It is an algebraic singular foliation in the sense of Definition 2.28 in chapter "What Is a Singular Foliation?" for the ring $\hat{\mathcal{O}}_m$.

Exercise 3.21 Let $(M, \mathcal{F})$ be a singular foliation. Assume all vector fields in $\mathcal{F}$ are zero at a given point $m \in M$. Show that the quotient space[23]

$$\frac{\mathcal{F}}{\mathcal{F} \cap \mathcal{I}_m^2 \mathfrak{X}(M)},$$

is a Lie algebra isomorphic to $\mathfrak{g}_m^{lin}(\mathcal{F})$.

The linear isotropy Lie algebra captures the "linear approximation" of the foliation at a given point. As a consequence, for foliations vanishing quadratically at a point, this Lie algebra is trivial.

Example 3.22 Let $\mathcal{F} = \mathcal{I}_0^n \mathbb{R}^d$. Then $\mathfrak{g}_0(\mathcal{F})$ will have dimension

$$d \times \binom{n+d-1}{n},$$

while the linear one will be trivial for $n \geq 2$.

When $\mathcal{F}$ admits real analytic generators, one can prove that the linear holonomy contains all the semi-simplicity of $\mathfrak{g}_m(\mathcal{F})$, i.e.:

Proposition 3.23 (The Semi-simple Part Is Linear) *For a real analytic singular foliation $\mathcal{F}$, the kernel of the linearization map $\mathfrak{g}_m(\mathcal{F}) \to \mathfrak{g}_m^{lin}(\mathcal{F})$ is a nilpotent Lie algebra.*

Proof We refer to Theorem 1.10 in [24] for the complete proof. The main ingredient of the proof is the Artin-Rees Lemma, which is valid for Noetherian rings (i.e., the ring of germs of analytic functions but not the ring of smooth functions). □

Whether the theorem holds also in the smooth category is still an open problem:

Question 3.24 *Is it possible to omit the assumption "locally real analytic" in Proposition 3.23?*

---

[23] Where $\mathcal{I}_m \subset \mathcal{C}^\infty(M)$ is the ideal of all functions vanishing in m.

Let us finish with a Levi-Malcev type of theorem, in the smooth setting.
Consider the following series of Lie algebras[24]

$$\mathcal{F}(m) \to j_m^\infty(\mathcal{F}) \to \ldots \to j_m^N(\mathcal{F}) \to j_m^{N-1}(\mathcal{F}) \to \ldots \to j_m^1(\mathcal{F}) = \mathfrak{g}_m^{lin}(\mathcal{F}),$$

Here $j_m^N(\mathcal{F}) = \frac{\mathcal{F}(m)}{I_m^{N+1}\mathfrak{X}(M)}$ are N-jets of vector fields on $\mathcal{F}(m)$ and $j_m^\infty(\mathcal{F})$ their projective limit, i.e., the space of Taylor expansions of elements in $\mathcal{F}(m)$. For $N \in \mathbb{N}$, the kernel of $j_m^N(\mathcal{F}) \to \mathfrak{g}_m^{lin}(\mathcal{F})$ is a nilpotent Lie ideal. The linear isotropy Lie algebra $\mathfrak{g}_m^{lin}(\mathcal{F})$ might still contain a solvable ideal. By dividing out the maximal solvable ideal $\mathfrak{r}$, we obtain a semisimple Lie algebra $\mathfrak{g}_m^{lin}(\mathcal{F})^{ss} = \frac{\mathfrak{g}_m^{lin}(\mathcal{F})}{\mathfrak{r}}$, which can be added on the right in the above filtration to obtain a series of surjections:

$$\mathcal{F} \to j_m^\infty(\mathcal{F}) \to \ldots \to j_m^N(\mathcal{F}) \to j_m^{N-1}(\mathcal{F}) \to \ldots \to \mathfrak{g}_m^{lin}(\mathcal{F}) \to \mathfrak{g}_m^{lin}(\mathcal{F})^{ss}.$$

Using classical techniques, and using nilpotency of the kernels as above, one can prove that singular foliations satisfies a sort of Levi-Malcev-theorem:

Proposition 3.25 (Cerveau, 1977 [6]–[24]) *There is a formal Lie algebra section of* $j_m^\infty(\mathcal{F}) \to \mathfrak{g}_m^{lin}(\mathcal{F})^{ss}$ *i.e., a Lie algebra homomorphism from the semi-simple part of the linear isotropy Lie algebra at m to the Lie algebra $j_m^\infty(\mathcal{F})$.*

3.6 The Holonomy Lie Algebroid of a Leaf

So far, we have attached several Lie algebras to a point m of a foliated manifold $(M, \mathcal{F})$. There is a more general object that captures the dynamics along a leaf, namely the holonomy Lie algebroid.

As for the isotropy Lie algebra, the simplest way to define it is purely algebraic, it consists in seeing it as a quotient. But this is only valid upon the assumption that L is embedded. We will start by assuming that it is embedded, then extend the construction to the general case using almost Lie algebroid. Also, we leave it to the reader to see how to adapt the result of this section to the complex or real analytic settings.

Let L be a leaf of a singular foliation $\mathcal{F}$ on a manifold M. To start with, let us assume that L is an embedded submanifold. Consider the quotient space

$$\frac{\mathcal{F}}{\mathcal{I}_L \mathcal{F}} \tag{11}$$

---

[24] $\mathcal{F}(m)$ stands here for vector fields in $\mathcal{F}$ that vanish at m, and are defined in a neighborhood of m.

Since L is an embedded leaf, $\mathcal{I}_L$ is what we used to call a foliated ideal, i.e.,

$$\mathcal{F}[\mathcal{I}_L] \subset \mathcal{I}_L$$

so that the quotient described in Eq. (11) inherits a Lie algebra bracket. There is also a $\mathcal{C}^\infty(M)$-module structure on the quotient (11), but since $\mathcal{I}_L$ acts by zero, it is in fact a $\mathcal{C}^\infty(M)/\mathcal{I}_L \simeq \mathcal{C}^\infty(L)$-module. Altogether, these two structures turn the quotient (11) into a Lie-Rinehart algebra over $\mathcal{C}^\infty(L)$. If one can show that it is in fact a *projective* $\mathcal{C}^\infty(L)$-module, then by the Serre-Swan theorem, there exists a vector bundle $A_L \to L$ such that

$$\Gamma_c(A_L) \simeq \frac{\mathcal{F}}{\mathcal{I}_L \mathcal{F}}, \tag{12}$$

as $\mathcal{C}^\infty(L)$-module. The Lie bracket then equips A_L with a Lie algebroid structure, whose anchor is given for all $\tilde{a} \in \Gamma(A_L)$ by

$$\rho(\tilde{a}) = \hat{a}_{|L}$$

where $\hat{a} \in \mathcal{F}$ is any element whose class modulo $\mathcal{I}_L \mathcal{F}$ corresponds to $\tilde{a}$ in Eq. (12).

But how do we prove that $\mathcal{F}/\mathcal{I}_L \mathcal{F}$ is a projective $\mathcal{C}^\infty(L)$-module? And can we deal with the non-embedded case? To address both issues, we will, again, use almost Lie algebroids. From now on, we make no assumption on L. Let (A, ρ) be an anchored bundle defined in a neighborhood $\mathcal{U}$ of a point $\ell \in L$. We denote by $(\mathcal{U} \cap L)_\ell$ the connected component of ℓ in $\mathcal{U} \cap L$. Without any loss of generality, we can assume this submanifold is now embedded. Theorem 1.9 implies that the Strong kernel of ρ has the same dimension at every point in $L \cap \mathcal{U}$. Since any element of the strong kernel admits a section through it which is valued in the Strong kernel at every other point, this guarantees that, altogether, the strong kernels of points in to $(L \cap \mathcal{U})_\ell$ assemble to a sub-vector bundle of $A_{|L}$. We denote by A_L the quotient. We leave it to the reader to check that the isomorphism (12) holds true in the following form: every point $\ell \in L$ has a neighborhood $\mathcal{V}$ in M such that

$$\Gamma_c(A_L|_\mathcal{U}) \simeq \frac{\mathcal{F}|_\mathcal{U}}{\mathcal{I}_{(L \cap \mathcal{U})_\ell} \mathcal{F}|_\mathcal{U}} \tag{13}$$

where $\mathcal{F}|_\mathcal{U}$ is the restriction of the singular foliation to $\mathcal{U}$, and $\mathcal{I}_{(L \cap \mathcal{U})_\ell}$ the ideal of functions vanishing on $(L \cap \mathcal{U})_\ell$. Since the right-hand side of the previous equation does not make any reference to the anchored bundle (A, ρ), it means that A_L does not depend on the choice of a particular anchored bundle. As a consequence, $A_L \to L$, together with its anchor, are defined on the whole leaf L. Moreover, the Lie bracket that the right-hand side of the previous equation equips A_L with a Lie algebroid structure, whose anchor map $A_L \to TL$ is surjective at every point.

Definition 3.26 (Holonomy Lie Algebroid of a Leaf) Let L be a leaf of a foliated manifold $(M, \mathcal{F})$. We call *holonomy Lie algebroid of $\mathcal{F}$ along L* the unique Lie algebroid $(A_L, \rho, [\cdot, \cdot])$ such that every point $\ell \in L$ admits a neighborhood $\mathcal{U}$ on which there is an isomorphism of both Lie algebra and $\mathcal{C}^\infty(L)$-module structures as in Eq. (13).

Also, there is a short exact sequence:

$$\begin{array}{ccccc} \mathfrak{g}_L(\mathcal{F}) & \hookrightarrow & A_L & \xrightarrow{\rho} & TL \\ \downarrow & & \downarrow & & \downarrow \\ L & = & L & = & L \end{array}$$

where $\mathfrak{g}_L(\mathcal{F}) = \bigsqcup_{\ell \in L} \mathfrak{g}_\ell(\mathcal{F})$ is a bundle of Lie algebras over L whose fiber at $\ell \in L$ is the isotropy Lie algebra $\mathfrak{g}_\ell(\mathcal{F})$.

Exercise 3.27 Show that the Lie algebroid A_L acts on the normal bundle $TM_{|L}/TL$ of L in M. This action is exploited by Androulidakis and Zambon in [3]. Show that if a leaf is a point, we recover the action described in Exercise 3.13.[25]

Exercise 3.28 Find an example where there is no section $TL \to A_L$ of the anchor map which is a Lie algebra morphism. Show that if the leaf is flat,[26] then such a section exists.

4 Bisubmersions over a Singular Foliation

4.1 Definitions

In order to understand singular foliations, the most crucial and intriguing object is certainly the so-called holonomy groupoid. This groupoid is constructed from an object called bisubmersion. Both concepts were introduced by Androulidakis and Skandalis in [2].

A word of caution. The notion of bisubmersion seems to be extremely basic. At first look, it seems to be only a "Lie-groupoid-like but without a product" object, that is, it

---

[25] *Hint:* If the leaf L is embedded, Exercise 3.13 can be imitated as follows: Show that the normal bundle of M at m is the dual bundle of a bundle whose sections are $\mathcal{I}_L/\mathcal{I}_L^2$, where $\mathcal{I}_L$ functions vanishing on L. Then show that the Lie algebra $\mathcal{F}$ acts on the vector space $\mathcal{I}_L/\mathcal{I}_L^2$, and that this action is zero for any element in $\mathcal{I}_L\mathcal{F}$. Conclude that A_L acts on dual of the normal bundle and therefore on the normal bundle. Use sheaves for the general case.

[26] See Exercise 8.16 in chapter "What Is a Singular Foliation?" and the discussion following it.

seems to be a very poor structure. However, as we will see in the coming lines, although its definition is extremely short, it is indeed a very subtle and rich notion. In particular, it is easy to make wrong statements about them, and many true statements are hard to prove.

There is an analogy, that has its limits, but may sound familiar to some readers: bisubmersions (more precisely those called atlases) are to the holonomy groupoid what Lie groupoids representing a stack are to the differential stack in question. Androulidakis and Zambon also pointed us out that bisubmersions can also be thought of as the plots for the holonomy groupoid, and that equivalence is like equivalence of plots in diffeology—an enriching perspective.

Let us give the definition. From now on, we work in the smooth setting, and leave it to the reader to adapt to real analytic or complex ones.

Definition 4.1 (Androulidakis-Skandalis' Bisubmersions) Let M be a manifold equipped with a singular foliation $\mathcal{F}$. A bisubmersion over $(M, \mathcal{F})$ is a triple (W, s, t) where:

1. W is a manifold,
2. $s, t \colon W \to M$ are submersions,[a] respectively called *source* and *target*,

such that

1. the pull-back singular foliations $s^{-1}(\mathcal{F})$ and $t^{-1}(\mathcal{F})$ are equal,[b]
2. and any vector field $Z \in s^{-1}(\mathcal{F}) = t^{-1}(\mathcal{F})$ is of the form[c] $Z = X + Y$ with $X \in \Gamma(\ker(Ts))$ and $Y \in \Gamma(\ker(Tt))$.

Also, we will use the name *bi-pull-back singular foliation* and the notation $^s\mathcal{F}^t$ for the singular foliation on W given by:

$$^s\mathcal{F}^t := s^{-1}(\mathcal{F}) = t^{-1}(\mathcal{F}) = \Gamma(\ker(Ts)) + \Gamma(\ker(Tt)). \qquad (14)$$

Last, vector fields in $\Gamma(\ker(Ts)) \cap \Gamma(\ker(Tt))$ shall be said to be is *bi-vertical*.

---

[a] We do not assume s and t to be surjective. See Sect. 5.2 in chapter "What Is a Singular Foliation?" for a definition of the pull-back singular foliation.
[b] Notice that this implies that $\Gamma(\ker(Ts)) + \Gamma(\ker(Tt)) \subset s^{-1}(\mathcal{F}) = t^{-1}(\mathcal{F})$.
[c] In general, this decomposition is not unique at all. This second condition can be equivalently stated as $\Gamma(\ker(Ts)) + \Gamma(\ker(Tt)) = s^{-1}(\mathcal{F}) = t^{-1}(\mathcal{F})$.

When we will need to insist on the names of all structures, such a bisubmersion shall be denoted by

$$M \xleftarrow{s} W \xrightarrow{t} M.$$

Quite often, we will simply say "a bisubmersion W over a foliated manifold $(M, \mathcal{F})$".

Warning!
Notice that it does not make much sense to say "Let $M \xleftarrow{s} W \xrightarrow{t} M$ be a bisubmersion", without mentioning over which singular foliation $\mathcal{F}$ it is a bisubmersion. We will always say "Let $M \xleftarrow{s} W \xrightarrow{t} M$ be a bisubmersion over $\mathcal{F}$" or "over $(M, \mathcal{F})$".

Exercise 4.2 Let W be a bisubmersion over $(M, \mathcal{F})$. Show that if $\dim(W) = \dim(M)$, then $\mathcal{F} = 0$.

Exercise 4.3 Show that, for any bisubmersion, compactly supported[27] bivertical vector fields form a $\mathcal{C}^\infty(W)$-module which is closed under Lie bracket. Notice that it may not be a singular foliation, for it may not be finitely generated, see Exercise 4.4.

Exercise 4.4 Let X be a compactly supported vector field on a manifold M, and $\mathcal{F}_X = \{fX | f \in \mathcal{C}^\infty(M)\}$ the singular foliation it generates as in Exercise 4.6 in chapter "What Is a Singular Foliation?". Show that $W := \mathbb{R} \times M$ equipped with the following source and target maps

$$s(u, m) := m \text{ and } t(u, m) := \phi_u^X(m) \text{ for all } m \in M, u \in \mathbb{R}$$

is a bisubmersion for $\mathcal{F}_X$. Show that if $M = \mathbb{R}$ and X is a vector field with support $[-1, 1]$, bivertical vector fields do not form a locally finitely generated module.

Exercise 4.5 Let M be a manifold, and $\mathcal{F} = \mathfrak{X}_c(M)$ be the singular foliation of all compacted supported vector fields. Show that $W := M \times M$, equipped with source and targets defined to be the projections on the first and second components, respectively, is a bisubmersion over $\mathcal{F}$ with no non-trivial bi-vertical bivector fields.

Here is a technical lemma: the reader may jump to Definition 4.9 directly. The importance of this lemma will appear later on, and may not be obvious at first sight.

---

[27] Again, one could avoid the use of compactly supported vector fields by using sheaves.

Lemma 4.6 *Consider a bisubmersion* $M \xleftarrow{s} W \xrightarrow{t} M$ *over a singular foliation* $\mathcal{F}$, *and let* ${}^s\mathcal{F}^t$ *be the bi-pull-back singular foliation as in Eq.* (14). *Choose* $x \in W$. *Let* $\underline{X}_1, \ldots, \underline{X}_{r_s}$ *be local generators of* $\mathcal{F}$ *in a neighborhood of* $s(x)$ *and* $\underline{Y}_1, \ldots, \underline{Y}_{r_t}$ *be local generators of* $\mathcal{F}$ *in a neighborhood of* $t(x)$. *Let* $k = 2\operatorname{rk}(\ker Ts) = 2\operatorname{rk}(\ker Tt)$. *Then* x *admits a neighborhood* $\mathcal{U}$ *on which* ${}^s\mathcal{F}^t$ *is generated by vector fields* $X_1, \ldots, X_{r_s}, Y_1, \ldots, Y_{r_t}, Z_1, \ldots, Z_k$ *satisfying the following properties:*

1. $X_1, \ldots, X_{r_s}$ *belong to* $\Gamma(\ker(Tt))$ *and are s-related to* $\underline{X}_1, \ldots, \underline{X}_{r_s}$,
2. $Y_1, \ldots, Y_{r_t}$ *belong to* $\Gamma(\ker(Ts))$ *and are t-related to* $\underline{Y}_1, \ldots, \underline{Y}_{r_t}$,
3. $Z_1, \ldots, Z_k \in \Gamma(\ker(Ts)) \cap \Gamma(\ker(Tt))$ *are bi-vertical vector fields,*
4. *the vector fields* $X_1, \ldots, X_{r_s}, Z_1, \ldots, Z_k$ *generate* $\ker(Tt)$ *at every point of* $\mathcal{U}$, *and the vector fields* $Y_1, \ldots, Y_{r_t}, Z_1, \ldots, Z_k$ *generate* $\ker(Ts)$ *at every point of* $\mathcal{U}$.

Proof By definition of the pull-back singular foliation $s^{-1}(\mathcal{F})$, there exists for every $i \in \{1, \ldots, r_s\}$ a vector field $X_i^! \in s^{-1}(\mathcal{F})$ which is s-related to $\underline{X}_i$. By definition of a bisubmersion, there exists $X_i^s \in \Gamma(\ker(Ts))$ and $X_i^t \in \Gamma(\ker(Tt))$ such that $X_i^! = X_i^s + X_i^t$. By construction, $X_i := X_i^t \in \Gamma(\ker(Tt))$ is s-related to $\underline{X}_i$ and t-related to 0. *Item 1 is therefore satisfied.*

The same argument, with s replaced by t, yields vector fields $Y_1, \ldots, Y_{r_t} \in \Gamma(\ker(Ts))$ that are t-related to $\underline{Y}_1, \ldots, \underline{Y}_{r_t}$ and s-related to 0. *Item 2 is therefore satisfied.*

Now, it is an elementary property of $s^{-1}(\mathcal{F})$ that for any vector field $Z \in s^{-1}(\mathcal{F})$, there exists functions $(g^i)_{i=1}^{r_s}$, defined in a neighborhood $\mathcal{U}$ of x in W, such that $Z - \sum_{i=1}^{r_s} g^i X_i \in \Gamma(\ker(Ts))$. In particular, for $Z \in \Gamma(\ker(Tt))$, it means that

$$Z - \sum_{i=1}^{r_s} g^i X_i$$

belongs to $\Gamma(\ker(Ts)) \cap \Gamma(\ker(Tt))$, i.e., is bi-vertical.

Consider now a local trivialization $Z'_1, \ldots, Z'_{k/2}$ of $\ker(Tt)$ on $\mathcal{U}' \subset \mathcal{U}$. By the above discussion, there exists local functions

$$\left(g_i^j\right)_{i=1,\ldots,r_s}^{j=1,\ldots,k/2},$$

defined on some neighborhood $\mathcal{U}'' \subset \mathcal{U}'$ of x, such that

$$Z_j := Z'_j - \sum_{i=1}^{r_s} g_i^j X_i \in \Gamma(\ker(Ts)) \cap \Gamma(\ker(Tt)), \quad j \in \{1, \ldots, k/2\},$$

as claimed in *Item 3*. The family of vector fields defined on $\mathcal{U}''$:

$$\left(X_1, \ldots, X_{r_s}, Z_1, \ldots, Z_{k/2}\right)$$

generates $\Gamma(\ker(Tt))$, *as required in item 4*. The same argument, with s replaced by t, yields vector fields $Z_{k/2+1} \ldots, Z_k \in \Gamma(\ker(Ts)) \cap \Gamma(\ker(Tt))$, *therefore satisfying Item 3*, such that the family

$$\left(Y_1, \ldots, Y_{r_t}, Z_{k/2+1}, \ldots, Z_k\right)$$

generates $\Gamma(\ker(Ts))$, *as required in item 4*.

Altogether, the families $X_1, \ldots, X_{r_s}$, $Y_1, \ldots, Y_{r_t}$ and, $Z_1, \ldots, Z_k$ satisfy all the requirements of Lemma 4.6. □

Exercise 4.7 Consider a bisubmersion over $\mathcal{F}$ as before.

1. Show that $\Gamma(\mathrm{Ker}(Ts))$ is generated by vector fields t-related to vector fields in $\mathcal{F}$.
2. Show also that any element of $\mathcal{F}$ is t-related to a section of $\Gamma(\mathrm{Ker}(Ts))$.

Hint: Use Lemma 4.6.

Remark 4.8 The vector fields $Z_1, \ldots, Z_k$ in Lemma 4.6 are *not*, in general, generators of bi-vertical vector fields. Otherwise, bi-vertical vector fields would be finitely generated, which is not true in view of Exercise 4.4. □

We claimed that bisubmersions may be thought of as an equivalent of a Lie groupoid, as a sort of "Lie-groupoid-without-a-product". To justify the analogy, as for Lie groupoids, we now define units and bisections.

Definition 4.9 (Some Important Notions: Units and Bisections) A *bisection* of a bisubmersion $M \xleftarrow{s} W \xrightarrow{t} M$ over $\mathcal{F}$ is a submanifold $\Sigma \subset W$ such that

1. $s(\Sigma) \subset M$ and $t(\Sigma) \subset M$ are open subsets,[a]
2. the restriction of s to Σ is a diffeomorphism from Σ to $s(\Sigma)$, and the restriction of t to Σ is a diffeomorphism from Σ to $t(\Sigma)$.

Let $\mathcal{U} \subset M$ be an open subset. A map $\epsilon : \mathcal{U} \longrightarrow W$ is said to be a *unit map*[b] if it is a section of both s and t, i.e., $s \circ \epsilon = t \circ \epsilon = \mathrm{id}_\mathcal{U}$.

---

[a] We will speak of *global bisections* when $s(\Sigma) = t(\Sigma) = M$, and *local bisections* otherwise.
[b] Again, we will speak of *global units* when $\mathcal{U} = M$, and *local units* otherwise.

Remark 4.10 The image $\epsilon(\mathcal{U})$ of a (local/global) unit map ϵ of a bisubmersion is a (local/global) bisection. Notice that, unlike for groupoids, the unit map may not exist. Moreover, even if it exists, it may not be unique. Examples will be given below. □

The next exercise explains how an anchored bundle can be recovered out of the data of a bisubmersion with unit. An even more general result will be proven in Lemma 4.17.

Exercise 4.11 Consider a bisubmersion $M \xleftarrow{s} W \xrightarrow{t} M$ over $\mathcal{F}$. Assume a unit map $\epsilon: M \to W$ exists.

1. Show that the normal bundle $N = TW/T(\epsilon(M))$ of $\epsilon(M)$ into W is canonically isomorphic to $\ker(Ts) \subset TW|_{\epsilon(M)}$ and to $\ker(Tt) \subset TW|_{\epsilon(M)}$
2. Show that the vector bundle morphism $Ts - Tt : TW|_{\epsilon(M)} \to TM$ goes to the quotient to give a vector bundle morphism $\rho_W: N \to TM$ (over the natural diffeomorphism $\epsilon(M) \to M$).
   From now on, we use the diffeomorphism $\epsilon: M \to \epsilon(M)$ in order to consider N as a vector bundle over M (and not as a vector bundle over $\epsilon(M)$) that we denote by A. Also, we consider ρ_W as a vector bundle morphism from A to TM (over the identity of M).
3. (*Not easy!*) Show that the pair (A, ρ_W) is an anchored bundle over $\mathcal{F}$ (i.e., $\rho_W(\Gamma(A)) = \mathcal{F}$). *Hint:* Use Lemma 4.6 and more precisely its consequence, Exercise 4.7.

The results of this exercise will soon be generalized, see Lemma 4.17 below.

Lemma 4.12 *Consider $M \xleftarrow{s} W \xrightarrow{t} M$ a bisubmersion over $\mathcal{F}$. For every $x \in W$, there exists at least one bisection through x.*

Proof There exists vector subspaces $V \subset T_xW$ of dimension equal to $\dim(M)$ whose intersection with $\ker(T_xs)$ and with $\ker(T_xt)$ are both reduced to zero. In particular, both $T_xs: V \to T_{s(x)}M$ and $T_xt: V \to T_{t(x)}M$ are linear invertible maps. Any submanifold of W through x admitting V as its tangent space admits a restriction to a neighborhood of x which is a bisection. □

Exercise 4.13 Prove the following statements.

1. Show that there exists infinitely many local bisections through a point $x \in W$, except maybe if $\mathcal{F} = 0$ in a neighborhood of $s(x)$.
2. Show that for any two local bisections Σ_0, Σ_1 through $x \in W$, there exists a neighborhood $\mathcal{U}$ of $s(x)$ and a smooth family $(\Sigma_\epsilon)_{\epsilon \in [0,1]}$ of bisections through x such that $s(\Sigma_\epsilon) = \mathcal{U}$ for all $\epsilon \in [0, 1]$.

As for Lie groupoids, every bisection Σ induces a diffeomorphism:

$$\underline{\Sigma}\colon s(\Sigma) \longrightarrow t(\Sigma),$$

that makes the following diagram commutative:

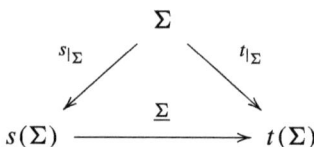

Exercise 4.14 Show that units of a bisubmersion are in one-to-one correspondence with bisections Σ such that $\underline{\Sigma}$ is the identity map of M.

Here is an important result, very similar to what happens for Lie groupoids.[28]

Proposition 4.15 (Bisections Induce Symmetries) *Let $\mathcal{F}$ be a singular foliation on a manifold M and let $M \xleftarrow{s} W \xrightarrow{t} M$ be a bisubmersion over $\mathcal{F}$. For every local bisection Σ, the induced (local) diffeomorphism*

$$\underline{\Sigma}\colon s(\Sigma) \longrightarrow t(\Sigma)$$

is an isomorphism of singular foliations from[a] $(s(\Sigma), \mathcal{F}_{s(\Sigma)})$ *to* $(t(\Sigma), \mathcal{F}_{t(\Sigma)})$. *In particular, when Σ is a global bisection, $\underline{\Sigma}$ is a symmetry of $(M, \mathcal{F})$.*

---

[a] Also called "local symmetry of $\mathcal{F}$" in Sect. 3.1 in chapter "What Is a Singular Foliation?". Recall that $\mathcal{F}_{s(\Sigma)}$ and $\mathcal{F}_{s(\Sigma)}$ stand for the restrictions of $\mathcal{F}$ to $s(\Sigma)$ and $t(\Sigma)$, respectively.

We prove Proposition 4.15 for a local bisection. The global case follows as a particular case. Let $^s\mathcal{F}^t$ be the bi-pull-back singular foliation on W as in Eq. (14). Denote by $\left(^s\mathcal{F}^t\right)_\Sigma$ its restriction to Σ (see Sect. 5.3 in chapter "What Is a Singular Foliation?"). Recall that $\left(^s\mathcal{F}^t\right)_\Sigma$, i.e., the subspace of $\mathfrak{X}(\Sigma)$ obtained by considering the restriction to Σ of all vector fields in $^s\mathcal{F}^t$ that are tangent to Σ, need not be a singular foliation in general. However, it is so in the present situation:

---

[28] See Proposition 2.8 in [2]. All concepts and results of this section come from [2], we will not give the exact correspondence from now on.

Lemma 4.16 *Let Σ be a bisection of W.*

1. *The restriction $\left({}^s\mathcal{F}^t\right)_\Sigma$ of ${}^s\mathcal{F}^t$ to Σ is a singular foliation on Σ.*
2. *Moreover, both $s|_\Sigma$ and $t|_\Sigma$ are diffeomorphisms of singular foliations from $(\Sigma, {}^s\mathcal{F}_\Sigma{}^t)$ to $(s(\Sigma), \mathcal{F}_{s(\Sigma)})$ and $(\Sigma, {}^s\mathcal{F}_\Sigma{}^t)$ to $(t(\Sigma), \mathcal{F}_{t(\Sigma)})$, respectively.*

Proof The following decomposition holds $\forall \sigma \in \Sigma$:

$$T_\sigma \Sigma \oplus \ker(T_\sigma s) = T_\sigma W.$$

Since $\ker(T_\sigma s) \subset T_\sigma{}^s\mathcal{F}^t$, we therefore have

$$T_\sigma \Sigma + T_\sigma{}^s\mathcal{F}^t = T_\sigma W.$$

In terms of the concepts introduced in Sect. 5.3 in chapter "What Is a Singular Foliation?", it means that Σ intersects cleanly the bi-pull-back singular foliation ${}^s\mathcal{F}_\Sigma{}^t$. Proposition 5.15 in chapter "What Is a Singular Foliation?" therefore applies: $\left({}^s\mathcal{F}^t\right)_\Sigma \subset \mathfrak{X}(\Sigma)$ is a singular foliation on Σ. *This proves item 1.*

Since the restriction $s|_\Sigma$ of s to Σ is a diffeomorphism, one can consider the push-forward singular foliation

$$(s|_\Sigma)_* \left(\left({}^s\mathcal{F}^t\right)_\Sigma\right).$$

It is by construction a singular foliation on $s(\Sigma)$. We first show the inclusion $\mathcal{F}_{s(\Sigma)} \subseteq (s_\Sigma)_*(({}^s\mathcal{F}^t)_\Sigma)$. Let $u \in \mathcal{F}_{s(\Sigma)}$. There exists a unique vector field $u^\Sigma \in \mathfrak{X}(\Sigma)$ such that $Ts_\Sigma(u^\Sigma) = u$. Let v be a vector field in $s^{-1}(\mathcal{F})$ that s-related to u. For every $\sigma \in \Sigma$, the difference $u_\sigma^\Sigma - v_\sigma$ is valued in $\ker(T_\sigma s)$. In view of the decomposition

$$T_\sigma W = T_\sigma \Sigma \oplus \ker(T_\sigma s),$$

there exists a vector field in $Z \in \Gamma(\ker(Ts))$ such that $u_\sigma^\Sigma - v_\sigma = Z_\sigma$ for all $\sigma \in \Sigma$. Consider $u' := v + Z$. The vector field u' belongs to $s^{-1}(\mathcal{F})$ by definition. Also, by construction, its restriction to Σ is tangent to Σ. Last, it coincides with u^Σ on Σ. This proves that $u^\Sigma \in \left({}^s\mathcal{F}^t\right)_\Sigma$. Therefore, the desired inclusion holds.

Let us show the opposite inclusion $(s|_\Sigma)_* \left({}^s\mathcal{F}^t\right)_\Sigma \subseteq \mathcal{F}_{s(\Sigma)}$: let v be a vector field in ${}^s\mathcal{F}^t$ that happens to be tangent to Σ. We show that there exists another vector field $\tilde{v}$ which coincides with v on Σ and is s-related to a vector field in $\mathcal{F}$. By construction, $v = \sum_i g_i v_i$ where the g_i are smooth functions on W and $v_i \in {}^s\mathcal{F}^t$ are s-related to elements $u_i \in \mathcal{F}$. Let $\tilde{g}_i$ be functions on W that coincide with g_i on Σ and are constant along s-fibers, i.e., $\tilde{g}_i = s^* f_i$ for some smooth function f_i on M. The vector field $\tilde{v} = \sum_i \tilde{g}_i v_i$ is tangent to Σ (since it coincides with v on Σ) and is s-related with $\sum_i f_i u_i \in \mathcal{F}$. Hence $s_*(v|_\Sigma) = \sum_i f_i u_i|_{s(\Sigma)}$ and in particular $(s_\Sigma)_*(v|_\Sigma) \in \mathcal{F}_{s(\Sigma)}$. In turn, this implies $(s_\Sigma)_* \left({}^s\mathcal{F}_\Sigma{}^t\right) \subset$

$\mathcal{F}_{s(\Sigma)}$. The same argument holds for $(t_\Sigma)_* \left({}^s\mathcal{F}_\Sigma{}^t\right) \subset \mathcal{F}_{t(\Sigma)}$. This completes the *proof of item 2*. □

Proof of Proposition 4.15 From Lemma 4.16, it follows that both $s_{|\Sigma}$ and $t_{|\Sigma}$ below are diffeomorphisms of singular foliations:

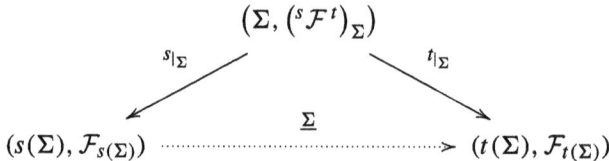

Hence the horizontal line $\underline{\Sigma}$ is a diffeomorphisms of singular foliations. This proves the claim. □

Let us now associate an anchored bundle over $\mathcal{F}$ to any bisection Σ, that we will assume to be global for the sake of simplicity. The vector bundle morphism:

$$TW_{|\Sigma} \longrightarrow TM$$
$$u \mapsto Tt(u) - T\underline{\Sigma} \circ Ts(u)$$

admits $T\Sigma$ in its kernel. Therefore, it goes to the quotient to yield a vector bundle morphism ρ_Σ

$$\begin{array}{ccc} TW_{|\Sigma}/T\Sigma = N_\Sigma & \xrightarrow{\rho_\Sigma} & TM \\ \downarrow & & \downarrow \\ \Sigma & \xrightarrow{t} & M \end{array}$$

where N_Σ stands for the normal bundle of Σ in W. In view of the decomposition:

$$T_\sigma W = T_\sigma \Sigma \oplus \ker(T_\sigma s) \text{ for all } \sigma \in \Sigma,$$

the restriction $\ker(Ts)_{|\Sigma}$ of $\ker(Ts)$ to Σ is, as a vector bundle, canonically isomorphic to the normal bundle N_Σ. Under this isomorphism, the following diagram is commutative

$$\begin{array}{ccc} N_\Sigma & \xrightarrow{\simeq} & \ker(Ts)_{|\Sigma} \\ & \searrow_{\rho_\Sigma} & \downarrow{Tt} \\ & & TM \end{array} \qquad (15)$$

Now, since $t : \Sigma \to M$ is a diffeomorphism on its image, N_Σ can be considered as a vector bundle $t^{-1} N_\Sigma$ over the open subset $t(\Sigma) \subset M$ (rather than over Σ). Under this identification, ρ_Σ becomes a vector bundle morphism $t^{-1} N_\Sigma \to TM$ over the identity map of $t(\Sigma) \subset M$ that we denote by the same letter ρ_Σ and call the *anchor of* Σ.

Lemma 4.17 *Let Σ be a bisection of a bisubmersion W over a foliated manifold $(M, \mathcal{F})$. The pair $(t^{-1} N_\Sigma, \rho_\Sigma)$ is an anchored bundle for the restriction of $\mathcal{F}$ to $t(\Sigma)$.*

Proof Choose any point $\sigma \in \Sigma$. In some neighborhoods, there exists vector fields as in Lemma 4.6. By the fourth item, $Y_1, \ldots, Y_{r_t}$ and $Z_1, \ldots, Z_k$ generate the kernel of Tt, and therefore its restriction to Σ. They identify therefore to sections of the normal bundle that generate it. By Eq. (15), their images under ρ_Σ are the local generators $\underline{Y}_1, \ldots, \underline{Y}_{r_t}$ of $\mathcal{F}$. This completes the proof. □

The end of this subsection is required for what follows, but is more technical/involved. We recommend the reader unfamiliar with the notion to look directly at Sect. 4.2.

Consider two local bisections Σ_0, Σ_1 of a bisubmersion W that contain the same point $w \in W$. In general, the induced local symmetries $\underline{\Sigma_0}, \underline{\Sigma_1}$ of $(M, \mathcal{F})$ do not coincide, even locally. However, we will see that they differ by an inner symmetry admitting $s(w)$ as a very-fixed point,[29] (equivalently, they differ by an inner symmetry admitting $t(w)$ as a very-fixed point). We refer to Sect. 3.1 in chapter "What Is a Singular Foliation?" for the vocabulary about inner symmetries, and very-fixed points of those. We also refer to this section[30] for the subtleties about the precise meaning of "smoothness" for a time-dependent vector field $(X_\epsilon)_{\epsilon \in I}$.

Proposition 4.18 *Consider a bisubmersion W over a foliated manifold $(M, \mathcal{F})$. Let $w \in W$. For any two local bisections Σ_0, Σ_1 of W through w, the local diffeomorphism $\underline{\Sigma_0} \circ \underline{\Sigma_1}^{-1}$ is an inner symmetry of $\mathcal{F}$ with very-fixed point[31] $t(w)$, at least in a neighborhood of $t(w)$.*

Said differently, there exists $(X_\epsilon)_{\epsilon \in [0,1]}$, a smooth time-dependent vector field in $\mathcal{F}$, such that

$$X_\epsilon|_x = \sum_{i=1}^{r} f_i(x, \epsilon) \, X_i|_x, \tag{16}$$

---

[29] See Definition 3.10 in chapter "What Is a Singular Foliation?".
[30] The parameter is denoted by t in that section and by ϵ here, since t is the target map.
[31] See Definition 3.10 in chapter "What Is a Singular Foliation?".

with $X_1, \ldots, X_r$ being local generators of $\mathcal{F}$, and $f_1, \ldots, f_r$ being smooth functions that vanish on $\{m\} \times [0, 1]$, whose time-1 flow $\Phi_1^{X_\epsilon}$ coincides with $\underline{\Sigma_0} \circ \underline{\Sigma_1}^{-1}$ in a neighborhood of $t(x)$.

The proof starts with a Lemma. Let $(\Sigma_\epsilon)_{\epsilon \in I}$ be a family of global bisections. We say that such a family is *smooth* if the sections $\Sigma_\epsilon^{-1} \colon M \to W$ that inverts $s_{|\Sigma_\epsilon} \colon \Sigma_\epsilon \to W$ depends smoothly on the parameter ϵ.

Lemma 4.19 *If there exists a smooth interpolation by bisections between two given bisections Σ_0 and Σ_1, then $\underline{\Sigma_1}^{-1} \circ \underline{\Sigma_0}$ is a local inner symmetry of $\mathcal{F}$.*

Proof For any small enough $\epsilon \in \mathbb{R}$ and any $w \in \Sigma_\epsilon$, the path

$$\eta \mapsto s_{|\Sigma_\eta}(s(m))$$

is a path in a fixed s-fiber and goes through m at $\epsilon = \eta$. Its derivative:

$$u_\epsilon := \left. \frac{\partial s_{|\Sigma_\eta}(s(m))}{\partial \eta} \right|_{\eta = \epsilon}$$

is therefore a section of $\ker(Ts)_{|\Sigma_\epsilon}$, that naturally identifies to a section of the anchored bundle $(A_{\Sigma_\epsilon}, \rho_{\Sigma_\epsilon})$ of Lemma 4.17. Its image through ρ_{Σ_ϵ} is a vector field that we denote by X_ϵ. In view of Lemma 4.17, X_ϵ belongs to $\mathcal{F}$. It is by construction a smooth time-dependent vector field in $\mathcal{F}$ as defined in Sect. 3.1 in chapter "What Is a Singular Foliation?". The relation $Tt(u_\epsilon) = X_\epsilon$ holds by construction, and implies that $\eta \mapsto t \circ s_{|\Sigma_\eta}(s(w))$ is an integral curve of the vector field X_ϵ, that relates $\underline{\Sigma_0}(s(w))$ to $\underline{\Sigma_1}(s(w))$ by construction. This proves the claim. □

Proof of Proposition 4.18 Upon replacing Σ_0, Σ_1 by neighborhoods of w in Σ_0, Σ_1 that we denote by the same symbols, one can assume that a smooth family of local bisections interpolating between Σ_0 and Σ_1 exists. Moreover, those can be chosen such that *(i)* we have $w \in \Sigma_\epsilon$ for all $\epsilon \in [0, 1]$, and *(ii)* there exists a neighborhood $\mathcal{U}$ of $t(w)$ contained in $t(\Sigma_\epsilon)$ for all value of $\epsilon \in [0, 1]$. This last point allows to use Lemma 4.19 (adapted to local bisections). It therefore shows that $\phi = \underline{\Sigma_0} \circ \underline{\Sigma_1}^{-1}$ in an inner symmetry of $\mathcal{F}$. Since all the bisections Σ_ϵ are through w, the section u_ϵ that appears in Lemma 4.19 vanishes at $t(w)$. As a consequence (see Exercise 3.13 in chapter "What Is a Singular Foliation?"), the smooth time-dependent vector field $(X_\epsilon)_{\epsilon \in [0,1]}$, whose time-1 flow is ϕ, is of the type given in Eq. (16). The construction in Lemma 4.19 yields therefore a smooth time-dependent vector field in $\mathcal{F}$ whose time 1-flow satisfies the required assumptions that makes $t(w)$ a very fixed point. □

For any two points in a foliated manifold $(M, \mathcal{F})$. Let us denote by $\mathrm{Sym}_{\mathcal{F}}(m, m')$ the set of local symmetries of $\mathcal{F}$, defined from a neighborhood of m to a neighborhood of m'. We denote by $\underline{\mathrm{Sym}}_{\mathcal{F}}(m, m')$ and call germs of such local symmetries the quotient of $\mathrm{Sym}_{\mathcal{F}}(m, m')$ through the equivalence relation that identifies two diffeomorphisms that coincide in a neighborhood of m. Of course, in general, $\mathrm{Sym}_{\mathcal{F}}(m, m')$ may be empty : it is however, non-empty if m and m' are in the same leaves: see Sect. 7 in chapter "What Is a Singular Foliation?".

Let us introduce an equivalence relation on $\underline{\mathrm{Sym}}_{\mathcal{F}}(m, m')$. We identify two elements $\bar{\Phi}, \bar{\Psi}$ if one of the equivalent conditions below holds:

(i) $\Phi^{-1} \circ \Psi$ coincides in a neighborhood of m with an inner symmetry of $\mathcal{F}$ with very-fixed point m.
(ii) $\Psi^{-1} \circ \Phi$ coincides in a neighborhood of m' with an inner symmetry with very-fixed point m'.

Exercise 4.20 Show that both conditions (i) and (ii) above are indeed equivalent, and that the latter is an equivalence relation. *Hint:* Look at Exercise 3.12 in chapter "What Is a Singular Foliation?".

We now denote the quotient of $\underline{\mathrm{Sym}}_{\mathcal{F}}(m, m')$ by the previous equivalence relation by $\mathrm{OutSym}_{\mathcal{F}}(m, m')$. We call this set the *germs of outer symmetries from m to m'*.

Let W be a bisubmersion for $(M, \mathcal{F})$. It follows from Proposition 4.15 that to any pair (x, Σ), with Σ a bisection of W through x, one can associate the germ of $\underline{\Sigma} \in \underline{\mathrm{Sym}}_{\mathcal{F}}(s(x), t(x))$. By Proposition 4.18, the class of $\underline{\Sigma}$ in $\mathrm{OutSym}_{\mathcal{F}}(s(x), t(x))$ does not depend on the choice of Σ. In view of Lemma 4.12, we have therefore defined a map that makes the following diagram commute:

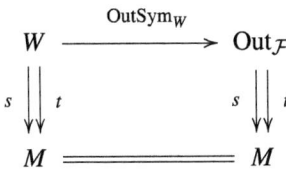

where $\mathrm{OutSym}_{\mathcal{F}} = \bigsqcup_{m,m' \in M} \mathrm{OutSym}_{\mathcal{F}}(m, m')$ and where the s and t are defined for any element in $\mathrm{OutSym}_{\mathcal{F}}(m, m')$ to be m and m', respectively.

Definition 4.21 (Outer-Germ Map) We call *Outer-germ map* and denote by Out_W the map

$$W \longrightarrow \mathrm{OutSym}_{\mathcal{F}}$$

defined by associating to any $w \in W$ the class, modulo inner symmetries with very-fixed points,[a] of the germ of symmetries[b] of $\mathcal{F}$ induced by an arbitrary[c] local bisection through w.

---

[a] See Definition 3.10 in chapter "What Is a Singular Foliation?".
[b] See Sect. 3.1 in chapter "What Is a Singular Foliation?".
[c] The discussion before the statement explains why any two bisections through w induce the same class in the quotient space.

Exercise 4.22 We intend to prove that the map Out_W is constant along integral curves of bi-vertical vector fields. Let $X \in \mathfrak{X}(W)$ be a bi-vertical vector field on a bisubmersion W over $(M, \mathcal{F})$. Let Φ_t^X be the flow of X and let $\epsilon \mapsto \gamma(\epsilon) = \Phi_\epsilon^X(m)$ be an integral curve of X starting from $x \in W$, defined for all ϵ is some interval I containing 0.

1. Let Σ be a bisection of W through $\gamma(0)$. Show that for all $\epsilon \in I$, there exists a neighborhood Σ' of x in Σ on which Φ_η^X is well-defined for all $\eta \in [0, \epsilon]$.
2. Show that $\Sigma_\eta := \Phi_\epsilon^X(\Sigma')$ is a bisection for all $\eta \in [0, \epsilon]$, and that $\underline{\Sigma_\eta}$ is constant for all $\eta \in [0, \epsilon]$.
3. Conclude that the map Out_W is constant along integral curves of bi-vertical vector fields.

Exercise 4.23 Let $(M, \mathcal{F})$ be a foliated manifold, W and W' two bisubmersions over $\mathcal{F}$. Show that if $x \in W$ and $x' \in W'$ are related by $\mathrm{Out}_W(x) = \mathrm{Out}_W(x')$, then there exists local bisections Σ and Σ' through x and x' such that $\underline{\Sigma} = \underline{\Sigma}'$. (*Hint*: We will have more tools to solve this exercise after right and left actions are defined. This exercise will be then repeated as Exercise 4.52.)

4.2 Examples of Bisubmersions

4.2.1 Basic Examples and Non-examples

The first exercise shows that Lie groupoids are bisubmersions - a non-trivial fact.

Exercise 4.24 Let Γ be a Lie groupoid over M. Let A be its Lie algebroid and $\rho: A \to TM$ its anchor map. The basic singular foliation is (by definition) the singular foliation

$\mathcal{F} := \rho\,(\Gamma(A))$ (see Sect. 4.2 in chapter "What Is a Singular Foliation?"). The purpose of this exercise is to show that any Lie groupoid Γ is a bisubmersion for its basic singular foliation $\mathcal{F}$.

Let us denote by $\overrightarrow{a}$ and $\overleftarrow{a}$ the right and left invariant vector fields on Γ associated to $a \in \Gamma(A)$ as in [28]. There are miscellaneous conventions: we chose them such that $\overrightarrow{a}$ is s-related to 0 and t-related to $\rho(a)$, while $\overleftarrow{a}$) is t-related to 0 and s-related to $\rho(a)$.

1. Show that the $\mathcal{C}^\infty(\Gamma)$-module $\mathcal{G}$ generated by vector fields of the form $\overrightarrow{a} + \overleftarrow{b}$ for some $a, b \in \Gamma(A)$ is a singular foliation on Γ.
2. In passing: describe its leaves. We may assume that Γ is source-connected for simplicity.
3. Use the Lie groupoid axioms to show that $\Gamma(\ker(Tt))$ (resp. $\Gamma(\ker(Ts))$) coincides with the $\mathcal{C}^\infty(\Gamma)$-module generated by $\{\overrightarrow{a} \mid a \in \Gamma(A)\}$ (resp. $\{\overleftarrow{a} \mid a \in \Gamma(A)\}$).
4. Show that $\mathcal{G} = \Gamma(\ker(Ts)) + \Gamma(\ker(Tt))$.
5. Show that $\mathcal{G} \subset s^{-1}(\mathcal{F})$ and $\mathcal{G} \subset t^{-1}(\mathcal{F})$. *Hint*: prove that $\mathcal{G}$ is generated by vector fields s-related (resp. t-related) to vector fields in $\mathcal{F}$.
6. Show that $s^{-1}(\mathcal{F}) \subset \mathcal{G}$ and $t^{-1}(\mathcal{F})) \subset \mathcal{G}$. *Hint*: Show that every $X \in \mathcal{F}$ is s-related (resp. t-related) to a vector field in $\Gamma(\ker(Tt))$, equivalently $\Gamma(\ker(Ts))$.
7. Conclude that Γ is a bisubmersion for its basic singular foliation $\mathcal{F}$.
8. Show that $\mathrm{Out}_\Gamma \colon \Gamma \to \mathrm{OutSym}_\mathcal{F}$ is a groupoid morphism (the groupoid structure on $\mathrm{OutSym}_\mathcal{F}$ is defined in the next Sect. 5.1).

It is also very important to have non-examples in mind.

Exercise 4.25 Here are non-examples of bisubmersions. Let $\mathcal{F}$ be a singular foliation on M.

1. Show that $W := M \times M$, equipped with the projections onto the first and second components as source and target, is *not* a bisubmersion over $\mathcal{F}$, unless $\mathcal{F} = \mathfrak{X}_c(M)$ is the singular foliation of all compactly supported vector fields on M.
2. Show that $W := M$ equipped with the identity map as source and target is *not* a bisubmersion over $\mathcal{F}$, unless $\mathcal{F} = 0$.
3. Give an example of a manifold W, equipped with two surjective submersions $s, t \colon W \to M$, that do satisfy $s^{-1}(\mathcal{F}) = t^{-1}(\mathcal{F})$, and is still not a bisubmersion over $\mathcal{F}$.

4.2.2 The Crucial Example: A Bisubmersion for Every Finitely Generated Singular Foliation

There is a very natural bisubmersion over any finitely generated singular foliation $\mathcal{F}$ on M. Let $X_1, \ldots, X_r$ be generators of $\mathcal{F}$. For the sake of simplicity, we assume that they are complete vector fields, and leave it to the reader to generalize. Consider the following triple:

1. The manifold $\mathbb{R}^r \times M$.
2. The map $s\colon \mathbb{R}^r \times M \to M$ given by the projection on the second factor.

$$s\colon \begin{array}{c} \mathbb{R}^r \times M \to M \\ ((\lambda_1 \ldots, \lambda_r), m) \to m \end{array}. \tag{17}$$

3. The map $t\colon \mathbb{R}^r \times M \to M$ given by:

$$t\colon \begin{array}{c} \mathbb{R}^r \times M \to M \\ ((\lambda_1 \ldots, \lambda_r), m) \mapsto \phi^{X_1}_{\lambda_1} \circ \cdots \circ \phi^{X_r}_{\lambda_r}(m) \end{array}. \tag{18}$$

If one do not assume that the vector fields $X_1, \ldots, X_r$ are complete vector fields, then the previous map t still makes sense, but the initial manifold $\mathbb{R}^r \times M$ has to be replaced by a neighborhood of $(0, \ldots, 0) \times M$ (i.e., of the zero section of the trivial bundle $\mathbb{R}^r \times M \to M$). The map s is always a surjective submersion. The map t is also a surjective submersion at least, again, in a neighborhood of the zero section.

The following proposition is very important and very non-trivial. Here, our inspiration is not [2] but we adapted a proof by Claire Debord [10]. We will soon state and prove a slightly different but more general statement (see Proposition 4.35 below).

Proposition 4.26 (A Crucial Example) *Let $\mathcal{F}$ be a finitely generated singular foliation. There is a neighborhood $\mathcal{V}$ of the zero section in $\mathbb{R}^r \times M \to M$ such that $M \xleftarrow{s} \mathcal{V} \xrightarrow{t} M$, with s, t as in Eqs. (17)–(18), is a bisubmersion over $\mathcal{F}$.*

Remark 4.27 If $\mathcal{F}$ is generated, as a module over smooth functions, by vector fields which are integrable by quadrature ($\simeq$ such that the flow can be computed "by hand"), then Proposition 4.26 means that there is an *explicit* bisubmersion for $\mathcal{F}$.

Proof Recall from (18) that the target map t is given by $((\lambda_1 \ldots, \lambda_r), m) \mapsto \phi^{X_1}_{\lambda_1} \circ \cdots \circ \phi^{X_r}_{\lambda_r}(m)$ and from Eq. (17) that the source map s is given by $((\lambda_1 \ldots, \lambda_r), m) \mapsto m$. The vector field on $\mathbb{R}^r \times M$ given by

$$Z_i\colon ((\lambda_1 \ldots, \lambda_r), m) \mapsto (0, X_i|_m)$$

is s-related to $X_i \in \mathcal{F}$ for every $i = 1, \ldots, r$. Hence, $Z_i \in s^{-1}(\mathcal{F})$. As a consequence, $\lambda_i Z_i \in s^{-1}(\mathcal{F})$ so that its flow $\phi_1^{\lambda_i Z_i} = \phi_{\lambda_i}^{Z_i}$ is a symmetry[32] of $s^{-1}(\mathcal{F})$ by Corollary 7.13 in chapter "What Is a Singular Foliation?". Hence, $\phi_{\lambda_1}^{Z_1} \circ \cdots \circ \phi_{\lambda_r}^{Z_r}$ is a symmetry of $s^{-1}(\mathcal{F})$—at least on the open set where it makes sense. Now, notice that

$$s \circ \phi_{\lambda_1}^{Z_1} \circ \cdots \circ \phi_{\lambda_r}^{Z_r}(\lambda_\bullet, m) = \phi_{\lambda_1}^{X_1} \circ \cdots \circ \phi_{\lambda_r}^{X_r}(m) = t(\lambda_\bullet, m)$$

at least on the open set where it makes sense.[33] Since there is a symmetry Φ of $s^{-1}(\mathcal{F})$ such that $s \circ \Phi = t$, we have $s^{-1}(\mathcal{F}) = t^{-1}(\mathcal{F})$ at least in a neighborhood of $\{0\} \times M$.

The rest of the proof uses the same techniques as in [10]. We have to prove that any vector field in $s^{-1}(\mathcal{F}) = t^{-1}(\mathcal{F})$ is the sum of a vector field in $\mathrm{Ker}(Ts)$ and a vector field in $\mathrm{Ker}(Tt)$ at least in a neighborhood of $\{0\} \times M$. Since we have already proven that $s^{-1}(\mathcal{F}) = t^{-1}(\mathcal{F})$, and since $\Gamma(\ker Ts) \subset t^{-1}(\mathcal{F})$ while $\Gamma(\ker Tt) \subset s^{-1}(\mathcal{F})$, we already have the inclusion

$$t^{-1}(\mathcal{F}) \supset \Gamma(\ker Ts) + \Gamma(\ker Tt).$$

We need to show the converse inclusion. For every $i \in \{1, \ldots, r\}$, let $\mathbf{A}_i(\lambda_i, m) := (a_{ij}^l)_{jl}(\lambda_i, m)$ be a $r \times r$ matrix with coefficients in $C^\infty(M)$ such that $\mathbf{A}_i(0, m) = \mathrm{Id}$ for all $m \in M$ and

$$\begin{pmatrix} \left(\phi_{\lambda_i}^{X_i}\right)_*(X_1) \\ \vdots \\ \left(\phi_{\lambda_i}^{X_i}\right)_*(X_r) \end{pmatrix}(m) = \mathbf{A}_i(\lambda_i, m) \begin{pmatrix} X_1 \\ \vdots \\ X_r \end{pmatrix}(m). \tag{19}$$

Such a matrix exists by Proposition 7.11 Sect. 7 in chapter "What Is a Singular Foliation?". A direct computation gives for every $i = 1\ldots, r$:

$$T_{(\lambda,m)}t\left(\frac{\partial}{\partial \lambda_i}\right) = \frac{d}{d\epsilon}\bigg|_{\epsilon=0} \left(\phi_{\lambda_1}^{X_1} \circ \cdots \circ \phi_{\lambda_i+\epsilon}^{X_i} \left(\phi_{\lambda_{i+1}}^{X_{i+1}} \circ \cdots \circ \phi_{\lambda_r}^{X_r}(m)\right)\right)$$

$$= T\left(\phi_{\lambda_1}^{X_1} \circ \cdots \circ \phi_{\lambda_{i-1}}^{X_{i-1}}\right)\left(X_i\big|_{\phi_{\lambda_i}^{X_i} \circ \phi_{\lambda_{i+1}}^{X_{i+1}} \circ \cdots \circ \phi_{\lambda_r}^{X_r}(m)}\right)$$

$$= T\left(\phi_{\lambda_1}^{X_1} \circ \cdots \circ \phi_{\lambda_{i-1}}^{X_{i-1}}\right)\left(X_i\big|_{\phi_{-\lambda_{i-1}}^{X_{i-1}} \circ \cdots \circ \phi_{-\lambda_1}^{X_1}(t(\lambda_\bullet, m))}\right)$$

---

[32] (at least whenever it is defined—hence a local symmetry).
[33] Here and below, we use the shorthand $\lambda_\bullet = (\lambda_1, \ldots, \lambda_r)$. Also, we will not recall at every line that expressions are only defined on an open subset containing the zero section M.

$$= \left(\phi_{\lambda_1}^{X_1} \circ \cdots \circ \phi_{\lambda_{i-1}}^{X_{i-1}}\right)_* X_i \bigg|_{t(\lambda_\bullet, m)}$$

$$= \left(\phi_{\lambda_1}^{X_1}\right)_* \circ \cdots \circ \left(\phi_{\lambda_{i-1}}^{X_{i-1}}\right)_* X_i \bigg|_{t(\lambda_\bullet, m)}$$

which, in view of Eq. (19) gives

$$T_{(\lambda_\bullet, m)} t \begin{pmatrix} \frac{\partial}{\partial \lambda_1} \\ \vdots \\ \frac{\partial}{\partial \lambda_r} \end{pmatrix} = \mathfrak{M}(\lambda_\bullet, m) \begin{pmatrix} X_1 \\ \vdots \\ X_r \end{pmatrix} (t(\lambda_\bullet, m))$$

where $\mathfrak{M}(\lambda_\bullet, m)$ is the $r \times r$ matrix whose i-th line is the i-th line of the following matrix:

$$\mathfrak{M}(\lambda_\bullet, m) = \mathbf{A}_{i-1}\left(\lambda_{i-1}, \phi_{-\lambda_{i-2}}^{X_{i-2}} \circ \cdots \circ \phi_{-\lambda_1}^{X_1} \circ t(\lambda_\bullet, m)\right)$$
$$\times \mathbf{A}_{i-2}\left(\lambda_{i-2}, \phi_{-\lambda_{i-3}}^{X_{i-3}} \circ \cdots \circ \phi_{-\lambda_1}^{X_1} \circ t(\lambda_\bullet, m)\right)$$
$$\times \cdots$$
$$\cdots \times \mathbf{A}_1(\lambda_1, t(\lambda, m))$$

for all $1 \leq i \leq r$. Here $\times$ stands for the product of $r \times r$ matrices. Despite its extraordinary description, the only point that matters in that this matrix exists and that, since $\mathfrak{M}(0, \ldots, 0, m) = \mathrm{Id}$ for all $m \in M$, it is invertible on an open subset $\mathcal{V}$ of $\mathbb{R}^r \times M$ that contains $\{0\} \times M$. In turn, this invertibility implies there exists for every $i \in 1, \ldots, r$ a linear combination of the vector fields $\frac{\partial}{\partial \lambda_1}, \ldots, \frac{\partial}{\partial \lambda_r}$, depending smoothly on the point in $\mathcal{V}$, which is t-related to X_i. Said otherwise, there exists for every $i \in 1, \ldots, r$ a vector field $Z_i \in \mathfrak{X}(\mathcal{V})$ which lies inside $\ker(Ts)$ and is t-related with X_i. By construction, $t^{-1}(\mathcal{F})$ is generated by $Z_1, \ldots, Z_r$ and vector fields in $\ker(Tt)$. This implies that, after restriction to $\mathcal{V}$:

$$s^{-1}(\mathcal{F}) = t^{-1}(\mathcal{F}) \subseteq \Gamma(\ker Ts) + \Gamma(\ker Tt).$$

This completes the proof. □

4.2.3 The Previous Example Made More Abstract: Anchored Bundles as Bisubmersions

Proposition 4.26 can be made more abstract by using an anchored bundle[34] over $\mathcal{F}$. The idea is to mimic the construction of the so-called parallel A-paths in integration of Lie

---

[34] We refer to Sect. 1.1 for results are the existence of anchored bundles. In short: in the smooth case, it exists if and only if $\mathcal{F}$ is finitely generated, and it always exists locally.

algebroids as in [7]. Let $\mathcal{F}$ be a singular foliation on M. Let (A, ρ) be an anchored bundle such that $\rho(\Gamma(A)) = \mathcal{F}$. We denote by $\pi: A \to M$ the projection onto the base manifold.

Definition 4.28 (Anchored Paths) Let (A, ρ) be an anchored bundle over $\mathcal{F}$. We say that a path $a: I \to A$ is *anchored* if

$$\frac{d\gamma(t)}{dt} = \rho_{\gamma(t)}(a(t))$$

where $\gamma = \pi \circ a : I \to M$ is the projection of $a(t)$ onto M.

Notice that the path $\gamma(t) = \pi \circ a(t)$ in Definition 4.28 can not "jump" from one leaf of $\mathcal{F}$ to another leaf.

Exercise 4.29 Show that for any anchored path $t \mapsto a(t)$, there exists a smooth time-depending section $(a_t)_{t \in I}$ of A such that $a_t(\gamma(t)) = a(t)$. Show that the $\gamma(t)$ is an integral curve of a smooth time-dependent vector field in $\mathcal{F}$.

Let us choose an affine connection[35] on A:

$$\nabla: \mathfrak{X}(M) \times \Gamma(A) \to \Gamma(A)$$
$$(X, a) \mapsto \nabla_X a$$

Definition 4.30 (Parallel Anchored Paths) Let $A \xrightarrow{\pi} M$ be an anchored bundle and ∇ be a connection on A. We say that an anchored path $a(t)$ is ∇-parallel if it satisfies:

$$\nabla_{\dot\gamma(t)} a(t) = 0$$

where $\gamma = \pi \circ a : I \to M$ is the projection of $a(t)$ onto M.

Here is a result which is purely a differential geometry result.

Lemma 4.31 *Every element $a \in A$ is the starting point of a ∇-parallel anchored path $t \mapsto a^\nabla(t)$, defined on some open interval containing $0 \in \mathbb{R}$, that we call the geodesic*

---

[35] They always exist in the smooth setting, and always exist locally in the complex or real-analytic settings.

starting at $a \in A$. Moreover, there is a unique linear vector field Ξ on A whose integral curves are these geodesics, i.e.,

$$a^\nabla(t) = \Phi_t^\Xi(a) \text{ for all } a \in A,$$
$$\text{and all } t \in \mathbb{R} \text{ for which it is defined.}$$

We call Ξ *the* geodesic vector field.

Proof The affine connection ∇ can be seen a sub-vector bundle $H^\nabla \subset TA$ in direct sum with the kernel of $T\pi$. In particular, for any $a \in A$, $T_a\pi : H^\nabla \longrightarrow T_{\pi(a)}M$ is one-to-one. We denote its inverse map by $u \mapsto H_a^\nabla(u)$ for any $u \in T_{\pi(a)}M$. We define Ξ by

$$\Xi|_a := H_a^\nabla\left(\rho_{\pi(a)}(a)\right).$$

Equivalently, Ξ is the only section of H^∇ such that $T_a\pi(\Xi|_a) = \rho(a)$ at every point $a \in A$. It is routine to check that the integral curves of Ξ are precisely the above defined geodesics. □

Exercise 4.32 Show that the geodesic starting from a point of the zero section is a constant path.

Exercise 4.33 Show that for $A = TM$ and $\rho = \mathrm{id}$, the geodesics defined above are the usual geodesics of an affine connection on M.

Exercise 4.34 Let (A, ρ) be an anchored bundle for a singular foliation $\mathcal{F}$. Let ∇ be a connection on A. Choose $m \in M$.

1. Show that for any $a \in A_m$, the geodesic path starting from a remains inside the leaf L_m through M, i.e., $t \mapsto \pi \circ \Phi_\Xi^t(a) \in L_m$ for all t for which it is defined.
2. Show that there is a neighborhood $\mathcal{V}$ of m in L_m such that for any $m' \in \mathcal{V}$, m and m' are the starting and ending point of the base path of some geodesic.
3. Here is an open question: are any two points in the same leaf the starting and end point of the base path of a geodesic?

For a given $a \in A$, the geodesic path $t \mapsto a^\nabla(t) = \Phi_t^\Xi(a)$ may not be defined for all t, but it follows from Exercise 4.32 there is a neighborhood $\mathcal{U}_A \subset A$ of the zero section where it is defined for all $t \in [0, 1]$. We denote by Φ_1^Ξ the map from $\mathcal{U}_A$ to A that sends a to $\Phi_1^\xi(a) = a^\nabla(1)$. Now, let us consider the triple made of

1. the neighborhood $\mathcal{U}_A$ of the zero section in A,
2. the projection $A \xrightarrow{\pi} M$ that we rename s,
3. the composition t of $a \mapsto \Phi_1^\Xi(a)$ with the projection : $A \xrightarrow{\pi} M$.

Notice that, by construction:

$$t = s \circ \Phi_1^\Xi \qquad (20)$$

Proposition 4.35 (The Same as Proposition 4.26 but More Abstract: Anchored Bundles Are "Fundamental" Bisubmersions) *Let (A, ρ) be any anchored bundle such that $\mathcal{F} = \rho(\Gamma(A))$, and ∇ be a connection on A. There is a neighborhood $\mathcal{A}$ of the zero section in A on which $M \xleftarrow{s} \mathcal{A} \xrightarrow{t} M$, with s, t as above, is a bisubmersion of $\mathcal{F}$.*

We call *fundamental bisubmersion* of $\mathcal{F}$ a bisubmersion $M \xleftarrow{s} \mathcal{A} \xrightarrow{t} M$ of the form given in Proposition 4.35, associated to some anchored bundle (A, ρ) and some connection ∇. We start with a lemma about the geodesic vector field Ξ of Lemma 4.31.

Lemma 4.36 *The geodesic vector field Ξ belongs to $s^{-1}(\mathcal{F})$.*

Proof By construction (see Sect. 5.2 in chapter "What Is a Singular Foliation?"), $s^{-1}(\mathcal{F})$ is generated by:

1. vertical vector fields, i.e., vector fields tangent to the fiber of $s = \pi : A \to M$,
2. the horizontal lifts $H^\nabla(X)$ of vectors $X \in \mathcal{F}$.

As a consequence, a vector field which is locally of the form $\sum_{i=1}^r f_i H^\nabla(\rho(e_i))$, where $e_1, \ldots, e_r$ is a local trivialization of A and $f_1, \ldots, f_r$ smooth real-valued functions on A, belongs to $s^{-1}(\mathcal{F})$. By construction, for any local trivialization $e_1, \ldots, e_r$ of A, and any $a \in A$:

$$\Xi_{|a} = \sum_{i=1}^r \langle e_i^*, a \rangle H^\nabla(\rho(e_i))$$

where $e_1^*, \ldots, e_r^*$ is the dual trivialization of A^*. This proves the claim. □

Here is an immediate consequence of Lemma 4.36.

Lemma 4.37 *There is a neighborhood of the zero section in A where $t^{-1}(\mathcal{F}) = s^{-1}(\mathcal{F})$.*

Proof Lemma 4.36 and Corollary 7.13 in chapter "What Is a Singular Foliation?" imply that the flow of Φ_t^Ξ is a symmetry of $s^{-1}(\mathcal{F})$. In particular, Φ_1^Ξ is a symmetry[36] of $s^{-1}(\mathcal{F})$ on some neighborhood $\mathcal{U}_A$ where this time 1-flow is well-defined:

$$\left(\Phi_1^\Xi\right)^{-1}\left(s^{-1}(\mathcal{F})\right) = s^{-1}(\mathcal{F}).$$

---

[36] And even an inner-symmetry, but this is not important here.

Since the target map consists precisely in composing Φ_1^Ξ with s (see (20)), we have:

$$t^{-1}(\mathcal{F}) = (s \circ \Phi_1^\Xi)^{-1}(\mathcal{F}) = \left(\Phi_1^\Xi\right)^{-1}\left(s^{-1}(\mathcal{F})\right) = s^{-1}(\mathcal{F}).$$

This proves therefore that $s^{-1}(\mathcal{F}) = t^{-1}(\mathcal{F})$. □

Proof of Proposition 4.35 In view of Lemma 4.37, we have to prove that any vector field $Z \in s^{-1}(\mathcal{F}) = t^{-1}(\mathcal{F})$ decomposes as $X + Y$ with X a vector field tangent to the fiber of s, i.e., the fiber of the canonical projection $\mathcal{A} \to M$, and the fiber of t. We are allowed to replace $\mathcal{A}$ by a smaller neighborhood of the zero section.

To start with, there is a natural anchored bundle structure over the singular foliation $s^{-1}(\mathcal{F})$ given by the vector bundle $s^!A \oplus s^!A \to \mathcal{A}$ equipped with the anchor[37]

$$\rho_{A \oplus A} : (s^!a_1, s^!a_2) \mapsto H(\rho(a_1)) + a_2^v$$

where $H : \mathfrak{X}(M) \to \mathfrak{X}(\mathcal{A})$ is the Ehresmann connection on $\mathcal{A}$ associated to ∇, and where a^v is the vertical vector field[38] on $\mathcal{A}$ associated to $a \in \Gamma(A)$. Since $\Xi \in s^{-1}(\mathcal{F})$, Proposition 2.11 allows lifting Φ_1^Ξ to an isomorphism of anchored bundles

$$\begin{array}{ccc} s^!A \oplus s^!A & \xrightarrow{\Psi_1} & s^!A \oplus s^!A \\ \downarrow & & \downarrow \\ \mathcal{A} & \xrightarrow{\Phi_1^\Xi} & \mathcal{A} \end{array}$$

Since $0 \oplus s^!A$ has by construction an image through the anchor map $\rho_{A \oplus A}$ which lies in $\ker(Ts)$, and since Φ_1^Ξ intertwines $\ker(Ts)$ and $\ker(Tt)$, the sub-vector bundle $\Psi_*(s^!A \oplus 0) \subset s^!A \oplus s^!A$ has an image through the anchor map that belongs to $\ker(Tt)$. If we can show that for some open neighborhood $\mathcal{A}' \subset \mathcal{A}$ of the zero section, the vector bundles $\Psi_*(0 \oplus s^!A)$ and $0 \oplus s^!A$ are in direct sum inside $s^!A \oplus s^!A$, then the result follows, since for any vector field Z in $s^\mathcal{F}$, it suffices to decompose a section $\alpha \in \Gamma(s^!A \oplus s^!A)$ such that

$$\rho_{A \oplus A}(\alpha) = Z$$

---

[37] For $\phi : M \to N$ a map and $E \to N$ a vector bundle, we denote by $\phi^!E$ the pull-back vector bundle $\phi^!E_m \simeq E_{\phi(m)}$ and by $\phi^!e \in \Gamma(\phi^!E)$ the pull-back of a section $e \in \Gamma(E)$.

[38] For very point a, there is an injection $A_{s(a)} \hookrightarrow T_a A$ due to the vector bundle structure that the fibers of $A \to M$ possesses. This allows to see sections of A as vector fields on A valued on $\ker(Ts)$.

under the form $\alpha = \alpha_1 + \alpha_2$ with α_1, α_2 in $0 \oplus s^!A$ and $\Psi_*(0 \oplus s^!A)$ respectively. By construction $Z = X_1 + X_2$ with $X_1 = \rho_{A \oplus A}(\alpha_1)$ and $X_2 = \rho_{A \oplus A}(\alpha_2)$ to have a decomposition of Z as the sum of an element in $\Gamma(\ker(Ts))$ and an element in $\Gamma(\ker(Tt))$.

Let us show this point. The vector field Ξ vanishes on the zero section, so that each point of the zero section is a fixed point of Φ_1^Ξ. Also, it satisfies, on any point m of the zero section, and for any $a \in \Gamma(A)$, the relation

$$[\Xi, a^v]_{|m} = \rho(a_m) \tag{21}$$

where the right-hand side, which belongs to $T_m M$, is to be seen as an element in $T_m A$ with the help of the zero section $T_m A \hookrightarrow T_m A$. Now, for proving Proposition 2.11, we lifted Ξ to a fiberwise linear vector field on $s^!A \oplus s^!A$. Since m is a fixed point, upon identifying linear vector fields with linear endomorphism of $A_m \oplus A_m$, Eq. (21) implies that this lift can be chosen to be given by the matrix:

$$\begin{pmatrix} 0 & \mathrm{id}_{A_m} \\ 0 & 0 \end{pmatrix}.$$

Its flow is therefore given by the matrix

$$\begin{pmatrix} \mathrm{id}_{A_m} & \mathrm{id}_{A_m} \\ 0 & \mathrm{id}_{A_m} \end{pmatrix}.$$

Hence, the restriction of $\Phi_1(0 \oplus s^!A)$ to M is the diagonal sub-bundle of $A \oplus A$. Since the vector sub-bundle $0 \oplus A$ is in direct sum with the diagonal sub-bundle of $A \oplus A$, the desired property holds true in a neighborhood $\mathcal{A}'$ of the zero section. This concludes the proof. □

Remark 4.38 When the anchored bundle is the trivial bundle $A = \mathbb{R}^r \times M \to M$ with anchor

$$\rho((t_1, \ldots, t_r), m) = \sum_{i=1}^{r} t_i X_i(m)$$

as in the proof of Proposition 1.5, and the connection is the trivial connection:

$$\nabla_X(f_1, \ldots, f_r) = (X[f_1], \ldots, X[f_r]) \text{ for any } f_1, \ldots, f_r \in C^\infty(M)$$

then the parallel path stating from $((t_1, \ldots, t_r), m) \in A$ is the path:

$$t \mapsto \left((t_1, \ldots, t_r), \phi_t^{\sum_{i=1}^{r} t_i X_i}(m)\right)$$

In this case, the bisubmersion in Proposition 4.35 is similar (but different) to the bisubmersion described in Proposition 4.26.

Remark 4.39 The fundamental bisubmersion of Proposition 4.35 could more be equivalently defined, in a more symmetric and therefore pleasant way, by using the maps $s = \pi \circ \Phi^\Xi_{-1/2}$ and $t = \pi \circ \Phi^\Xi_{+1/2}$ with $\pi : A \to M$ the projection onto the base manifold. Proposition 4.35 still stands with this definition.

4.2.4 Discussion on the Notion of Bisubmersion

As already mentioned, the word "bisubmersion" used alone does not make sense if used alone: only the expression "bisubmersion over the singular foliation $\mathcal{F}$" makes sense. To clarify this point, we introduce the following notion (maybe not interesting by itself, but practical for pedagogical purposes).

Definition 4.40 We call *twin-submersion* the data of

1. two manifolds X, M,
2. two surjective submersions $s, t : X \to M$.

A *unit map for twin-submersions* is a smooth map $\epsilon : M \to X$ which is a section of both s and t.

Here is a natural question:

Question 4.41 *Given a twin-submersion (X, M, s, t), when is it a bisubmersion for some singular foliation $\mathcal{F}$?*

We answer this question through the following exercises.

Exercise 4.42 Show that a twin-submersion can not be a bisubmersion for two different singular foliations on M.
Hint: Show that $\mathcal{F} \neq \mathcal{F}'$ implies $s^{-1}(\mathcal{F}) \neq s^{-1}(\mathcal{F}')$.

Exercise 4.43 Let (X, M, s, t) be a twin-submersion.

1. Show that the following two conditions are equivalent.
   (i) $\mathcal{G} := \Gamma(\ker(Ts)) + \Gamma(\ker(Tt))$ is stable under Lie bracket.
   (ii) $[\Gamma(\ker(Ts)), \Gamma(\ker(Tt))] \subset \Gamma(\ker(Ts)) + \Gamma(\ker(Tt))$

Show that the previous equivalent two conditions are in turn equivalent to:
(iii) ker(Tt) is generated by s-projectable[39] vector fields and ker(Ts) is generated by t-projectable[40] vector fields.
Hint: Show that $\mathcal{G}$ is generated by vector fields X which are s-projectable. Then write $X = X^s + X^t$ with X^s, X^t in the kernels of Ts and Tt, respectively. Show that X^t is s-projectable and belongs to the kernel of Tt.

2. We now assume that the fibers of s and t are connected. Show that if one of the equivalent conditions above hold, then there exists singular foliations $\mathcal{F}_s$ and $\mathcal{F}_t$ on M such that

$$\mathcal{G} = s^{-1}(\mathcal{F}_s) = t^{-1}(\mathcal{F}_t)$$

3. Let us now assume that there exists a unit map $\epsilon \colon M \hookrightarrow X$, i.e a left inverse of both s and t. Show that $\mathcal{F}_s = \mathcal{F}_t$.

4. Show that if a twin-submersion admits a unit map and has connected s- and t-fibers, then there exists a singular foliation with respect to which it is a bisubmersion if and only if

$$[\Gamma(\ker(Ts)), \Gamma(\ker(Tt))] \subset \Gamma(\ker(Ts)) + \Gamma(\ker(Tt)).$$

4.2.5 More Examples of Bisubmersions

By composing the source or the anchor with a symmetry of a bisubmersion over $(M, \mathcal{F})$, one still obtains a bisubmersion over $\mathcal{F}$. The following definition therefore makes sense.

> **Definition 4.44 (Composing Bisubmersions with Symmetries)** Let $M \xleftarrow{s} W \xrightarrow{t} M$ be a bisubmersion over $\mathcal{F}$, and $\phi \colon M \to M$ a symmetry of $\mathcal{F}$. Then
>
> $$M \xleftarrow{\phi \circ s} W \xrightarrow{t} M \text{ and } M \xleftarrow{s} W \xrightarrow{\phi^{-1} \circ t} M$$
>
> are bisubmersions of $\mathcal{F}$ again. We call them the *right and left composition by the symmetry* ϕ, respectively.

An analogous construction can be done for local bisubmersions over $\mathcal{F}$. In fact, any bisubmersion over $\mathcal{F}$ is of the type constructed in Proposition 4.35, up to composition by a symmetry, at least locally. This is the topic of the next Exercise.

---

[39] I.e. s-related to a vector field on M, see Definition 5.4 in chapter "What Is a Singular Foliation?".
[40] I.e. t-related to a vector field on M, see Definition 5.4 in chapter "What Is a Singular Foliation?".

Exercise 4.45 Let $M \xleftarrow{s} W \xrightarrow{t} M$ be a bisubmersion of $\mathcal{F}$. Choose Σ a bisection. Let N_Σ be the normal bundle of Σ in W as in Lemma 4.17.

Show that there is a neighborhood of any bisubmersion Σ in W isomorphic, as a bisubmersion, to the left composition by $\underline{\Sigma}$ of a fundamental bisubmersion (see Proposition 4.35) of the anchored bundle (A_Σ, ρ). (*Hint*: use Theorem 4.62 below, together with Exercise 4.64).

4.3 Left and Right "Actions" of Anchored Bundles

Let $\mathcal{F}$ be a singular foliation on M. So far, we have seen two classes of objects "over $\mathcal{F}$".

1. bisubmersions $M \xleftarrow{s} W \xrightarrow{t} M$ of $\mathcal{F}$.
2. anchored bundles (A, ρ) over $\mathcal{F}$.

Let us assume that we are given both. What is the relation between them? Proposition 4.35 was already a partial answer to that question, but there is more. If W is a Lie groupoid with Lie algebroid (A, ρ), then W would be a bisubmersion and (A, ρ) would be an anchored bundle for the same singular foliation $\mathcal{F} = \rho(\Gamma(A))$ (see exercise 4.24). In that case, left and right Lie algebroids actions of A on W can be defined with the help of the notion of left and right invariant vector fields. Those can also be defined through left and right Lie groupoid action. Despite the lack of product on a bisubmersion, there is a very similar construction. Or course, it is not an "action" *stricto sensu*, but the following object can be defined, following [26].

Definition 4.46 (Right and Left Action) Let $(M, \mathcal{F})$ be a foliated manifold. Let $M \xleftarrow{s} W \xrightarrow{t} M$ and (A, ρ) be a bisubmersion and an anchored bundle over $(M, \mathcal{F})$, respectively.

We call, respectively, right and left actions of (A, ρ) on the bisubmersion W vector bundle morphisms:[a]

$$L: s^!A \longrightarrow \ker(Tt) \text{ and } R: t^!A \longrightarrow \ker(Ts)$$

making the following diagrams commutative:

$$\begin{array}{ccc} s^!A & \xrightarrow{L} & \ker(Tt) \\ {\scriptstyle \rho}\downarrow & \swarrow{\scriptstyle Ts} & \\ TM & & \end{array} \quad \text{and} \quad \begin{array}{ccc} \ker(Ts) & \xleftarrow{R} & t^!A \\ & \searrow{\scriptstyle Tt} & \downarrow{\scriptstyle \rho} \\ & & TM \end{array}$$

---

[a] $s^!A, t^!A$ are the pull-back vector bundles of the vector bundle $A \to M$ to W through s and t, respectively.

The notion is interesting for many reasons. To start with, it really exists:

> **Proposition 4.47 (Right and Left Actions Exist)** *Let $(M, \mathcal{F})$ be a foliated manifold. Let $M \xleftarrow{s} W \xrightarrow{t} M$ of $\mathcal{F}$ and (A, ρ) be a bisubmersion and an anchored bundle over $(M, \mathcal{F})$ respectively. Right and left-actions of (A, ρ) on W exist.*

Proof We will prove it for the right action R. Let $x \in W$, and let $e_1, \ldots, e_r$ be a local trivialization of A near $t(x)$. Then $\rho(e_1), \ldots, \rho(e_r)$ are generators of $\mathcal{F}$. In view of Lemma 4.6, there exists a neighborhood $\mathcal{U}$ of x and vector fields $X_1, \ldots, X_r$ in $\Gamma_{\mathcal{U}}(\ker(Ts))$ such that X_i is t-related to $\rho(e_i)$. We define a vector bundle morphism $R_{\mathcal{U}}$ from $t^!A|_{\mathcal{U}}$ to $\ker(Ts)|_{\mathcal{U}}$ that satisfies the desired property in a neighborhood of x by mapping the sections[41] $t^!e_1, \ldots, t^!e_r$ to $X_1, \ldots, X_r$. The proposition then follows from the fact that given a partition of unity $(\mathcal{U}_i, \chi_i)_{i \in I}$ of W such that such a vector bundle morphism $R_{\mathcal{U}_i}$ as in the statement exists on $\mathcal{U}_i$, the linear combination

$$R := \sum_{i \in I} \chi_i R_{\mathcal{U}_i}$$

is a vector bundle morphism[42] from $t^!A$ to $\ker(Ts)$ that satisfies all desired properties. □

Remark 4.48 When the bisubmersion W is a Lie groupoid and A is its Lie algebroid (see exercise 4.24), then the usual right and left actions of the Lie algebroid are instances of the previous left and right actions. They are not the unique ones in general.

For any left or right actions and any $a \in \Gamma(A)$, a vector field on W is defined by $L(s^!a)$ or $R(t^!a)$ respectively. We denote these vector fields by $\vec{a}$ and $\overleftarrow{a}$ respectively.
The maps:

$$\Gamma(A) \to \mathfrak{X}(W)$$
$$a \mapsto \vec{a}$$
$$a \mapsto \overleftarrow{a}$$

satisfy for any $a \in \Gamma(A)$ the following conditions by construction:

1. the vector field $\vec{a}$ (resp. $\overleftarrow{a}$) s-projects (resp. t-projects) to $\rho(a) \in \mathfrak{X}(M)$

---

[41] Recall that for $\phi : N \to M$ a map, $E \to M$ a vector bundle, and e a section of E, we denote by $\phi^!e$ the section of $\pi^!E = E \times_M N$ whose value at $n \in M$ is $(e|_{\phi(n)}, n)$.

[42] (now defined on the whole manifold W).

2. the vector field $\vec{a}$ (resp. $\overleftarrow{a}$) is tangent to the fibers of t (resp. s).
3. For any $F \in C^\infty(M)$, we have $\overrightarrow{Fa} = t^*F \, \vec{a}$ and $\overleftarrow{Fa} = s^*F \, \overleftarrow{a}$.
4. For any $a, b \in \Gamma(A)$, the vector field $[\vec{a}, \overleftarrow{b}]$ is bi-vertical.

Remark 4.49 For a different choice for the right action, the corresponding vector fields $\vec{a}$ will differ by bivertical vector fields. Hence, the class of $\overleftarrow{a}, \vec{a}$ modulo bivertical vector fields does not depend on the choice of a right action. The same holds for the left action.

Lemma 4.50 *Let W be a bisubmersion over $(M, \mathcal{F})$. Any $w \in W$ admits a neighborhood $\mathcal{V}$ such that there exists* (i) *anchored bundles (A_R, ρ_R) and (A_L, ρ_L) over $\mathcal{F}$, defined on $s(\mathcal{V})$ and $t(\mathcal{V})$, and* (ii) *a choice of right and left actions R and L near w such that both R and L*

$$L : s^!A_L \xrightarrow{\simeq} \ker(Tt)|_\mathcal{V} \text{ and } R : t^!A_R \xrightarrow{\simeq} \ker(Ts)|_\mathcal{V}$$

are vector bundle isomorphisms over $\mathcal{V}$. Moreover, if $s(w) = t(w)$, then one can assume (A_R, ρ_R) and (A_L, ρ_L) coincide near $s(w) = t(w)$.

Proof Let Σ be a local bisection through w and N_Σ be the normal bundle of Σ in W. In view of Lemma 4.17, this normal bundle can be seen as an anchored bundle for M on $s(\Sigma)$ or $t(\Sigma)$: it suffices to identify the henceforth obtained vector bundle with $\ker(Tt)$ or to $\ker(Ts)$ through s and t and to equip it with Ts and Tt, respectively. The proof of Proposition 4.47 can then be adapted so that R and L are the identity maps on $\ker(Ts)|_\Sigma$ or to $\ker(Tt)|_\Sigma$. Hence, both R and L have to be invertible on a neighborhood of Σ in W. This completes the proof. □

Now, assume that the anchored bundle (A, ρ) is equipped with a bracket making it an almost Lie algebroid: such a bracket exists by Proposition 2.4. It is interesting to notice that for every choice of such an almost Lie algebroid bracket on (A, ρ), the vector fields that measures the default of the left and right actions to preserve the brackets, i.e., the vector fields

$$\overleftarrow{[a,b]_A} - [\overleftarrow{a}, \overleftarrow{b}] \text{ and } \overrightarrow{[a,b]_A} - [\vec{a}, \vec{b}] \text{ with } a, b \in \Gamma(A),$$

are bi-vertical vector fields.[43] Of course, it does not make sense to hope for a Lie algebra morphism, since $\Gamma(A)$ is not a Lie algebra. However, the following exercise describes a natural and canonical Lie algebra morphism.

---

[43] We also saw that the vector field $[\vec{a}, \overleftarrow{b}]$ is always bi-vertical.

Exercise 4.51 Let W be a bisubmersion over $(M, \mathcal{F})$. Show that:

1. t-projectable[44] sections of $\ker(Ts)$ form a Lie algebra, denoted by $\Gamma(\ker(Ts))^t$.
2. Bivertical vector fields form a Lie ideal (denoted Bivert) of the previous Lie algebra.
   Let (A, ρ) be an anchored bundle over $\mathcal{F}$. Let $\text{Ker}(\rho) \subset \Gamma(A)$ be the subspace of all $a \in \Gamma(A)$ such that $\rho(a) = 0$.
3. Show that $\text{Ker}(\rho) \subset \Gamma(A)$ is a left and right ideal for the bracket of $\Gamma(A)$ and that the quotient space $\Gamma(A)/\text{Ker}(\rho)$ is a Lie algebra isomorphic to $\mathcal{F}$.
4. Show that for any choice of a right-action, $\overrightarrow{a}$ is bivertical if $a \in \text{Ker}(\rho) \subset \Gamma(A)$.
5. Show that the induced map:

$$\mathcal{F} \simeq \frac{\Gamma(A)}{\text{Ker}(\rho)} \longrightarrow \frac{\Gamma(\ker(Ts))^t}{\text{Bivert}}$$

is a Lie algebra morphism.
6. Show that this Lie algebra morphism does not depend on the choice of the right action R (see Remark 4.49).
7. Show that this Lie algebra morphism is in fact an isomorphism.

We now use the notion of left and right action to show the following interesting point about Out_W. Let $(M, \mathcal{F})$ be a foliated manifold, W and W' two bisubmersions over $\mathcal{F}$. We intend in Exercise 4.52 below to show that if $x \in W$ and $x' \in W'$ are related by $\text{Out}_W(x) = \text{Out}_W(x')$, then there exists local bisections Σ and Σ' through x and x' respectively such that $\underline{\Sigma} = \underline{\Sigma'}$:

Exercise 4.52 Let Σ_0 and Σ'_0 be bisections of W and W' through x and x'. By assumption $\Phi := \Sigma_0 \circ (\Sigma'_0)^{-1}$ is an inner symmetry admitting $m = t(x) = t'(x')$ as a very fixed point, i.e., there exists a smooth time-dependent vector field $(X_\epsilon)_{\epsilon \in [0,1]}$ in $\mathcal{F}$ whose time ϵ-flow Φ_ϵ is defined for all $\epsilon \in [0, 1]$ and whose time 1-flow is Φ (at least in a neighborhood of X). Moreover, by Exercise 3.13 in chapter "What Is a Singular Foliation?", $X_\epsilon = \rho(a_\epsilon)$ where $(a_\epsilon)_{\epsilon \in [0,1]}$ is a smooth[45] time dependent section of $\Gamma(A)$ that we can assume to satisfy $a_\epsilon(m) = 0$ for every $\epsilon \in [0, 1]$.

1. Choose a right action on W. Let Ψ_ϵ be the flow of the time dependent vector field $(\overleftarrow{a_\epsilon})_{\epsilon \in [0,1]}$. Show that there exists a neighborhood Σ of x in Σ_0 such that $\Sigma_\epsilon := \Psi_\epsilon(\Sigma)$ is well-defined for every $\epsilon \in [0, 1]$, and is a bisection of W for all $\epsilon \in [0, 1]$.
2. Show that the bisection Σ_ϵ contains x for every $\epsilon \in [0, 1]$.

---

[44] I.e. t-related (see Definition 5.4 in chapter "What Is a Singular Foliation?") to a vector field in $\mathfrak{X}(M)$.

[45] I.e., smoothly depending on the parameter $\epsilon \in I$.

3. Show that $\underline{\Sigma}_\epsilon = \Phi_\epsilon \circ \underline{\Sigma}$ for every $\epsilon \in [0, 1]$.
4. Conclude that the bisections $\Sigma_1 := \Psi_1(\Sigma)$ and Σ'_0 induce diffeomorphisms that coincide in a neighborhood of $s(x) = s'(x')$.
5. Conclude.

Last, right and left actions are extremely practical to check that bisubmersions satisfy the following property.

Definition 4.53 Let I be an interval containing zero. Let $(M, \mathcal{F})$ be a foliated manifold, and $\mathcal{U} \subset M$ an open subset. We say that a bisubmersion $M \xleftarrow{s} W \xrightarrow{t} M$ represents *all small inner symmetries on* $\mathcal{U}$ if, for every smooth time-dependent vector field $(X_\epsilon)_{\epsilon \in I}$ whose time t-flow is defined on $\mathcal{U}$ for all $\epsilon \in I$, there exists for all $m \in M$ and all ϵ small enough a local bisection Σ_ϵ of W such that $\underline{\Sigma}_t$ and the flow $\phi_\epsilon^{X\bullet}$ of $(X_\epsilon)_{\epsilon \in I}$ coincide in a neighborhood of m.

Here is a simple condition to have this condition satisfied, and we will use right actions to prove it.

Proposition 4.54 *If a bisubmersion $M \xleftarrow{s} W \xrightarrow{t} M$ over $\mathcal{F}$ admits a unit map ϵ, defined on $\mathcal{U} \subset M$, then it represents all small inner symmetries on $\mathcal{U}$.*

Proof Let $(X_\epsilon)_{\epsilon \in I}$ be as in Definition 4.53. Let $m \in \mathcal{U}$, and let (A, ρ) be an anchored bundle over $\mathcal{F}$ near m. Let $(a_\epsilon)_{\epsilon \in A}$ be a smooth time-dependent section of A such that $\rho(a_\epsilon) = X_\epsilon$. Choose a right action, and consider the smooth time-dependent vector field $\vec{a}_\epsilon$ on W. Since this vector field tangent to the fibers of s and is t-related to X_ϵ, its flow $\Phi_t^{a\bullet}$ and the flow $\phi_t^{X\bullet}$ are related by $t \circ \Phi_t^{a\bullet} \circ \epsilon = \phi_t^{X\bullet}$ for every ϵ for which it is defined. The submanifold $\Sigma_\epsilon = \Phi_\epsilon^{a\bullet} \circ \epsilon(\mathcal{U})$ is a bisubmersion that satisfies the desired property. $\square$

Example 4.55 The bisubmersion of Proposition 4.26 represents all small inner symmetries on the $\mathcal{U}$ where it is defined. The fundamental bisubmersions in Proposition 4.35, associated to an anchored bundle (A, ρ) defined on $\mathcal{U}$, represents all small inner symmetries on $\mathcal{U}$.

4.4 Products and Inverse of Bisubmersions

Bisubmersions for a given singular foliation behave like Lie groupoids, but so far there is still no inverse and no product. The following notions present an analogy of those:[46]

---

[46] See Proposition 2.4 in [2].

Definition 4.56 (Product and Inverse) Let $\mathcal{F}$ be a singular foliation on a manifold M.

1. The *inverse* W^{-1} of a bisubmersion $M \xleftarrow{s} W \xrightarrow{t} M$ over $\mathcal{F}$ is the bisubmersion $M \xleftarrow{t} W \xrightarrow{s} M$.
2. The *composition* $W * W'$ of two bisubmersions $M \xleftarrow{s} W \xrightarrow{t} M$ and $M \xleftarrow{s'} W' \xrightarrow{t'} M$ over $\mathcal{F}$ is the fibered product

$$W \times_{s,M,t'} W' = \{(x, x') \in W \times W' \text{ s.t. } t(x) = s'(x')\}$$

equipped with the source $(x, x') \mapsto s(x)$ and target $(x, x') \mapsto t'(x')$.

Exercise 4.57 We leave it as an exercise to check that the product of bisubmersions for $\mathcal{F}$, defined as above, is a bisubmersion for $\mathcal{F}$ again.

We finish this discussion with some explanation of the names "inverse" and "compositions" of bisubmersions.

Proposition 4.58 *Let $(M, \mathcal{F})$ be a foliated manifold.*

1. *Consider a bisubmersion $M \xleftarrow{s} W \xrightarrow{t} M$ over a singular foliation $\mathcal{F}$. Then the outer-germ maps of W and W^{-1} are related as follows*

$$\mathrm{Out}_{W^{-1}}(x) = (\mathrm{Out}_W(x))^{-1} \text{ for all } x \in W.$$

2. *Consider a second bisubmersion $M \xleftarrow{s'} W' \xrightarrow{t'} M$ over a singular foliation $\mathcal{F}$, the outer-germ map of $W * W'$, W, and W' are related by:*

$$\mathrm{Out}_{W*W'}(x, x') = \mathrm{Out}_{W'}(x') \circ \mathrm{Out}_W(x)$$

for all $x \in W$, $x' \in W'$ such that $t(x) = s(x')$.

Proof Let us prove item 1, for every local bisection $\Sigma \subset W$ through x, Σ is also a bisection for W^{-1}. But the corresponding isomorphism of $\mathcal{F}$ being the inverse of the source, restricted to Σ, composed with the target, they give isomorphisms which are inverse one to the over.

Let us prove Item 2. For two local bisections Σ, Σ' through x and x', the fibered product $\Sigma \times_{s,M,t'} \Sigma'$ is a bisection of the product bisubmersion $W * W'$. Moreover, we have the following property:

$$\underline{\Sigma \times_{s,M,t'} \Sigma'} = \underline{\Sigma'} \circ \underline{\Sigma} \tag{22}$$

This completes the proof of the claim. $\square$

4.5 Equivalence of Bisubmersions (and Their Compositions)

There is[47] a "Morita equivalence-like" equivalence relation on the set of all bisubmersions over $\mathcal{F}$. Its definition is very natural for the reader used to Morita equivalence of Lie groupoid. There is also a natural notion of morphism, that, surprisingly, will be (more or less) the same as an equivalence.

Definition 4.59 (Equivalence of Bisubmersions) Consider two bisubmersions $M \xleftarrow{s} W \xrightarrow{t} M$ and $M \xleftarrow{s'} W' \xrightarrow{t'} M$ over $\mathcal{F}$.

1. A *morphism* from the first one to the second one is a map $W \to W'$ making the following diagram commutative:

$$\begin{array}{ccccc} M & \xleftarrow{s} & W & \xrightarrow{t} & M \\ \| & & \downarrow & & \| \\ M & \xleftarrow{s'} & W' & \xrightarrow{t'} & M \end{array}$$

2. An *equivalence between them* is a third bisubmersion $M \xleftarrow{s''} P \xrightarrow{t''} M$ of $\mathcal{F}$ equipped with two surjective submersions $P \to W$ and $P \to W'$ making the following diagram commutative:

$$\begin{array}{c} W \\ {}^{s} \nearrow \uparrow^{\pi'} \nwarrow^{t} \\ M \xleftarrow{s''} P \xrightarrow{t''} M \\ {}_{s'} \searrow \downarrow^{\pi} \swarrow_{t'} \\ W' \end{array} \tag{23}$$

---

[47] We follow here Section 2 in [2].

Exercise 4.60 Show that the so-called equivalence defined above is indeed an equivalence relation on bisubmersions over $\mathcal{F}$.

Exercise 4.61 Let $\Gamma \rightrightarrows M$ be a Lie groupoid with Lie algebroid (A, ρ). For any connection ∇ on A, its exponential map $\exp \colon \mathcal{U}_A \to \Gamma$ is a (iso)morphism of bisubmersions from the fundamental bisubmersion $\mathcal{U}_A \subset A$ to an open neighborhood of the unit manifold M in Γ. The following exercise details this point (and recalls the definition of the exponential map in the context of Lie algebroids).

1. For every $m \in M$ and every $a \in A_m$, there exists a unique curve $\epsilon \mapsto \gamma(\epsilon)$ such that for every ϵ for which it is defined:

$$s \circ \gamma(\epsilon) = m \text{ and } a(\epsilon) = \gamma^{-1}(\epsilon) \frac{d\gamma(\epsilon)}{d\epsilon}$$

is the geodesic of A starting from a. We define $\exp(a) \in \Gamma$ to be $\gamma(1)$.
2. Show that $t \circ \Phi_\epsilon^\Xi(a) = t(\gamma(\epsilon))$.
3. Conclude.

Here is an important and surprising theorem.

Theorem 4.62 (Equivalence of Bisubmersions) *Consider two bisubmersions $M \xleftarrow{s} W \xrightarrow{t} M$ and $M \xleftarrow{s'} W' \xrightarrow{t'} M$ for $\mathcal{F}$. The following statements are equivalent:*

(i) *Both bisubmersions are equivalent.*
(ii) *The following two conditions hold:*
   a. *any $x \in W$ admits a neighborhood $\mathcal{U}$ on which a morphism $\mathcal{U} \to W'$ exists,*
   b. *and any $x' \in W'$ admits a neighborhood $\mathcal{U}'$ on which a morphism $\mathcal{U}' \to W$ exists.*
(iii) *Both bisubmersions induce the same outer-germs,[a] i.e.*

$$\mathrm{Out}_W(W) = \mathrm{Out}_{W'}(W').$$

---

[a] Equivalently: for every $x \in W$ there exists $x' \in W'$ such that $\mathrm{Out}_W(x) = \mathrm{Out}_{W'}(x')$ and conversely for every $x' \in W'$ there exists $x \in W$ such that $\mathrm{Out}_{W'}(x') = \mathrm{Out}_W(x)$.

Proof Let us prove (i) ⇒ (ii). Consider an equivalence as in Eq. (23). For every $x \in W$, a local section σ of $\pi: W'' \to W$ can be defined in an open neighborhood of x. The composition of σ with the projection $\pi': P \to W'$ is a morphism as in (ii) item (a). Similarly, for every $x \in W'$ the composition of the projection $\pi: P \to W$ with a local section of $\pi': P \to W'$ is a morphism as in (ii) item (b). This shows that (i) implies (ii).

Let us show (ii) ⇒ (iii). The image of any local bisection Σ of W through a morphism $\phi: \mathcal{U} \to W'$ (with $\mathcal{U}$ an open subset of W) is a bisection of W', and $\phi(\Sigma) = \Sigma$. In particular, for any x, x' as in item (ii), we have $\mathrm{Out}_W(x) = \mathrm{Out}_{W'}(x')$.

Let us prove that (iii) implies (i). First, let us use the convenient notations $\mathcal{G}$ and $\mathcal{G}'$ for the bi-pull-back singular foliations on the bisubmersions W and W', respectively.

Let $x \in W$ and x' be such that $\mathrm{Out}_W(x) = \mathrm{Out}_{W'}(x')$. By exercise 4.52, there exists local bisections Σ and Σ' through x and x' such that $\underline{\Sigma} = \underline{\Sigma}'$. In particular, the diffeomorphism

$$\phi \colon \Sigma \to \Sigma'$$
$$x \mapsto \left(s'_{|\Sigma'}\right)^{-1} \circ s_{|\Sigma} \colon \Sigma(x)$$

makes the following diagram a commutative diagram:

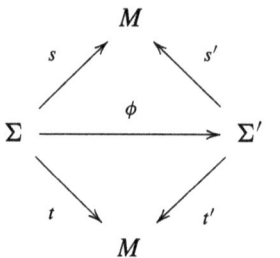

Let $\{\underline{X}_i, i = 1, \ldots, r\}$ be a set of generators of $\mathcal{F}$ in a neighborhood of $s(x)$ and let $\{\underline{Y}_i, i = 1, \ldots, r\}$ be a set of generators of $\mathcal{F}$ in a neighborhood of $t(x)$. Let

$$(X; Y; Z) = (X_1, \ldots, X_{r_s}; Y_1, \ldots, Y_{r_t}; Z_1, \ldots, Z_k)$$

and

$$(X'; Y'; Z') = (X'_1, \ldots, X'_{r_s}; Y'_1, \ldots, Y'_{r_t}; Z'_1, \ldots, Z'_{k'})$$

be generators of $\mathcal{G}$ and $\mathcal{G}'$ as in Lemma 4.6. Without any loss of generality, one can assume $k = k'$: it suffices for instance to add $k' - k$ times the vector field 0 if $k' > k$. Consider the manifold

$$P''_x := \mathrm{Gr}_\phi(\Sigma) \times \mathbb{R}^{r_s} \times \mathbb{R}^{r_t} \times \mathbb{R}^k,$$

where $\mathrm{Gr}_\phi(\Sigma) \subset \Sigma \times \Sigma'$ is the graph of ϕ.

From now on, we use for any n-tuple of vector fields $\xi_\bullet = (\xi_1, \ldots, \xi_n)$ the notation

$$\exp(\lambda_\bullet \xi_\bullet) := \Phi_{\lambda_1}^{\xi_1} \circ \cdots \circ \Phi_{\lambda_n}^{\xi_n}$$

for any $\lambda_\bullet = (\lambda_1, \cdots, \lambda_n) \in \mathbb{R}^n$ such that the flows make sense. There is an open subset of P_x'' containing $\mathrm{Gr}_\phi(\Sigma) \times (0, \ldots, 0) \times (0, \ldots, 0) \times (0, \ldots, 0)$, that we still denote by P_x'' with a slight abuse of notations, on which the following two maps are well-defined:

$\pi:$
$$P_x'' \to W$$
$$\big((\sigma, \sigma'), (\lambda_1, \ldots, \lambda_{r_s}), (\mu_1, \ldots, \mu_{r_t}), (\nu_1, \ldots, \nu_k)\big) \mapsto \exp(\lambda_\bullet X_\bullet) \circ \exp(\mu_\bullet Y_\bullet) \circ \exp(\nu_\bullet Z_\bullet)(\sigma)$$

$\pi':$
$$P_x'' \to W'$$
$$\big((\sigma, \sigma'), (\lambda_1, \ldots, \lambda_{r_s}), (\mu_1, \ldots, \mu_{r_t}), (\nu_1, \ldots, \nu_k)\big) \mapsto \exp(\lambda_\bullet X'_\bullet) \circ \exp(\mu_\bullet Y'_\bullet) \circ \exp(\nu_\bullet Z'_\bullet)(\sigma')$$

Above (σ, σ') is an element of the graph of ϕ, i.e., $\sigma' = \phi(\sigma)$.
Consider the point inside P_x'' given by

$$O_\sigma := ((\sigma, \sigma'), \underbrace{0, \ldots, 0}_{r_s + r_t + k \text{ times}})$$

It is not hard to see that the image of the differential at O_σ of π (resp. π') is generated by $T_\sigma \Sigma$ and $T_\sigma \mathcal{G}$ (resp. $T_{\sigma'} \Sigma'$ and $T_{\sigma'} \mathcal{G}'$). Both π, π' are therefore surjective submersions in a neighborhood of O_σ that we denote by P_x'' again. Moreover, we claim that the commutativity of diagram (23) holds, i.e.,

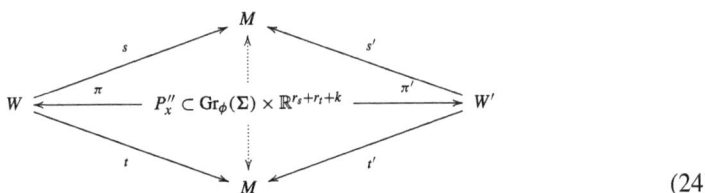

$$(24)$$

This is an easy consequence of the following facts, valid for every index

1. the vector fields X_i and X'_i are s-related to the same vector field in $\mathcal{F}$ (namely $\underline{X_i}$) and t-related with the same vector field on $\mathcal{F}$ (namely, zero).
2. the vector fields Y_i and Y'_i are t-related to the same vector field in $\mathcal{F}$ (namely $\underline{Y_i}$) and s-related with the same vector field on $\mathcal{F}$ (namely, zero).
3. the vector fields Z_i and Z'_i are s- and t-related to the same vector field in $\mathcal{F}$ (namely zero).
4. the commutativity holds true on $\mathrm{Gr}_\phi(\Sigma) \times 0^{r_s+r_t+k}$ by definition of a morphism of bisubmersions (which means that $s(\sigma) = s'(\sigma')$ and $t(\sigma) = t'(\sigma')$ for all $(\sigma, \sigma') \in \mathrm{Gr}_\phi(\Sigma)$).

This shows that every point in $x \in W$ admits a neighborhood $\mathcal{U}_x$ on which an equivalence of bisubmersion exists between that neighborhood and some open subset of W'. There exists a countable family $(x_i)_{i \in \mathbb{N}}$ such that the open subsets $\mathcal{U}_{x_i}$ cover W. The disjoint union $P_1 := \coprod_{i \in \mathbb{N}} P_{x_i}$ is an equivalence between W and an open subset of W'. Now, by the same reasoning, an equivalence P_2 between W' and an open subset of W can be constructed. Their disjoint union $P_1 \coprod P_2$ is an equivalence between the bisubmersions W and W'.

This completes the proof of the claim. □

Exercise 4.63 Consider two bisubmersions $M \xleftarrow{s} W \xrightarrow{t} M$ and $M \xleftarrow{s'} W' \xrightarrow{t'} M$ for $\mathcal{F}$. Using Theorem 4.62, show that if there exists a surjective morphism of bisubmersions from W to W', then W and W' are equivalent.

Exercise 4.64 Consider two bisubmersions $M \xleftarrow{s} W \xrightarrow{t} M$ and $M \xleftarrow{s'} W' \xrightarrow{t'} M$ for $\mathcal{F}$ of the same dimension. Show that if $x \in W$ and $x' \in W'$ are equivalent, then there exist neighborhoods $\mathcal{U}, \mathcal{U}'$ of x and x' and an isomorphism of bisubmersions $\psi : \mathcal{U} \to \mathcal{U}'$, i.e., a morphism which is a diffeomorphism onto its image. *Hint:* By the proof of Theorem 4.62, there exists an equivalence $y \in W''$ such that $\pi(y) = x$ and $\pi'(y) = x'$. Consider a local bisection Σ though y, i.e., a submanifold of W'' on which π, π' are diffeomorphisms onto their images. Such local bisections exist. The induced map $\underline{\Sigma}$ does the job.

Theorem 4.62 can be restated as in the following manner, which is interesting by itself:

Corollary 4.65 *Consider two bisubmersions $M \xleftarrow{s} W \xrightarrow{t} M$ and $M \xleftarrow{s'} W' \xrightarrow{t'} M$ for $\mathcal{F}$. For any two points $x \in W$ and $x' \in W'$, the following statements[48] are equivalent:*

(i) *A neighborhood $\mathcal{U}$ of x is W is equipped with a morphism of bisubmersions $\phi : \mathcal{U} \to W'$ mapping x to x'.*
(ii) *A neighborhood $\mathcal{U}'$ of x' is W' is equipped with a morphism of bisubmersions $\phi' : \mathcal{U}' \to W$ mapping x' to x.*
(iii) *There exist local bisections Σ through x and Σ' through x' that induce the same germ of isomorphisms of $\mathcal{F}$ (i.e., $\underline{\Sigma} = \underline{\Sigma}'$ near $s(x)$).*
(iv) $\mathrm{Out}_W(x) = \mathrm{Out}_{W'}(x')$.

---

[48] Notice that all these statements imply $s(x) = s(x')$ and $t(x) = t(x')$.

5   Holonomy Groupoid of a Singular Foliation

We now introduce the holonomy groupoid of Androulidakis and Skandalis, defined in the [2], Section 3. We shall use a presentation which may seem quite different from the original one, but the difference is only a difference of presentation.

5.1   The Holonomy Groupoid Without Its Topology

An important point about the holonomy groupoid is that it is a topological groupoid, and even a diffeological groupoid (although we will not develop this point here). To start with, we present a construction strongly inspired by a construction given by Garmendia and Villatoro [16] of Androulidakis-Skandalis' holonomy groupoid. But this construction will be incomplete, because it only describes it as a set—without a topology.[49] We will call it for the moment the "flowing groupoid" of a singular foliation, and later on prove that it is isomorphic, as a groupoid (in the category of sets), to the properly defined holonomy groupoid (which will be a topological groupoid, i.e., a groupoid in the category of topological spaces—and even more).

Let $(M, \mathcal{F})$ be a foliated manifold. In Sect. 3.1 in chapter "What Is a Singular Foliation?", we defined two sets, both equipped with two projections onto M called source[50] and target and denoted by s and t:

1. The set $\mathrm{Sym}_{\mathcal{F}}$ equipped with two maps $s : \mathrm{Sym}_{\mathcal{F}} \to M$ and $t : \mathrm{Sym}_{\mathcal{F}} \to M$ called source and target, respectively, such that $s^{-1}(m) \cap t^{-1}(n)$ is for all[51] $m, n \in M$ made of germs of local isomorphisms ϕ of $\mathcal{F}$ mapping a m to n.
2. The set $\mathrm{OutSym}_{\mathcal{F}}$ obtained by taking the quotient of the previous set modulo the equivalence relation $\Phi \sim \Psi$ if $\Phi \circ \Psi^{-1}$ coincides in a neighborhood of m an inner symmetry having m as a very fixed point.[52]

---

[49] Villatoro and Garmendia do equip it with a topology, but using an infinite dimensional manifold structure that we do not wish to introduce here.

[50] Conventions on source and target of a groupoid are not the same in non-commutative geometry and in Poisson geometry. We use the following convention: for $\Gamma \rightrightarrows M$ a groupoid over M, the product of two elements γ_1 and γ_2 is defined if $t(\gamma_1) = s(\gamma_2)$ and the product $\gamma_1 \cdot \gamma_2$ admits the source of γ_1 as its source and the target of γ_2 as its target.

[51] For a given m, n, it may of course be empty. For m, n on the same leaf, it is never empty in view of Theorem 1.21 in chapter "What Is a Singular Foliation?". For m, n on the regular part, provided all regular leaves have the same dimension, it is also non-empty.

[52] See Definition 3.10 in chapter "What Is a Singular Foliation?" for a definition. We saw in that section this condition is equivalent to: $\Psi \circ \Phi^{-1}$ coincides in a neighborhood of n with an inner symmetry having n as a very fixed point.

So far, we have not spoken of any structure on these sets. We do it now.

1. The set $\mathrm{Sym}_{\mathcal{F}}$ admits a natural groupoid structure: source and target are already defined, the composition consists in composing germs of local isomorphisms of $\mathcal{F}$, the inverse is obtained by inverting such an isomorphism, and the unit map consists in mapping $m \in M$ to the germ of the identity map.
2. The set $\mathrm{OutSym}_{\mathcal{F}}$ also admits a groupoid structure, obtained as a quotient of the previous one. Let us describe now. Consider the bundle of groups $\mathrm{K}_{\mathcal{F}}$ over M whose fiber at $m \in M$ are the germs of inner symmetries admitting m as a very fixed point. This bundle of groups is by construction included into $\mathrm{Sym}_{\mathcal{F}}$. It follows from Exercise 3.12 in chapter "What Is a Singular Foliation?" that it is a normal bundle or groups, i.e.,

$$[\phi]^{-1} \, \mathrm{K}_{\mathcal{F}}|_m \, [\phi] \in \mathrm{K}_{\mathcal{F}}|_n$$

for any germ $[\phi]$ of a local isomorphism ϕ of $\mathcal{F}$ such that $\phi(m) = n$. Since the quotient $\mathrm{Sym}_{\mathcal{F}}$ of $\mathrm{K}_{\mathcal{F}}$ is precisely $\mathrm{OutSym}_{\mathcal{F}}$, the latter inherits a groupoid structure.

By construction, there is a short exact sequence of groupoids:

$$\begin{array}{ccccc} \mathrm{K}_{\mathcal{F}} & \hookrightarrow & \mathrm{Sym}_{\mathcal{F}} & \twoheadrightarrow & \mathrm{OutSym}_{\mathcal{F}} \\ \downarrow & & \downdownarrows & & \downdownarrows \\ M & = & M & = & M \end{array} \qquad (25)$$

Now, there is a second natural groupoid $\mathrm{InnerSym}_{\mathcal{F}} \subset \mathrm{Sym}_{\mathcal{F}}$ of $\mathrm{Sym}_{\mathcal{F}} \rightrightarrows M$ obtained by considering all germs of local inner symmetries of $(M, \mathcal{F})$, see Definition 3.7 in chapter "What Is a Singular Foliation?". As a set, it consists in taking all smooth time dependent vector fields $(X_\epsilon)_{\epsilon \in [0,1]}$ in $\mathcal{F}$ (recall that those were defined in Sect. 3.1 in chapter "What Is a Singular Foliation?"), then to take germs of their time-1 flows at every point where they are well-defined.

Lemma 5.1 *Let $(M, \mathcal{F})$ be a foliated manifold. $\mathrm{InnerSym}_{\mathcal{F}} \rightrightarrows M$ is a subgroupoid of $\mathrm{Sym}_{\mathcal{F}} \rightrightarrows M$.*

Proof Lemma 3.8 in chapter "What Is a Singular Foliation?" can be adapted to show that $\mathrm{InnerSym}_{\mathcal{F}}$ is indeed a groupoid. Corollary 7.13 in chapter "What Is a Singular Foliation?" states that any local inner symmetry is a symmetry of $\mathcal{F}$, hence we have a groupoid inclusion $\mathrm{InnerSym}_{\mathcal{F}} \subset \mathrm{Sym}_{\mathcal{F}}$. □

Since InnerSym$_\mathcal{F} \rightrightarrows M$ contains $\mathcal{K}_\mathcal{F}$, Eq. (25) allows defining the flowing groupoid[53]

Definition 5.2 Let $(M, \mathcal{F})$ be a foliated manifold. We call *flowing groupoid*[54] of $\mathcal{F}$ and denote by

$$\text{Flow}_\mathcal{F} \rightrightarrows M$$

the image of InnerSym$_\mathcal{F}$ through the projection Sym$_\mathcal{F} \longrightarrow$ OutSym$_\mathcal{F}$.

We will see in Proposition 5.20 that the flowing groupoid <u>coincides</u> with the holonomy groupoid, but it is however important for pedagogical reasons, and also because it is the most practical manner to construct the holonomy groupoid of a given singular foliation, see, e.g., Exercises 5.3 and 5.4 below. By construction, there is a short exact sequence of groupoids:

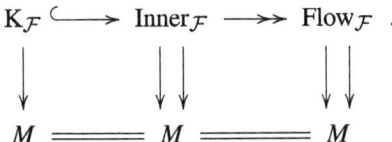

Exercise 5.3 Let $\mathcal{F}$ be the singular foliation on $M = \mathbb{R}^n$ with $n \geq 2$ made of all vector fields vanishing at 0 as in Sect. 4.3 in chapter "What Is a Singular Foliation?". Show that the flowing groupoid is the disjoint union of the pair groupoid of $M\backslash\{0\}$ with the group of invertible matrices of positive determinant (over the origin).

Exercise 5.4 Let $\mathcal{F}$ be the singular foliation on $M = \mathbb{R}^n$ with $n \geq 2$ made of all vector fields vanishing at 0 at order k as in Sect. 4.3 in chapter "What Is a Singular Foliation?". Show that the flowing groupoid is the disjoint union of the pair groupoid of $M\backslash\{0\}$ with the quotient of the group of formal diffeomorphism admitting the origin as a fixed point, whose derivatives at 0 vanish up to order $k-1$ by the group of formal diffeomorphism admitting the origin as a fixed point, whose derivatives at 0 vanish up to order $k-1$. Show that this group is in fact isomorphic to a vector space, equipped with the addition as a group product, if $k \geq 2$.

---

[53] Our construction of the flowing groupoid is equivalent to the one completed by [16] using $\mathcal{F}$-cuts to the leaves.

[54] The flowing groupoid will be soon isomorphic (as a set) to the holonomy groupoid, see Proposition 5.20! So we do not insist too much on the notion.

5.2 Atlases

Let $(M, \mathcal{F})$ be a smooth singular foliation. Let us introduce a very particular type of bisubmersions,[55] the quotient of which is going to be, by definition, the holonomy groupoid (now equipped with a topology).

Definition 5.5 (Atlases: A Groupoid-Like Bisubmersions) Let $(M, \mathcal{F})$ be a foliated manifold. We say that a bisubmersion $M \xleftarrow{s} W \xrightarrow{t} M$ over $\mathcal{F}$ is an *atlas of* $\mathcal{F}$ when

1. W is equivalent[a] to its inverse[b] W^{-1},
2. the composition[c] $W * W$ is equivalent to W.
3. W admits local unit maps[d]

Also, atlases are said to be *equivalent* when they are equivalent as bisubmersions over $\mathcal{F}$.

[a] See Definition 4.59.
[b] See Sect. 4.4.
[c] See Sect. 4.4.
[d] See Definition 4.9.

Exercise 5.6 We show in this exercise that the third assumption in Definition 5.5 is a consequence of the two first ones, and can therefore be omitted. Consider $M \xleftarrow{s} W \xrightarrow{t} M$ an atlas of $\mathcal{F}$. Let $m \in M$ be a point. Let $x \in W$ be such that $t(x) = m$. Let Σ be a germ of a bisection of W through x.

1. Consider the inverse bisection $W^{-1} := M \xleftarrow{t} W \xrightarrow{s} M$, and $\Psi : W \to W^{-1} \simeq W$ a morphism of bisubmersions defined in a neighborhood of x. Let $\Sigma' = \Psi(\Sigma)$. Show that Σ' is a bisection of W' through $x' = \Psi(x)$ and that $\underline{\Sigma'} = \underline{\Sigma}^{-1}$.
2. Show that the graph $\Sigma'' := \mathrm{Gr}(\Psi|_\Sigma) = \{(y, \Psi(y)| y \in \Sigma\}$ of the restriction $\Psi|_\Sigma$ of Ψ is a bisection through (x, x') for the product bisection $W * W$.
3. Let $\Xi : W * W \to W$ be a morphism of bisubmersions defined in a neighborhood of (x, x'). Show that $\Xi(\Sigma'')$ is a bisection of W through $\Xi(x, x')$ whose induced diffeomorphism is the identity map in a neighborhood of m.
4. Use Exercise 4.14 to conclude the argument.

[55] Following Section 1.3.2 in [2].

Here is a consequence of Theorem 4.62, more precisely of the equivalence of item (i) and (iii) in that Theorem.

Proposition 5.7 *A bisubmersion $M \xleftarrow{s} W \xrightarrow{t} M$ over $\mathcal{F}$ is an atlas of $\mathcal{F}$ if and only if the image $\mathrm{Out}_W(W)$ of the outer-germ map is a sub-groupoid of $\mathrm{OutSym}_{\mathcal{F}} \rightrightarrows M$.*

Also, two atlases W and W' are equivalent if and only if the sub-groupoids $\mathrm{Out}_W(W) \rightrightarrows M$ and $\mathrm{Out}_{W'}(W') \rightrightarrows M$ coincide.

The following result, presented as an exercise, can be practical to compute atlases.

Exercise 5.8 Let $M \xleftarrow{s} W \xrightarrow{t} M$ of $\mathcal{F}$ be a bisubmersion over a singular foliation $\mathcal{F}$ that admits a (local) unit $\epsilon \colon M \hookrightarrow W$. Show that W is an atlas if and only if $W * W^{-1}$ is equivalent to W. *Hint:* Consider the following commutative diagrams

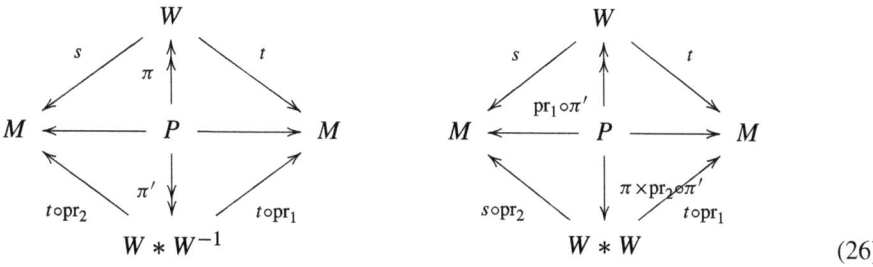

(26)

and use Theorem 4.62.

We are in particular interested in a particular subclass of atlases, that we call inner atlases.

Definition 5.9 (An Inner Atlas) Let $(M, \mathcal{F})$ be a foliated manifold. We say that an atlas W for $\mathcal{F}$ is an inner atlas if

$$\mathrm{Out}_W(W) = \mathrm{Flow}_{\mathcal{F}}.$$

Definition 5.9 can be restated as meaning that

1. for any local bisection Σ through $w \in W$, the local symmetry $\underline{\Sigma}$ coincides, in a neighborhood of every point where it is defined, with an inner symmetry of $\mathcal{F}$, and,
2. any inner symmetry of $\mathcal{F}$ is induced by a bisection of W, at least near any point where it is defined.

Even more explicitly, an atlas is an inner atlas if and only if:

1. For any local diffeomorphism of the type $\Phi := \phi_1^{X_1} \circ \cdots \circ \phi_n^{X_n}$ with $X_1, \ldots, X_n$ in $\mathcal{F}$, and any $m \in M$ such that $\Phi(m)$ is well-defined, there exists $x \in W$ and a bisection Σ through x such that $\underline{\Sigma} = \Phi$ is a neighborhood of m.
2. Conversely, for any bisection Σ through x, there exists $X_1, \ldots, X_n$ in $\mathcal{F}$ such that $m = \Phi(s(x))$ is well-defined and $\underline{\Sigma} = \Phi$ is a neighborhood of m.

Notice that we have not assumed in Definition 5.9 that the inner atlas is an atlas: it is a priori simply a bisubmersion. However, it follows from the definition that it has to be an atlas:

Lemma 5.10 *Any inner atlas for $\mathcal{F}$ is an atlas for $\mathcal{F}$.*

Proof By equivalence of item (i) and (iii) in Corollary 4.65, a bisubmersion is an atlas if and only if its germs of bisections are stable under inverse and product. It is of course the case for all local diffeomorphisms as in the second item of Definition 5.9. This completes the proof. □

Lemma 5.11 *Any two inner atlases for a singular foliation $\mathcal{F}$ are equivalent.*

Proof This is an immediate consequence of the equivalence of (i) and (iii) in Theorem 4.62. □

Exercise 5.12 Let W be an inner atlas with source s and target t for $\mathcal{F}$. Show that two points $\ell_1, \ell_2 \in M$ are in the same leaf of $\mathcal{F}$ if and only if there exists $w \in W$ such that $s(w) = \ell_1$ and $t(w) = \ell_2$. The converse is not true, see Exercise 5.15.

Proposition 5.13 *Any Lie groupoid with source-connected fibers is an inner atlas for its basic singular foliation.*

Proof Let $\Gamma \rightrightarrows M$ be a Lie groupoid with s- and t-connected fibers. Let $(A \to M, \rho, [\cdot, \cdot])$ be its algebroid and $\mathcal{F} = \rho(\Gamma(A))$ its basic singular foliation. Recall that Γ is a bisubmersion of $\mathcal{F}$ (see Exercise 4.24). The space Γ is an atlas of $\mathcal{F}$, since the inverse map $\text{inv} \colon \Gamma \to \Gamma$ is an equivalence between

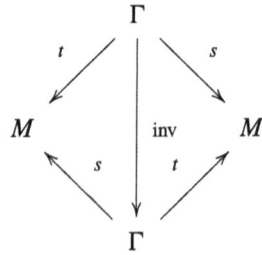

while the product μ gives an equivalence of bisubmersions:

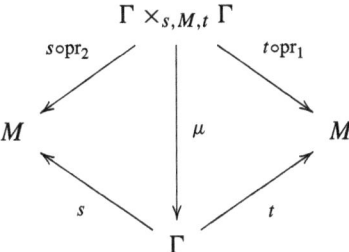

This proves that Γ is an atlas. To show that it is an inner atlas, we proceed as follows. Since fibers of s are connected, there exists, for any $\gamma \in \Gamma$ with source m, a time dependent section a_t of A such that the time 1-flow Φ_t of $\overrightarrow{a_t} \in \mathfrak{X}(\Gamma)$ maps m (seen as an element in Γ through the unit map ϵ) to γ. Let $\mathcal{U} \subset M$ be a neighborhood of m on which $\Phi_t \circ \epsilon$ is well-defined. Then $\Phi_1 \circ \epsilon(\mathcal{U})$ is a bisection of Γ through γ. Its induced diffeomorphism $\Phi_1 \circ \epsilon(\mathcal{U})$ is the time 1-flow of the vector field $\rho(a_t)$, which is a vector field on $\mathcal{F}$ that depends smoothly on t. It is therefore an inner symmetry of $\mathcal{F}$. $\square$

Here is a non-example of inner atlas.

Example 5.14 Let $\mathcal{F}$ be a singular foliation on a manifold M. Let W be an inner atlas with source s and target t. Assume that there exists a finite group G acting on M by symmetries of $\mathcal{F}$, denoted by $\phi_g : M \to M$ for a given $g \in G$. Then

$$\coprod_{g \in G} \{g\} \times W$$

is an atlas for $\mathcal{F}$, when equipped with the source $(g, w) \mapsto s(w)$ and $t(g, w) \mapsto \phi_g(t(w))$. If ϕ_g is not an inner symmetry for at least one $g \in G$, it is however not an inner atlas, since the symmetry associated to the bisection $m \mapsto (\{g\}, \epsilon(m))$, with ϵ a local unit map for W, is not an inner symmetry of $\mathcal{F}$.

Exercise 5.15 Apply the construction of Example 5.14 to vector fields on $\mathbb{R}^2$ vanishing quadratically at zero, and to a finite sub-group of the group of rotations. Show that the converse of the statement in Exercise 5.12 is not true.

The most important of all atlases (that will be proven soon to be an inner atlas) was introduced by Androulidakis and Skandalis under the name of *path holonomy atlas of $\mathcal{F}$ associated to an anchored bundle and a connection*. It is obtained by the following procedure:

1. Take a family $(A_i \to \mathcal{U}_i, \rho_i)_{i \in I}$ of anchored bundles such that the open subsets $(\mathcal{U}_i)_{i \in I}$ where they are defined cover M. Assume each one of them is equipped with a connection ∇_i.
2. For every $i \in I$, there exists a neighborhood $\mathcal{A}_i$ of the zero section of $A_i \to \mathcal{U}_i$ on which there is a bisubmersion over $\mathcal{F}$ (see Proposition 4.35).
3. Then consider the disjoint union[56] for all $n \geq 0$ and all $i_1, \ldots, i_n \in I$ of all direct products

$$\mathcal{A}_{i_1}^\star \times_M \cdots \times_M \mathcal{A}_{i_n}^\star \quad (n \text{ times})$$

where $\star$ means that we consider $\mathcal{A}$ or its inverse $\mathcal{A}^{-1}$.

Denote by W_{path} the disjoint union of all manifolds as in the third item.

It deserves to be noticed that every connected component of W_{path} the manifold is finitely dimensional, although the dimension is not bounded. We have the following lemma:

Lemma 5.16 W_{path} is an atlas for $\mathcal{F}$.

Proof There is a natural isomorphism between W_{path} and W_{path}^{-1}. Also $W_{path} * W_{path}$ injects into W_{path} by construction. □

Following [2], we call *path holonomy atlas of* $\mathcal{F}$ such an atlas.

Proposition 5.17 *Any path holonomy atlas of $\mathcal{F}$ is an inner atlas of $\mathcal{F}$.*

Proof This is a direct consequence of Example 4.55, which shows that each one of the fundamental bisubmersion $\mathcal{A}_i$ contains all small inner symmetries on $\mathcal{U}_i$. In turn, since any inner-symmetry Φ is, in a neighborhood of any point where it is defined, a composition

$$\Phi = \phi_1 \circ \cdots \circ \phi_n$$

where for every index k, we have $\phi_k = \underline{\Sigma_k} \subset \mathcal{A}_{i_k}$ a bisection as in Definition 4.53 for some $i_k \in I$. Now, $\Sigma := \Sigma_n \times_M \cdots \times_M \Sigma_1$ is a bisection of the bisubmersion $\mathcal{A}_{i_1} \times_M \cdots \times_M \mathcal{A}_k$. By construction, and in view of Eq. (22), we have

$$\underline{\Sigma} = \underline{\Sigma_n \times_M \cdots \times_M \Sigma_1} = \underline{\Sigma_1} \times \cdots \times \underline{\Sigma_n} = \phi_1 \ldots \phi_n = \Phi.$$

This completes the proof. □

---

[56] These sets are called *longitudinal charts* in [2].

5.3 Holonomy Groupoid

We can now at last define the holonomy groupoid of Androulidakis and Skandalis [2]. We start with a general result.

Proposition 5.18 (From Atlases to Groupoids) *Let $M \xleftarrow{s} W \xrightarrow{t} M$ of $\mathcal{F}$ be an atlas of $\mathcal{F}$. Consider the equivalence relation on W given by $x \sim x'$ if and only if x and x' have neighborhoods which are equivalent as bisubmersions of $\mathcal{F}$.*

The equivalence classes of this relation form a topological[a] groupoid over M.

---

[a] A subset of the groupoid is open if and only if its inverse image in the atlas (which is a manifold) is open.

Proof The inverse of an equivalence class represented by $x \in W$ is represented by any point in W^{-1} which equivalent to x. It is easy to check that it is well-defined. The same applies to product. Given two compatible x_1, x_2, there exists $x_3 \in X$ such that x_3 and (x_1, x_2) are equivalent. This construction goes to the quotient w.r.t. the equivalence relation on X and defines the groupoid product. All these maps are continuous. $\square$

Theorem 4.62 implies that equivalent atlases of $\mathcal{F}$ induce canonically homeomorphic quotient groupoids. In particular, Definition 3.26 makes sense.

Definition 5.19 (Holonomy Groupoid of a Singular Foliation) We call holonomy groupoid of $\mathcal{F}$ the topological[a] groupoid associated to a fundamental atlas of $\mathcal{F}$. We denote it by $\mathrm{Hol}(\mathcal{F}) \rightrightarrows M$.

---

[a] In [2], this groupoid is equipped with much more than just a topology: a C^*-algebra of "smooth functions" is even introduced. We will not go as far here.

To start with, let us state that the flowing groupoid, introduced for pedagogical reason, is a notion we can now get rid of.

Proposition 5.20 *For any singular foliation, the flowing groupoid[57] and the holonomy groupoid are canonically isomorphic.*

Proof The flowing groupoid can be redefined as being the image through $\text{Out}_{\mathcal{F}}$ of any inner atlas. The quotient that defines the holonomy groupoid consists in identifying two points in an inner atlas whose image in $\text{Sym}_{\mathcal{F}}$ coincide. The result follows. □

Remark 5.21 Let us assume that the singular foliation $(M, \mathcal{F})$ comes from a Lie groupoid $\Gamma \rightrightarrows M$ whose and t-fibers are connected, i.e., let us assume that $\mathcal{F} = \rho(\Gamma(A))$ with A the Lie algebroid of $\Gamma \rightrightarrows M$. In Proposition 5.13, we saw that Γ is an inner atlas for $\mathcal{F}$. There is therefore a surjective groupoid morphism Φ as follows:

$$\begin{array}{ccc} \Gamma & \xrightarrow{\Phi} & \text{Hol}(\mathcal{F}) \\ \downarrow\downarrow & & \downarrow\downarrow \\ M & \xrightarrow{=} & M. \end{array}$$

Let us study the fibers of this map Φ. For that purpose, let us consider

$$\mathcal{K} := \{a \in \Gamma(A) \text{ s.t. } \rho(a)\}.$$

By construction $\mathcal{K}$ is the $\mathcal{C}^{\infty}(M)$-module of sections of A which are valued in the kernel of ρ for all point $m \in M$. Any section $a \in \mathcal{K}$ gives, by right action on Γ, a vector field $\vec{a}$ in Γ. The $\mathcal{C}^{\infty}(\Gamma)$ module that these vector fields generate is the module of bi-vertical vector fields. Assume for simplicity that it is a singular foliation (i.e., that it is locally finitely generated). Then two points in Γ that are in the same leaf of the bi-vertical singular foliation are in the same fiber of Φ. See e.g., Exercise 4.22. The quotient $\Gamma/\sim$ of the Lie groupoid Γ by the equivalence class defined by the bi-vertical singular foliation is a topological groupoid, and Φ induces a continuous groupoid morphism $\Gamma/\sim \longrightarrow \text{Hol}(\mathcal{F})$.

5.4 About Smoothness of the Holonomy Groupoid: Two Theorems by Claire Debord

Recall that a *Lie groupoid* is a groupoid $\Gamma \rightrightarrows M$ such that Γ and M are manifolds, the source, and target are smooth surjective submersions, and all structural maps (unit, product, inverse) are smooth.[58]

---

[57] See Definition 5.2.

[58] Or real analytic, or holomorphic, depending on the setting.

The holonomy groupoid is certainly not a smooth groupoid in general. It is a topological groupoid, and the topology may be quite horrible -very far away from a manifold topology. However, the following theorem was proven by Claire Debord [12].

> **Theorem 5.22 (Along a Leaf)** *[12] The orbits of the holonomy groupoid* $\mathrm{Hol}(\mathcal{F})$ *of a singular foliation* $\mathcal{F}$ *are the leaves of* $\mathcal{F}$. *Moreover, its restriction to any leaf* L:
>
> 1. *is a Lie groupoid,*
> 2. *whose Lie algebroid is the holonomy Lie algebroid*[a] *of the leaf* L.
>
> ---
> [a]See Sect. 3.6.

Proof The first sentence was established in Exercise 5.12. The hard part of the proof relies on a theorem that bounds below the periods of a periodic orbit of a vector field in a neighborhood of a point. This lower bound forbids a bisubmersion to have "too many quotients", it makes the quotient "discrete-like". We refer to [12]. □

Smoothness of the holonomy groupoid happens in a second situation. This theorem is also due to Claire Debord [11].

> **Theorem 5.23 (Projective Case)** *The holonomy groupoid of a Debord*[a] *foliation is a Lie groupoid whose Lie algebroid is the Lie algebroid in Proposition 4.13 in chapter "What Is a Singular Foliation?".*
>
> ---
> [a]See Sect. 4.2.4 in chapter "What Is a Singular Foliation?".

5.5   The Fundamental Groupoid

An alternative approach to the construction of the holonomy groupoid is by anchored paths divided by a certain equivalence relation. This is how the Lie groupoid integrating a Lie algebroid has been constructed in [7,32]. In the context of singular foliations, this approach has been taken in [16], Section 1.1.2 in [25] and Section 4.6 of [23].

Le $\mathcal{F}$ be a singular foliation on M and A an anchored bundle over $\mathcal{F}$. We start by introducing the right notion of homotopy between anchored paths (cf. Definition 4.28). For this, we introduce part of the longitudinal complex[59] $(\Lambda^\bullet_{\mathcal{C}^\infty(M)}\mathcal{F})*$ of $\mathcal{F}$ (See Section 4.1. in [23]). Its cochains are given by skew-symmetric $\mathcal{C}^\infty(M)$-multilinear maps from $\mathcal{F}^{\times k}$ to $\mathcal{C}^\infty(M)$. For us, only the first differential

$$d^\mathcal{F} : \mathcal{F}^* = \mathrm{Hom}_{\mathcal{C}^\infty(M)}(\mathcal{F}, \mathcal{C}^\infty(M)) \to (\Lambda^2_{\mathcal{C}^\infty(M)}\mathcal{F})^*,$$

defined by the usual formula

$$(d^\mathcal{F}\alpha)(X, Y) = X(\alpha(Y)) - Y(\alpha(X)) - \alpha([X, Y])$$

will play a role. For an anchored bundle A, we will denote by ρ^* the induces map from $\mathcal{F}^*$ to $\Gamma(A)$ and given an anchored morphism $\phi : A_1 \to A_2$, we will denote by ϕ^*, the induced map $\Gamma(\Lambda A_2^*) \to \Gamma(\Lambda A_1^*)$. With those conventions, we can finally define anchored homotopies:

Definition 5.24 Let $A \xrightarrow{\pi} M$ be an anchored bundle and $\gamma, \tilde{\gamma} : I \to A$ anchored paths. We say that γ is A-homotopic to $\tilde{\gamma}$, if there exists an anchored bundle morphism:

$$H : T[0, 1]^2 \to A$$

such that:

- $H(t, 0) = \gamma(t), H(t, 1) = \tilde{\gamma}(t)$.
- $H(0, s)$ and $H(1, s)$ are constant and equal to zero.
- The following diagram commutes:

$$\begin{array}{ccc} \Omega^1(T[0, 1]^2) & \xleftarrow{H^* \circ \rho^*} & \mathcal{F}^* \\ \downarrow d & & \downarrow d^\mathcal{F} \\ \Omega^2(T[0, 1]^2) & \xleftarrow{H^* \circ \rho^*} & (\Lambda^2_{\mathcal{C}^\infty(M)}\mathcal{F})^* \end{array}$$

Then we call H an anchored homotopy between γ and $\tilde{\gamma}$.

---

[59] $(\Lambda^\bullet_{\mathcal{C}^\infty(M)}\mathcal{F})^*$ is *not*, in general, isomorphic to the exterior tensor product $(\Lambda^\bullet_{\mathcal{C}^\infty(M)}\mathcal{F}^*)$ of $\mathrm{Hom}_{\mathcal{C}^\infty(M)}(\mathcal{F}, \mathcal{C}^\infty(M))$.

In order to define the fundamental groupoid of a singular foliation, we just need to make paths composable, since in general the concatenation of smooth paths need not be smooth. The lazy solution to this problem is by using lazy paths (cf. e.g., [35] also referred to as sitting instants initially [9]), i.e., paths which are constant near the boundary.

Definition 5.25 (Fundamental Groupoid) Let M be a manifold and $\mathcal{F}$ a singular foliation. Let A be any anchored bundle over $\mathcal{F}$. The fundamental groupoid of $\mathcal{F}$ is the quotient

$$\Phi_1(M, \mathcal{F}) = \frac{\text{lazy anchored paths in } A}{A - \text{Homotopies}}$$

The composition of the fundamental groupoid is given by the concatenation of lazy paths.

As usual for such constructions, the source, and target maps are the projections on the endpoints of a path and the unit elements are given by the constant paths equal to 0. The orbits of the fundamental groupoid of $\mathcal{F}$ are exactly the leaves of the foliation $\mathcal{F}$. Also, for any regular leaf L, $\Phi_1(M, \mathcal{F})_L$ is just the fundamental groupoid of the leaf.

$\Phi_1(M, \mathcal{F})_L$ can be seen as a universal cover of the holonomy groupoid in the following sense. There is a groupoid morphism:

$$\Phi_1(M, \mathcal{F}) \longrightarrow \text{Hol}(\mathcal{F})$$

such that for every $m \in M$, the restriction of the map above to $s^{-1}(m)$ is a submersion and a local diffeomorphism. Since, by construction, the fundamental groupoid is source-simply-connected, it means that $\Phi_1(M, \mathcal{F})$ is obtained by taking, for every leaf L, the source-simply-connected groupoid integrating the holonomy Lie algebroid A_L. See [16, 23, 25] for several variations on this topic.

6 Geometric Resolutions of a Singular Foliation

6.1 Definition and Universality of Existence of Geometric Resolutions

6.1.1 Introduction

Let us work within the smooth setting for the moment, and use Definition 2.1 in chapter "What Is a Singular Foliation?" of singular foliations.[60] We saw in Sect. 1.1 that for every

---

[60] Recall that the index $\Gamma_c(E)$ means "compactly supported sections of E" and $\mathfrak{X}_c(M)$ means "compactly supported vector fields on M".

finitely generated singular foliation $\mathcal{F}$, there exists an anchored vector bundle[61]

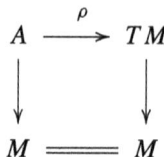

such that
$\rho(\Gamma_c(A)) = \mathcal{F}$. Now, consider the kernel[62] $\ker(\rho)$ of

$$\rho\colon \Gamma_c(A) \longrightarrow \mathfrak{X}_c(M).$$

The space $\ker(\rho) \subset \Gamma_c(A)$ is again a $\mathcal{C}^\infty(M)$-submodule of $\Gamma_c(A)$. If it is finitely generated, then there exists a second vector bundle[63] $B \to M$ and a vector bundle morphism $d\colon B \longrightarrow A$ such that

$$d\,(\Gamma_c(B)) = \ker(\rho)$$

In particular, we have $\rho \circ d = 0$, and

$$\begin{array}{ccccc} B & \xrightarrow{d^{(2)}} & A & \xrightarrow{\rho} & TM \\ \downarrow & & \downarrow & & \downarrow \\ M & = & M & = & M \end{array}$$

is a complex of vector bundles which is exact at the level of sections, i.e., that the sequence

$$\Gamma_c(B) \xrightarrow{d^{(2)}} \Gamma_c(A) \xrightarrow{\rho} \mathcal{F}$$

is exact. The procedure continues when the kernel of

$$d^{(2)}\colon \Gamma_c(B) \longrightarrow \Gamma_c(A)$$

---

[61] We can even assume A to be a trivial bundle, see the discussion in Sect. 1.1.
[62] Notice that this kernel consist in compactly supported sections of A which for every $m \in M$ are valued in the Strong kernel of ρ at m.
[63] We can even assume B to be a trivial bundle, see the discussion in Sect. 1.1.

is again finitely generated as a $C^\infty(M)$-module: there exists a vector bundle $C \to M$ and a vector bundle morphism $d\colon C \to B$ such that $d^{(3)}(\Gamma_c(C)) = \ker(d^{(2)})$. By construction,

$$\begin{array}{ccccccc} C & \xrightarrow{d^{(3)}} & B & \xrightarrow{d^{(2)}} & A & \xrightarrow{\rho} & TM \\ \downarrow & & \downarrow & & \downarrow & & \downarrow \\ M & = & M & = & M & = & M \end{array}$$

is a complex of vector bundles and the following is an exact complex:

$$\Gamma_c(C) \xrightarrow{d^{(3)}} \Gamma_c(B) \xrightarrow{d^{(2)}} \Gamma_c(A) \xrightarrow{\rho} \mathcal{F} \ .$$

Here is a natural question.

Question 6.1 *When can the construction of the complex of vector bundles described above be continued "up to infinity" (i.e., can one be certain that the kernels are finitely generated)?*

Does it stop at some point? (i.e., can we manage that the kernel of $d^{(k)}$ is trivial for k large enough?

Assume it can be constructed, what kind of geometric information is encoded in that complex?

6.1.2 Definitions

Let us start with by precise definition and a precise vocabulary (valid in the smooth, real analytic or complex settings).

Definition 6.2 An *anchored complex of vector bundles* consists of a triple $(E_{-\bullet}, d^{(\bullet)}, \rho)$, where

1. $E_{-\bullet} = (E_{-i})_{i \geq 1}$ is a family of vector bundles over M, indexed by negative integers.
2. for every $i \geq 1$, $d^{(i+1)} \in \mathrm{Hom}(E_{-i-1}, E_{-i})$ is a vector bundle morphism over the identity of M called the *differential map*.
3. $\rho\colon E_{-1} \longrightarrow TM$ is a vector bundle morphism over the identity of M called the *anchor map*.

such that

$$\begin{array}{ccccccccccccc} \longrightarrow & E_{-i-1} & \xrightarrow{d^{(i+1)}} & E_{-i} & \xrightarrow{d^{(i)}} & E_{-i+1} & \longrightarrow & \cdots & \xrightarrow{d^{(2)}} & E_{-1} & \xrightarrow{\rho} & TM \\ & \downarrow & & \downarrow & & \downarrow & & & & \downarrow & & \downarrow \\ = & M & = & M & = & M & = & \cdots & = & M & = & M \end{array}$$

form a complex, i.e., such that

$$d^{(i)} \circ d^{(i+1)} = 0 \text{ and } \rho \circ d^{(2)} = 0$$

Let us fix some vocabulary about such complexes:

1. The integer $-i$ is called the *degree* of the vector bundle E_{-i}. The choice of negative numbers may seem surprising: it will be justified when introducing Lie ∞-algebroid structures.
2. An anchored complex of vector bundle $(E_{-\bullet}, d^{(\bullet)}, \rho)$ is said to be of *length* $n \in \mathbb{N}$ if $E_{-i} = 0$ for $i \geq n+1$ and *finite length* if $E_{-i} = 0$ except for finitely many indices.
3. We shall speak of *anchored complex of trivial bundles* when all the vector bundles $(E_{-i})_{i \geq 1}$ are trivial vector bundles. We do not assume TM to be a trivial bundle.

There are two main cohomologies that one can associate to an anchored complex of vector bundles.

1. **Cohomology at the level of sections.** An anchored complex of vector bundles $(E_{-\bullet}, d^{(\bullet)}, \rho)$ induces a complex of sheaves of modules over functions. More explicitly, for every open subset $\mathcal{U} \subset M$, there is a complex:

$$\cdots \longrightarrow \Gamma_{\mathcal{U}}(E_{-i-1}) \xrightarrow{d^{(i+1)}} \Gamma_{\mathcal{U}}(E_{-i}) \xrightarrow{d^{(i)}} \Gamma_{\mathcal{U}}(E_{-i+1}) \longrightarrow \cdots \longrightarrow \Gamma_{\mathcal{U}}(E_{-1}) \xrightarrow{\rho} \mathfrak{X}(\mathcal{U}).$$

In particular, $\mathrm{Im}\left(d^{(i+1)}\right) \subseteq \ker d^{(i)}$ for every $i \in \mathbb{N}$, so that the quotient spaces:

$$H^{-i}(E_\bullet, \mathcal{U}) = \begin{cases} \frac{\ker \rho}{\mathrm{Im}(d^{(2)})} & \text{for } i = 1 \\[6pt] \frac{\ker d^{(i)}}{\mathrm{Im}(d^{(i+1)})} & \text{if } i \geq 2 \end{cases}$$

is a module over functions on $\mathcal{U}$ that we call *i-th cohomology of* $(E_{-\bullet}, d^{(\bullet)}, \rho)$ *at the level of sections*.
2. **Cohomology at an arbitrary point** $m \in M$.
An anchored complex of vector bundles $(E_{-\bullet}, d^{(\bullet)}, \rho)$ at an arbitrary point $m \in M$, restricts to a complex of vector spaces

$$\cdots \longrightarrow E_{-i-1}|_m \xrightarrow{d^{(i+1)}_{|m}} E_{-i}|_m \xrightarrow{d^{(i)}_{|m}} E_{-i+1}|_m \longrightarrow \cdots \xrightarrow{d^{(2)}_{|m}} E_{-1} \xrightarrow{\rho_{|m}} T_m M.$$

In particular, $\mathrm{Im}(\mathrm{d}^{(i+1)}) \subseteq \ker \mathrm{d}^{(i)}_{|m}$ for every $i \in \mathbb{N}$, and we call the quotient vector spaces:

$$H^{-i}(E_\bullet, m) = \begin{cases} \dfrac{\ker \rho_{|m}}{\mathrm{Im}\left(\mathrm{d}^{(2)}_{|m}\right)} & \text{for } i = 1 \\[2ex] \dfrac{\ker \mathrm{d}^{(i)}_{|m}}{\mathrm{Im}\left(\mathrm{d}^{(i+1)}_{|m}\right)} & \text{if } i \geq 2 \end{cases}$$

the i-th *cohomology of* $(E_{-\bullet}, \mathrm{d}^{(\bullet)}, \rho)$ *at the point* m.

Warning!

It is important to notice that $H^{-i}(E_\bullet, m)$ may be non-zero at a point m even if $H^{-i}(E_\bullet, \mathcal{U})$ is zero in every open neighborhood $\mathcal{U}$ of m. For instance, for a Debord singular foliation $\mathcal{F}$. Let A stand for its associated Lie algebroid A and ρ the anchor. The complex of vector bundles defined by $E_{-1} = A$ and $E_{-i} = 0$ for $i \geq 2$, together with a trivial d and the anchor ρ, is exact at the level of sections on any open set. But this is not exact when evaluated at a singular point of $\mathcal{F}$.

The converse, however, is not possible, as we will see in Proposition 6.3 below.

A complex of vector bundles needs not to be exact at a point, even when it is exact at the level of sections. But if it is exact at a point, then it is also exact, locally, at the level of sections, as we now see.

Proposition 6.3 *Let $i \in \mathbb{N}$. Every $m \in M$ such that $H^{-i}(E_\bullet, m) = 0$ has an open neighborhood $\mathcal{U}$ for which $H^{-i}(E_\bullet, \mathcal{U}) = 0$.*

Proof Below, $\mathrm{rk}(E)$ stands for the rank of a vector bundle $E \to M$, and $\mathrm{rk}(d)$ stands for the rank (= dimension of the image) of a vector bundle morphism d. Also, for $i = 1$, in this proof, it must be understood that $\mathrm{d}^{(1)}$ is the anchor ρ. Assume that $\ker \mathrm{d}^{(i)}_{|m} = \mathrm{Im}\left(\mathrm{d}^{(i+1)}_{|m}\right)$ for some $m \in M$. This implies that

$$\mathrm{rk}\left(\mathrm{d}^{(i+1)}_{|m}\right) = \dim\left(\ker \mathrm{d}^{(i)}_{|m}\right)$$
$$= \mathrm{rk}(E_{-i}) - \mathrm{rk}\left(\mathrm{d}^{(i)}_{|m}\right). \tag{27}$$

It is a general fact that the functions from M to $\mathbb{N}$ given by

$$m' \mapsto \mathrm{rk}\left(\mathrm{d}^{(i)}_{|m'}\right) \text{ and } m' \mapsto \mathrm{rk}\left(\mathrm{d}^{(i+1)}_{|m'}\right)$$

are lower semi-continuous functions. Hence, there exists an open neighborhood $\mathcal{U}$ of m such that for all $m' \in \mathcal{U}$

$$\mathrm{rk}\left(\mathrm{d}^{(i+1)}_{|m'}\right) \geq \mathrm{rk}\left(\mathrm{d}^{(i+1)}_{|m}\right)$$

$$= \mathrm{rk}(E_{-i}) - \mathrm{rk}\left(\mathrm{d}^{(i)}_{|m}\right), \quad \text{by Eq. (27)}$$

$$\geq \mathrm{rk}(E_{-i}) - \mathrm{rk}\left(\mathrm{d}^{(i)}_{|m'}\right), \quad \text{by lower semi-continuity of } m' \mapsto \mathrm{rk}\left(\mathrm{d}^{(i)}_{|m'}\right)$$

$$= \dim\left(\ker \mathrm{d}^{(i)}_{|m'}\right).$$

Since we are considering a complex of vector bundles, we also have

$$\mathrm{Im}\left(\mathrm{d}^{(i+1)}_{|m'}\right) \subseteq \ker \mathrm{d}^{(i)}_{|m'}$$

for all $m' \in M$. This contradicts the previous inequality, unless if $\ker \mathrm{d}^{(i)}_{|m'}$ coincides with $\mathrm{Im}\left(\mathrm{d}^{(i+1)}_{|m'}\right)$ for all $m' \in \mathcal{U}$. This proves the claim. □

Definition 6.4 Let $(E_{-\bullet}, \mathrm{d}^{(\bullet)}, \rho)$ and $(E'_{-\bullet}, (\mathrm{d}')^{(\bullet)}, \rho')$ be anchored complexes of vector bundles.

1. An *anchored chain map* or *anchored complex of vector bundle morphisms* between the anchored complexes of vector bundles $(E_{-\bullet}, \mathrm{d}^{(\bullet)}, \rho)$ and $(E'_{\bullet}, (\mathrm{d}')^{(\bullet)}, \rho')$ is a collection of vector bundle morphisms (of degree zero) $\varphi_\bullet : E_{-\bullet} \longrightarrow E'_{-\bullet}$ such that the following diagram commutes

$$\begin{array}{ccccccccc}
\cdots & \xrightarrow{\mathrm{d}^{(i)}} & E_{-i} & \xrightarrow{\mathrm{d}^{(i)}} & E_{-i+1} & \xrightarrow{} & \cdots & \xrightarrow{\mathrm{d}^{(2)}} & E_{-1} & \xrightarrow{\rho} & TM \\
 & & \downarrow{\varphi_i} & & \downarrow{\varphi_{i-1}} & & & & \downarrow{\varphi_1} & & \| \\
\cdots & \xrightarrow{\mathrm{d}'^{(i)}} & E'_{-i} & \xrightarrow{} & E'_{-i+1} & \xrightarrow{} & \cdots & \xrightarrow{\mathrm{d}'^{(2)}} & E'_{-1} & \xrightarrow{\rho'} & TM
\end{array} \quad (28)$$

that is, $\rho' \circ \varphi_1 = \rho$ and $\mathrm{d}'^{(i)} \circ \varphi_i = \varphi_{i-1} \circ \mathrm{d}^{(i)}$ for every $i \geq 2$.

2. A *homotopy* between two anchored complexes of vector bundle morphisms $\varphi_\bullet, \psi_\bullet : E_{-\bullet} \longrightarrow E'_{-\bullet}$ is the datum $\{h_i : E_{-i} \longrightarrow E'_{-i-1}\}_{i \geq 1}$ of vector bundle

morphisms (of degree -1) that satisfies $\psi_1 - \varphi_1 = d'^{(2)} \circ h_1$ and for each $i \geq 2$, $\psi_i - \varphi_i = d'^{(i+1)} \circ h_i + h_{i-1} \circ d^{(i)}$

$$\begin{array}{ccccccccc}
\cdots & \longrightarrow & E_{-i-1} & \xrightarrow{d^{(i+1)}} & E_{-i} & \xrightarrow{d^{(i)}} & E_{-i+1} & \longrightarrow & \cdots \\
& & {\scriptstyle \psi_{i+1}-\varphi_{i+1}}\downarrow & \overset{h_i}{\swarrow} & {\scriptstyle \psi_i-\varphi_i}\downarrow & \overset{h_{i-1}}{\swarrow} & {\scriptstyle \psi_{i-1}-\varphi_{i-1}}\downarrow & & \\
\cdots & \longrightarrow & E'_{-i-1} & \xrightarrow{d'^{(i+1)}} & E'_{-i} & \xrightarrow{d'^{(i)}} & E'_{-i+1} & \longrightarrow & \cdots
\end{array} \quad (29)$$

(a) When there exists a homotopy between two anchored complexes of vector bundle morphisms $\varphi_\bullet, \psi_\bullet : E_{-\bullet} \longrightarrow E'_{-\bullet}$, we say that φ, ψ are *homotopic* and we write $\varphi \sim \psi$.

(b) (E, d, ρ) and (E', d', ρ') are said to be *homotopy equivalent*, if there exist anchored chain maps $\varphi_\bullet : E_{-\bullet} \longrightarrow E'_{-\bullet}$ and $\psi_\bullet : E'_{-\bullet} \longrightarrow E_{-\bullet}$ such that $\varphi \circ \psi \sim \mathrm{id}_{E'_{-\bullet}}$ and $\psi \circ \varphi \sim \mathrm{id}_{E_{-\bullet}}$.

Exercise 6.5 Check that

1. $\sim$ is an equivalence relation on the class of complexes of vector bundle morphisms.
2. "homotopy equivalence" on the class of anchored complexes of vector bundles is an equivalence relation.

Lemma 6.6 *Let $(E_{-\bullet}, d^{(\bullet)}, \rho)$ and $(E'_{-\bullet}, (d')^{(\bullet)}, \rho')$ be homotopy equivalent anchored complexes of vector bundles of finite length n and n' respectively. The alternating sum of the ranks of the vector bundles $(E_{-i})_{i \in \mathbb{N}}$ and $(E'_{-i})_{i \in \mathbb{N}}$ respectively, are equal, i.e.,*

$$\sum_{i=1}^{n}(-1)^i \mathrm{rk}(E_{-i}) = \sum_{i=1}^{n'}(-1)^i \mathrm{rk}(E'_{-i}).$$

Here $\mathrm{rk}(E)$ stands for the rank of a vector bundle $E \to M$.

Proof Note first that the restriction of both complexes to a point $m \in M$ give two finite length complexes of vector spaces of finite dimension. The result is an immediate consequence of the fact that in every degree the cohomology group of two equivalent complexes of vector spaces are isomorphic. It follows by taking the alternating sum of their dimensions and using the Rank-nullity theorem. □

For the next definition, we invite the reader to consider singular foliations as defined in Definition 2.18 or 2.22 in chapter "What Is a Singular Foliation?".

Definition 6.7 (Geometric Resolution of a Singular Foliation) Let $\mathcal{F} \subseteq \mathfrak{X}(M)$ be a singular foliation on a smooth, real analytic or complex manifold M. An anchored complex of vector bundles $(E_{-\bullet}, d^{(\bullet)}, \rho) :=$

$$\cdots \xrightarrow{} E_{-i-1} \xrightarrow{d^{(i+1)}} E_{-i} \xrightarrow{d^{(i)}} E_{-i+1} \xrightarrow{} \cdots \xrightarrow{d^{(2)}} E_{-1} \xrightarrow{\rho} TM$$
$$\downarrow \qquad \downarrow \qquad \downarrow \qquad \qquad \downarrow \qquad \downarrow$$
$$M = M = M = \cdots = M = M$$

is said

1. to *terminate in* $\mathcal{F}$ if $\rho\left(\Gamma(E_{-1})\right) \subseteq \mathcal{F}$
2. to be *over* $\mathcal{F}$ if $\rho(\Gamma(E_{-1})) = \mathcal{F}$
3. to be a *geometric resolution of* $\mathcal{F}$ if the following complex of sheaves[a] is exact:

$$\cdots \longrightarrow \Gamma(E_{-i-1}) \xrightarrow{d^{(i+1)}} \Gamma(E_{-i}) \xrightarrow{d^{(i)}} \Gamma(E_{-i+1}) \longrightarrow \cdots \longrightarrow \Gamma(E_{-1}) \xrightarrow{\rho} \mathcal{F} \longrightarrow 0.$$

A geometric resolution $(E_{-\bullet}, d^{(\bullet)}, \rho)$ is said to be *minimal at a point* $m \in M$ if for each $i \geq 2$ the linear map $d^{(i)}_{|m} : E_{-i}|_m \longrightarrow E_{-i+1}|_m$ vanishes.

---
[a] See discussion below.

Let us recall what we mean precisely by the sheaf condition above, and explain how this condition simplifies in the smooth case. "Exact as sheaves" means that for any $i \in \mathbb{N}$ and any $m \in M$, there is a neighborhood $\mathcal{V}$ of m such that for any $\mathcal{U}$ in $\mathcal{V}$ the complex:

$$\cdots \longrightarrow \Gamma_\mathcal{U}(E_{-i-1}) \xrightarrow{d^{(i+1)}} \Gamma_\mathcal{U}(E_{-i}) \xrightarrow{d^{(i)}} \Gamma_\mathcal{U}(E_{-i+1}) \longrightarrow \cdots.$$

is exact. In the smooth setting, this is equivalent to demand that the complex be exact at the level of global compactly supported sections. In the smooth setting, therefore, the notion of geometric resolution is much easier, and Definition 6.7 can be rewritten as follows.

Definition 6.8 (Geometric Resolution of a Smooth Singular Foliation, an Equivalent Definition) Let $\mathcal{F} \subseteq \mathfrak{X}_c(M)$ be a singular foliation on a smooth manifold M. A *geometric resolution* of the singular foliation $\mathcal{F}$ is a complex of vector bundles $(E_{-\bullet}, \mathrm{d}^{(\bullet)}, \rho) :=$

$$\longrightarrow E_{-i-1} \xrightarrow{\mathrm{d}^{(i+1)}} E_{-i} \xrightarrow{\mathrm{d}^{(i)}} E_{-i+1} \longrightarrow \cdots \longrightarrow E_{-1} \xrightarrow{\rho} TM$$
$$\downarrow \qquad \downarrow \qquad \downarrow \qquad \qquad \downarrow \qquad \downarrow$$
$$M = M = M = \cdots = M = M$$

such that the following complex is exact:

$$\cdots \longrightarrow \Gamma_c(E_{-i-1}) \xrightarrow{\mathrm{d}^{(i+1)}} \Gamma_c(E_{-i}) \xrightarrow{\mathrm{d}^{(i)}} \Gamma_c(E_{-i+1}) \longrightarrow \cdots \longrightarrow \Gamma_c(E_{-1}) \xrightarrow{\rho} \mathcal{F}.$$

Remark 6.9 When a geometric resolution $(E_{-\bullet}, \mathrm{d}^{(\bullet)}, \rho)$ is minimal at a point $m \in M$ then one has, $H^{-i}(E_\bullet, m) = E_{-i}|_m$ for all $i \geq 2$.

Remark 6.10 A singular foliation $\mathcal{F}$ is Debord if and only if there exists a geometric resolution of length 1.

6.1.3 Universality

We conclude this section with a theorem which says that, given a foliated manifold $(M, \mathcal{F})$, in the category where

1. objects are anchored complexes of vector bundles that terminate within $\mathcal{F}$.
2. arrows are homotopy classes of morphisms,

geometric resolutions are terminal[64] (a.k.a. "universal") objects. For the reader not familiar with categories, this gives the following result, that extends Theorems 1.7.

---

[64] Also called "final": it is an object γ such that for any other object α there is one and only one arrow from α to γ. It is a general property of category theory that for any two terminal (a.k.a. final, a.k.a. universal) objects γ, γ' there exist a unique arrow $\gamma \to \gamma'$ and $\gamma' \to \gamma$ which are inverse one to the other.

Theorem 6.11 (Geometric Resolutions Are Terminal Objects) *Let $\mathcal{F} \subseteq \mathfrak{X}(M)$ be a singular foliation on a smooth manifold M that admits a geometric resolution $(E_{-\bullet}, \mathrm{d}^{(\bullet)}, \rho)$.*

1. *For any anchored complex of vector bundles $(E'_{-\bullet}, \mathrm{d}^{(\bullet)}, \rho')$ that terminates within $\mathcal{F}$, there exists a chain map of anchored vector bundles*

$$(E'_{-\bullet}, \mathrm{d}^{(\bullet)}, \rho') \longrightarrow (E_{-\bullet}, \mathrm{d}^{(\bullet)}, \rho)$$

    and any two such chain maps are homotopy equivalent.
2. *In particular, two geometric resolutions of the same singular foliations are homotopy equivalent.*

The same results hold in the complex and real analytic setting, but in a neighborhood of a point only.

Proof These are reinterpretations of classical results of algebraic topology, see Section 3.2 in [23] for more explanations: for any commutative algebra $\mathcal{O}$, resolutions of an $\mathcal{O}$-module, say $\mathcal{F}$, by projective $\mathcal{O}$-modules are universal objects in the category where objects are complexes of $\mathcal{O}$-modules ending in $\mathcal{F}$ and arrows are homotopy classes of chain maps. □

The fact that two geometric resolutions of $\mathcal{F}$, when they exist, are homotopy equivalent, has many consequences on the topic "whatever is canonically invariant under homotopy equivalence is canonically attached to the singular foliation - provided it admits geometric resolutions." It is the case, for instance, of the alternating sums of the ranks by Lemma 6.6.

Corollary 6.12 ([23], Proposition 2.5) *Let $\mathcal{F} \subseteq \mathfrak{X}(M)$ be a singular foliation on a smooth, real analytic or complex connected manifold M that admits a geometric resolution $(E_{-\bullet}, \mathrm{d}^{(\bullet)}, \rho)$ of finite length. Then*

1. *all the regular leaves have the same[65] dimension r,*
2. *the alternating sum of the ranks of $E_{-\bullet}$ is equal to the dimension of the regular leaves, i.e.*

$$r = \sum_{i \geq 1} (-1)^{i+1} \mathrm{rk}(E_{-i}).$$

---

[65] This point is in fact automatic in the complex and real analytic settings, it is only non-trivial in the smooth case, see Remark 3.22 in chapter "What Is a Singular Foliation?".

If two geometric resolutions are homotopy equivalent, their restrictions to a point $m \in M$ are also homotopy equivalent. In consequence, the complexes obtained by evaluation at this point have isomorphic cohomologies. This proves the first part of the following corollary, we leave the rest to the reader.

Corollary 6.13 *Let $\mathcal{F} \subseteq \mathfrak{X}(M)$ be a singular foliation on a smooth, real analytic or complex manifold M that admits a geometric resolution. Then, for every $m \in M$, the cohomologies $H^{-i}(E_{-\bullet}, m) \simeq H^{-i}(E'_{-\bullet}, m)$ are canonically isomorphic.*

It therefore makes sense to denote these spaces of cohomologies by $H^{-i}(\mathcal{F}, m)$, erasing the dependency[66] on the geometric resolution. We call them *isotropy spaces of $\mathcal{F}$ at m*. In particular, the dimensions $d_1, \ldots, d_i, \ldots$ of these spaces are canonically attached to $\mathcal{F}$. Also, Proposition 6.3 implies that the following items are equivalent:

1. m is a regular point,
2. $H^{-1}(\mathcal{F}, m) = 0$
3. $H^{-i}(\mathcal{F}, m') = 0$ for every $i \geq 1$ and every m' in a neighborhood of m.

Remark 6.14 The integers $d_1, \ldots, d_i, \ldots$ were constructed without making any use of the Lie bracket of vector fields, so that they are, as a matter of fact, attached to $\mathcal{F}$ seen as a module over functions, and not to $\mathcal{F}$ seen as a singular foliation. We suggest interpreting them as follows:

1. d_1 is the rank of $\mathcal{F}$ at m minus the dimension of the leaf through m,
2. d_2 is the rank at m of the module of relations between the previous generators near m,
3. d_3 is the rank of the module over relations between the generators of the module of relations between the generators of $\mathcal{F}$,
4. ...and so on.

6.2 Existence of Geometric Resolutions (Noetherianity, Syzygies and a Flatness Theorem by Tougeron-Malgrange)

We defined and studied geometric resolutions in Sect. 6.1, but do they exist?

Here are some cases where geometric resolutions of a singular foliation always exist at least locally, and are of finite length.

---

[66] Some readers may recognize that $H^{-i}(E_{-\bullet}, m)$ is the so-called Tor-functor (see [39]) in the category of $\mathcal{O}$-modules, with $\mathcal{O}$ the sheaf of functions on M, applied to the $\mathcal{O}$-modules $\mathcal{F}$ and $\mathbb{K}$. On the second one, the action is given by $F \cdot \lambda = F(m)\lambda$ for all $F \in \mathcal{O}$ and $\lambda \in \mathbb{K}$.

Proposition 6.15 (A Particular Case of Syzygy Theorem) *Let $\mathcal{F}$ be a real analytic or holomorphic singular foliation on a manifold M of dimension d. Any point in M admits a neighborhood on which a geometric resolution by trivial vector bundles exists.*

Moreover, its length can be chosen to be less than $d + 1$.

Proof This is in fact a general theorem for coherent sheaves on a complex or real analytic manifold, and has nothing to do with singular foliation. It is called Syzygy theorem, and is related to Oka's coherence theorem. We invite the reader to refer to the classical literature on the matter, e.g., Griffiths and Harris's *Principles of Algebraic Geometry* [15] page 696. □

Remark 6.16 For an algebraic singular foliation on an affine variety W as in Sect. 2.4 in chapter "What Is a Singular Foliation?" over $\mathbb{C}$ or $\mathbb{R}$, since the ring of functions $\mathcal{O}_W$ is Noetherian, a geometric resolution of $\mathcal{F}$ by trivial vector bundles has to exist, for the following reason:

1. $\mathcal{F}$ being finitely generated over $\mathcal{O}_W$ by definition, there exists an integer d_1 and a surjective $\mathcal{O}_W$-linear map $\rho : \mathcal{O}_W^{d_1} \to \mathcal{F}$.
2. The kernel of ρ being an $\mathcal{O}_W$-module, by Noetherianity, there exists an integer d_2 and a surjective $\mathcal{O}_W$-linear map $d^{(2)} : \mathcal{O}_W^{d_2} \to \mathrm{Ker}(\rho)$.
3. The kernel of $d^{(2)}$ being an $\mathcal{O}_W$-module, by Noetherianity, there exists an integer d_3 and a surjective $\mathcal{O}_W$-linear map $d^{(3)} : \mathcal{O}_W^{d_2} \to \mathrm{Ker}(d^{(2)})$
4. ...and so on up to infinity...

Now, the module $\mathcal{O}_W^{d_i}$ can be considered as a module of sections of a trivial vector bundle E_{-i}, and the maps ρ and $d^{(\bullet)}$ can be considered a vector bundle morphisms. This completes the construction. In general, the procedure described above never stops. But for $W = \mathbb{C}^d$, we can make it stop after at most $d + 1$ steps. This is a deep result in algebraic geometry called Syzygy theorem (See, e.g., Theorem 1.1. in [13]). The same results hold for $W = \mathbb{R}^N$, and for a polynomial singular foliation on W. See [23], Section 3.2, for more details.

Here is a technical but important question. Assume that we are given a real analytic foliated manifold $(M, \mathcal{F}_{ra})$ a geometric resolution $(E_\bullet, d^{(\bullet)}, \rho)$ of $\mathcal{F}_{ra}$. The real analytic manifold can be considered as a smooth manifold. The singular foliation can be considered as a smooth singular foliation $\mathcal{F}_{sm}$ (it suffices to consider the sheaf of $\mathcal{C}_c^\infty(M)$-module generated[67] by $\mathcal{F}$):

$$\mathcal{F}_{sm} := \mathcal{C}^\infty(M)\,\mathcal{F}_{ra}$$

Moreover, a real analytic geometric resolution $(E_{-\bullet}, d^{(\bullet)}, \rho)$, being a family real analytic vector bundles equipped with a family real analytic vector bundle morphisms, can be considered as a smooth complex of vector bundles over $\mathcal{F}_{sm}$: But it is not obvious that it is still a geometric resolution of $\mathcal{F}_{sm}$. It is not obvious that exactness of the complex of sheaves[68]

$$\cdots \longrightarrow \Gamma_{ra}(E_{-i-1}) \xrightarrow{d^{(i+1)}} \Gamma_{ra}(E_{-i}) \xrightarrow{d^{(i)}} \Gamma_{ra}(E_{-i+1}) \longrightarrow \cdots \longrightarrow \Gamma_{ra}(E_{-1}) \xrightarrow{\rho} \mathcal{F}_{ra}$$

implies the exactness of the complex of sheaves[69]

$$\cdots \longrightarrow \Gamma_{sm}(E_{-i-1}) \xrightarrow{d^{(i+1)}} \Gamma_{sm}(E_{-i}) \xrightarrow{d^{(i)}} \Gamma_{sm}(E_{-i+1}) \longrightarrow \cdots \longrightarrow \Gamma_{sm}(E_{-1}) \xrightarrow{\rho} \mathcal{F}_{sm}.$$

Notice that this is absolutely not true if we deal with global sections. If true, it can only be true at the level of sheaves, i.e., in a neighborhood of a point. Such results are called flatness. What we are asking is: are smooth functions flat over real analytic functions? The answer is "no", and counter-examples are easily found. However, it is true[70] that *germs of smooth functions at a point form a flat module over converging real analytic functions*. This is enough for our present purpose: because it means that given a smooth section $a \in \Gamma_{sm}(E_{-i-1})$ which is $d^{(i)}$-closed, defined on some open subset $\mathcal{U}$ of M, each point $m \in \mathcal{U}$ admits a neighborhood $\mathcal{V}_m$ where it is exact $a_{|\mathcal{V}_m} = d^{(i+1)} b_{\mathcal{V}_m}$. Using an open cover and partitions of unity,[71] we can glue the $b_{\mathcal{V}_m}$'s to obtain some $b \in \Gamma_{sm}(E_{-i-2})$ such that $d^{(i+1)} b = a$. Hence:

Proposition 6.17 ([23], Proposition 2.3) *A real analytic geometric resolution of a real analytic singular foliation $\mathcal{F}_{ra}$ is also a smooth geometric resolution of its induced smooth singular foliation $\mathcal{F}_{sm}$.*

---

[67] If the reader prefers to think in terms of global section compactly supported vector fields as in Definition 2.1 in chapter "What Is a Singular Foliation?", $\mathcal{F}_{sm}$ must be seen as compactly supported vector fields X such that every $m \in M$ admits a neighborhood $\mathcal{U}$ on which $X = \sum_{i=1}^k f_i X_i$ with $f_i \in \mathcal{C}^\infty(\mathcal{U})$ and X_i a local section of $\mathcal{F}_{ra}$.
[68] Γ_{ra} stands for the sheaf of real analytic sections.
[69] Γ_{sm} stands for the sheaf of smooth sections.
[70] See Theorem 4 in [36] by Tougeron—a result generally attributed to Malgrange.
[71] See the proof of Proposition 2.3 in [23].

Remark 6.18 In a similar fashion, an algebraic singular foliation on $\mathbb{R}^d$ can be seen as a smooth singular foliation.[72] Any geometric resolution as in Remark 6.16 can be seen as a smooth geometric resolution. This is follows from the consecutive use of

1. flatness[73] of real analytic functions over the ring of polynomials, which implies that a polynomial geometric resolution is a real analytic geometric resolution,
2. then Proposition 6.17, which says that any real analytic geometric resolution is also a smooth geometric resolution.

For similar reasons, an algebraic singular foliation on $\mathbb{C}^d$ can be seen as a holomorphic singular foliation.[74]

We say that a smooth singular foliation is *locally real analytic* if every point admits a local chart on which the generators of the singular foliation are real analytic, i.e., have real analytic coefficients. We *do not* assume that these charts are patched together by real analytic transition function: it is fine if those are smooth only. This is enough to guarantee the existence of real analytic geometric resolution near every point, by using the Syzygy-type argument of Proposition 6.15. Then these real analytic geometric resolutions can be considered as smooth ones, see Proposition 6.17. Now, in the smooth context, and only in this context, smooth geometric resolutions can be "glued"—in some sense, and after modification to homotopy equivalent ones: we refer to the discussion of Proposition 3.24 in [23]. Eventually, one obtains the following theorem (Theorem 2.4 in [23]).

Theorem 6.19 (Existence of Geometric Resolutions) *A locally real analytic singular foliation on a manifold of dimension d admits a geometric resolution of length $\leq d + 1$ on any relatively compact open subset of M.*

Below, we describe some examples of geometric resolutions of singular foliations.

Exercise 6.20 Show that the singular foliation on $M = \mathbb{R}$ generated by the vector field $\chi(t)\partial_t$ with $\chi(t) = e^{-1/t^2}$ for $t > 0$ and $\chi(t) = 0$ for $t \leq 0$ admits no geometric resolution. (*Hint*: Use Corollary 6.12).

---

[72] See Proposition 2.29 in chapter "What Is a Singular Foliation?".

[73] Polynomial functions form a valuation ring. A module over a valuation ring is flat if and only if it is torsion free (see, e.g., stack project, section 15.22), which is the case of real analytic functions on an open subset of $\mathbb{R}^d$.

[74] See the previous footnote: the argument extends to holomorphic function on any open connected subset of $\mathbb{C}^d$.

Example 6.21 Let M be a smooth, real analytic or complex manifold, with sheaf of functions $\mathcal{O}$. Let $\mathcal{I} \subset \mathcal{O}$ be a sheaf of ideals. We say that $\mathcal{I}$ admits a geometric resolution when there exists a complex of vector bundles

$$\cdots \xrightarrow{d} I_{-2} \xrightarrow{d} I_{-1} \xrightarrow{\epsilon} \mathbb{K} \times M$$
$$\downarrow = \quad \downarrow = \quad \downarrow = \quad \downarrow$$
$$\cdots \longrightarrow M \longrightarrow M \longrightarrow M$$

which are exact at the level of sections and such that the image of ϵ is $\mathcal{I}$. Now, $\mathcal{I}\mathfrak{X}$, i.e., the sheaf of vector fields whose coefficients are in $\mathcal{I}$, form a singular foliation. A geometric resolution of it is given by taking the tensor product[75] with TM, as follows:

$$\cdots \xrightarrow{d \otimes \mathrm{id}} I_{-2} \otimes TM \xrightarrow{d \otimes \mathrm{id}} I_{-1} \otimes TM \xrightarrow{\epsilon \otimes \mathrm{id}} TM$$
$$\downarrow = \quad \downarrow = \quad \downarrow = \quad \downarrow$$
$$\cdots \longrightarrow M \longrightarrow M \longrightarrow M$$

This type of geometric resolution is in fact quite common, as we now see in Examples 6.22 and 6.23.

Example 6.22 Let $\mathcal{F}_0 = \{X \in \mathfrak{X}(V) \mid X(0) = 0\}$ be the singular foliation made of all vector fields vanishing at the origin of a vector space V of dimension N over $\mathbb{C}$ or $\mathbb{R}$). We are precisely in the situation of Example 6.21, with $\mathcal{I}$ the ideal of polynomial functions vanishing at 0.

In view of the previous example, therefore, we have to find a geometric resolution of that ideal, then take the tensor product with the tangent bundle. This can be done as follows. The contraction by the Euler vector field

$$\vec{E} = \sum_{i=1}^{N} x_i \frac{\partial}{\partial x_i}$$

gives rise to a complex of trivial vector bundles

$$\cdots \longrightarrow \wedge^3 T^*V \xrightarrow{\iota_{\vec{E}}} \wedge^2 T^*V \xrightarrow{\iota_{\vec{E}}} T^*V \xrightarrow{\iota_{\vec{E}}} \mathbb{C} \times V =: \underline{\mathbb{C}}, \tag{30}$$

---

[75] Notice the use of $(\mathbb{K} \times M) \otimes TM \simeq TM$ in the last column.

whose complex on the level of sections is $(\Omega^\bullet(V), \iota_{\vec{E}})$. Here $(x_1, \ldots, x_N)$ are the canonical coordinates on V. Since the x_i's form a regular sequence, it is well known that $(\Omega^\bullet(V), \iota_{\vec{E}})$ is exact, and the previous complex is therefore a geometric resolution of the ideal of polynomial functions that vanish at 0. A geometric resolution of $\mathcal{F}_0$ is therefore given by the trivial bundles with fiber $E_{-i} := (\wedge^i V^* \otimes V) \times V \to V$, the differential $\iota_{\vec{E}} \times \mathrm{id}$, and the anchor given at a point $v \in V$:

$$E_{-1}|_v \simeq V^* \otimes V \to V \simeq T_v V$$
$$\alpha \otimes u \to (\iota_{\vec{E}}{}_{|_v} \alpha) u = \alpha(v)\, u.$$

Example 6.23 Example 6.22 can be further extended. Consider the singular foliation $\mathcal{I}\mathfrak{X}(V)$ of all vector fields on a finite dimensional vector space V over $\mathbb{K} = \mathbb{R}$ or $\mathbb{C}$ whose coefficients[76] are in $\mathcal{I}$. When the ideal $\mathcal{I}$ is generated by a regular sequence $\phi_1, \ldots, \phi_a$ of polynomial functions, then it admits a Koszul resolution.[77] This is constructed as follows. Let B be a vector space of dimension a, and α be the section of the trivial vector bundle $B^* \times V \to V$ given by

$$\alpha = \sum_{i=1}^{a} \phi_i\, e_i^*$$

with $e_1^*, \ldots, e_a^*$ a basis of V^*. Consider the trivial vector bundle $\wedge^\bullet B \times V \to V$ with fiber $\wedge^\bullet B$. The pair $(\Gamma(\wedge^\bullet B \times V), \iota_\alpha)$ is a complex of vector bundles.[78] Koszul theorem states that this construction is a geometric resolution of the ideal $\mathcal{I}$, which in turn gives a geometric resolution of $\mathcal{I}\mathfrak{X}(V)$ by taking the tensor product with TV as in Example 6.21.

Example 6.24 Here is a case of singular foliation on $\mathbb{K}^d$ which is of the form $\mathcal{I}\mathfrak{X}(V)$ for some vector space V and some ideal $\mathcal{I}$, but the construction of Example 6.23 does not apply because $\mathcal{I}$ is not generated by a regular sequence. Let $\mathcal{F}_2$ be the algebraic singular foliation made of vector fields vanishing at order at least 2 at the origin of $V = \mathbb{K}^2$ (see Sect. 4.3 in chapter "What Is a Singular Foliation?"). Here $\mathcal{I}$ is the ideal generated by the monomials x^2, xy, y^2. It is *not* generated by a regular sequence. However, it admits a geometric resolution

$$0 \longrightarrow \mathcal{O}(\mathbb{K}^2) \oplus \mathcal{O}(\mathbb{K}^2) \xrightarrow{d} \mathcal{O}(\mathbb{K}^2) \oplus \mathcal{O}(\mathbb{K}^2) \oplus \mathcal{O}(\mathbb{K}^2) \xrightarrow{\epsilon} \mathcal{I} \longrightarrow 0, \qquad (31)$$

---

[76] For $\mathbb{K} = \mathbb{C}$, when $\mathcal{I}$ is the ideal of functions on an affine variety W, this is precisely the singular foliation of vector fields that vanish on W, see Sect. 4.4 in chapter "What Is a Singular Foliation?".

[77] See Section 4.5 in [39], or the original (in French) [18].

[78] In degree $-i$, the vector bundle is $\wedge^i B \times V \to V$.

where for all $f, g, h \in \mathcal{O}(\mathbb{K}^2)$,

$$\epsilon(f, g, h) = x^2 f + xyg + y^2 h \text{ and } d(f, g) = (-yf, xf - yg, xg).$$

Here $\mathcal{O}(\mathbb{K}^2)$ stands for polynomial functions on $\mathbb{K}^2$. This corresponds to a complex of trivial vector bundles

$$0 \longrightarrow \mathbb{K}^2 \times V \xrightarrow{d} \mathbb{K}^3 \times V \xrightarrow{\epsilon} \mathbb{K} \times V. \tag{32}$$

Taking the tensor product with TV, which is a trivial bundle of rank 2, we see that a geometric resolution is therefore given by the following trivial bundles:

$$0 \longrightarrow \mathbb{K}^2 \otimes \mathbb{K}^2 \times V \xrightarrow{d \otimes \mathrm{id}} \mathbb{K}^3 \otimes \mathbb{K}^2 \times V \xrightarrow{\epsilon \otimes \mathrm{id}} \mathbb{K} \otimes \mathbb{K}^2 \simeq TV. \tag{33}$$

This geometric resolution can possibly be seen as a real analytic or a smooth singular foliation, as in Remark 6.18, when $\mathbb{K} = \mathbb{R}$ or a holomorphic one when $\mathbb{K} = \mathbb{C}$.

Example 6.25 Let $\varphi \in \mathcal{O}(V)$ be a polynomial function on a finite dimensional vector space V over $\mathbb{R}$ or $\mathbb{C}$. Consider the algebraic singular foliation on V of all polynomial vector fields that "kill".[79] φ:

$$\mathcal{F}_\varphi := \{X \in \mathfrak{X}(V) \mid X[\varphi] = 0\}.$$

The contraction by the exact 1 form $d\varphi \in \Omega^1(V)$ turns sections[80] of $\wedge^\bullet TV$, over V into a complex of vector bundles

$$\cdots \xrightarrow{\iota_{d\varphi}} \wedge^3 TV \xrightarrow{\iota_{d\varphi}} \wedge^2 TV \xrightarrow{\iota_{d\varphi}} TV \xrightarrow{\iota_{d\varphi}} \mathbb{C} \times V. \tag{34}$$

Consider the complex on the level of sections:

$$\cdots \xrightarrow{\iota_{d\varphi}} \mathfrak{X}^3(V) \xrightarrow{\iota_{d\varphi}} \mathfrak{X}^2(V) \xrightarrow{\iota_{d\varphi}} \mathfrak{X}(V) \xrightarrow{\iota_{d\varphi}} \mathcal{O}(V) \tag{35}$$

(where $\mathfrak{X}^i(V) := \Gamma(\wedge^i TV)$ stands for i-vector fields on V). Koszul theorem[81] states that this complex is exact in all degree, except in degree 0, if $\left(\frac{\partial \varphi}{\partial x_1}, \cdots, \frac{\partial \varphi}{\partial x_N}\right)$ is a regular

---

[79] This algebraic singular foliation was considered in Sect. 4.4 in chapter "What Is a Singular Foliation?".
[80] I.e., polynomial polyvector fields, which here identify to polynomial maps from V to $\wedge^\bullet V$.
[81] See Section 4.5 in [39], or the original article [18].

sequence. The theorem of Cohen-Macaulay[82] then says that $\left(\frac{\partial \varphi}{\partial x_1}, \cdots, \frac{\partial \varphi}{\partial x_N}\right)$ is a regular sequence if ϕ is weight homogeneous of non-zero degree with one isolated singularity at the origin. In all these cases, the truncated complex:

$$\cdots \xrightarrow{\iota_{d\varphi}} \mathfrak{X}^3(V) \xrightarrow{\iota_{d\varphi}} \mathfrak{X}^2(V) \xrightarrow{\iota_{d\varphi}} \ker(\iota_{d\varphi}) = \mathcal{F}_\varphi \qquad (36)$$

is an algebraic geometric resolution of $\mathcal{F}_\varphi$. It can therefore be seen as a real analytic or a smooth singular foliation, as in Remark 6.18.

In the case $\mathbb{K} = \mathbb{C}$, it can be seen as a holomorphic one, in view of the same remark.

6.3 Geometric Resolutions of Length ≤ 2 and Singular Foliations

As we saw in Sect. 4.2.4 in chapter "What Is a Singular Foliation?", a singular foliation $\mathcal{F}$ admits a geometric resolution $0 \to E_{-1} \to TM$ of length 1 if and only if it is Debord. E_{-1} then always admits an almost Lie algebroid bracket $[\cdot, \cdot]_{E_{-1}}$ which is automatically a Lie algebroid bracket. In conclusion, if a geometric resolution of length 1 exists, then it admits a Lie algebroid structure.

In this section, we discuss the case where a singular foliation $\mathcal{F}$ admits a geometric resolution of length 2. In this case, we claim that there are Lie algebra-like structures on it. This will be generalized in Sect. 7.

Let $(M, \mathcal{F})$ be a singular foliation that admits a geometric resolution of length 2, namely

$$(E_{-\bullet}, d^{(\bullet)}, \rho): \quad 0 \longrightarrow E_{-2} \xrightarrow{d^{(2)}} E_{-1} \xrightarrow{\rho} TM. \qquad (37)$$

Since Eq. (37) is a geometric resolution of $\mathcal{F}$, the pair $(E_{-1} \to M, \rho)$ is an anchored bundle over $\mathcal{F}$. Therefore, by Proposition 2.4, item 1, in the smooth case, E_{-1} can be endowed with an almost Lie algebroid structure $(E_{-1}, [\cdot, \cdot]_{E_{-1}}, \rho)$, and in the real analytic or complex cases, the almost Lie algebroid structure exists in a neighborhood of any point.

Let us assume from now that we are given on an almost Lie algebroid structure on E_{-1}. It is quite judicious to ask whether we can extend this bracket to sections of E_{-2}. If yes, what kind of structures will we have?

Since the complex (37) is a geometric resolution of $\mathcal{F}$, the complex

$$0 \longrightarrow \Gamma(E_{-2})_U \xrightarrow{d^{(2)}} \Gamma(E_{-1})_U \xrightarrow{\rho} \mathcal{F}_U \longrightarrow 0 \qquad (38)$$

is exact for all open subsets $U \subset M$.

---

[82] See e.g., [34] for detailed proofs or [29], Section 3, for a quick overview.

1. For all $a \in \Gamma_U(E_{-1}), b \in \Gamma_U(E_{-2})$:
   $$\rho([a, d^{(2)}b]_{E_{-1}}) = [\rho(a), \rho \circ d^{(2)}b] = 0, \text{ (by the anchor condition,}^{[83]} \text{ and since } \rho \circ d^{(2)} \equiv 0).$$

   In other words, the bracket $[a, d^{(2)}b]_{E_{-1}}$ is in the kernel of the anchor map ρ. By exactness of the complex (38), there exists a unique local section denoted by $\nabla_a b \in \Gamma(E_{-2})_U$ such that

   $$d^{(2)}(\nabla_a b) = [a, d^{(2)}b]_{E_{-1}}. \tag{39}$$

   The bilinear map:

   $$\Gamma(E_{-1})_U \otimes \Gamma(E_{-2})_U \to \Gamma(E_{-2})_U$$
   $$(a, b) \mapsto \nabla_a b$$

   does not depend on the chosen open subset U, i.d. is globally defined on M, and satisfies:
   (a) $d^{(2)}\nabla_a b = [d^{(2)}a, b]_{E_{-1}}$, $\forall a \in \Gamma(E_{-2})_U$, $b \in \Gamma(E_{-1})_U$, by construction, and
   (b) for all function $f \in \mathcal{O}(U)$: for all $a \in \Gamma(E_{-1}), b \in \Gamma(E_{-2})$, $\nabla_a(fb) = f\nabla_a b + \rho(a)[f]b$ and $\nabla_{fa} b = f\nabla_a b$,
2. For all $a, b, c \in \Gamma(E_{-1})$, we have

   $$\text{Jac}(a, b, c) := [a, [b, c]_{-1}]_{E_{-1}} + [b, [c, a]_{E_{-1}}]_{E_{-1}} + [c, [a, b]_{E_{-1}}]_{E_{-1}} \in \ker \rho.$$

   By using again exactness of the complex (38) there is a unique local section that we denote by $[a, b, c]_{E_{-1}} \in \Gamma(E_{-2})_U$ that satisfies

   $$d^{(2)}[a, b, c]_{E_{-1}} = \text{Jac}(a, b, c). \tag{40}$$

   It is easily checked that the map $(a, b, c) \mapsto [a, b, c]_{E_{-1}}$ is linear over functions, and therefore comes from a pointwise linear vector bundle morphism $\wedge^3 E_{-1} \to E_{-2}$. In particular, it is globally defined on M.

   The following lemma recapitulates the discussion above.

Lemma 6.26 *Let $(M, \mathcal{F})$ be a singular foliation that admits a geometric resolution of length 2 as in (37) such that $(E_{-1}, [\cdot, \cdot]_{E_{-1}}, \rho)$ admits an almost Lie algebroid bracket.*

---

[83] I.d. Eq. (4).

1. There is a bilinear map:[84]

$$\Gamma(E_{-1}) \otimes \Gamma(E_{-2}) \to \Gamma(E_{-2})$$
$$(a, b) \mapsto \nabla_a b$$

and a skew-symmetric trilinear map:

$$[\cdot, \cdot, \cdot]_{E_{-1}} : \Gamma(E_{-1}) \wedge \Gamma(E_{-1}) \wedge \Gamma(E_{-1}) \longrightarrow \Gamma(E_{-2})$$

2. such that for every function f:
   (a) $\nabla_a f b = f \nabla_a b + \rho(a)[f] b$ and $\nabla_{fa} b = f \nabla_a b$, for all $a \in \Gamma(E_{-1}), b \in \Gamma(E_{-2})$,
   (a) $[fa, b, c]_{E_{-1}} = f[a, b, c]_{E_{-1}}$ for all $a, b, c \in \Gamma(E_{-1})$,

We can now state our main result, which will be soon enlarged (see Theorem 7.34).

Proposition 6.27 *Let $(M, \mathcal{F})$ be a singular foliation that admits a geometric resolution of length 2 as in (37) such that $(E_{-1}, [\cdot, \cdot]_{E_{-1}}, \rho)$ admits an almost Lie algebroid bracket. Let ∇ and $[\cdot, \cdot, \cdot]_{E_{-1}}$ be as in Lemma 6.26. The 2-ary bracket on $\Gamma(E_{-1} \oplus E_{-2})$ defined by:*

$$[a, b]_2 = \begin{cases} [a, b]_{E_{-1}} & \text{for } a, b \in \Gamma(E_{-1}) \\ \nabla_a b & \text{for } a \in \Gamma(E_{-1}), b \in \Gamma(E_{-2}) \\ -\nabla_b a & \text{for } a \in \Gamma(E_{-2}), b \in \Gamma(E_{-1}) \\ 0 & \text{for } a, b \in \Gamma(E_{-2}) \end{cases}$$

together with the 3-ary bracket on $\Gamma(E_{-1} \oplus E_{-2})$ defined by $[a, b, c]_3 = [a, b, c]_{E_{-1}}$ if $a, b, c \in \Gamma(E_{-1})$ and zero otherwise, satisfies

1. *$[\cdot, \cdot, \cdot]_3$ is linear over functions, while for every function f and for all $a \in \Gamma(E_{-1}), b \in \Gamma(E_{-1})$ or $\Gamma(E_{-2})$,*

$$[a, fb] = f[a, b]_2 + \rho(a)[f] b$$

2. *for all $a \in \Gamma(E_{-2}), b \in \Gamma(E_{-1})$,*

$$d^{(2)}[a, b]_2 = [d^{(2)} a, b]_2, \tag{41}$$

---

[84] Here, $\Gamma(E)$ refers to local sections of the vector bundle E. In the smooth case, it suffices of course to work with global sections.

3. for all $a, b, c \in \Gamma(E_{-1})$

$$d^{(2)}[a,b,c]_3 + [a,[b,c]_2]_2 + [b,[c,a]_2]_2 + [c,[a,b]_2]_2 = 0$$

4. for all $a, b \in \Gamma(E_{-1})$ and $c \in \Gamma(E_{-2})$

$$[a, b, d^{(2)}c]_3 + [a,[b,c]_2]_2 + [b,[c,a]_2]_2 + [c,[a,b]_2]_2 = 0.$$

Definition 6.28 The structure $(E_{-1} \oplus E_{-2}, d, \rho, [\cdot,\cdot]_2, [\cdot,\cdot,\cdot]_3)$ described in the third item of Proposition 6.27 is known as a *Lie 2-algebroid*.

A generalization to arbitrary singular foliation admitting a geometric resolution will be discussed in the Sect. 7.

7 Universal Q-manifolds (a.k.a. Universal Lie ∞-Algebroids)

Beyond the world of manifolds is the universe of "manifolds up to homotopy", which are known under various names: some or more or less equivalent, and some are mostly dual notions:

- ♣ Lie ∞-algebroids, also called Lie algebroids up to homotopy.
- ◇ Q-manifolds, also called dg-manifolds (dg= differential graded) of positive degree.

The notions ♣ and ◇ are in fact "equivalent" in the sense that they are dual[85] one to the other. To explain where the notion of Q-manifold comes from, and the duality with Lie ∞-algebroids, let us start with some basic points about Lie algebras. We invite the reader familiar with Lie ∞-algebroids and NQ-manifolds to go directly to Theorem 7.34.

7.1 Two Dual Point of Views on Lie Algebras

For V a vector space, the exterior algebra $\wedge^\bullet V = \bigoplus_{k=0}^\infty \wedge^k V$ is a graded algebra with respect to a product that we denote by $\wedge$.

Definition 7.1 A *co-Lie algebra* is a vector space V equipped with a degree $+1$ derivation

$$\delta: \wedge^\bullet V \longrightarrow \wedge^{\bullet+1} V$$

such that $\delta^2 = 0$.

---

[85] The reader familiar to these concepts knows that it is a bit more subtle: one needs to choose a splitting.

Before explaining this definition, let us start with a few comments.

1. We write $\delta \colon \wedge^\bullet V \mapsto \wedge^{\bullet+1} V$ to mean that for every $k \geq 0$, δ maps $\wedge^k V$ to $\wedge^{k+1} V$, i.e., it is of degree $+1$.
2. By a degree $+1$ derivation, we mean that

$$\delta(\alpha \wedge \beta) = \delta(\alpha) \wedge \beta + (-1)^k \alpha \wedge \delta(\beta)$$

for all $\alpha \in \wedge^k V$ and $\beta \in \wedge^\bullet V$. The signs are exactly those of the de Rham differential (which is also a degree $+1$ derivation).
3. For any degree $+1$ derivation, δ^2 is easily seen to be a degree $+2$ derivation.[86]
4. A degree $+1$ derivation of $\wedge^\bullet V$ is entirely determined by its restriction to V, which is a map $\mu \colon V \longrightarrow \wedge^2 V$ that we call the *co-Lie-bracket*. This comes from the derivation property:

$$\delta(v_1 \wedge \cdots \wedge v_k) = \sum_{i=1}^{k} (-1)^{i+1} v_1 \wedge \cdots \wedge v_{i-1} \wedge \mu(v_i) \wedge v_{i+1} \wedge \cdots \wedge v_k \qquad (42)$$

for every $\in \mathbb{N}$ and $v_1, \ldots, v_k \in V$.
5. Conversely, any linear map $\mu \colon V \longrightarrow \wedge^2 V$ extends to a unique degree $+1$ derivation by using (42).

Proposition 7.2 (Lie Algebras Are Dual to Co-Lie Algebra) *There is a one-to-one correspondence[a] between finite dimensional Lie algebras and finite dimensional co-Lie algebras.*

---

[a]More precisely, to a Lie algebra structure on $\mathfrak{g}$ corresponds a co-Lie algebra structure on $\mathfrak{g}^*$, and to a co-Lie algebra structure on V corresponds a co-Lie algebra structure on V^*.

Proof The correspondence goes as follows.

1. Given a Lie algebra $(\mathfrak{g}, [\cdot, \cdot])$, the dual of the Lie algebra bracket $[\cdot, \cdot] \colon \wedge^2 \mathfrak{g} \to \mathfrak{g}$ is a map $\mu \colon \mathfrak{g}^* \to (\wedge^2 \mathfrak{g})^*$. Since the dimensions are finite, there is a canonical isomorphism $(\wedge^2 \mathfrak{g})^* \simeq \wedge^2 \mathfrak{g}^*$, and we still denote by μ the map $\mu \colon \mathfrak{g}^* \to \wedge^2 \mathfrak{g}^*$. Using the derivation

---

[86] Of course, this is not true for degree 0 derivation, otherwise the formula $(fg)'' = f''g + fg''$ would be true.

property in Eq. (42), one extends μ to a degree $+1$ derivation δ of $\wedge^\bullet \mathfrak{g}^*$. It is a routine to check that if the Jacobi identity holds for $[\cdot,\cdot]$, then it implies that $\delta^2 = 0$.

2. Conversely, given a co-Lie algebra, the dual of the co-Lie bracket $\mu: V \to \wedge^2 V$ is a linear map $[\cdot,\cdot]: \wedge^2 V^* \to V^*$. It is a routine to check that $\delta^2 = 0$ implies that the Jacobi identity holds for the bracket $[\cdot,\cdot]$.

Moreover, the two maps above are inverse one to the other. □

Remark 7.3 The degree $+1$ derivation corresponding to a finite-dimensional Lie algebra over the field $\mathbb{K}$ is the Chevalley-Eilenberg differential computing the Lie algebra cohomology (valued in $\mathbb{K}$).

To put it all in a nutshell:

What is a finite dimensional Lie algebra? Two dual answers.	
Direct notion:	Dual notion:
A vector space $\mathfrak{g}$ *and* a linear map $[\cdot,\cdot]: \wedge^2 \mathfrak{g} \to \mathfrak{g}$	A vector space V *and* a degree $+1$ derivation $\delta: \wedge^\bullet V \to \wedge^{\bullet+1} V$
such that	
The Jacobi identity holds	$\delta^2 = 0$

7.2 Graded Symmetric Algebras

Throughout of this section we are working on a field $\mathbb{K} \in \{\mathbb{R}, \mathbb{C}\}$.

Let us introduce some terminology. Consider a graded vector space:

$$Z_\bullet := \oplus_{i \in \mathbb{Z}} V_i.$$

An element in $V_\bullet$ is said to be *homogeneous* when there exists $i \in \mathbb{Z}$ such that it belongs to Z_i.

Definition 7.4 (Graded Symmetric Algebras) Let $Z_\bullet := \oplus_{i \in \mathbb{Z}} Z_i$ be a graded vector space. We call *graded symmetric algebra of* $Z_\bullet$ the quotient of the tensor algebra $\oplus_{k \geq 0} Z^{\otimes k}$ by the ideal generated by all elements of the form

$$\{x \otimes y - (-1)^{ij} y \otimes x \mid i, j \in \mathbb{Z}, x \in V_i, y \in V_j\}.$$

We denote it by $S(Z_\bullet)$. We denote its product by $\odot$.

Let us state a few basic points about graded symmetric algebras and their terminology. To start with, the graded symmetric algebra comes with two different "degrees" that we have to distinguish: linear combinations of elements of the form

$$z_1 \odot \cdots \odot z_k$$

with $z_1 \in Z_{i_1}, \ldots, z_k \in Z_{i_k}$ shall be said of

1. *polynomial degree k*, because they are products of k elements,
2. and of *degree* $i_1 + \cdots + i_k$, because the individual degrees of the terms in the product add up to this integer.[87]

With respect to both the polynomial degree and the degree, $S(Z_\bullet)$ is "graded" in the sense that the degree of the product of two terms is the sum of their degrees. But with respect to the degree, it is also graded commutative, i.e.,

$$P \odot Q = (-1)^{ij} Q \odot P$$

for any $P, Q \in S(Z_\bullet)$ of degrees i and j, respectively.

For homogeneous elements $z_1, \ldots, z_k \in Z_\bullet$ and permutation $\sigma \in \mathfrak{S}_k$, the *Koszul sign*, denoted by $\epsilon(\sigma, z_1, \ldots, z_k)$ or simply by $\epsilon(\sigma)$ when there is no ambiguity, is the sign induced by the permutation of the v_i in the graded symmetric algebra:

$$z_{\sigma(1)} \odot \cdots \odot z_{\sigma(k)} = \epsilon(\sigma, z_1, \ldots, z_k) z_1 \odot \cdots \odot z_k. \tag{43}$$

---

[87] The "degree" is sometimes called "ghost degree" by theoretical physicists.

For $Z_\bullet = \oplus_{i\in\mathbb{Z}} Z_i$ a graded vector space, we call *graded dual* denote by $Z_\bullet^*$ the graded vector space:

$$Z_\bullet^* = \oplus_{i=-\infty}^{\infty} Z_{-i}^*,$$

with the understanding that elements in Z_{-i}^* are of degree $+i$. Notice that $Z_\bullet^*$ is strictly contained in the dual Z^*. According to our conventions $S(Z_\bullet^*)$ stands for the graded commutative symmetric algebra generated by $Z^* = \oplus_{i=1}^{\infty} Z_{-i}^*$. Elements in $Z_{-i_1}^* \odot \cdots \odot Z_{-i_k}^*$ are therefore of polynomial degree k and degree $i_1 + \cdots + i_k$. We shall define elements of polynomial degree zero to be elements in $\mathbb{K}$. In the present section, we will be interested in two kinds of symmetric algebras:

- ♣ those of the form $S(E_\bullet)$ with $E_\bullet := \oplus_{i \geq 1} E_{-i}$ a negatively graded vector space. Except $S^0(E_\bullet) = \mathbb{K}$, all its components are of negative degrees,
- ◇ and those of the form $S(V_\bullet)$ with $V_\bullet := \oplus_{i \geq 1} V_i$ a positively graded vector space. Except $S^0(V_\bullet) = \mathbb{K}$, all its components are of positive[88] degrees.

If each one of the vector spaces E_{-i} and V_i are of finite dimension, and if E_{-i} is the dual of V_i for all $i \geq 1$, then there is a natural duality between:

- ♣ elements of polynomial degree k and degree $-i$ in $S(\oplus_{i \geq 1} E_{-i})$,
- ◇ elements of polynomial degree k and degree $+i$ in $S(\oplus_{i \geq 1} V_i)$,

given for all homogeneous $v_1, \ldots, v_k \in V_\bullet$ and $e_k, \ldots, e_1 \in E_\bullet$ by the pairing:[89]

$$\langle v_1 \odot \cdots \odot v_k \,|\, e_k \odot \cdots \odot e_k \rangle = \sum_{\sigma \in \mathfrak{S}_k} \epsilon(\sigma, v_1, \ldots, v_k) \prod_{i=1}^{k} \langle v_{\sigma(i)}, e_i \rangle.$$

7.2.1 Lie ∞-Algebras

We are now ready for the following question.[90]

> **Question 7.5 (Towards Lie ∞-Algebras)** *Let $V_\bullet = \oplus_{i \geq 1} V_i$ be a positively graded vector space with each V_i of finite dimension. Let E_{-i} be the dual of V_i.*
> *Assume $S(V_\bullet)$ comes equipped with a degree $+1$ derivation δ such that $\delta^2 = 0$. What kind of structures do we obtain on the dual spaces $E_\bullet := \oplus_{i \geq 1} E_{-i}$?*

---

[88] Since it might be confusing for speakers of Latin languages, let us be precise: here, positive means ≥ 0, and negative means ≤ -1.

[89] Some authors would divide by $k!$.

[90] It is complicated to make a history of the notion presented in this section: let us just say that no idea here is ours, and that the notion of Lie ∞-algebra originated in [33].

We saw in the Sect. 7.1 that if $V_i = 0$ for $i \geq 2$, such derivations are in one-to-one correspondence with Lie algebras structures on $E_{-1} := V_1^*$. Let us go back to the general case. The derivation δ is entirely determined by its restriction to $V_\bullet$. Decomposing according to the polynomial degree, we see that $\delta = \sum_{k \geq 1} \delta^{(k)}$, with $\delta^{(k)} \colon V_\bullet \mapsto S^k V_\bullet$ a degree $+1$ map. By duality, there is a one-to-one correspondence between:

- ♣ the datum $\left(E_\bullet = \sum_{i=1}^{\infty} E_{-i}, (\ell_k)_{k \geq 1}\right)$ made of a collection of vector spaces $E_\bullet = (E_{-i})_{i \geq 1}$ of finite dimension together with a family of degree $+1$ linear maps $(\ell_k \colon S^\bullet(E) \longrightarrow E)_{k \geq 1}$ called k-ary brackets,
- ◇ a sequence $\delta^{(k)}$ of linear maps $V_\bullet \longrightarrow S^k(V_\bullet)$, with $V_i = E_{-i}^*$ for all $i \geq 1$.

The relation between both is given for all $v \in V_\bullet, e_1, \ldots, e_k \in E_\bullet$ by:

$$\langle \delta^{(k)}(v), e_k \odot \cdots \odot e_1 \rangle = \langle v, \ell_k(e_k \odot \cdots \odot e_1) \rangle.$$

A direct computation gives that $\delta^2 = 0$ holds if and only if the ℓ_k's equip $E_{-\bullet}$ with a Lie ∞-algebra structure, the latter being defined as follows.[91]

> **Definition 7.6 (Lie ∞-Algebras)** A *negatively graded Lie ∞-algebra* is the datum $\left(E_\bullet, (\ell_k)_{k \geq 1}\right)$ made of a collection of vector spaces $E_\bullet = (E_{-i})_{i \geq 1}$ together with a family of degree $+1$ linear maps $(\ell_k \colon S^\bullet(E_\bullet) \longrightarrow E_\bullet)_{k \geq 1}$ called k-ary brackets, which fulfill the compatibility conditions the so-called *higher Jacobi identities*: for all homogeneous elements $e_1, \ldots, e_n \in E$
>
> $$\sum_{i=1}^{n} \sum_{\sigma \in \mathfrak{S}_{i,n-i+1}} \epsilon(\sigma) \ell_{n-i+1}\left(\ell_i(e_{\sigma(1)}, \ldots, e_{\sigma(i)}), v_{\sigma(i+1)}, \ldots, e_{\sigma(n)}\right) = 0. \quad (44)$$
>
> Here $\epsilon(\sigma)$ is the Kozsul sign associated to σ and $e_1, \ldots, e_n$, and $\mathfrak{S}_{i,n-i+1}$ stands for the set of (n, i)-shuffles.

In particular, for $n = 1$, Eq. (44) means $\ell_1^2 = 0$, so that $(E_\bullet, \ell_1)$ is a complex. Let us conclude this section by the following result.

---

[91] We prefer to avoid any historical comments. See, e.g. [30], Sections 2.1 and 2.2, for a pedagogical introduction.

Proposition 7.7 (Dual Point of Views) *Let $E_{-\bullet} = \oplus_{i \geq 1} E_{-i}$ and $V_\bullet = \oplus_{i \geq 1} V_{+i}$ be finite dimensional graded vector spaces in duality.[a] There is a one-to-one correspondence between*

- *Lie ∞-algebras brackets $(\ell_k)_{k \geq 1}$ on $E_\bullet$,*
- *degree $+1$ derivations squaring to zero of $S(V_\bullet)$.*

---

[a] I.e., E_{-i} is the dual of V_i for all $i \geq 1$.

To put it all in a nutshell, in view of Proposition 7.7, we have

What is a finite dimensional Lie ∞-algebra? Two dual answers	
Direct notion:	Dual notion:
A neg. graded vector space $E_\bullet = \oplus_{i=1}^\infty E_{-i}$ and a family of degree $+1$ linear maps $\ell_k \colon S^k(E_\bullet) \longrightarrow E_\bullet$ (called "brackets")	A pos. graded vector space $V_\bullet = \oplus_{i=1}^\infty V_i$ and a degree $+1$ derivation $\delta \colon S(V_\bullet) \longrightarrow S(V_\bullet)$
such that	
Eq. (44) ("higher Jacobi") holds for all n	$\delta^2 = 0$

7.3 (Positively Graded) NQ-manifold

We will now extend the previous discussion from Lie ∞-algebras to Lie ∞-algebroids,[92] and from $S(V_\bullet)$ equipped with a degree $+1$ differential to the so-called Q-manifolds.

---

[92] We prefer not to make an historical introduction: we acknowledge being inspired by several works by Pavol Severa [31], Theodore Voronov [38], as well as by Giuseppe Bonavolontà and Norbert Poncin's [5], see also the first sections of [8] and Section 3.4 in [23] for a similar presentation.

7.3.1 Graded Manifolds

Let us first define positively graded manifolds. In addition to the purely mathematical definition, we give as footnotes some vocabularies commonly used by mathematical physicists working with this object.

Definition 7.8 (Graded Manifolds: The Objects) Let M be a smooth, real analytic or complex manifold and $\mathbb{K} = \mathbb{R}$ or $\mathbb{C}$ depending on the chosen context. Let $\mathcal{O}$ be the corresponding sheaf of (smooth, real analytic, or holomorphic) functions on M.

A *positively graded manifold* over a manifold[a] M is a sheaf

$$\mathcal{E} : \mathcal{U} \mapsto \mathcal{E}_\mathcal{U}$$

of graded commutative algebras over $\mathbb{K}$ such that every $m \in M$ admits an open neighborhood $\mathcal{U} \subset M$ on which the sheaf[b] structure takes the form[c]

$$\mathcal{E}_\mathcal{U} = \mathcal{O}_\mathcal{U} \otimes_\mathbb{K} S(V_\bullet)$$

for some graded vector space[d] $V = \oplus_{i=1}^\infty V_{+i}$.

---

[a] It is customary to call M the *base* of the sheaf.
[b] It is customary to call sections of the sheaf $\mathcal{E}$ *graded functions*.
[c] It is convenient to denote a graded manifold as a pair $(M, \mathcal{E})$.
[d] It is customary to say that the graded manifolds in *concentrated in degrees* $[1, \ldots, N]$ if all $V_i = 0$ when $i \notin [1, \ldots, N]$.

Remark 7.9 In the smooth setting, it can be proven that there exists a globally and canonically defined graded vector bundle $V_\bullet \to M$ such that $\mathcal{E}$ is isomorphic to the sheaf of sections of the graded commutative algebra bundle $S(V_\bullet) \to M$, obtaining by considering pointwise the symmetric algebra of $V_\bullet$. Although $V_\bullet$ is canonical, the isomorphism of sheaves

$$\mathcal{E} \xleftarrow{\simeq} \Gamma(S(V))$$

is *not* canonical, and is called a *splitting*. For a statement adapted to the present situation, see [19]. Upon choosing such an isomorphism, a function $F \in \mathcal{E}_i$ is a formal sum

$$F = \sum_{k \geq 0} F^{(k)} \tag{45}$$

with $F^{(k)} \in \mathcal{E}$ an element of polynomial degree k and degree i. For degree reasons, the sum in (45) must be finite. To be more precise, the only possible values of k are between $1 \leq k \leq i$.

Remark 7.10 Remark 7.9 has consequences on vocabulary. Even in the holomorphic or real analytic settings, one calls sections in the sheaf $\mathcal{E}$ under the name of *(graded) functions*. But what are these sections "functions" on? In the smooth case, we can answer this question. Consider the negatively graded vector bundle $E_{-\bullet} = \oplus_{i \geq 1} E_{-i}$ with $E_{-i} = V_i^*$ for all $i \geq 1$. Sections of the sheaf $\mathcal{E}$ maybe considered as *functions on* $E_\bullet$.

Remark 7.11 If the manifold M is connected, then for every $i \geq 1$, the dimension of the component V_i of degree i in the graded vector space $V_\bullet$ that appear in Definition 7.8 does not depend on the point m. From now on, we will assume that M is connected and denote by r_i the rank of V_i.

7.3.2 Local Coordinates of a Graded Manifold

Among the somewhat cumbersome vocabulary that makes graded manifolds hard to deal with for the non-used reader is the notion of "local graded coordinates". Assume that M is connected and let r_i be as in Remark 7.11. For any $\mathcal{U} \subset M$ an open subset such that $(V_i)_\mathcal{U} \xrightarrow{\sim} \mathcal{U} \times \mathbb{K}^{r_i}$ for every $i \geq 1$, we call *graded coordinates* on the graded manifold $(M, \mathcal{E})$ the data made of

1. a system of coordinates $(x_1, \ldots, x_n)$ of M on $\mathcal{U}$.
2. and in all degree $i \geq 1$, a local trivialization $\left(\xi_i^{[1]}, \ldots, \xi_i^{[r_i]}\right)$ of V_i on $\mathcal{U}$.

After choosing graded coordinates as above, i.e., a list

$$\left(x_1, \ldots, x_n, \xi_1^{[1]}, \ldots, \xi_1^{[r_1]}, \ldots, \xi_i^{[1]}, \ldots, \xi_i^{[r_i]}, \ldots\right),$$

then any element of $\mathcal{E}(\mathcal{U})$ is a "polynomial" in finitely many of the variables

$$\left(\xi_i^{[j]}\right)_{j=1,\ldots,r_i,\ i \geq 1}$$

with coefficients in the algebra $\mathcal{O}(\mathcal{U})$ of smooth, real analytic or complex functions over U. For instance, an element in degree 3 decomposes as a sum of the form

$$\sum_{1\leq i<j<k\leq r_1} f_{ijk}(x_\bullet)\,\xi_1^{[i]}\odot\xi_1^{[j]}\odot\xi_1^{[k]} + \sum_{\substack{1\leq i\leq r_1\\ 1\leq j\leq r_2}} g_{ij}(x_\bullet)\,\xi_1^{[i]}\odot\xi_2^{[j]} + \sum_{1\leq i\leq r_3} h_i(x_\bullet)\,\xi_3^{[i]}$$

with f_{ijk}, g_{ij}, h_i being smooth, real analytic or holomorphic functions in the variables $x_\bullet = x_1, \ldots, x_n$.

Example 7.12 The sheaf of differential forms $(M, \mathcal{E} = \Omega(M))$ on a manifold M is a graded manifold since for every point $m \in M$, it takes the form $\mathcal{O}_\mathcal{U} \otimes_\mathbb{K} \wedge^\bullet T_m^* M$ on an open neighborhood $\mathcal{U}$ of m. Exterior forms can be seen as functions on the graded vector bundle $E_{-1} = TM$, concentrated in degree -1.

Example 7.13 Let k be a positive integer. A finite dimensional vector space E and its dual V can be seen as graded vector bundles of respective degree $-k$ and k over a point. E is a graded manifold over $M = \{\text{pt}\}$ and the sheaf $\mathcal{E}$ is isomorphic (as an algebra) to $\wedge V$ for k odd and $S(V)$ for k even. It must be understood that $\wedge^i V$ or $S^i(V)$ are then of degree ki.

We now define morphisms.

Definition 7.14 (Graded Manifolds: The Morphisms) A *morphism of graded manifolds* from (M', E') to $(M, \mathcal{E})$ is a pair made of a smooth or real analytic or holomorphic map $\phi: M' \longrightarrow M$ called the *base map* and a sheaf morphism over ϕ^*, i.e., a family of graded algebra morphisms:

$$\mathcal{E}_\mathcal{U} \to \mathcal{E}'_{\phi^{-1}(\mathcal{U})},$$

compatible with the restriction maps, such that

$$\Phi(f\alpha) = \phi^*(f)\Phi(\alpha). \tag{46}$$

for all $f \in \mathcal{O}_\mathcal{U}$ and $\alpha \in \mathcal{E}_\mathcal{U}$.

(continued)

Definition 7.14 (continued)

A *homotopy* between two morphisms of graded manifolds $\Phi, \Psi \colon (M', E') \longrightarrow (M, \mathcal{E})$ is a morphism of graded manifolds[a]

$$(M, \mathcal{E}) \longrightarrow (M', \mathcal{E}') \times ([0, 1], \Omega([0, 1]))$$

whose restrictions to the extremities of the interval $[0, 1]$ coincide with Φ and Ψ respectively.

---

[a]The direct product of graded manifolds considered above goes with some subtleties. The henceforth obtained sheaf is a sheaf over $M' \times [0, 1]$. The graded vector bundle as in Remark 7.9 is $p_1^! V_1 \oplus p_2^! T[0, 1]$ in degree 1 and $p_1^! V_i$ in degree $i \geq 2$, with p_1, p_2 the projections on the first and second component respectively. We refer to Section 3.4 in [23] for more details.

7.3.3 Vector Fields on Graded Manifolds

Vector fields on a manifold M are derivations of its sheaf of algebra of functions: this principle is valid in the smooth, real analytic or complex settings. For a graded manifold $(M, \mathcal{E})$, the equivalent of functions are the sections of the sheaf $\mathcal{E}$. Since it is not commutative but graded commutative, one has to consider graded derivations. Consider $k \in \mathbb{Z}$. A *graded derivation of degree k* of $\mathcal{E}$ is the data, for every $\mathcal{U} \subset M$ of a linear map

$$Q \colon (\mathcal{E}_{\mathcal{U}})_{\bullet} \longrightarrow (\mathcal{E}_{\mathcal{U}})_{\bullet + k},$$

compatible with all restriction maps, that increases the degree by $+k$ and satisfies:

$$Q[F\,G] = Q[F]\,G + (-1)^{ki} F\,Q[G]$$

for every $F \in (\mathcal{E}_{\mathcal{U}})_i, G \in \mathcal{E}_{\mathcal{U}}$. Since we think geometrically, we will simply say *vector fields of degree k* instead of graded derivations.

Definition 7.15 We call Let $(M, \mathcal{E})$ be a smooth graded manifold. For any open subset $\mathcal{U} \subset M$, we denote by $\mathfrak{X}_{\bullet}(\mathcal{E})$ the space of graded derivations of degree k and call its elements *vector fields of degree k*

In the complex or real analytic settings, graded derivations of degree k do not form not a sheaf, and one has to define $\mathfrak{X}_{\bullet}(\mathcal{E})$ using local graded derivations.

In all three cases, the sheaf

$$\mathcal{U} \longmapsto \mathfrak{X}_\bullet(\mathcal{E})_\mathcal{U} = \oplus_{k\in\mathbb{Z}} \mathfrak{X}_k(\mathcal{E})_\mathcal{U}$$

is a sheaf of $\mathcal{E}$-modules that we call *sheaf of graded vector fields*[93] on $(M, \mathcal{E})$.

Let us list some important facts on vector fields on $(M, \mathcal{E})$:

1. The sheaf $\mathfrak{X}_\bullet(\mathcal{E}) := \oplus_{k\in\mathbb{Z}} \mathfrak{X}_k(\mathcal{E})$ of vector fields on $(M, \mathcal{E})$ is a graded $\mathcal{E}$-module. Also, $\mathfrak{X}_\bullet(\mathcal{E})$ of is a sheaf of graded Lie algebras. The graded Lie bracket

$$[P, Q] = P \circ Q - (-1)^{kl} Q \circ P \qquad (47)$$

   of two vector fields P, Q of degree k, l respectively is a vector field of degree $k + l$. It is easily checked that the bracket (47) fulfills
   (a) $[P, Q] = -(-1)^{jk}[Q, P]$ (graded skew-symmetry)
   (b) $(-1)^{jl}[P, [Q, R]] + (-1)^{jk}[Q, [R, P]] + (-1)^{kl}[R, [P, Q]] = 0$, (graded Jacobi identity)
   for all graded vector fields $P, Q, R \in \mathfrak{X}(\mathcal{E})$ of degrees j, k and l, respectively.

2. Let $E_\bullet$ be a graded vector bundle over M as in Remark 7.9, i.e., such that there exists a splitting

$$\mathcal{E} \simeq \Gamma(S(V))$$

   with $V_i = E^*_{-i}$ for all $i \geq 0$. Any section $e \in \Gamma(E_{-i})$ corresponds to a vector field[94] $\iota_e \in \mathfrak{X}_{-i}(\mathcal{E})$ defined by contraction with e.

3. Let $(\mathcal{U}, x_1, \ldots, x_n)$ local coordinates on M and $(\xi_i^{[j]})_{j=1,\ldots,r_i}$ with $i \geq 1$ be local coordinates on $\mathcal{E}$, defined as above.
   Let $(e_i^{[j]})_{j=1,\ldots,r_i}, i \geq 1$ be the dual basis of $(\xi_i^{[j]})_{j=1,\ldots,r_i}, i \geq 1$. Then for every pair i, j, it is customary to consider the contraction $\iota_{e_i^{[j]}}$ of item as being the "partial derivative":

$$\frac{\partial}{\partial \xi_i^{[j]}}$$

The notation comes from the computational fact that, in local coordinates, the effect of $\iota_{e_i^{[j]}}$ is similar to the one of the partial derivative with respect to $\xi_i^{[j]} \in \Gamma(V_i)$.

---

[93] We will simply say "vector fields on $\mathcal{E}$" most of the time.
[94] Such vector fields are called *vertical* because they are linear over functions on the base manifold. In this case, the polynomial degree is -1 and its degree is $-i$.

Elements in the following list are vector fields of degree k:

$$\left(\xi_{i_1}^{[j_1]} \odot \cdots \odot \xi_{i_l}^{[j_l]} \frac{\partial}{\partial x_j}\right)_{i_1+\cdots+i_l=k} \quad \text{and} \quad \left(\xi_{i_1}^{[j_1]} \odot \cdots \odot \xi_{i_l}^{[j_l]} \frac{\partial}{\partial \xi_i^{[j]}}\right)_{i_1+\cdots+i_l-i=k}$$

that form a "basis" for $\mathfrak{X}_k(\mathcal{E})(\mathcal{U})$ in the sense that any vector field $Q \in \mathfrak{X}_k(\mathcal{E})(\mathcal{U})$ is an infinite[95] sum of the form

$$Q = \sum_{i_1\cdots i_l=k} \frac{1}{l!} \, {}^j Q_{i_1\cdots i_l}^{j_1\cdots j_l} \, \xi_{i_1}^{[j_1]} \odot \cdots \odot \xi_{i_l}^{[j_l]} \frac{\partial}{\partial x_j}$$

$$+ \sum_{i_1\cdots i_l-i=k} \frac{1}{l!} \, {}^{ij} Q_{i_1\cdots i_l}^{j_1\cdots j_l} \, \xi_{i_1}^{[j_1]} \odot \cdots \odot \xi_{i_l}^{[j_l]} \frac{\partial}{\partial \xi_i^{[j]}}.$$

for some smooth, real analytic or holomorphic functions $Q_{i_1\cdots i_l}^{j_1\cdots j_l}$ on $\mathcal{U}$. These functions are unique, in one chooses an ordering on the indices. For example, a vector field Q of degree $+1$ can be written in these notations as

$$Q = \sum_{\substack{1 \le u \le r_1 \\ j=1,\ldots,n}} {}^j Q_1^u \, \xi_1^{[u]} \frac{\partial}{\partial x_j} + \sum_{i_1+\cdots+i_l-i=1} \frac{1}{l!} \, {}^{ij} Q_{i_1\cdots i_l}^{j_1\cdots j_l} \, \xi_{i_1}^{[j_1]} \odot \cdots \odot \xi_{i_l}^{[j_l]} \frac{\partial}{\partial \xi_i^{[j]}}.$$

We can define NQ-manifolds.

Definition 7.16 (NQ-manifolds) A NQ-manifold[a] is a positively graded manifold $(M, \mathcal{E})$ endowed with a degree $+1$ homological vector field $Q \in \mathfrak{X}_1(\mathcal{E})$ such that $Q^2 = 0$.
It shall be denoted as a triple $(M, \mathcal{E}, Q)$.

---

[a] Also called in this context positively graded Q-manifold or positively graded dg-manifold.

Example 7.17 When trying to understand NQ-manifolds, the novice reader may be surprised by phrases like "A finite dimensional Lie algebra is a NQ-manifold with

---

[95] The sum containing $\frac{\partial}{\partial x_j}$ is finite, but the second one is only finite if finitely many of the E_{-i}'s are non-zero.

base manifold a point and concentrated in degree 1". Let us explain the meaning of such words. Given a finite dimensional Lie algebra $(\mathfrak{g}, [\cdot, \cdot])$ of dimension d. The pair $(M = \{\text{pt}\}, \mathcal{E} = \wedge^\bullet \mathfrak{g}^*)$ is a graded manifold over $M = \{\text{pt}\}$, which is generated by a vector space of degree -1 (hence "concentrated in degree -1"). When equipped with the Chevalley-Eilenberg differential which is a derivation of degree $+1$, i.e., a vector field of degree $+1$, and squares to zero. It is therefore a NQ-manifold. Let us write it in coordinates: Fix a basis $(e_i)_{i=1,\ldots,d}$ of $\mathfrak{g}$ and let $\xi^i \in \mathfrak{g}^*$ with $i = 1, \ldots, d$ be the dual basis. Consider the Christoffel symbols $\lambda_{ij}^k \in \mathbb{K}$ defined by

$$[e_i, e_j] = \sum_{k=1}^{d} \lambda_{ij}^k e_k.$$

The Chevalley-Eilenberg operator is given by the degree $+1$ "vector field":

$$Q = \frac{1}{2} \sum_{i,j,k=1}^{d} \lambda_{ij}^k \xi^i \wedge \xi^j \frac{\partial}{\partial \xi^k}.$$

Example 7.18 Given a complex of vector bundles $(E_{-\bullet}, \mathrm{d}^{(\bullet)})$ over M. There is a natural dg-manifold given by its sheaf of sections $(M, \mathcal{E} = \Gamma(S(E^*))$ and whose homological vector field $Q \in \mathfrak{X}_{+1}$ is obtained by dualization $\mathrm{d}^* \colon S^1(E^*) \longrightarrow S^1(E^*)$ of the differential map $\mathrm{d} \colon E \longrightarrow E$ map and $\rho^* \colon T^*M \to E_{-1}^*$ of the anchor $\rho \colon E_{-1} \to TM$

$$\Gamma(E_{-1}^*) \xrightarrow{\mathrm{d}^*} \Gamma(E_{-2}^*) \xrightarrow{\mathrm{d}^*} \cdots \tag{48}$$

that we extend to a derivation on $\mathcal{E}$ squaring to zero.

Example 7.19 Let M be a manifold. The sheaf $(M, \mathcal{E} = \Omega(M))$, equipped with De Rham differential, is a NQ-manifold. In terms of coordinates, the homological vector field reads

$$\sum_{i=1}^{n} dx_i \frac{\partial}{\partial x_i}.$$

Let us introduce some vocabulary that will need to use.

Definition 7.20 Let $(M, \mathcal{E}', Q')$ and $(M, \mathcal{E}, Q)$ be two NQ-manifolds coming with splittings $\mathcal{E} \simeq \Gamma(S(V_\bullet))$ and $\mathcal{E}' \simeq \Gamma(S(V'_\bullet))$. A linear map $\Phi \colon \mathcal{E} \to \mathcal{E}'$ is said to be of *polynomial degree/degree* $j \in \mathbb{Z}$ provided that for all functions $\alpha \in \mathcal{E}$ of polynomial degree/degree i, $\Phi(\alpha)$ is of polynomial degree/degree $i + j$.

Remark 7.21 Assume that we are in the context of Definition 7.20, then two points are worth noticing.

1. Any map $\Phi \colon \mathcal{E} \to \mathcal{E}'$ of degree i decomposes w.r.t the polynomial degree as follows:

$$\Phi = \sum_{r \in \mathbb{Z}} \Phi^{(r)}$$

with $\Phi^{(r)} \colon \mathcal{E} \to \mathcal{E}'$ a map of polynomial degree r.

2. When $\Phi \colon \mathcal{E} \to \mathcal{E}'$ is a graded morphism of algebras, its degree is zero, which implies that $\Phi^{(r)} = 0$ for all $r < 0$, i.e., it can not have components of negative polynomial degree. Furthermore, for all $n, r \in \mathbb{N}$ and all $\xi_1, \ldots, \xi_k \in \Gamma(V)$ one has the following decomposition:

$$\Phi^{(r)}(\xi_1 \odot \cdots \odot \xi_n) = \sum_{i_1 + \cdots + i_n = r} \Phi^{(i_1)}(\xi_1) \odot \cdots \odot \Phi^{(i_n)}(\xi_n). \tag{49}$$

In particular, Φ is uniquely determined by the values on $\Gamma(V)$.

In the context of Lie algebroids, the following idea is attributed to Vaintrob [37].

Definition 7.22 (Morphisms) Let $(M, \mathcal{E}, Q)$ and $(M', \mathcal{E}', Q')$ be two NQ-manifolds with sheaves of functions $\mathcal{E}$ and $\mathcal{E}'$ respectively. A *morphism of NQ-manifold over M* is a morphism of graded manifolds $\Phi \colon \mathcal{E} \to \mathcal{E}'$ over which intertwines Q and Q', i.e.,

$$\Phi \circ Q = Q' \circ \Phi. \tag{50}$$

7.4 Negatively Graded Lie ∞-algebroids and Their Morphisms

We denote by $\mathcal{O}$ the sheaf of smooth, real analytic or holomorphic functions on M, and by $\Gamma(E)$ the sheaf of sections of the graded bundle $E_{-\bullet}$.

Definition 7.23 (Lie ∞-Algebroid) A *Lie ∞-algebroid*[a] $\left(E_{-\bullet}, (\ell_k)_{k \geq 1}, \rho\right)$ is a collection of vector bundles $E = (E_{-i})_{i \geq 1}$ over M endowed with a sheaf of Lie ∞-algebra structures $(\ell_k)_{k \geq 1}$ over the sheaf of sections of E together with a vector

(continued)

Definition 7.23 (continued)
bundle morphism $\rho \colon E_{-1} \to TM$, called the *anchor map*, such that the k-ary-brackets

$$\ell_k \colon \underbrace{\Gamma(E_\bullet) \times \cdots \times \Gamma(E_\bullet)}_{k\text{-times}} \longrightarrow \Gamma(E_\bullet)$$

are all $\mathcal{O}$-multilinear except when $k = 2$ and at least one of the arguments is of degree -1. The 2-ary bracket satisfies the Leibniz identity

$$\ell_2(x, fy) = \rho(x)[f]y + f\ell_2(x, y), \quad f \in \mathcal{O}, \ x \in \Gamma(E_{-1}), y \in \Gamma(E_\bullet). \tag{51}$$

---

[a]"Negatively graded Lie ∞-algebroid" would be a more precise name, but since they are all negatively graded, we just say "Lie ∞-algebroid". We do not like the confusing notation L_∞-algebroid, although it is often used.

Exercise 7.24 Show that Definition 7.23 implies that $\rho(\ell_2(x, y)) = [\rho(x), \rho(y)]$ for all $x, y \in \Gamma(E_{-1})$ and that $\rho \circ \ell_1 = 0$.

Remark 7.25 Definition 7.23 implies the following facts

1. The sequence of morphisms of vector bundles

$$\cdots \xrightarrow{\ell_1} E_{-2} \xrightarrow{\ell_1} E_{-1} \xrightarrow{\rho} TM$$

is a complex of vector bundles that we call the *linear part*. A Lie ∞-algebroid is said to be *acyclic* if its linear part has no cohomology at the level of sections in degree ≤ -1.
2. The 2-ary bracket restricts to an almost-Lie algebroid structure on E_{-1}. Hence, by Lemma 2.3, the image[96] $\mathcal{F} := \rho(\Gamma(E_{-1})) \subseteq \mathfrak{X}(M)$ is a singular foliation on M called the *basic singular foliation* of $\bigl(E_\bullet, (\ell_k)_{k\geq 1}, \rho\bigr)$. We say, then, that the Lie ∞-algebroid $\bigl(E_\bullet, (\ell_k)_{k\geq 1}, \rho\bigr)$ is *over* $\mathcal{F}$.

Here is the most important theorem of the present section [5, 38].

---

[96] As usual, one can take global compactly supported sections in the smooth case, otherwise, the image has to be taken in the sense of sheaves.

Theorem 7.26 (Duality NQ-manifolds: Lie ∞-Algebroids) *Let M be a manifold, $E_\bullet$ and $V_\bullet$ negatively graded and, respectively, positively graded vector bundles over M such that $E_{-i} = V_i^*$ for all $i \geq 1$. There is a one-to-one correspondence between:*

(i) negatively graded Lie ∞-algebroids $\left(E_\bullet, (\ell_k)_{k\geq 1}, \rho\right)$,
(ii) NQ-manifolds $(M, \mathcal{E}, Q)$ with $\mathcal{E} = \Gamma(S(V_\bullet))$.

To make Theorem 7.26 complete, let us describe the duality relation through which both structures correspond one to the other:

1. for all $f \in \mathcal{O}, e \in \Gamma(E_{-1})$

$$\langle Q(f), e \rangle = \rho(e)[f], \tag{52}$$

2. for all $\xi \in \Gamma(E^*)$ and $e \in \Gamma(E)$:

$$\langle Q^{(0)}(\xi), e \rangle = (-1)^{|\xi|} \langle \xi, \ell_1(e) \rangle. \tag{53}$$

3. for all homogeneous elements $e_1, e_2 \in \Gamma(E)$ and $\xi \in \Gamma(E^*)$

$$\langle Q^{(1)}(\xi), e_1 \odot e_2 \rangle = \rho(e_1)[\langle \xi, e_2 \rangle] - \rho(e_2)[\langle \xi, e_1 \rangle] - \langle \xi, \ell_2(e_1, e_2) \rangle, \tag{54}$$

with the understanding that the anchor ρ vanishes on E_{-i} when $i \geq 1$.
4. for every $k \geq 3$, the k-ary brackets $\ell_k \colon \Gamma(S_\mathbb{K}^k(E)) \to \Gamma(E)$ and the polynomial degree $k-1$ component $Q^{(n-1)} \colon \Gamma(E^*) \longrightarrow \Gamma(S_\mathbb{K}^k(E^*))$ of Q are dual to each other.

Here $\langle \cdot, \cdot \rangle$ stands for the duality pairing between sections of a vector bundle and sections of its dual.

Definition 7.27 A Lie ∞-algebroid and a NQ-manifold that corresponds one another as in Theorem 7.26 are said to be *dual* one to the other.

Example 7.28 ([37]) Let $(A, [\cdot, \cdot]_A, \rho)$ be a Lie algebroid concentrated in degree -1. The graded manifold $(M, \mathcal{E} = \Gamma(\wedge A^*))$ carries a dg-manifold structure $Q \in \mathfrak{X}(A)$ which is given by

$$\langle Q[f], a \rangle = \rho(a)[f]$$
$$\langle Q[\xi], a \wedge b \rangle = \rho(a)[\langle \xi, b \rangle] - \rho(b)[\langle \xi, a \rangle] - \langle \xi, [a, b]_A \rangle$$

for $f \in \mathcal{O}, \xi \in \Gamma(A^*)$ and $a, b \in \Gamma(A)$. This is sufficient to extend Q by derivation on $\mathcal{E}$. One can check that $Q^2 = 0$ because of Jacobi identity. In particular, the Lie algebroid of vector fields on M corresponds to the De Rham complex.

Definition 7.29 (Lie ∞-Morphisms Are Defined Through Duality) Let $(E'_\bullet, (\ell'_k)_{k \geq 1}, \rho')$ and $(E_\bullet, (\ell_k)_{k \geq 1}, \rho)$ be Lie ∞-algebroids over M and, respectively, M'. Let $(M', \mathcal{E}', Q')$ and respectively, $(M, \mathcal{E}, Q)$ be their dual NQ-manifolds.[a] A *Lie ∞-algebroids morphism* or *Lie ∞-morphism* from $(E'_\bullet, (\ell'_k)_{k \geq 1}, \rho')$ to $(E_\bullet, (\ell_k)_{k \geq 1}, \rho)$ is a morphism[b] of NQ-manifolds from $(M, \mathcal{E}, Q)$ to $(M', \mathcal{E}', Q')$.

---

[a] See Definition 7.27.
[b] Notice that it goes backward.

7.4.1 Homotopic Lie ∞-algebroids

Having defined objects (Lie ∞-algebroids) and arrows (morphisms of Lie ∞-algebroids), we still have to define homotopy between morphisms.

Definition 7.30 (Homotopies Between Morphisms) Let $(E'_\bullet, (\ell'_k)_{k \geq 1}, \rho')$ and $(E_\bullet, (\ell_k)_{k \geq 1}, \rho)$ be smooth Lie ∞-algebroids over M and, respectively, M'. We assume $E_\bullet$ and $E'_\bullet$ to be of finite length. Let $(M', \mathcal{E}', Q')$ and, respectively, $(M, \mathcal{E}, Q)$ be their dual NQ-manifolds.[a] Two Lie ∞-morphisms $\Phi, \Psi \colon (M, \mathcal{E}, Q) \longrightarrow (M', \mathcal{E}', Q')$ are said to be *homotopic*[b] (and we then write $\Phi \sim \Psi$) if there is a NQ-manifold morphism[c]

$$(M, \mathcal{E}, Q) \longrightarrow (M, \mathcal{E}', Q') \times ([0,1], \Omega^\bullet([0,1]), d^{dR}),$$

whose restriction to the extremities of the interval are Φ and Ψ, respectively.

---

[a] See Definition 7.27.
[b] In the real analytic settings, one can assume forms on $[0, 1]$ are real analytic, and in the complex case, $[0, 1]$ can be replaced by a neighborhood of $[0, 1]$ in $\mathbb{C}$. One has then to consider the equivalence relation generated by this relation, which is not necessary in the smooth case.
[c] We leave it to the reader to define the direct product of two NQ-manifolds (for graded manifolds, see Definition 7.14).

When $M = M'$, Lie ∞-algebroid morphisms can be also defined using Taylor coefficients[97]

Remark 7.31 Equivalently, an equivalence $\Phi \sim \Psi$, consists of:

1. a piecewise-smooth path $t \mapsto \Phi_t$ valued in NQ-manifold morphisms such that

$$\Phi_0 = \Phi \quad \text{and} \quad \Phi_1 = \Psi,$$

2. a piecewise-smooth path $t \mapsto H_t$ valued in Φ_t-derivations of degree -1, such that the following equation:

$$\frac{d\Phi_t}{dt} = Q' \circ H_t + H_t \circ Q \tag{55}$$

holds for every $t \in [0, 1]$ where it is defined.

We refer to Section 3.4.4 in [23] or to Section 1.2.5. in [22] for more details - there is a subtlety if the length of $E_\bullet$ or $E'_\bullet$ is not bounded.

Exercise 7.32 Show that Definition 7.30 implies that for every pair of homotopic Lie ∞-morphisms $\Phi, \Psi \colon (M', \mathcal{E}', Q') \longrightarrow (M, \mathcal{E}, Q)$, there exists an $\mathcal{O}$-linear map $H \colon \mathcal{E} \longrightarrow \mathcal{E}'$ of degree -1 such that:

$$\Psi - \Phi = Q' \circ H + H \circ Q. \tag{56}$$

Exercise 7.33 Show that, in the smooth setting, homotopy of Lie ∞-algebroid morphisms is an equivalence relation and also this equivalence relation is compatible with composition of Lie ∞-morphisms.

7.5 NQ-manifolds and Singular Foliations

For any Lie ∞-algebroid over M, the image of the anchor map is a singular foliation called its basic foliation. Now we will analyze the opposite direction, i.e., given a singular foliation $\mathcal{F} \subseteq \mathfrak{X}(M)$, we will try to find a Lie ∞-algebroid over M whose basic foliation is $\mathcal{F}$? We will of course construct this structure on a geometric resolution, studied in Sect. 6. We will show that it is as unique as can be, i.e., unique up to equivalence. This extends a similar discussion that we had about almost-Lie algebroids, see Theorem 7.37.

---

[97] See, e.g., [8] for a pedagogical introduction.

The next theorem is obtained by proving that every graded almost Lie algebroid over a geometric resolution can be extended to a (unique up to homotopy) Lie ∞-algebroid structure. It appeared first in an explicit form in the PhD of Sylvain Lavau [20], followed by a referred version by C.L.G., Sylvain Lavau and Thomas Strobl in [23], but the authors acknowledge it was discussed several years earlier by Ralph Mayer and, even more extensively, by Chenchang Zhu. Also, Theodore Voronov and his collaborators discussed this result in a slightly different context. The construction was then reinterpreted by Yaël Frégier and Rigel Juarez-Ojeda in [14] using semi-models category. Last, it has been generalized later for arbitrary Lie-Rinehart algebras over a commutative unitary algebra in [22].

> **Theorem 7.34 (Universal Lie ∞-Algebroid: Existence)** *Let $\mathcal{F}$ be a singular foliation on a manifold M.*
>
> 1. *In the smooth setting, any geometric resolution $(E_{-\bullet}, d^{(\bullet)}, \rho)$ comes equipped with a Lie ∞-algebroid structure whose linear part[a] is $(E_{-\bullet}, d^{(\bullet)}, \rho)$.*
> 2. *In the real analytic or holomorphic setting, this still holds true, but in a neighborhood of any point of M only.*
>
> ---
> [a] Defined in Remark 7.25.

Proof The proof goes by recursion. Section 6.3 gave a proof when a geometric resolution of length 2 exists. In the general case, the first step of the proof consists in making E_{-1} an almost Lie algebroid, then in showing that there exists a family of graded symmetric bilinear maps

$$\ell_2(\Gamma(E_{-i}), \Gamma(E_{-j})) \longrightarrow \Gamma(E_{-i-j+1})$$

that give back the almost Lie algebroid bracket if $i = j = 1$, and satisfy for all $a \in \Gamma(E_{-i}), b \in \Gamma(E_{-j})$, and for every function f,

$$d^{(i+j-1)}\ell_2(a,b) = \ell_2(d^{(i)}(a), b) + (-1)^i \ell_2(a, d^{(j)}(b))$$

and $\ell_2(a, fb) = f\ell_2(a,b)$ unless $i = 1$, in which case this is replaced by:

$$\ell_2(a, fb) = f\ell_2(a,b) + \rho(a)[f]b.$$

Then we look at the graded Jacobiator and continue by recursion. □

Definition 7.35 (Universal Lie ∞-Algebroid: Definition) We call *universal Lie ∞-algebroid*[a] *of a singular foliation* a Lie ∞-algebroid whose linear part is a geometric resolution of $\mathcal{F}$.

---

[a] Its dual is called a *universal N Q-manifold of $\mathcal{F}$*.

Here is an immediate consequence of Theorem 7.34 and the existence results of Sect. 6.

Corollary 7.36 *A real analytic or holomorphic singular foliation admits a universal Lie ∞-algebroid in a neighborhood of any of its point.*

A locally real analytic singular foliation admits a universal Lie ∞-algebroid on any relatively compact open subset.

The name "universal" is justified: it is indeed a universal object in the category of Lie ∞-algebroids whose anchor map is valued in $\mathcal{F}$. The arrows of that category are defined to be homotopy classes of morphisms.

Theorem 7.37 (Universal Lie ∞-Algebroids Deserve the Name "Universal")
Let $\mathcal{F}$ be a singular foliation over smooth M. Given,

(a) a Lie ∞-algebroid $\left(E'_\bullet, (\ell'_k)_{k\geq 1}, \rho'\right)$ that terminates in $\mathcal{F}$, i.e, $\rho'(\Gamma(E'_{-1})) \subseteq \mathcal{F}$,
(b) a universal Lie ∞-algebroid $\left(E_\bullet, (\ell_k)_{k\geq 1}, \rho\right)$ of $\mathcal{F}$,

then

1. there exists a Lie ∞-algebroid morphism from $\left(E'_\bullet, (\ell'_k)_{k\geq 1}, \rho'\right)$ to $\left(E_\bullet, (\ell_k)_{k\geq 1}, \rho\right)$.
2. and any two such morphisms are homotopic.

In the real analytic or complex settings, the same holds, but in a neighborhood of a point only.

As for geometric resolutions (see Theorem 6.11), or as for almost Lie algebroids (see discussion after Proposition 2.7), Theorem 7.37 means that, given a singular foliation $\mathcal{F}$, in the category where

1. objects are Lie ∞-algebroid morphisms that terminates in $\mathcal{F}$,
2. and arrows are homotopy classes of Lie ∞-algebroid morphisms,

universal Lie ∞-algebroids of $\mathcal{F}$ are terminal (a.k.a. final, or universal) objects.[98] Here is an immediate corollary of this result, which is valid for any pair of terminal objects in any category.

Corollary 7.38 (The Universal Lie ∞-Algebroid Is as Unique as It Can Be) *Two universal Lie ∞-algebroids of a smooth singular foliation are homotopy equivalent. Moreover, the homotopy equivalence between them is unique up to homotopy.*

In the real analytic and complexes cases, the same holds, but in a neighborhood of a point only.

Theorems 7.34–7.37 are proven in [20–23] by a finite and constructive recursion.[99] This does not mean that it is easy to construct it. These theorems are therefore constructive. Here are some examples of universal Lie ∞-algebroids of singular foliations.

Example 7.39 For a regular foliation $\mathcal{F}$ on a manifold M, the Lie algebroid $T\mathcal{F} \subset TM$ is a universal Lie ∞-algebroid of $\mathcal{F}$.

Example 7.40 Let $\mathcal{F} \subseteq \mathfrak{X}(M)$ be a Debord foliation, i.e., $\rho(\Gamma(A)) \simeq \mathcal{F} \subset \mathfrak{X}(M)$ for some Lie algebroid $(A \to M, [\cdot, \cdot]_A, \rho)$ whose anchor map $\rho \colon A \to TM$ is injective on an open dense subset of M. The latter Lie algebroid is a universal Lie ∞-algebroid of $\mathcal{F}$.

Example 7.41 The Lie 2-algebroid constructed in Proposition 6.27 over a singular foliation $\mathcal{F}$ admitting a geometric resolution of length 2 is a universal Lie ∞-algebroid of $\mathcal{F}$.

---

[98] As usual, add "in a neighborhood of a point" in the real analytic and complex cases—or redefine morphisms and homotopies as in Lavau's PhD, (see the last chapter in [20]) as being local ones that glue up to homotopy on the intersections.

[99] Provided that the geometric resolution $(E_\bullet, (\ell_k)_{k \geq 1}, \rho)$ and an almost Lie-algebroid bracket on E_{-1} are given for Theorem 7.34, provided a morphism of geometric resolution is given for Theorem 7.37.

Example 7.42 We go back to Example 6.25. A universal Lie ∞-algebroid of $\mathcal{F}_\varphi \subset \mathfrak{X}(V)$ is given on the free resolution $\left(E_{-\bullet} = \wedge^{\bullet+1} V, \mathrm{d} = \iota_{\mathrm{d}\varphi}, \rho = -\iota_{\mathrm{d}\varphi}\right)$ by defining the following n-ary brackets:

$$\{\partial_{I_1}, \cdots, \partial_{I_n}\}_n := \sum_{i_1 \in I_1, \ldots, i_n \in I_n} \epsilon(i_1, \ldots, i_n) \varphi_{i_1 \cdots i_n} \partial_{I_1^{i_1} \bullet \cdots \bullet I_n^{i_n}}; \tag{57}$$

and the anchor map given for all $i, j \in \{1, \ldots, n\}$ by

$$\rho\left(\frac{\partial}{\partial x_i} \wedge \frac{\partial}{\partial x_j}\right) := \frac{\partial \varphi}{\partial x_j} \frac{\partial}{\partial x_i} - \frac{\partial \varphi}{\partial x_i} \frac{\partial}{\partial x_j}. \tag{58}$$

Above, for every multi-index $J = \{j_1, \ldots, j_n\} \subseteq \{1, \ldots, d\}$ of length n, ∂_J stands for the n-vector field $\frac{\partial}{\partial x_{j_1}} \wedge \cdots \wedge \frac{\partial}{\partial x_{j_n}}$ and $\varphi_{j_1 \cdots j_n} := \frac{\partial^n \varphi}{\partial x_{j_1} \cdots \partial x_{j_n}}$. Also, $I_1 \bullet \cdots \bullet I_n$ is a multi-index obtained by concatenation of n multi-indices $I_1, \ldots, I_n$. For every $i_1 \in I_1, \ldots, i_n \in I_n$, $\epsilon(i_1, \ldots, i_n)$ is the signature of the permutation which brings $i_1, \ldots, i_n$ to the first n slots of $I_1 \bullet \cdots \bullet I_n$. Last, for $i_s \in I_s$, we define $I_s^{i_s} := I_s \setminus i_s$. We refer to [23], Example 3.101 or to Section 3.2.1 in [22] for a proof.

Question 7.43 *For the algebraic*[100] *vector fields on $\mathbb{C}^n$ tangent to an affine variety $W \subset \mathbb{C}^d$, what is the universal Lie ∞-algebroid like?*

7.6 The Isotropy Lie ∞-Algebra of a Singular Foliation at a Point

Let M be a real analytic, smooth, or complex manifold.

Given a point $m \in M$, there is a functor:[101]

$$\mathrm{Isotropy}_m : \{ \text{Lie } \infty\text{-algebroids on } M \} \longrightarrow \{ \text{Lie } \infty\text{-algebras} \} \tag{59}$$

that we describe in the next lines. Then, we apply this functor to the universal Lie ∞-algebroids at an arbitrary point m, and explain why the henceforth obtained Lie ∞-algebras deserve to be called isotropy Lie ∞-algebras by relating them to AS-isotropy Lie algebras (see Sect. 3).

7.6.1 Isotropy Functor: Lie ∞-Algebroid + Point $\mapsto$ Lie ∞-Algebra

Let $\left(E_{-\bullet}, (\ell_k)_{k \geq 1}, \rho\right)$ be a Lie ∞-algebroid with anchor ρ on a manifold[102] M.

---

[100] But seen as holomorphic, as in Remark 6.16.
[101] It is implicitly assume here that the grading of the Lie ∞-structures go from -1 to $-\infty$.
[102] It suffices the Lie ∞-algebroid structure to be defined in a neighborhood of m, it is important for the real analytic and complex settings.

Consider the graded vector space $ev(E,m)_\bullet$ given by

$$ev(E,m)_i = \begin{cases} E_{-i}|_m & \text{for } i \geq 2 \\ \ker(\rho_m) & \text{for } i = 1 \\ 0 & \text{for } i \leq 0 \end{cases}$$

> **Proposition 7.44 (Specialization of a Lie ∞-Algebroid at a Point)** *For every point $m \in M$, the k-ary bracket of $(E_{-\bullet}, (\ell_k)_{k\geq 1}, \rho)$ restrict to $ev(E,m)_\bullet$, and equipped the latter with a Lie ∞-algebra structure.*
> *This restriction is functorial.[a]*
>
> ---
> [a] I.e., a (maybe local near m) morphism of Lie ∞-algebroids over the identity of M induces a morphism of Lie ∞-algebras, and homotopic morphisms are mapped to homotopic morphisms.

Proof We denote the evaluation of a section $s \in \Gamma(E_\bullet)$ at m by $s(m)$ or $s|_m$ depending on the context. We denote the induced brackets on $ev(E,m)_\bullet$. For every $k \geq 1$, we set

$$\{x_1, \ldots, x_k\}_k := \ell_k(s_1, \ldots, s_k)|_m \tag{60}$$

for all $x_1, \ldots, x_k \in ev(E,m)$, where $s_1, \ldots, s_k \in \Gamma(E)$ are sections of E such that $s_i(m) = x_i$ with $i = 1, \ldots, k$.

For $k \neq 2$, this is well-defined, since ℓ_k is linear over functions. But it is not so immediate that the 2-ary bracket is also well-defined. Let $(e_1^i, \ldots e_{\text{rk}(E_{-i})}^i)$ be a local trivialization of E_{-i} on a neighborhood $\mathcal{U}$ of the point $m \in M$. For $x_1 \in \ker(\rho_m)$ and $x_2 \in E_{-i}|_m$, write

$$x_1 = \sum_{k=1}^{\text{rk}(E_{-1})} \lambda_k e_k^1(m), \quad x_2 = \sum_{k=1}^{\text{rk}(E_{-i})} \mu_k e_k^i(m)$$

for some scalars (λ_i) in $\mathbb{K}$. The scalars (λ_k), (μ_k) extend to functions (f_k), (g_k) on $\mathcal{U}$. Therefore, we have

$$\{x_1, x_2\}_2 = \ell_2(s_1, s_2)|_m$$

with

$$s_1 = \sum_{k=1}^{\text{rk}(E_{-1})} f_k e_k^1, \quad s_2 = \sum_{k=1}^{\text{rk}(E_{-i})} g_k e_k^i.$$

If $\widetilde{s}_2$ is another extension of x_2, then $(s_2 - \widetilde{s}_2)(m) = 0$ and this is equivalent to $(g_k - \widetilde{g}_k)(m) = 0$ for $k = 1, \ldots, \text{rk}(E_{-i})$. It follows that

$$\ell_2(s_1, s_2 - \widetilde{s}_2)|_m = \sum_{k=1}^{\text{rk}(E_{-i})} \ell_2\left(s_1, (f_k - \widetilde{g}_k) e_k^i\right)\bigg|_m$$

$$= \sum_{k=1}^{\text{rk}(E_{-i})} \cancel{(f_k - \widetilde{g}_k)(m)} \ell_2\left(s_1, e_k^i\right)\bigg|_m + \cancel{\rho(s_1)|_m [f_k - \widetilde{g}_k] e_k^i}$$

$$= 0.$$

This proves the claim.

Also, the induces brackets $\{\cdots\}_k$ have values in $ev(E, m)_\bullet$ for degree reasons, except maybe for the 2-ary bracket when applied to elements of degree -1 (i.e., elements of the kernel of ρ_m): in that case the bracket is still in the kernel of ρ_m since

$$\rho_m(\{x_1, x_2\}_2(m)) = \rho_m(\ell_2(s_1, s_2)|_m)$$
$$= \rho(\ell_2(s_1, s_2))|_m$$
$$= [\rho(s_1), \rho(s_2)]|_m = 0$$

In the last line we have used the fact that the Lie bracket of two vector fields that vanish at m is a vector field that vanishes again at m. Functoriality is left to the reader. □

Remark 7.45 This proof above is elementary, but let us explain how the proof would work when working with NQ-manifolds, exploiting the duality od Definition 7.27. Let $(M, \mathcal{E}, Q)$ be the NQ-manifold dual to the Lie ∞-algebroid structure. It is easy to check that the ideal of $\mathcal{E} = \Gamma(S(E_\bullet^*))$ generated by $\mathcal{I}_m + Q(\mathcal{I}_m)$, with $\mathcal{I}_m$ the ideal of functions vanishing at m, is a Q-ideal. Hence, the quotient $\mathcal{E}/((\mathcal{I}_m + Q(\mathcal{I}_m))\mathcal{E})$ comes equipped with a derivation of degree $+1$ squaring to 0. It suffices then to check that this quotient is canonically isomorphic to the graded symmetric algebra of the dual of $ev(E, m)_\bullet$. This gives a very aesthetic proof of Theorem 7.44, and also gives functoriality in a trivial manner.

7.6.2 Cohomology Functor: Lie ∞-Algebra ↦ DGLA

Now, there is a second functor

$$\text{Cohom}: \{\text{ Lie } \infty\text{-algebras }\} \longrightarrow \{\text{Differential Graded Lie algebras }\} \tag{61}$$

that consists in noticing that, for a given Lie ∞-algebra on $F_{-1} \oplus F_{-2} \oplus \cdots$ with brackets $(\ell_k)_{k \geq 1}$, the high Jacobi identities (44) applied for $n = 1, 2, 3$ imply that

1. the 1-ary bracket squares to zero, and therefore turns $(E_\bullet, d)$ into a complex, whose cohomology we denote by $\oplus_{i \geq 1} H^{-i}$,
2. the 2-ary bracket goes to the quotient to define a degree $+1$ graded symmetric bilinear map $H^{-i} \times H^{-j} \to H^{-i-j+1}$ that we denote by $[\cdot, \cdot]$.
3. the Jacobiator of the 2-ary bracket ℓ_2 being, the Jacobiator of $[\cdot, \cdot]$ is zero.

Hence, $(H^{-\bullet}, [\cdot, \cdot])$ is a graded Lie algebra.[103] This map behaves well with respect to morphisms and homotopies.

7.6.3 The Isotropy DLGA at a Point of a Singular Foliation

Now, in the particular case we are interested in, the cohomology of the complex $(ev(E, m)_m, \{\cdot\}_1)$ is the cohomology obtained by evaluating a geometric resolution at m. We saw these cohomologies in Sect. 6.1, more precisely in Corollary 6.13. We called them isotropy spaces of $\mathcal{F}$ at m, and denote them by $H^\bullet(\mathcal{F}, m)$ and showed that it does not depend on the chosen point. For any singular foliation that admits a (local) geometric resolution $(E_{-\bullet}, d^{(\bullet)}, \rho)$ therefore, by

1. constructing a universal Lie ∞-algebroid structure on $(E_{-\bullet}, d^{(\bullet)}, \rho)$,
2. then applying the functor Isotropy$_m$ of Eq.(59)
3. then applying the functor Cohom of Eq. (61)

one obtains a differential graded Lie algebra (=DGLA). Moreover, the whole construction is "canonical" in the sense that Theorem 7.37 states that two different choices of a universal Lie ∞-algebroid would lead to equivalent DGLA. Now, Proposition 4.12 in [23] states that the situation is in fact even better: the final DGLA is independent of the choice of a universal Lie ∞-algebroid. This justifies the next definition.

Definition 7.46 We call the differential Lie algebra structure on $H^\bullet(\mathcal{F}, m) := \oplus_{i \geq 1} H^{-i}(\mathcal{F}, m)$ described above the *isotropy DGLA* of the singular foliation at the point m.

---

[103] With a graded symmetric degree -1 Lie bracket, which is not the most convention may be used to. To get a degree 0 graded skew-symmetric bracket, one has to define a new degree by declaring H^{-i} to be of degree $-i + 1$ and to replace $[\cdot, \cdot]$ by $(a, b) \mapsto (-1)^i [a, b]$ for any $b \in E_{-i}$.

The name is justified by the coincidence of its first term with Androulidakis and Skandalis isotropy Lie algebra of Sect. 3.

Proposition 7.47 *For every point m in a foliated manifold that admits a (maybe local) geometric resolution, the restriction of the isotropy differential graded Lie algebra to its component of degree -1 is a Lie algebra that coincides with the Androulidakis and Skandalis isotropy Lie algebra $\mathfrak{g}_m(\mathcal{F})$.*

Proof For $m \in M$, we construct a Lie algebra is isomorphism $\zeta \colon \frac{\ker(\rho_m)}{\operatorname{im}(\mathrm{d}_m^{(2)})} \to \mathfrak{g}_m$ as follows: For an element $u \in \ker(\rho_m)$, let $\tilde{u}$ be an extension of u to a local section on E_{-1}. By construction, one has $\rho(\tilde{u}) \in \mathcal{F}(m)$. Let $\tilde{\rho}_m$ be the surjective linear map defined by

$$\tilde{\rho}_m \colon \ker(\rho_m) \longrightarrow \mathfrak{g}_m, \quad u \longmapsto [\rho(\tilde{u})].$$

Since any other extension $\tilde{u}'$ for u differs from the first one by a section in $\mathcal{I}_m \Gamma(E_{-1})$, the map $\tilde{\rho}_m$ is well-defined. Surjectivity is due to the fact that every vector field of $\mathcal{F}$ vanishing at $m \in M$ is of the form $\rho(e)$ with e a (local) section of E_{-1} whose value at m belongs to $\ker(\rho_m)$. In addition, it is not hard to see that $\tilde{\rho}_m$ is a morphism of brackets.

It remains to show that $\ker(\tilde{\rho}_m) = \operatorname{im}(\mathrm{d}_m^{(2)})$: let $u \in \ker(\tilde{\rho}_m) \subset \ker(\rho_m)$ and $\tilde{u}$ be a local section of E_{-1} that extends u. By definition of u, the class of $\rho(\tilde{u})$ is zero in $\mathfrak{g}_m$, therefore, there exists some functions $f_i \in \mathcal{I}_m$ and $X_i \in \mathcal{F}, i = 1, \ldots, k$, local generators such that

$$\rho(\tilde{u}) = \sum_{i=1}^{k} f_i X_i.$$

This implies that

$$\rho(\tilde{u} - \sum_{i=1}^{k} f_i e_i) = 0.$$

where for $i = 1, \ldots, k$, e_i is a (local) section of E_{-1} whose image through ρ is X_i. Since $(E_{-\bullet}, \mathrm{d}^\bullet, \rho)$ is a geometric resolution, there exists a (local) section $q \in \Gamma(E_{-2})$ such that

$$\tilde{u} = \sum_{i=1}^{k} f_i e_i + \mathrm{d}^{(2)} q \tag{62}$$

By evaluating Eq. (62) at m, we find out that $u \in \operatorname{im}(\mathrm{d}_m^{(2)})$. Conversely, for $v \in E_{-2|m}$, choose a (local) section q of E_{-2} through v. Therefore, $\mathrm{d}^{(2)} q \in \ker \rho$, is a (local) extension of $\mathrm{d}_m^{(2)} v \in \operatorname{im}(\mathrm{d}_m^{(2)})$. The image of $\mathrm{d}_m^{(2)} v$ through $\tilde{\rho}_m$ is zero. This proves that $\ker(\tilde{\rho}_m) = \operatorname{im}(\mathrm{d}_m^{(2)})$. □

7.6.4 The Isotropy Lie ∞-Algebra

Now, there are more structures that a "simple" differential graded Lie algebra on $H^\bullet(\mathcal{F}, m)$. The 2-ary bracket of that structure is in fact the 2-ary bracket of a Lie ∞-algebra structure whose 1-ary bracket is zero. To obtain it, we proceed as follows. For any singular foliation that admits a (local) geometric resolution $(E_{-\bullet}, d^{(\bullet)}, \rho)$ therefore, by

1. We must manage to replace it by another geometric resolution which is minimal[104] at m,
2. then constructing a universal Lie ∞-algebroid structure on $(E_{-\bullet}, d^{(\bullet)}, \rho)$,
3. then applying the functor Isotropy$_m$ of Eq.(59)
4. and then we stop there (we do *not* apply functor Cohom of Eq. (61)).

If the underlying complex $(E_{-\bullet}, \ell_1, \rho)$ of (E, Q) is minimal at m then, for every $i \geq 2$, the vector space $H^{-i}(\mathcal{F}, m)$ is canonically isomorphic to $E_{-i}|_m$. Also, $H^{-1}(\mathcal{F}, m)$ is canonically isomorphic to $\ker(\rho_m)$. Therefore, the outcome of the construction is a Lie ∞-algebra over the "complex" $H^\bullet(\mathcal{F}, m)$ whose differential is zero. The 2-ary bracket is the isotropy differential graded Lie algebra bracket. Proposition 4.12 in [23] shows[105] that this Lie ∞-algebra is well-defined up to a strict isomorphism on which there are even more constraints. The following definition then makes sense.

> **Definition 7.48** For any point m of a foliated manifold $(M, \mathcal{F})$ that admits a geometric resolution (at least, near m), the graded vector space $H^\bullet(\mathcal{F}, m)$ carries a Lie ∞-algebra structure called the *isotropy Lie ∞-algebra of the singular foliation* $\mathcal{F}$ *at* m.

This structure is used in Proposition 1.11 in chapter "What Is a Singular Foliation?".

References

1. I. Androulidakis, O. Mohsen, R. Yuncken, A pseudodifferential calculus for maximally hypoelliptic operators and the Helffer-Nourrigat conjecture, arXiv.2201.12060 (2022)
2. I. Androulidakis, G. Skandalis, The holonomy groupoid of a singular foliation. J. Reine Angew. Math. **626**, 1–37 (2009)
3. I. Androulidakis, M. Zambon, Holonomy transformations for singular foliations. Adv. Math. **256**, 348–397 (2014)

---

[104] See Definition 6.7.
[105] See also [22] Section 2.1.3.

4. H. Bursztyn, H. Lima, E. Meinrenken, Splitting theorems for Poisson and related structures. J. Reine Angew. Math. **754**, 281–312 (2019)
5. G. Bonavolontà, N. Poncin, On the category of Lie n-algebroids. J. Geom. Phys. **73**, 70–90 (2013)
6. D. Cerveau, Distributions involutives singulières. Ann. Inst. Fourier (Grenoble) **29**(3), xii, 261–294 (1979)
7. M. Crainic, R.L. Fernandes, Integrability of Lie brackets. Ann. Math. (2) **157**(2), 575–620 (2003)
8. R. Caseiro, C. Laurent-Gengoux, Modular class of Lie ∞-algebroids and adjoint representations. J. Geom. Mech. **14**(2), 273–305 (2022)
9. A. Caetano, R.F. Picken, An axiomatic definition of holonomy. Int. J. Math. **5**(6), 835–848 (1994)
10. C. Debord, Local integration of Lie algebroids. Banach Center Publ. **54**, 21–33 (2000)
11. C. Debord, Holonomy groupoids of singular foliations. J. Differential Geom. **58**(3), 467–500 (2001)
12. C. Debord, Longitudinal smoothness of the holonomy groupoid. C. R., Math. Acad. Sci. Paris **351**(15–16), 613–616 (2013)
13. D. Eisenbud, *The Geometry of Syzygies. A Second Course in Commutative Algebra and Algebraic Geometry*, volume 229 of Grad. Texts Math. (Springer, New York, 2005)
14. Y. Fregier, R.A. Juarez-Ojeda, Homotopy theory of singular foliations, arXiv.1811.03078 (2019)
15. P. Griffiths, J. Harris, *Principles of Algebraic Geometry*. Pure and Applied Mathematics. A Wiley-Interscience Publication, vol. XII (John Wiley & Sons, New York, 1978), 813 p. £ 29.60; SPSSlashDollar 58.00
16. A. Garmendia, J. Villatoro, Integration of Singular Foliations via Paths. Int. Math. Res. Not., rnab177 (2021)
17. J. Huebschmann, Multi derivation Maurer-Cartan algebras and sh Lie-Rinehart algebras. J. Algebra **472**, 437–479 (2017)
18. J.-L. Koszul, Homologie et cohomologie des algèbres de Lie. Bull. Soc. Math. France **78**, 65–127 (1950)
19. A. Kotov, V. Salnikov, The category of z-graded manifolds: What happens if you do not stay positive. Differential Geom. Appl. **93**, 102109 (2024)
20. S. Lavau, Lie ∞-algebroids and singular foliations, PhD, University of Lyon, arXiv1703.07404 (2018)
21. S. Lavau, A short guide through integration theorems of generalized distributions. Differ. Geom. Appl. **61**, 42–58 (2018)
22. C. Laurent-Gengoux, R. Louis, Lie-Rinehart algebras $\simeq$ acyclic lie ∞-algebroids. J. Algebra **594**, 1–53 (2022)
23. C. Laurent-Gengoux, S. Lavau, T. Strobl, The universal Lie ∞-algebroid of a singular foliation. Doc. Math. **25**, 1571–1652 (2020)
24. C. Laurent-Gengoux, L. Ryvkin, The neighborhood of a singular leaf. J. Éc. Polytech. Math. **8**, 1037–1064 (2021)
25. C. Laurent-Gengoux, L. Ryvkin, The holonomy of a singular leaf. Sel. Math. **28**(2), 45 (2022)
26. R. Louis, On symmetries of singular foliations. J. Geom. Phys. **189**, 104833 (2023)
27. R. Louis, A series of Nash resolutions of a singular foliation. arXiv:2301.08706 (2024)
28. K. Mackenzie, *Lie Groupoids and Lie Algebroids in Differential Geometry*, volume 124 of London Mathematical Society Lecture Note Series (Cambridge University Press, Cambridge, 1987)
29. A. Pichereau, Poisson (co)homology and isolated singularities. J. Algebra **299**(2), 747–777 (2006)
30. L. Ryvkin, Observables and symmetries of n-plectic manifolds. BestMasters. Master Thesis. Wiesbaden: Springer Spektrum; Bochum: Univ. Bochum (2016)

31. P. Ševera, Some title containing the words "homotopy" and "symplectic", e.g. this one, in *Proceedings of the 4th Conference on Poisson Geometry, Luxembourg, June 7–11, 2004* (Université du Luxembourg, Luxembourg, 2005), pp. 121–137
32. P. Ševera, Letters to Alan Weinstein about Courant algebroids. arXiv:1707.00265 (2017)
33. J. Stasheff, Differential graded Lie algebras, quasi-Hopf algebras and higher homotopy algebras, in *Quantum Groups. Proceedings of Workshops, held in the Euler International Mathematical Institute, Leningrad, USSR, Fall 1990* (Springer, Berlin, 1992), pp. 120–137
34. B. Sturmfels, *Algorithms in Invariant Theory*. Texts Monogr. Symb. Comput. (Springer, Wien, 1993)
35. U. Schreiber, K. Waldorf, Parallel transport and functors. J. Homotopy Relat. Struct. **4**(1), 187–244 (2009)
36. J.-C. Tougeron, Idéaux de fonctions différentiables. I. Ann. Inst. Fourier (Grenoble) **18**(fasc., fasc. 1), 177–240 (1968)
37. A.Yu. Vaintrob, Lie algebroids and homological vector fields. Russ. Math. Surv. **52**(2), 428–429 (1997)
38. Th.Th. Voronov, Q-manifolds and higher analogs of Lie algebroids, in *XXIX Workshop on Geometric Methods in Physics*, volume 1307 of AIP Conf. Proc. (Amer. Inst. Phys., Melville, NY, 2010), pp. 191–202
39. C.A. Weibel, *An Introduction to Homological Algebra*, volume 38 of Camb. Stud. Adv. Math. (Cambridge University Press, Cambridge, 1994)

State of the Art and Open Questions

Camille Laurent-Gengoux, Ruben Louis, and Leonid Ryvkin

We list in this section several open problems and questions—some are vague, some are precise, some seem major, some seem mere anecdotes.

1 Existence of Lie Algebroids Generating a Singular Foliation

Let us present the most intriguing open question regarding singular foliations.

Let us start by making the terminology precise. So far, it was part of the definition of a "Lie algebroid $(A \to M, \rho, [\cdot\,,\cdot])$" that $A \to M$ had to be a finite rank vector bundle by M, i.e., that $A \to M$ is a vector bundle modeled over a finite dimensional vector space. In this section, however, let us distinguish:

C. Laurent-Gengoux (✉)
Université de Lorraine, CNRS, IECL, Metz, France
e-mail: camille.laurent-gengoux@univ-lorraine.fr

R. Louis
Department of Mathematics, Jilin University, Changchun, Jilin, China

Institut für Mathematik, Georg-August-Universität Göttingen, Göttingen, Germany
e-mail: ruben.louis@mathematik.uni-goettingen.de

L. Ryvkin
Université Claude Bernard Lyon 1, Villeurbanne, France

University of Göttingen, Göttingen, Germany
e-mail: ryvkin@math.univ-lyon1.fr

1. *finite rank Lie algebroids*, i.e., Lie algebroids as defined so far, with $A \to M$ a finite rank vector bundle,
2. *infinite rank Lie algebroids*, which have precisely the same definition, except that $A \to M$ is now a vector bundle of infinite rank.

As we saw in Sect. 4.2 in chapter "What Is a Singular Foliation?", for any finite rank Lie algebroid $(A \to M, \rho, [\cdot, \cdot])$, the image of the anchor map $\mathcal{F} = \rho\left(\Gamma(A)\right)$ is a singular foliation on M.

Exercise 1.1 Let $(A \to \mathcal{U}, [\cdot, \cdot], \rho)$ be an infinite rank Lie algebroid on a smooth manifold. Check that[1] $\mathcal{F} = \rho\left(\Gamma_c(A)\right)$ is

1. a $\mathcal{C}^\infty(M)$-submodule of $\mathfrak{X}_c(M)$,
2. involutive, i.e., $[\mathcal{F}, \mathcal{F}] \subset \mathcal{F}$.

Extend the result to the complex or real analytic contexts.

Example 1.2 Here is an example for which $\mathcal{F}$ is not locally finitely generated as a $\mathcal{C}^\infty(M)$-module but still comes from an infinite rank Lie algebroid:

(a) $M = \mathbb{R}$,
(b) A is the trivial vector bundle with generators indexed $(e_i)_{i \in \mathbb{N}}$ indexed by $\mathbb{N}$,
(c) the anchor map is $\rho(e_i) = \frac{1}{x^i} e^{-\frac{1}{x^2}} \frac{\partial}{\partial x}$, for all $i \in \mathbb{N}$,
(d) the Lie bracket is defined by $[e_i, e_j] = (i-j) e^{-\frac{1}{x^2}} e_{i+j+1}$. for all $i, j \in \mathbb{N}$

Here is a simple open question, that—as far as we know—first appeared in a printed version in Androulidakis and Zambon's [3]. We were told by several mathematicians, e.g., Rui Loja Fernandes and Georges Skandalis, that the question was already circulating orally in the early 2000s.

> **Question 1.3 (Lie Algebroid? [3])** *Let $\mathcal{F}$ be a singular foliation on a manifold M. Does every point m admit a neighborhood $\mathcal{U}$ on which there exists a finite rank Lie algebroid $(A \to \mathcal{U}, [\cdot, \cdot], \rho)$ such that $\mathcal{F} = \rho\left(\Gamma(A)\right)$?*

---

[1] The index c means compactly supported.

Here is a slightly more general formulation of the question:

Question 1.4 (Lie Algebroid (Version II)? [3]) *Is any finitely generated singular foliation the image through the anchor map of a finite rank Lie algebroid?*

In addition to the local problem, there is also a global "gluing" problem. This one only makes sense on the smooth setting.

Question 1.5 *If a smooth singular foliation is the image of the Lie algebroid two on open subsets U_1 and U_2, is it the image of a Lie algebroid on $U_1 \cup U_2$?*

Even if we assume both Lie algebroid structures to be defined on the restrictions to U_1 and U_2 of the same vector bundle on $U_1 \cap U_2$, Question 1.5 remains non-trivial.

Example 1.6 Singular foliations whose number of local generators are not globally bounded can not be, globally, the image through the anchor map of a finite rank Lie algebroid. Hence, the Androulidakis and Zambon's "non-finitely-many-generators" singular foliation (see Example 2.36 in chapter "What Is a Singular Foliation?") is not the image through the anchor map of a finite rank Lie algebroid on the whole manifold $M = \mathbb{R}^2$.

Exercise 1.7 The purpose of this exercise is to show that any finitely generated singular foliation is the image through the anchor map of an infinite rank Lie algebroid.

1. Let $X_1, \ldots, X_d$ be vector fields on a manifold M, and let $\mathfrak{g}_{free}^d$ be the free Lie algebra with d-generators $e_1, \ldots, e_d$. Show that exists a unique Lie algebra morphism $\rho \colon \mathfrak{g}_{free}^d \to \mathfrak{X}(M)$ such that $\rho(e_i) = X_i$.
2. Assume now that $X_1, \ldots, X_d$ are generators of a singular foliation $\mathcal{F}$. Use the previous Lie algebra morphism to construct a Lie algebroid structure on the trivial bundle $\mathfrak{g}_{free}^d \times M \to M$ such that the image of its anchor map is $\mathcal{F}$.

Quite a few singular foliations are the image through the anchor map of a Lie algebroid: symplectic foliations of Poisson structures (see Sect. 4.6 in chapter "What Is a Singular Foliation?") for instance, or orbits of a Lie algebra action (see Sect. 4.2 in chapter "What Is a Singular Foliation?"). Here is an example of a singular foliation of rank 6 for which no Lie algebroid is known. Notice that this question is understandable by any master degree student—but is still open!

Question 1.8 (A Frustrating Example) *Is the singular foliation of vector fields on $\mathbb{R}^2$ vanishing quadratically[a] at the origin 0 the image through the anchor map of a finite rank Lie algebroid?*

---

[a] See Sect. 4.3 in chapter "What Is a Singular Foliation?".

Here are other examples of singular foliations for which no finite rank Lie algebroid is known, except in some particular cases:

1. vector fields on $\mathbb{C}^n$ tangent to a given affine variety $W \subset \mathbb{C}^n$,
2. vector fields on $\mathbb{C}^n$ vanishing at every point of an affine variety $W \subset \mathbb{C}^n$,
3. vector fields $X \in \mathfrak{X}(\mathbb{C}^n)$ such that $X[\varphi] = 0$ for some polynomial function $\varphi \in \mathbb{C}[x_1, \ldots, x_n]$ (see Example 6.25 in chapter "Canonical Geometric and Algebraic Structures Hidden Behind a Singular Foliation").

Exercise 1.9 Show that any singular foliation $\mathcal{F}$ whose rank can be only $k = 1, 2$ at a given point comes from a Lie algebroid. (*Hint*: construct an almost Lie algebroid of rank k over $\mathcal{F}$ and show that its Jacobiator has to be trivial).

Exercise 1.10 Let $\varphi \in \mathbb{C}[x, y, z]$ be a polynomial function on $\mathbb{C}^3$. Check that the following bivector field:

$$\{x, y\} = \frac{\partial \varphi}{\partial z}, \ \{y, z\} = \frac{\partial \varphi}{\partial x}, \ \{z, x\} = \frac{\partial \varphi}{\partial y}$$

is a Poisson bivector field, and that φ is a Casimir function. Consider the corresponding Lie algebroid on $A = T^*\mathbb{C}^3$. Show that the image of its anchor map is a sub-singular foliation of the singular foliation $\mathcal{F}_\varphi$ of all vector fields $X \in \mathfrak{X}(\mathbb{C}^3)$ such that $X[\varphi] = 0$. Show that if φ is weight homogeneous with an isolated singularity at zero, then $\rho(\Gamma(A)) = \mathcal{F}_\varphi$. *Hint*: This is done by Anne Pichereau in [32].

Discussion Question 1.3 may be misleading, in the sense that "behind" a singular foliation there is a Lie ∞-algebroid.[2] The Lie algebroid, even if there is one, is certainly not unique (one could take the direct product with any Lie algebra for instance). But the universal Lie ∞-algebroid is unique (up to homotopy, see Corollary 7.38 in chapter

---

[2] (= Q-manifold = dg-manifold).

"Canonical Geometric and Algebraic Structures Hidden Behind a Singular Foliation"), so that any homotopy invariant information obtained out of a universal Lie ∞-algebroid is canonically attached to the singular foliation.

Moreover, the universal Lie ∞-algebroid itself gives some hints about a possible Lie algebroid that whose image through the anchor map would be the singular foliation. It is shown in [22] that some singular foliations of rank r are not the image through the anchor map of a Lie algebroid of rank r. In fact, the following result is shown in Example 4.32 in [22]:

Proposition 1.11 (No Minimal Rank Lie Algebroid) *The singular foliations of all vector fields X on $\mathbb{C}^4$ such that $X[\phi] = 0$ with $\phi(z_1, z_2, z_3, z_4) = z_1^3 + z_2^3 + z_3^3 + z_4^3$:*

1. *has rank 6 at the origin,*
2. *but can not be the image through the anchor map of a Lie algebroid of rank 6 on a given neighborhood of the origin.*

This relatively elementary result uses the universal Lie ∞-algebroid. Proposition 4.29 of [22] states that if a Lie algebroid of rank r exists in a neighborhood of a leaf reduced to a point, say m, then the isotropy Lie ∞-algebra[3] at m admits a minimal model whose 3-ary bracket vanishes. Now, there are cohomological obstructions to such a cancellation. Here is the exact statement:

Proposition 1.12 ([22], Proposition 4.29) *A singular foliation, defined in a neighborhood of $0 \in \mathbb{R}^n$ and of rank r at this point, which admits a geometric resolution, and for which the 3-ary bracket of any minimal model of the Lie ∞-isotropy Lie algebra at 0 is not exact as a Chevalley-Eilenberg cocycle for the isotropy Lie algebra at 0, can not be the image through the anchor map of a Lie algebroid of rank r.*

Let us state a striking corollary of this statement. Let $X_1, \ldots, X_r$ be generators of a singular foliation $\mathcal{F}$. There exists (see Exercise 2.35 in chapter "What Is a Singular Foliation?") Christoffel coefficients, i.e., functions c_{ij}^k (with $i, j, k = 1, \ldots, r$ satisfying

$$[X_i, X_j] = \sum_{k=1}^{r} c_{ij}^k X_k$$

---

[3] See Sect. 7.6 in chapter "Canonical Geometric and Algebraic Structures Hidden Behind a Singular Foliation".

but those are not unique, since there are relations between the generators. Without any loss of generality, we can assume

$$c_{ij}^k = -c_{ij}^k, \tag{1}$$

and, since the Jacobi identity holds, we have:

$$0 = [X_i, [X_j, X_k]] + c.p._{(i,j,k)}$$

$$= \sum_{a=1}^{r} \left(X_i[c_{jk}^a] + \sum_{b=1}^{r} c_{ij}^b c_{bk}^a + c.p.(i,j,k) \right) X_a$$

If for every $a \in \{1, \ldots, r\}$,

$$X_i[c_{jk}^a] + \sum_{b=1}^{r} c_{ij}^b c_{bk}^a + c.p.(i,j,k) = 0 \tag{2}$$

then there exists a Lie algebroid of rank r whose image through the anchor map is $\mathcal{F}$: the Lie algebroid on a trivial bundle of rank r whose bracket is given by

$$[e_i, e_j] = \sum_{k=1}^{r} c_{ij}^k e_k$$

and whose anchor is $\rho(e_k) = X_k$ for all k. Proposition 1.12 (i.e., Proposition 4.29 in [22]) explains that, if the isotropy Lie ∞-algebra at a point satisfies the cohomological condition linked to its 3-ary bracket, then there is no way that coefficients c_{ij}^k could be found that satisfy both Eqs. (1) and (2).

Another Relation Between the Universal Lie ∞-Algebroid and a Lie Algebroid over $\mathcal{F}$ Here is a known result, which makes more precise the claims made in Theorem 7.37 in chapter "Canonical Geometric and Algebraic Structures Hidden Behind a Singular Foliation".

Proposition 1.13 ([19], Proposition 2.3) *If a Lie algebroid A over a singular foliation $\mathcal{F}$ exists, and if $\mathcal{F}$ admits a geometric resolution, then there exists a universal Lie ∞-algebroid $(\mathcal{E}, Q)$ constructed with the help of a geometric resolution $(E_{-i})_{i \geq 1}$ that satisfies the additional conditions:*

1. $E_{-1} = A$
2. *the 2-ary bracket, restricted to A, is the bracket of A*
3. *the restriction to $E_{-1} = A$ of all the n-ary brackets are 0 for $n \geq 3$.*

This proposition makes the next question a natural one:

Question 1.14 *If a singular foliation (i) admits a geometric resolution and (ii) is the image through the anchor map of a Lie algebroid, does it admit a universal Lie ∞-algebroid for which all n-ary brackets are zero for $n \geq 3$?*

Is the answer to Question 1.3 yes or no? Karandeep Jandu Singh's [36] goes in the opposite direction, by showing that the 3-ary brackets need happen not to be zero for some singular foliation associated to the symplectic Lie algebra. However, this does not answer the question negatively, since other choices could be made in the construction at early steps.

We have the following conjecture, which rather goes in the direction "the Lie algebroid seems to exist" but certainly does not prove it, and leaves room for counter-examples. If true, then finding negative answers to Question 1.4 will be a hard task.

Conjecture 1.15 (The Isotropy Lie ∞-Algebra at a Point Is Formal) *At every point, the isotropy Lie ∞-algebra of a singular foliation that admits a geometric resolution is homotopy equivalent to a finite dimensional differential graded Lie algebra.*

If true, the conjecture implies, for instance, that it is not possible to have a homotopy Lie ∞-algebra of a singular foliation at a point isomorphic to the so-called string Lie 2-algebra. i.e., the Lie 2-algebra of the form

$$\mathfrak{g} = \mathfrak{g}_{-2} \oplus \mathfrak{g}_{-1}$$

with $\mathfrak{g}_{-1}$ a semi-simple Lie algebra, $\mathfrak{g}_{-2} = \mathbb{R}$, the 1-ary bracket equal to 0, the 2-ary bracket equal to the Lie algebra bracket on $\mathfrak{g}_{-1}$, and the 3-ary bracket:

$$\wedge^3 \mathfrak{g}_{-1} \to \mathfrak{g}_{-2}$$

given by the Cartan 3-form. The statement comes from the fact that the latter is not homotopy equivalent to a finite dimensional differential graded Lie algebra.

2 About Geometric Resolutions: When Do They Exist?

When do geometric resolutions exist, at least in the neighborhood of every point? In the complex or real analytic setting, they exist locally for coherent sheaves, by the classical Hilbert's syzygy theorem.

In the smooth case, it is relatively simple to construct singular foliations that do not admit geometric resolutions. For instance, we can look at the following function $f : \mathbb{R} \to \mathbb{R}$ vanishing on $\{x \leq 0\}$ in $\mathbb{R}$.

$$f(x) = \begin{cases} e^{\frac{-1}{x^2}} & \text{if } x > 0 \\ 0 & \text{if } x \leq 0 \end{cases}$$

Then $\mathcal{F} = \mathcal{C}_c^\infty(M) \cdot f \cdot \partial_x$ defines a singular foliation, which has regular leaves of dimension 0 (=the points $\{x\}$ for $x < 0$) and one regular leaf of dimension 1 (=the half-line $\{x > 0\}$). Hence, by Corollary 6.12 in chapter "Canonical Geometric and Algebraic Structures Hidden Behind a Singular Foliation", this foliation can not admit a geometric resolution.

However, we saw in the discussion around Theorem 7.37 in chapter "Canonical Geometric and Algebraic Structures Hidden Behind a Singular Foliation" that geometric resolutions of a singular foliation exist for locally real analytic singular foliations on a relatively compact open subset. This point was proven in [22], Theorem 2.4. Can one make something more general? A very natural condition to be imposed on a singular foliation $\mathcal{F}$ is that it forms a subspace of $\mathfrak{X}_c(M)$ which is <u>closed</u> with respect to the Fréchet topology. Any locally real analytical singular foliation is closed, however not every closed singular foliation is locally real analytic, hence is a counter-example.

Example 2.1 Consider $M = \mathbb{R}^2$ and the ideal $I = (f) \subset \mathcal{C}^\infty(M)$ generated by $f(x, y) = y^2 - e^{-\frac{1}{x^2}}$. By Example 4.8 in [37] this ideal is closed, hence the foliation $\mathcal{F} = f \cdot \mathfrak{X}_c(M)$ is also closed. It can however not be real analytically generated near the origin, since there the union of all 0-dimensional are made of two curves intersecting flatly (i.e., with all derivatives colinear). At a consequence, vector fields of the form $f(x, y)\frac{\partial}{\partial x}$ is not locally real analytic, although it is Fréchet closed.

Since locally real-analytic singular foliations locally admit geometric resolutions and hence, and since Fréchet-closed singular foliations being a generalization of locally real analytic ones, the following question is very natural.

Question 2.2 (Closed Singular Foliations Admit Geometric Resolutions?) *Does any Fréchet-closed singular foliation locally admit a geometric resolution?*

If the answer is "yes", then Fréchet-closed singular foliation admit a universal Lie ∞-algebroids in a neighborhood of every point by Theorem 7.37 in chapter "Canonical Geometric and Algebraic Structures Hidden Behind a Singular Foliation".

3 Molino-Atiyah Classes

Let us first recall the construction of the Molino class of a regular foliation $\mathcal{F}$. As its name indicates, the Molino class is a class in some cohomology: we first describe the cohomology to which it belongs.

Let $\mathcal{F}$ be a regular foliation on M, with tangent bundle $T\mathcal{F} \subset TM$. Notice that $\mathcal{F} = \Gamma_c(T\mathcal{F})$. We assume the reader is familiar with the language of Lie algebroids: the presentation is influenced by Zhou Chen, Mathieu Stiénon, Ping Xu's [7].

1. The tangent bundle $T\mathcal{F}$ is a Lie sub-algebroid of the tangent Lie algebroid TM, whose anchor map is the inclusion $T\mathcal{F} \hookrightarrow TM$.
2. Consider the *normal bundle* $N_\mathcal{F} := TM/T\mathcal{F}$. Denote by $u \mapsto \bar{u}$ the natural projection $TM \longrightarrow N_\mathcal{F} = TM/T\mathcal{F}$. The normal bundle comes equipped with a $T\mathcal{F}$-connection,[4] called the *Bott connection*, and defined by:

$$\nabla_X^{Bott} \bar{u} = \overline{[X, u]}$$

for all $X \in \mathcal{F}$ and $u \in \Gamma(TM)$.

3. It follows from the Jacobi identity for vector fields on M that the Bott connection is a flat connection. As a consequence $X \mapsto \nabla_X^{Bott}$ turns $N_\mathcal{F}$ into a Lie algebroid representation of $T\mathcal{F}$.
4. The dual of a Lie algebroid representation of $T\mathcal{F}$, and the tensor or symmetric products of two Lie algebroid representations of $T\mathcal{F}$ being Lie algebroid representations of $T\mathcal{F}$ again, the vector bundle $S^2 N_\mathcal{F}^* \otimes N_\mathcal{F}$ (i.e., the vector bundle of symmetric bilinear maps from the normal bundle to itself) is a Lie algebroid representation of $T\mathcal{F}$.

The Molino class is a cohomology class of degree 1 for the Chevalley-Eilenberg cohomology[5] of $T\mathcal{F}$ valued in the module $S^2 N_\mathcal{F}^* \otimes N_\mathcal{F}$. By construction, this class has to be represented by a vector bundle morphism:

$$\alpha \colon T\mathcal{F} \otimes S^2 N_\mathcal{F} \longrightarrow N_\mathcal{F},$$

---

[4] For the reader non-familiar with Lie algebroids, items (2) and (3) mean that the restriction of the normal bundle to a leaf is equipped with a natural flat connection.
[5] For the reader non-familiar with Lie algebroid: the restriction of $S^2 N_\mathcal{F}^* \otimes N_\mathcal{F}$ to a leaf comes with a flat connection.

which has to satisfy (in-order to be a closed-cocycle) for all $X, Y \in \mathcal{F}$ and $u, v \in \Gamma(N_\mathcal{F})$:

$$\alpha([X, Y], u, v) = \nabla_X^{Bott}\alpha(Y, u, v) - \alpha(Y, \nabla_X^{Bott}u, v) - \alpha(Y, u, \nabla_X^{Bott}v) - (X \longleftrightarrow Y). \tag{3}$$

Let us now construct the Molino class for a regular foliation.

1. Consider a TM-connection[6] ∇ on $N_\mathcal{F}$:

$$(X, \overline{u}) \mapsto \nabla_X \overline{u}$$

whose restriction to $\mathcal{F} \times \Gamma(N_\mathcal{F})$ is the Bott connection, i.e., such that for all $X \in \mathcal{F}$:

$$\nabla_X \overline{u} = \nabla_X^{Bott} \overline{u}.$$

   (a) Such connections always exist.
   (b) Without any loss of generality, we can assume that its torsion is zero. The *torsion* is the vector bundle morphism defined by

$$T^\nabla : \wedge^2 TM \to TM \\ (X, Y) \mapsto \overline{\nabla_X \overline{Y} - \nabla_Y \overline{X} - [X, Y]}.$$

   From now on, we will assume the torsion to be zero.

2. Consider the curvature κ^∇ of such a connection ∇. By construction, κ^∇ is a vector bundle morphism

$$\kappa^\nabla : TM \wedge TM \otimes N_\mathcal{F} \longrightarrow N_\mathcal{F}.$$

3. Since the Bott connection is flat, for any $X_1, X_2 \in \mathcal{F}$, $\kappa^\nabla(X_1, X_2) = 0$ if both X_1 and X_2 are in $\mathcal{F}$. This means that for every $X \in \mathcal{F}$, the vector bundle morphism $i_X \kappa^\nabla : Y \longrightarrow \kappa^\nabla(X, Y)$ vanishes as soon as $Y \in \mathcal{F}$. It therefore can be seen as a vector bundle morphism $\overline{i_X \kappa^\nabla} : N_\mathcal{F} \otimes N_\mathcal{F} \longrightarrow N_\mathcal{F}$.

4. The map $X \mapsto \overline{i_X \kappa^\nabla}$ can therefore be seen as a vector bundle morphism from $T\mathcal{F}$ to $N_\mathcal{F} \otimes N_\mathcal{F} \longrightarrow N_\mathcal{F}$,

5. i.e., it can be seen as a vector bundle morphism $T\mathcal{F} \otimes N_\mathcal{F} \otimes N_\mathcal{F} \longrightarrow N_\mathcal{F}$.

6. We leave it to the reader to check that the vanishing of the torsion implies that $X \mapsto \overline{i_X \kappa^\nabla}$ is symmetric in the two last variables $N_\mathcal{F}$, and is indeed a vector bundle morphism

$$\alpha^\nabla : T\mathcal{F} \otimes S^2 N_\mathcal{F} \longrightarrow N_\mathcal{F}.$$

---

[6] i.e., a linear connection is the usual sense.

State of the Art and Open Questions

Now, it is a direct computation that the Bianchi identity implies that α satisfies (3) above, and is therefore a cocycle of the Chevalley-Eilenberg cohomology of $T\mathcal{F}$ is the module $S^2 N_\mathcal{F}^* \otimes N_\mathcal{F}$, called the *Molino cocycle of the torsion-free connection* ∇. It can be shown that different choices of connections ∇ would give the same class in cohomology. This completes the definition of the *Molino class*, which is also called *Atiyah class* (see [7, 23] for a unification of the Molino class with the Atiyah class in complex geometry -hence the double name. See [29] for an original construction -in French- by Pierre Molino). The previous discussion has established the following fact:

Proposition 3.1 ([29]) *The Molino class is the obstruction to the existence of an extension of the Bott connection whose curvature 2-form is zero as soon as one element tangent to the foliation is applied to it.*

Question 3.2 (Molino Class and Meaning?) *What is the equivalent of the Molino (also called Atiyah class) for a singular foliation? And what is its geometrical meaning?*

Let us state a few points.

1. The Bott connection has a natural extension to the singular case, see [5, 24]:
   (a) The formula $(X, \overline{u}) \mapsto \overline{[X, u]}$ defines a flat Lie-Rinehart connection of $\mathcal{F}$ on the $\mathcal{C}^\infty(M)$-module $\mathfrak{X}(M)/\mathcal{F}$.
   (b) The adjoint representation "up to homotopy" of any universal Lie ∞-algebroid of $\mathcal{F}$ is a flat Lie ∞-algebroid connection on a geometric resolution of the $\mathcal{C}^\infty(M)$-module $\mathfrak{X}(M)/\mathcal{F}$. It can also be understood as a generalization of the Bott connection.
2. Geometrically, the vanishing of the Molino class of a regular foliation has several consequences. For any leaf L, and any $x \in L$, the holonomy:

$$Hol_{\mathcal{F}, L, l} : \pi_1(x, L) \longrightarrow Diff_0(N_\mathcal{F}|_x)$$

   valued in germs at 0 of diffeomorphisms of the normal bundle. If the Molino class vanishes, the holonomy is linearizable, i.e., the group morphism $Hol_{\mathcal{F}, L, l}$ can be assumed to be valued in linear invertible endomorphisms of $N_\mathcal{F}|_x$. See, e.g., Theorem 8.5 in [21].
   It is not obvious to see what the equivalent of the previous points for a singular leaf is.

Question 3.3 *Assuming it has been defined, what is the geometrical meaning of the vanishing of the (to be constructed) Molino class for a singular foliation?*

A recent article by Seokbong Seol [34] may lead to conjecture that if the Molino class (to be defined) vanishes, then the foliation has to be a regular foliation.

4 Miscellaneous

Here is a "potpourri" of several questions, mostly anecdotal at first sight, but to which we have no immediate answer.

Yahya Turki [38] suggested the following notion: we say that a bivector field $\pi \in \Gamma(\wedge^2 TM)$ is *foliated* if $\pi^\sharp(\Omega^1(M))$ is closed under the Lie bracket, i.e., is a singular foliation.

Example 4.1 Poisson bivector fields, but also twisted-Poisson bi-vector fields, are examples. Yahya Turki [38] gave examples of foliated bivector fields that are not of this type, but proved that they are twisted Poisson near any one of their regular points (= points in a neighborhood of which $\pi^\sharp$ has constant rank).

Question 4.2 (Foliated Bivector Fields) *Let π be a foliated bivector field. Can a Lie algebroid structure with anchor map $\pi^\sharp$ be constructed on T^*M?*

It is known that T^*M comes equipped with a Lie algebroid structure with anchor $\pi^\sharp \colon T^*M \longrightarrow TM$ when π is twisted Poisson [35], so the question makes sense.

Sébastien Michéa asked if for any smooth Poisson structure π on $\mathbb{R}^n$, there is another structure π' on $\mathbb{R}^n$ which coincides with π in a neighborhood of 0 and vanishes outside a compact subset. The corresponding question for singular foliations is much easier:

Exercise 4.3 Given a smooth singular foliation $\mathcal{F}$ on $\mathbb{R}^n$, show that there exist another singular foliation $\mathcal{F}'$ on $\mathbb{R}^n$ which coincides with $\mathcal{F}$ in a neighborhood of 0 and vanishes outside a compact subset.

Here is however, a more delicate question:

Can a smooth Debord foliation on a half-plane (well-defined in the boundary) extend to a Debord foliation on the whole plane?

5 Linearization

Can we enlarge the classical theorems (i.e., Conn's [11, 12] or Zung's [39], Crainic-Fernandes' [8]) about linearizations of Poisson structures or Lie algebroid actions or Lie groupoid actions to the context of singular foliations and of its holonomy groupoid?

The previous linearization theorems all share the same logic. There are first relatively easy results whose patterns are:

State of the Art and Open Questions

$$\text{Fixed point} + \text{Semi-simple} \implies \text{Formally Linearizable}.$$

For instance, it is not so complicated to show that if a Lie algebroid $(A, \rho, [\cdot, \cdot])$ admits a point m where $\rho_m = 0$ and the isotropy Lie algebra $\mathfrak{g}_m = A_m$ is semi-simple, then the Lie algebroid is formally equivalent to the transformation Lie algebroid $\mathfrak{g}_m \times T_m M \to T_m M$ for some action of $\mathfrak{g}_m$ by linear endomorphisms $T_m M$. Dominique Cerveau [6] has a result of this type for singular foliations: it says that if the isotropy Lie algebra at a point is a semi-simple Lie algebra, then the singular foliation is formally equivalent to the one associated to a linear action of this Lie algebra. See [20] Proposition 1.12 for an enlarged version.

Beyond these (relatively easy) results, there are then much more difficult results whose patterns are:

$$\text{Fixed point} + \text{Compact (and semi-simple)} \implies \text{Locally Linear}$$

The difficulty consists in going from "formal" to "local". Recall (see Exercise 3.13 in chapter "Canonical Geometric and Algebraic Structures Hidden Behind a Singular Foliation") that if the leaf through a point m is reduced to $\{m\}$, or equivalently if $T_m \mathcal{F} = 0$, then there is a natural action of the isotropy Lie algebroid $\mathfrak{g}_m(\mathcal{F})$ on $T_m M$.

Question 5.1 (Extend Zung's Linearization [39] to Singular Foliations) *Let $\mathcal{F}$ be a singular foliation on a smooth manifold M. Let m be point such that $T_m \mathcal{F} = 0$.*

Assume the isotropy Lie algebra $\mathfrak{g}_m(\mathcal{F})$ of $\mathcal{F}$ at m is semi-simple of compact type. Then, is there a saturated[a] neighborhood of m on which $\mathcal{F}$ is isomorphic to a saturated neighborhood of 0 for the singular foliation associated to the natural representation of the isotropy Lie algebra $\mathfrak{g}_m(\mathcal{F})$ on $T_m M$?

---
[a] i.e., a leaf that has non-empty intersection with that neighborhood is contained in it.

In other words, we want to prove that the short exact sequence

$$\mathcal{I}_m \mathcal{F} \hookrightarrow \mathcal{F} \twoheadrightarrow \mathfrak{g}_m(\mathcal{F})$$

splits with a section σ:

$$\mathcal{I}_m \mathcal{F} \hookrightarrow \mathcal{F} \xrightarrow[\sigma]{} \mathfrak{g}_m(\mathcal{F})$$

which is a Lie algebra morphism, at least in a neighborhood of m. Since any action of a semi-simple Lie algebra of compact type is linearizable near a fixed point, this section σ may be seen as being an action of $\mathfrak{g}_m(\mathcal{F})$ on the vector space $T_m M$.

We could of course enlarge these questions to neighborhood of leaves. Again, the formal case is relatively easy: for instance it has been proven that [20] that Levi-Malcev style theorems hold: those are formal linearization theorems in a neighborhood of a leaf. Of course, only the semi-simple part of the holonomy Lie algebroid (defined in that article) is formally linearizable. For Lie algebroids or Poisson structures, several authors e.g., [9, 13, 15] have proven recently several linearizations or normal form theorems in neighborhood of leaves of Lie algebroids or singular foliations: Pretty much any one of these theorems admit a natural generalization for singular foliations.

There are similar questions about the holonomy groupoid [1]. Recall that it is a topological groupoid, although it is not a Lie groupoid (but each fiber is a variety [14]). The topology is the push-forward topology of any atlas of bisections that define it. It makes sense, therefore, to speak of a singular foliation $\mathcal{F}$ whose holonomy groupoid $\mathrm{Hol}(\mathcal{F})$ is proper: it is a singular foliation for which

$$(s, t) \colon \mathrm{Hol}(\mathcal{F}) \longrightarrow M \times M$$

is a proper map.

Definition 5.2 We say that a singular foliation $\mathcal{F}$ is *proper* if $\mathrm{Hol}(\mathcal{F})$ is a proper topological groupoid.

Example 5.3 Consider a proper groupoid $\Gamma \rightrightarrows M$, e.g., the action groupoid associated to an action of a compact group on a manifold. Then the basic singular foliation[7] is a proper singular foliation (since G itself is an atlas, see Proposition 5.13 in chapter "Canonical Geometric and Algebraic Structures Hidden Behind a Singular Foliation").

Proper groupoids have very strong linearization properties. Here is a theorem by Nguyen Tien Zung:[8]

Theorem 5.4 ([39]) *Consider a proper Lie groupoid* $\Gamma \rightrightarrows M$. *Every fixed point[9]* $m \in M$ *admits a saturated neighborhood* $\mathcal{U}$ *on which the restriction of* Γ *is isomorphic, as a Lie groupoid, to a transformation groupoid of the action of the compact isotropy group* G_m *on the tangent space* $V = T_m M$.

---

[7] i.e., $\mathcal{F} = \rho(\Gamma(A))$ with $(A, \rho, [\cdot, \cdot])$ the Lie algebroid of $G \rightrightarrows M$.
[8] Recall that for every fixed point $m \in M$ of a Lie groupoid (i.e., any point for which $t(s^{-1}(m)) = \{m\}$), the isotropy group at m acts naturally by linear automorphisms of the tangent space $T_m M$.
[9] I.e., the Γ orbit through m is $\{m\}$.

It is therefore very natural to guess that the following variation of the previous question should be true. The action of holonomy Lie algebra $\mathfrak{g}_m(\mathcal{F})$ at m on the tangent space $T_m M$ is defined in Exercise 3.13 in chapter "Canonical Geometric and Algebraic Structures Hidden Behind a Singular Foliation".

> **Question 5.5 (Extend Zung's Linearization to SF)** *Consider a proper singular foliation.*
>
> *Is it true that every point* $m \in M$ *such that*[a] $T_m \mathcal{F} = 0$ *admits a saturated neighborhood*[b] $\mathcal{U}$ *on which the restriction of* $\mathcal{F}$ *is isomorphic, as a singular foliation, to a saturated neighborhood of* 0 *for the singular foliation associated to the action of the holonomy Lie algebra* $\mathfrak{g}_m(\mathcal{F})$ *on the tangent space* $T_m M$?
>
> *If one replaces* m *by a leaf* L, *then the question is still valid, upon replacing the action of* $\mathfrak{g}_m(\mathcal{F})$ *on* $T_m M$ *by the natural action*[c] *of the holonomy Lie algebroid of* L *on the normal bundle of* L *in* M.

---

[a] I.e., the leaf through m is $\{m\}$.
[b] I.e., a leaf intersecting this neighborhood entirely belongs to that neighborhood.
[c] Considered by Androulidakis and Zambon [3].

For regular foliations, properness of the holonomy Lie groupoid implies, for instance, that every leaf has a saturated neighborhood on which the holonomy map is by linear automorphisms of a finite group, so that the answer to the question is affirmative in this case. Also, for singular foliations coming from Poisson manifolds of compact type [10], the answer to Question 5.5 is "yes". Notice also that it has been proven in [33] that singular foliations arising from a compact Lie groupoid can be made a regular foliation by finitely many blow-up operations of its most singular leaves. Can one generalize to a singular foliation whose holonomy groupoid is compact?

6 Longitudinal Differential Operators

For $(M, \mathcal{F})$ a singular foliation, we call *longitudinal differential operator* any linear combination of operators of the form

$$C^\infty(M) \to C^\infty(M)$$
$$F \mapsto X_1 \circ \cdots \circ X_k[F] \qquad (4)$$

with $X_1, \ldots, X_k \in \mathcal{F}$. We denote by Diff($\mathcal{F}$) the algebra of longitudinal differential operators.

For a regular foliation,[10] longitudinal differential operators coincide with the universal enveloping algebra of the Lie algebroid $T\mathcal{F}$. This is not the case for a generic singular foliation.

Let us first define the universal enveloping algebra of a singular foliation.[11]

1. To start with, consider the universal enveloping algebra $U(\mathcal{F})$ of the *Lie* algebra $\mathcal{F}$.
2. Now, divide $U(\mathcal{F})$ by the ideal generated by:

$$X \cdot (fY) - (fX) \cdot Y - (X[f]Y) \tag{5}$$

where $X, Y \in \mathcal{F}$ and $f \in C^\infty(M)$. We denote the quotient by $\mathcal{U}(\mathcal{F})$ and call it the *universal enveloping algebra of the singular foliation* $\mathcal{F}$.

There is a natural algebra morphism from $U(\mathcal{F})$ to Diff($\mathcal{F}$) defined by

$$\begin{aligned} U(\mathcal{F}) &\to \text{Diff}(\mathcal{F}) \\ X_1 \cdot \ldots \cdot X_k &\mapsto X_1 \circ \cdots \circ X_k \end{aligned} \tag{6}$$

The previously defined map (6) being equal to 0 on the ideal generated by the expressions (5), it goes to the quotient to define a surjective algebra morphism

$$\mathcal{U}(\mathcal{F}) \twoheadrightarrow \text{Diff}(\mathcal{F}) .$$

$$P \longrightarrow \underline{P}$$

that we call *realization of* $\mathcal{U}(\mathcal{F})$.

Exercise 6.1 Show that the realization of $\mathcal{U}(\mathcal{F})$ is injective if $\mathcal{F}$ is a Debord singular foliation. Show that it is not injective in general. *Hint:* The singular foliation of all vector fields on X on $\mathbb{R}^3$ such that $X[x^2 + y^2 + z^2] = 0$ provides a counter-example.

Elements in Diff($\mathcal{F}$) of the form (4) are called *monomials of degree k*, and we say that a longitudinal differential operator is of degree $\leq k$ if it is a sum of monomials of degree $\leq k$. The degree defines an increasing filtration on the algebra Diff($\mathcal{F}$), making it a filtered algebra $\left(\text{Diff}^{\leq k}(\mathcal{F})\right)_{k \geq 0}$. Similarly, the algebra $\mathcal{U}(\mathcal{F})$ comes equipped with a

---

[10] Or, more generally, a Debord foliation.
[11] One recognizes here the construction of the universal enveloping algebra of a Lie-Rinehart algebra by, see e.g., [16], [27] or [17]. See also [25].

filtration $(\mathcal{U}^{\leq k}(\mathcal{F}))_{k\geq 0}$ defined in a same manner. The realization (6) is a morphism of filtered algebras.

Let us address the following question (which is not an open question, as we will see).

Question 6.2 *What is the symbol of a longitudinal differential operator?*

Androulidakis and Skandalis in [2] gave an answer that is too involved to be dealt with here, using the C^*-algebra of half densities of the holonomy Lie groupoid. Later on, Mohsen, then Androulidakis, Mohsen and Yuncken's [4, 28] gave a second elaborate answer involving representations, more subtle than the one considered below. In the process, they showed that the Nash blowup invented by Omar Mohsen (see Sect. 5.8 in chapter "What Is a Singular Foliation?") provides an answer to Question 6.2 sufficient for several purposes, including dealing with pseudo-differential operator calculus developed by Androulidakis and Skandalis in [2], using the holonomy Lie groupoid and bisubmersions. We present our interpretation of this answer.

To start with, we will define two "symbols":

1. the symbol of an element in the universal enveloping algebra $\mathcal{U}(\mathcal{F})$ of $\mathcal{F}$,
2. the symbol of a longitudinal differential operator, i.e., an element in $\mathrm{Diff}(\mathcal{F})$.

Before defining these two symbols, let us very briefly recall the definition of the symbol in the context of Lie algebroids as presented in [17]. Let $B \to M$ be a Lie algebroid. We denote by $\mathcal{U}(B)$ and call the universal algebra of the Lie algebroid B the quotient of the universal algebra $U(\Gamma(B))$ of the Lie algebra of sections of B by the ideal generated by

$$X \cdot (fY) - (fX) \cdot Y - (\rho(X)[f]Y).$$

with $X, Y \in \Gamma(B)$ and $f \in C^\infty(M)$. It is again a filtered algebra, with $\mathcal{U}^{\leq k}(B)$ being the subspace generated by monomials of degree $\leq k$. Now, recall that

1. upon choosing a Lie algebroid connection, there is a grading preserving coalgebra isomorphism $\mathcal{U}(B) \simeq \Gamma(S(B))$ from the universal enveloping algebra of B to the symmetric algebra of B. See, e.g., [23].
2. since this isomorphism preserves the grading, it induces a $C^\infty(M)$-linear map from $\mathcal{U}^{\leq k}(B)$ to[12] $S^k(B)$.

---

[12] We insist that it is $S^k(B)$, not $S^{\leq k}(B)$: we project on the top component, i.e., the space generated by monomials of degree k.

3. This map does not depend on the choice of a Lie algebroid connection.[13]

Now, a section of $S^k(B)$ may be seen as a function on B^*, which is fiberwise polynomial and homogeneous of degree k. For every $P \in \mathcal{U}^{\leq k}(B)$, we denote by σ_P this function and call it the *symbol* of P.

Let us now define the symbol as an element in $\mathcal{U}(\mathcal{F})$. For every leaf L of a singular foliation $\mathcal{F}$ on a manifold M, there exists a natural restriction map

$$\mathcal{F} \longrightarrow \Gamma_L(A_L)$$

where A_L is the holonomy Lie algebroid of the leaf L (see Sect. 3 in chapter "Canonical Geometric and Algebraic Structures Hidden Behind a Singular Foliation"). This map is $\mathcal{C}^\infty(M)$-linear and is a Lie algebra morphism, so that it induces an algebra morphism

$$U(\mathcal{F}) \to U(\Gamma(A_L))$$
$$P \mapsto P(L)$$

from the algebra of longitudinal operators $U(\mathcal{F})$ to the universal enveloping algebra $U(\Gamma(A_L))$. It goes to the quotient to induce an algebra morphism

$$\mathcal{U}(\mathcal{F}) \to \mathcal{U}(A_L)$$
$$P \mapsto P(L) \qquad (7)$$

from the universal enveloping algebra $\mathcal{U}(\mathcal{F})$ of the singular foliation $\mathcal{F}$ to the universal enveloping algebra $\mathcal{U}(A_L)$ of the Lie algebroid A_L.

For a given $P \in \mathcal{U}^{\leq k}(\mathcal{F})$ of degree $\leq k$, consider, for every leaf L, the symbol $\sigma_{P(L)}$. It is a fiberwise homogeneous of degree k smooth function on A_L^*.

Remark 6.3 For any leaf L, the realization of P, restricted to L, is a differential operator in the usual sense. For a regular leaf, $A_L = TL$ and $\sigma_L(P)$ is simply the usual symbol [30] of this differential operator.

We call *symbol of P* the collection $(L, \sigma_{P(L)})$ indexed by leaves of $\mathcal{F}$ of the functions $\sigma_L(P)$. We denote it by σ_P.

From now on, we will assume that the regular leaves of $(M, \mathcal{F})$ are all the same dimension, so that the Nash blowup

---

[13] Only the top component does not depend on the connection, the Lower components do. This independence means that we could work with local connections, so that what we say here extends to the real analytic or holomorphic settings.

$$\left(\mathrm{Bl}(M, \mathcal{F}), \pi^! \mathcal{F}\right)$$

makes sense, see Sect. 5.8 in chapter "What Is a Singular Foliation?". Recall from Theorem 5.64 in chapter "What Is a Singular Foliation?" that the Nash blowup is a Debord singular foliation, with associated Lie algebroid (the Nash blowup Lie algebroid of Definition 5.65 in chapter "What Is a Singular Foliation?") the canonical quotient bundle A_{Bl} of the Grassmann bundle, restricted to $\mathrm{Bl}(M, \mathcal{F})$. From now on, we denote by $(A_{\mathrm{Bl}}, \rho_{\mathrm{Bl}})$ this Lie algebroid. Consider the lift

$$\mathcal{F} \to \Gamma_{\mathrm{Bl}(M,\mathcal{F})}(A_{\mathrm{Bl}})$$

mapping $X \in \mathcal{F}$ to the unique section of $\Gamma_{\mathrm{Bl}(M,\mathcal{F})}(A_{\mathrm{Bl}})$ corresponding to $\pi^! X$. This lift is (1) a $\mathcal{C}^\infty(M)$-module morphism and (2) a Lie algebra morphism. This implies that it lifts to an algebra morphism

$$\mathrm{Diff}(\mathcal{F}) \longrightarrow \mathcal{U}(A_{\mathrm{Bl}}),$$

where $\mathcal{U}(A_{\mathrm{Bl}})$ stands for the universal algebra of the Nash-blowup Lie algebroid A_{Bl}. We denote by $D \longrightarrow \pi^! D$ this algebra morphism. Since $\pi^! D$ is an element of degree $\leq k$ of the universal Lie algebroid A_{Bl}, its symbol is a (perfectly well-defined) element of $\Gamma(S^k A_{\mathrm{Bl}})$, or, equivalently, a function on A^*_{Bl}, which is fiberwise polynomial and homogeneous of degree k. Since its restriction to the regular part is entirely determined by the image of D through the realization map, and since this symbol depends continuously on the base point by construction, the next definition makes sense.

Definition 6.4 Let D be a longitudinal differential operator of degree $\leq k$. We call *symbol of D of degree k* the symbol of the element $\pi^! D$ in the universal algebra $\mathcal{U}(A_{\mathrm{Bl}})$ of the Nash blowup algebroid A_{Bl}.

By construction, this symbol that we denote σ_D is a fiberwise homogeneous of degree k polynomial function on the dual A_{Bl} of the Nash blowup Lie algebroid of Definition 5.65 in chapter "What Is a Singular Foliation?".

We now have two symbols: one for the algebra of longitudinal differential operators $\mathrm{Diff}(\mathcal{F})$ and one for the universal enveloping algebra $\mathcal{U}(\mathcal{F})$.

To relate these two symbols, we need to give some explanation about the Helffer-Nourrigat cone [4] (see also Exercise 2.9 in chapter "Canonical Geometric and Algebraic Structures Hidden Behind a Singular Foliation"). Let us first explain a general phenomenon. Given a vector bundle $E \to M$ and a map $\Pi \colon N \to M$, consider the pull-back bundle $\pi^! E \to N$. Now, let $T \subset \Pi^* E$ be a subvector bundle over N then there is a natural fiberwise injective vector morphism

$$\begin{array}{ccc} \left(\frac{\pi^! E}{T}\right)^* & \xrightarrow{j} & E^* \\ \downarrow & & \downarrow \\ N & \longrightarrow & M \end{array}$$

It consists in identifying for every point $n \in N$ the dual of the quotient space $\left(\frac{\pi^! E_n}{T_n}\right)^*$ with the annihilator $T_n^\perp \subset E^*_{\pi(n)}$ of T_n, and to inject the later in E^*. Moreover, the image of j can be described as follows. For every $m \in M$, its intersection with E_m is a union of vector spaces of dimension $\mathrm{rk}(E) - \mathrm{rk}(T)$. We call *image cone* this image.

We apply this construction to

1. $E := A$, with (A, ρ) an anchored bundle over $\mathcal{F}$,
2. $N := \mathrm{Bl}(M, \mathcal{F})$ the Nash blowup[14] of Sect. 5.8 in chapter "What Is a Singular Foliation?", computed with respect to (A, ρ),
3. T the canonical bundle τ_A^{-r} on the Grassmannian $\mathrm{Grass}_{-r}(A)$, restricted to $\mathrm{Bl}(M, \mathcal{F})$.

In this case, Theorem 5.64 in chapter "What Is a Singular Foliation?" identifies the quotient $\pi^! E/T$ with the Nash blowup Lie algebroid[15] A_{Bl} of Definition 5.65 in chapter "What Is a Singular Foliation?" associated to the Nash blow up of $\mathcal{F}$. We therefore obtain a fiberwise injective[16] vector bundle morphism:

$$\begin{array}{ccc} j \colon A^*_{\mathrm{Bl}} & \longrightarrow & A^* \\ \downarrow & & \downarrow \\ \mathrm{Bl}(M, \mathcal{F}) & \xrightarrow{\pi} & M \end{array}.$$

We leave it to the reader to check[17] that the image cone $j(A^*_{\mathrm{Bl}})$ in this case coincides with the Helffer-Nourrigat cone $\mathrm{HN}(\mathcal{F})$ (computed with respect to (A, ρ)) defined in Exercise 2.9 in chapter "Canonical Geometric and Algebraic Structures Hidden Behind a Singular Foliation", i.e.,

$$\mathrm{HN}_A(\mathcal{F}) := \overline{\cup_{m \in M_{\mathrm{reg}}} (\rho^*(T^*_m M))} \subset A^*.$$

Moreover, for every leaf L of $\mathcal{F}$, the image of j takes values in the annihilator of the strong kernel of A, which can be identified with A^*_L (the dual of the holonomy Lie algebroid). The

---

[14] Of course, it may not be a manifold, but this defect has no practical consequence here.

[15] That we recall to be the Lie algebroid of the Mohsen's groupoid, see Theorem 5.64 in chapter "What Is a Singular Foliation?".

[16] But of course not injective.

[17] This can be proven as follows: Since π is proper, both sets are closed. It therefore suffices that the image of j on $\pi^{-1}(M_{\mathrm{reg}})$ (i.e., the regular part) coincides with the image of ρ^* (i.e., the annihilator of the kernel of ρ).

previous vector bundle morphism, therefore, induces a vector bundle morphism[18] (which is still fiberwise injective)

$$q\colon A_{\mathrm{Bl}}^* \longrightarrow \coprod_{L\in\text{Leaves}} A_L^*$$
$$\downarrow \qquad\qquad\qquad \downarrow$$
$$\mathrm{Bl}(M,\mathcal{F}) \xrightarrow{\pi} M$$

whose image, again, is the Helffer-Nourrigat cône $\mathrm{HN}(\mathcal{F})$, described as in Question 3 in Exercise 2.9 in chapter "Canonical Geometric and Algebraic Structures Hidden Behind a Singular Foliation".

Proposition 6.5 *Let $D \in \mathrm{Diff}^{\leq k}(\mathcal{F})$ be a longitudinal differential operator of degree $\leq k$ and let $P \in \mathcal{U}^{\leq k}(\mathcal{F})$ be any element whose realization is D. The symbol σ_D is the pull-back through the projection q of the restriction of the symbol σ_P of P to the Helffer-Nourrigat cone bundle of $\mathcal{F}$. In Equation:*

$$\sigma_D = q^* \, \sigma_P|_{\mathrm{HN}(\mathcal{F})}.$$

Proof It holds true on the regular part. The result follows by density. □

A remarkable consequence of this statement is that the restriction to $\mathrm{HN}(\mathcal{F})$ of the symbol of an element in P depend on its realization $\underline{P}$ only.

Now, recall that a differential operator is said to be elliptic if its symbol vanishes only at the origin. For a Lie algebroid [30], an element in the universal Lie algebra is said to be elliptic if its symbol vanishes only along the zero section.

Question 6.6 *Let $\mathcal{F}$ be a singular foliation. What is a longitudinally elliptic differential operator?*

Here is, in our opinion, the correct answer [28].

Definition 6.7 A longitudinal differential operator D of degree k is said to be *longitudinally elliptic* if its symbol[19] σ_D is a strictly positive function, except on the zero section.

In particular, this implies that D is longitudinally elliptic if and only if $\pi^!(D)$ is elliptic for the Nash blowup Lie algebroid A_{Bl} in the sense of [30, 31], which seems to us to be a

---

[18] Here, we consider $\coprod_{L\in\text{Leaves}} A_L^*$ as a "singular" vector bundle over M: the ranks of the fibers vary.

[19] Recall that the latter is a function on the dual A_{Bl}^* of the Nash blowup Lie algebroid.

convincing justification of the notion. Now, choose some $P \in \mathcal{U}(\mathcal{F})$ whose realization is D. The symbol σ_P does *not* need to be strictly positive outside the Helffer-Nourrigat cone by Proposition 6.5, so that our definition does *not* imply that $P(L) \in \mathcal{U}(A_L)$ (see Eq. (7)) is elliptic for all leaves (although it certainly has to be elliptic on regular leaves).

7 Cohomologies of a Singular Foliation

We already saw that the derived cohomological spaces $\mathrm{Tor}_{\mathcal{C}^\infty(M)}(\mathcal{F}, \mathbb{K})$ come equipped with a Lie ∞-algebra structure, whose cohomology permits to solve some elementary problems. But these are cohomologies associated to points or to leaves. Our next question is rather vague:

> **Question 7.1 (Relevant Cohomologies?)** *What are the interesting global cohomology theories for singular foliations?*

Here are several candidates.[20] Also, for any $\mathcal{C}^\infty(M)$-module $\mathcal{E}$, the notation $\mathcal{E} \wedge_\mathcal{O} \mathcal{E}$ below stands for the wedge product over $\mathcal{O}$, i.e., we allow

$$X \wedge FY = FX \wedge Y \text{ for all } X, Y \in \mathcal{E}, F \in \mathcal{O}$$

1. *Longitudinal cohomology of a singular foliation*, see [22], Section 4.1. Let us describe it. To make the notation easier, we write $\mathcal{O}$ instead of $\mathcal{C}^\infty(M)$:

Chains in degree k	Differential on chains of degree k
Skew symmetric and $\mathcal{O}$-multilinear maps $\underbrace{\mathcal{F} \wedge_\mathcal{O} \cdots \wedge_\mathcal{O} \mathcal{F}}_{k-\text{times}} \to \mathcal{O}$ i.e., $\mathrm{Hom}^\bullet_\mathcal{F}(\mathcal{F}, \mathcal{O}) := \mathrm{Hom}_\mathcal{O}(\wedge^\bullet_\mathcal{O} \mathcal{F}, \mathcal{O})$	$\forall \omega \in \mathrm{Hom}^k_\mathcal{F}(\mathcal{F}, \mathcal{O})$ and all $X_0, \ldots, X_k \in \mathcal{F}$ $\delta \omega (X_0, \ldots, X_k) =$ $\sum_{i=0}^k (-1)^i X_i \left[\omega(X_0, \ldots, \widehat{X_i}, \ldots, X_k) \right]$ $+ \sum_{i<j} (-1)^{i+j+1}$ $\omega([X_i, X_j], X_0, \ldots, \widehat{X_i}, \ldots, \widehat{X_j}, \ldots, X_k)$
Chains in degree 0	Differential on chains of degree 0
In degree 0, chains are simply elements of $\mathcal{O}$	$\forall F \in \mathcal{O},$ $\delta(F) \colon \mathcal{F} \to \mathcal{O}$ $X \mapsto X[F]$

---

[20] We describe them in the smooth context: for the real-analytic or holomorphic settings, one has to add a Čech-type differential for a good covering—as always in sheaf theory.

For a regular foliation, this cohomology is simply the De Rham cohomology along the leaves, *i.e.*, it is the complex $\left(\Gamma(\wedge^\bullet T^*\mathcal{F}), d_\mathcal{F}^{dR}\right)$ with $d_\mathcal{F}^{dR}$ being the De Rham differential restricted to $T\mathcal{F}$, but computed leaf by leaf.

2. The *basic cohomology* is the sub-complex of $(\Omega(M), d^{dR})$ made, in degree k, of all k-forms that vanish when k vector fields in $\mathcal{F}$ are applied. Equivalently, these are k-forms ω whose pull-back to any leaf L is zero.

   It has been studied by David Miyamoto [26].

3. The *universal cohomology of* $\mathcal{F}$ is the cohomology of the commutative differential graded algebra of functions[21] on any universal Q-manifold[22] of $\mathcal{F}$. This is more precisely defined as the cohomology of $\left(\Gamma(S(\oplus_{i\geq 1} E_i^*)), Q\right)$. The definition makes sense: it can be proven that since any two universal Lie ∞-algebroid of $\mathcal{F}$, say (E, Q) and (E', Q') are homotopy equivalent, the differential graded commutative algebras $\left(\Gamma(S(\oplus_{i\geq 1} E_i^*)), Q\right)$ and $\left(\Gamma(S(\oplus_{i\geq 1} (E_i')^*)), Q'\right)$ are homotopy equivalent in a unique up to homotopy manner, see e.g. [22] Section 3.4.4. In particular, their cohomologies are canonically isomorphic.[23]

   Universal cohomology is linked to longitudinal cohomology, since there is a map of differential graded commutative algebras:

   $$\text{Longitudinal cohomology of } \mathcal{F} \longrightarrow \text{Universal cohomology of } \mathcal{F}.$$

   See the discussion on universal and longitudinal cohomologies in [22], Section 4.1.

4. It is also interesting to consider the Chevalley-Eilenberg cohomology for the adjoint representation [5, 18] of any universal Lie ∞-algebroid of $\mathcal{F}$. This coincides with the cohomology of vector fields on the universal Q-manifolds, equipped with ad_Q, and plays a role in deformation theory.

The list will certainly be continued.

References

1. I. Androulidakis, G. Skandalis, The holonomy groupoid of a singular foliation. J. Reine Angew. Math. **626**, 1–37 (2009)
2. I. Androulidakis, G. Skandalis, Pseudodifferential calculus on a singular foliation. J. Noncommut. Geom. **5**(1), 125–152 (2011)
3. I. Androulidakis, M. Zambon, Smoothness of holonomy covers for singular foliations and essential isotropy. Math. Z. **275**(3–4), 921–951 (2013)

---

[21] One can also choose compactly supported functions.

[22] See Sect. 7 in chapter "Canonical Geometric and Algebraic Structures Hidden Behind a Singular Foliation".

[23] It is tempting to believe that because the geometric resolution has no cohomology, the universal cohomology has to be zero as well. This is not at all true.

4. I. Androulidakis, O. Mohsen, R. Yuncken. A pseudodifferential calculus for maximally hypoelliptic operators and the Helffer-Nourrigat conjecture (2022)
5. R. Caseiro, C. Laurent-Gengoux, Modular class of Lie ∞-algebroids and adjoint representations. J. Geom. Mech. **14**(2), 273–305 (2022)
6. D. Cerveau, Distributions involutives singulières. Ann. Inst. Fourier (Grenoble) **29**(3), xii, 261–294 (1979)
7. Z. Chen, M. Stiénon, X. Ping, From Atiyah classes to homotopy Leibniz algebras. Commun. Math. Phys. **341**(1), 309–349 (2016)
8. M. Crainic, R.L. Fernandes, A geometric approach to Conn's linearization theorem. Ann. Math. (2) **173**(2), 1121–1139 (2011)
9. M. Crainic, I. Mărcuţ, Reeb-Thurston stability for symplectic foliations. Math. Ann. **363**(1–2), 217–235 (2015)
10. M. Crainic, R.L. Fernandes, D. Martínez Torres, Poisson manifolds of compact types (PMCT 1). J. Reine Angew. Math. **756**, 101–149 (2019)
11. J.F. Conn, Normal forms for analytic Poisson structures. Ann. Math. 2(**119**), 577–601 (1984)
12. J.F. Conn, Normal forms for smooth Poisson structures. Ann. Math. 2(**121**), 565–593 (1985)
13. M. Crainic, I. Struchiner, On the linearization theorem for proper Lie groupoids. Ann. Sci. Éc. Norm. Supér. (4) **46**(5), 723–746 (2013)
14. C. Debord, Holonomy groupoids of singular foliations. J. Differential Geom. **58**(3), 467–500 (2001)
15. P. Frejlich, I. Mărcuţ, The normal form theorem around Poisson transversals. Pac. J. Math. **287**(2), 371–391 (2017)
16. J. Huebschmann, Duality for Lie-Rinehart algebras and the modular class. J. Reine Angew. Math. **510**, 103–159 (1999)
17. J. Huebschmann, On the history of Lie brackets, crossed modules, and Lie-Rinehart algebras. J. Geom. Mech. **13**(3), 385–402 (2021)
18. M. Jotz, R.A. Mehta, T. Papantonis, Modules and representations up to homotopy of Lie n-algebroids. J. Homotopy Relat. Struct. **18**(1), 23–70 (2023)
19. C. Laurent-Gengoux, R. Louis, Lie-Rinehart algebras $\simeq$ acyclic lie ∞-algebroids. J. Algebra **594**, 1–53 (2022)
20. C. Laurent-Gengoux, L. Ryvkin, The neighborhood of a singular leaf. J. Éc. Polytech. Math. **8**, 1037–1064 (2021)
21. C. Laurent-Gengoux, Y. Voglaire, Invariant connections and PBW theorem for Lie groupoid pairs. Pac. J. Math. **303**(2), 605–667 (2019)
22. C. Laurent-Gengoux, S. Lavau, T. Strobl, The universal Lie ∞-algebroid of a singular foliation. Doc. Math. **25**, 1571–1652 (2020)
23. C. Laurent-Gengoux, M. Stiénon, P. Xu, Poincaré-Birkhoff-Witt isomorphisms and Kapranov dg-manifolds. Adv. Math. **387**, 62 (2021). Id/No 107792
24. S. Lavau, The modular class of a singular foliation. J. Geom. Phys. **192**, 32 (2023). Id/No 104902
25. H.O. Maakestad, The enveloping algebra of a Lie algebra of differential operators. arXiv:1903.04285 (2019)
26. D. Miyamoto, The basic de Rham complex of a singular foliation. Int. Math. Res. Not. **2023**(8), 6364–6401 (2023)
27. I. Moerdijk, J. Mrcun, On the universal enveloping algebra of a Lie-Rinehart algebra. arXiv:0801.3929 (2008)
28. O. Mohsen, Blow-up groupoid of singular foliations. arXiv:2105.05201 (2021)
29. P. Molino, Classe d'Atiyah d'un feuilletage et connexions transverses projetables. C. R. Acad. Sci., Paris Sér. A **272**, 779–781 (1971)

30. V. Nistor, A. Weinstein, X. Ping, Pseudodifferential operators on differential groupoids. Pac. J. Math. **189**(1), 117–152 (1999)
31. M.J. Pflaum, H. Posthuma, X. Tang, The index of geometric operators on Lie groupoids. Indag. Math. New Ser. **25**(5), 1135–1153 (2014)
32. A. Pichereau, Poisson (co)homology and isolated singularities. J. Algebra **299**(2), 747–777 (2006)
33. H. Posthuma, X. Tang, K. Wang, Resolutions of proper Riemannian Lie groupoids. Int. Math. Res. Not. **2021**(2), 1249–1287 (2021)
34. S. Seol, The Atiyah class of dg manifolds of amplitude $+1$. arXiv:2312.16622 (2024)
35. P. Ševera, A. Weinstein, Poisson geometry with a 3-form background, in *Noncommutative Geometry and String Theory. Proceedings of the International Workshop, Keio Univ., Yokohama, Japan, March 16–22, 2001* (2001), pp. 145–154. Kyoto: Progress of Theoretical Physics
36. K.J. Singh, On the universal L_∞-algebroid of linear foliations. J. Lie Theory **33**(3), 925–952 (2023)
37. J.-C. Tougeron, Idéaux de fonctions différentiables. I. Ann. Inst. Fourier (Grenoble) **18**(fasc., fasc. 1), 177–240 (1968)
38. Y. Turki, A Lagrangian for Hamiltonian vector fields on singular Poisson manifolds. J. Geom. Phys. **90**, 71–87 (2015)
39. N.T. Zung, Proper groupoids and momentum maps: linearization, affinity, and convexity. Ann. Sci. Éc. Norm. Supér. (4) **39**(5), 841–869 (2006)

MIX
Papier aus verantwortungsvollen Quellen
Paper from responsible sources
FSC® C105338

If you have any concerns about our products,
you can contact us on
ProductSafety@springernature.com

In case Publisher is established outside the EU,
the EU authorized representative is:
**Springer Nature Customer Service Center GmbH
Europaplatz 3, 69115 Heidelberg, Germany**

Printed by Libri Plureos GmbH
in Hamburg, Germany